RADIATION PROTECTION
A Systematic Approach to Safety

In Two Volumes

INTERNATIONAL MEETING COMMITTEE (IMC)

TUVIA SCHLESINGER	Chairman
AHARON EISENBERG	General Secretary
YITZHAK G. GONEN	Scientific Secretary
ALEXANDER DONAGI	Treasurer
SHLOMO LETZTER	Exhibition Coordinator

INTERNATIONAL MEETING PROGRAM COMMITTEE (IMPC)

P. M. BRYANT	Chairman United Kingdom
Y. G. GONEN	Scientific Secretary Israel
A. DONAGI	Scientific Co-Secretary Israel
T. SCHLESINGER	Congress Chairman Israel
J. K. BASSON	South Africa
A. BOUVILLE	France
H. CEMBER	United States
M. FABER	Denmark
J. R. HORAN	Austria
A. OSIPENCO	Belgium
J. PÉNSKO	Poland
S. PRÊTRE	Switzerland
F. D. SOWBY	United Kingdom
E. STRAMBI	Italy

8235
24.3

£89.00
2 VOL

THE UNIVERSITY OF ASTON IN BIRMINGHAM LIBRARY

2142002-01-03
363.179 INT
RADIATION PROTECT...
INTERNATIONAL RAD...

LEND   363.179 INT
2142002-01-xx

3 0116 00114 2007

International Radiation Protec
Radiation protection: a system

14 OCT 1981	27 MAY 1987
	ASTON U.L.
6 FEB 1984	
ASTON U.L.	24 MAY 1995
	LONG LOAN
31 MAY 1984	10 MAY 1999
ASTON U.L.	LONG LOAN
26 OCT 1984	
ASTON U.L.	

This book is due for return not later than the last date stamped above, unless recalled sooner.

RADIATION PROTECTION

A Systematic Approach to Safety

Proceedings of the 5th Congress of the International Radiation Protection Society, Jerusalem, March 1980

In Two Volumes

Volume 2

PERGAMON PRESS
OXFORD · NEW YORK · TORONTO · SYDNEY · PARIS · FRANKFURT

U.K.	Pergamon Press Ltd., Headington Hill Hall, Oxford OX3 0BW, England
U.S.A.	Pergamon Press Inc., Maxwell House, Fairview Park, Elmsford, New York 10523, U.S.A.
CANADA	Pergamon of Canada, Suite 104, 150 Consumers Road, Willowdale, Ontario M2J 1P9, Canada
AUSTRALIA	Pergamon Press (Aust.) Pty. Ltd., P.O. Box 544, Potts Point, N.S.W. 2011, Australia
FRANCE	Pergamon Press SARL, 24 rue des Ecoles, 75240 Paris, Cedex 05, France
FEDERAL REPUBLIC OF GERMANY	Pergamon Press GmbH, 6242 Kronberg-Taunus, Hammerweg 6, Federal Republic of Germany

Copyright © 1980 Pergamon Press Ltd.

All Rights Reserved. No part of this publication may be reproduced, stored in a retrieval system or transmitted in any form or by any means: electronic, electrostatic, magnetic tape, mechanical, photocopying, recording or otherwise, without permission in writing from the publishers.

First edition 1980

British Library Cataloguing in Publication Data

International Radiation Protection Society.
Congress, 5th, Jerusalem, 1980
Radiation protection.
1. Ionizing radiation - Safety measures - Congresses
I. Title
614.8'39   RA569   80-49879
ISBN 0-08-025912-X

In order to make this volume available as economically and as rapidly as possible the authors' typescripts have been reproduced in their original forms. This method has its typographical limitations but it is hoped that they in no way distract the reader.

Printed in Great Britain by A. Wheaton & Co. Ltd, Exeter

CONTENTS

Volume 1

LIST OF CONTRIBUTORS   xxv

SIEVERT LECTURE

Some Non-Scientific Influences on Radiation Protection
Standards and Practice
Lauriston S. Taylor   3

OCCUPATIONAL EXPOSURE

Reducing Occupational Radiation Exposures at LWRs   19
D. Lattanzi, C. Neri, C. Papa and S. Paribelli

Radiation Protection Aspects in Decommissioning of a Fuel
Reprocessing Plant   27
P. Kotrappa, P. P. Joshi, T. K. Theyyunni, B. M. Sidhwa and
M. N. Nadkarni

The Practical Application of ICRP Recommendations Regarding
Dose-Equivalent Limits for Workers to Staff in Diagnostic
X-Ray Departments   31
J. R. Gill, P. F. Beaver and J. A. Dennis

Radiation Protection in the Dental Profession   35
B. Holyoak, J. K. Overend and J. R. Gill

The Development of Criteria for Limiting the Non-Stochastic
Effects of Non-Uniform Skin Exposure   39
M. W. Charles and J. Wells

Occupational Exposures Associated with Laser Heat Sealers   44
F. J. Bradley, L. Schuster, L. Cabasino, C. Tedford, D. Warren
and L. Fitzrandolph

Contents

Investigation of the Nature of a Contamination Caused by
Tritium Targets used for Neutron Production — 48
E. M. M. de Ras, J. P. Vaane and W. Van Suetendael

Les Problemes de Radioprotection Rencontres Dans un Laboratoire
de Marquage de Molecules au Carbone-11 — 52
H. Vialettes and A. Moreau

Some Radiation Protection Problems in a Cancer Hospital and
Associated Research Institute — 56
N. G. Trott, W. Anderson, R. Davis, R. P. Parker, D. M. Carden,
Nina Pearson and Elizabeth Harbottle

Problemes de Radioprotection Rencontres Dans un Laboratoire
de Controle Radiopharmaceutique - Irradiation des Extremites
des Mains — 60
J. J. Andre, J. Goubert, A. Moreau and J. P. Perotin

Health Physics Evaluation of an Acute Overexposure to a
Radiography Source — 64
J. K. Basson, A. P. Hanekom, F. C. Coetzee and D. C. Lloyd

Contamination of Persons Occupationally Exposed to Natural
Radioactivity in a Coal Fired Power Plant — 68
Alica Bauman and Djurdja Horvat

Occupational Thyroid Exposure Due to Internal Radioiodine
Contamination in Radiation Workers Handling Iodine-131(125) for
Non-Therapeutic Use — 72
H. R. Erlenbach

Personnel Dosimetry System Based on TLD at PNC Tokai Works — 76
H. Ishiguro and S. Fukuda

Mean and Individual Radiation Exposures of the Staff of the
Karlsruhe Nuclear Research Center, 1969-1978 — 80
Winfried Koelzer and H. Kiefer

Personnel Hazards from Medical Electron Accelerator Photoneutrons — 84
R. C. McCall, T. M. Jenkins, R. A. Shore and P. D. LaRiviere

Investigation Levels of Radioisotopes in the Body and in Urine.
Consequences of the Recent Recommendations on the Annual Limits
of Intake — 88
Y. Shamai, M. Tirkel and T. Schlesinger

Health Physics Documentation — 92
G. Stäblein

Analysis of Medical Occupational Exposure to Ionizing Radiation
on Taiwan During Past Two Decades — 95
P.-S. Weng and S.-Y. Li

ENVIRONMENTAL MODELLING AND DOSE ASSESSMENT

A Methodology for the Evaluation of Collective Doses Arising from
Radioactive Discharges to the Atmosphere — 101
J. Hallam and G. S. Linsley

Contents

Collective Doses Due to Radioactive Emissions from Nuclear Plants 105
 H. Bonka and H.-G. Horn

Evaluation de la Dose Collective a l'Echelle Europeenne due a des Rejets Atmospheriques 109
 A. Després, J. Le Grand, A. Bouville and Jean-Marie Guezengar

Influence of the Most Important Data on the Calculation of the Maximum Radiation Exposure in the Vicinity of Nuclear Facilities 113
 P. Schmidtlein, H. Bonka, D. Hesel and H.-G. Horn

Impact Sanitaire du Programme Nucleaire Francais en 1990 117
 C. Maccia and F. Fagnani

The Status of Radioactive Waste Management: Needs for Reassessments 121
 Merril Eisenbud

Radioecological Models for Estimating Short and Long-Term Effects of Releases in the Nuclear Fuel Cycle 129
 M. Nilsson and B. Persson

Evaluation des Doses Individuelles et Collectives Resultant de Rejets en Riviere 133
 H. Fabre, J. Le Grand and A. Bouville

Modelisation des Echanges de Produits Alimentaires en vue de l'Evaluation des Doses Collectives 137
 A. Garnier, J. Brenot, A. M. Obino and J. M. Herbecq

The Identification of Critical Groups 141
 G. J. Hunt and J. G. Shepherd

Assessment of the Radiation Exposure from the Radioactive Material Released from the Stack of a 2000 MWe Coal Fired Power Station 145
 W. C. Camplin and J. Hallam

Effect of the Foodchain in Radioactivities Released from Thermal Power Plants 149
 K. Okamoto

Hazards from Radioactivity of Fly Ash of Greek Coal Power Plants (CPP) 153
 C. Papastefanou and S. Charalambous

PATIENT EXPOSURE AND RADIOTHERAPY

Optimising Information Density in Nuclear Medicine Imaging Procedures 161
 L. D. Brown

A Study of Gonad Doses in X-Ray Radiographic Examinations of the Abdomen 165
 L. D. Brown

Diagnostic X-Ray Equipment Evaluation in Brazil 169
 Anna Maria Campos de Araujo, J. E. Peixoto and Vera R. G. Reis

Contents

Surveillance of X-Ray Machines in Israel ... 173
A. Donagi, J. Hai and M. Kuszpet

Projections of Organ Doses from Diagnostic Radiology in Radiation Protection and Control ... 175
J. J. Fletcher and H. Cember

Computer Aided Design of Fast Neutron Therapy Units ... 179
A. E. Gileadi, H. J. Gomberg and I. Lampe

Measurements of the Effect of "Thyroid Blocking" in Patients Investigated with ^{125}I-Fibrinogen ... 183
L. Jacobsson, S. Mattsson and B. Nosslin

Whole Body Electron Therapy Using the Philips SL75/10 Linear Accelerator ... 186
Y. Mandelzweig, M. Tatcher and M. Yudelev

Neutron Production and Leakage from Medical Electron Acclerators ... 190
R. C. McCall and W. P. Swanson

Patient Dose Evaluations from Medical X-Ray Exposure in Italy: An Analysis of Next Data ... 194
M. Andrea, P. F. Mario, Susanna Antonio, C. Angelo and S. Paolo

Retention of Molybdenum-99 in Adult Man ... 198
A. Parodo, F. Manca and G. Madeddu

Organ Doses in Diagnostic Radiology ... 202
M. Rosenstein

Internal Radiation Dosimetry of F-18-5-Fluorouracil ... 206
J. Shani, D. Young, T. Schlesinger and W. Wolf

The Current Contribution of Diagnostic Radiology to the Population Dose in Great Britain ... 210
B. F. Wall, S. Rae, G. M. Kendall, S. C. Darby, E. S. Fisher and S. V. Harries

RADIOACTIVE RELEASES AND ENVIRONMENTAL MONITORING

Detection of Low-Level Environmental Exposure Rates Due to Noble Gas Releases from the Muehleberg Nuclear Power Plant ... 217
J. Czarnecki, H. Völkle and S. Prêtre

Retention of ^{14}C in Nuclear Power Plants and Reprocessing Plants ... 221
D. Gründler and H. Bonka

Statistical Correlation of Environmental Tritium Values at Trombay ... 225
T. S. Iyengar, J. V. Deo, S. D. Soman and A. K. Ganguly

An Air Monitoring Programme in the Environment of a Major Nuclear Establishment: Operation and Results ... 229
A. Knight, B. M. R. Green, M. C. O'Riordan and G. S. Linsley

Six-Year Experiences in the Environmental Radioactivity Monitoring of Taiwan ... 233
Y.-M. Lin and C.-H. Cheng

Contents

The Pre-Operational Monitoring – How Useful are Recommendations of International Organizations and Various National Programs 237
M. Mihailović

Migration of Radionuclides Following Shallow Land Burial 240
J. Sedlet and N. W. Golchert

Principles and Results of Environmental Surveillance of the Austrian Research Center at Seibersdorf within the Last Twenty Years 244
F. Steger, E. Etzersdorfer and H. Sorantin

Six-Year Experiences in the Operation of a Low Level Liquid Waste Treatment Plant 248
S.-J. Wen, S.-L. Hwang and C.-M. Tsai

ASSESSMENT OF RISK

Risk Assessment Perspectives in Radiation Protection 255
W. D. Rowe

On the Characterization of Risk 262
Y. Tzur

Stochastic and Non-Stochastic Effects: A Conceptual Analysis 266
L. R. Karhausen

Risk Estimates of Stochastic Effects Due to Exposure to Radiation – A Stochastic Harm Index 268
Y. G. Gonen

Risk Ratios for Use in Establishing Dose Limits for Occupational Exposure to Radiation 272
P. E. Metcalf and B. C. Winkler

Comparaison des Couts Marginaux de Protection dans les Centrales Thermiques Classiques et Nucleaires 276
A. Oudiz and J. Lochard

Evaluation Quantitative et Comparative des Risques Lies Aux Centrales Nucleaires 280
S. Vignes, J. C. Nénot and M. Bertin

Radiation and the Perception of Risk in the USA 284
J. A. Hébert and R. L. Kathren

Physical Characteristics of the Japanese in Relation to *Reference Man* 288
G-I. Tanaka, H. Kawamura and E. Nomura

Competing Risk Theory and Radiation Risk Assessment 292
P. G. Groer

Contents

DOSE EQUIVALENT

Medical Irradiation and the Use of the "Effective Dose Equivalent" Concept
  B. R. R. Persson — 299

Is the Dose Equivalent Index a Quantity to be Measured?
  S. R. Wagner — 307

Dose Equivalent Conversion Factors for External Photon Irradiation
  R. Kramer and G. Drexler — 311

A Theoretical Approach for the Measurement of the Effective Dose Equivalent for External Radiations
  R. Jahr, R. Hollnagel and B. Siebert — 317

Biological Effectiveness of Fast Neutrons
  M. Coppola and G. Silini — 321

DOSIMETRY

A Systematic Approach to Personnel Neutron Monitoring
  R. V. Griffith and D. E. Hankins — 327

The Present State of Nuclear Accident Dosimetry
  B. Majborn — 335

Remarks on the Present TLD Concept in Personnel Monitoring
  D. F. Regulla and G. Drexler — 339

Phosphate Glass Dosimetry: A Potential Alternative in Personnel Monitoring
  E. K. A. Piesch and D. F. Regulla — 343

Track Structure Calculation of the Thermoluminescent Yields of Heavy Charged Particles
  Y. S. Horowitz and J. Kalef-Ezra — 347

Fast Neutron Dosimetry Using $CaSO_4$:Dy Thermoluminescent Dosimeters
  N. J. Azorín, C. R. Salvi, J. L. Rubio and C. A. Gutiérrez — 351

European Interlaboratory Test Programme for Luminescence Dosemeter Systems Used in Environmental Monitoring
  B. Burgkhardt, E. Piesch and H. Seguin — 355

Simultaneous Sensitisation and Re-Estimation in Thermoluminescent LiF
  M. W. Charles and Z. U. Khan — 359

Skin Dose Assessment in Routine Personnel Beta/Gamma Dosimetry
  P. Christensen — 363

Electret Dosimetry
  G. Dreyfus, J. Lewiner, D. Perino, W. Buttler and J. C. Magne — 367

Contents

Development of Neutron Dosimeters for Fast and Epithermal Neutrons — 371
Y. Eisen, Y. Shamai, E. Ovadia, Z. Karpinovitch, S. Faerman and T. Schlesinger

Arrangement of a TLD System to Measure the Dose to Patients Undergoing Irradiation — 375
C. Furetta, R. Pellegrini and C. Bacci

Polyurethane as a Base for a Family of Tissue Equivalent Materials — 380
R. V. Griffith

Development of a Personnel Neutron Dosimeter/Spectrometer — 384
R. V. Griffith, J. C. Fisher, L. Tommasino and C. Zapparoli

Neutron Sensitivity of Geiger-Müller Photon Dosemeters for Neutron Energies Between 100 keV and 19 MeV — 388
S. Guldbakke, R. Jahr, H. Lesiecki and H. Schölermann

Microdosimetric Approach for Lung Dose Assessments — 392
W. Hofmann, F. Steinhäusler, E. Pohl and G. Bernroider

Calibration and Application of the Multisphere Technique in Neutron Spectrometry and Dosimetry — 396
C. J. Huyskens and G. J. H. Jacobs

Spark Counting Technique with an Aluminium Oxide Film — 400
H. Kawai, T. Koga, H. Morishima, T. Niwa and Y. Nishiwaki

A Microcomputer Controlled Thermoluminescence Dosimetry System — 404
C. J. Huyskens and P. J. H. Kicken

Recent Development of Fluoro-Glass Dosimeter in Japan — 408
Y. Nishiwake and T. Omori

TSEE Dosimeter for Gamma-Rays and Fast Neutrons Using Ceramic BeO — 412
S. Ohtani

Beta Response of TL Personnel and Environmetnal Surveillance Dose Meters — 416
D. F. Regulla and L. Caldas

Dosimetric Applications of Cellular Electrophysiological Changes Under High- and Low-LET Irradiation in Health Physics — 420
F. Steinhäusler, W. Hofmann, P. Eckl and Johanna Pohl-Rüling

Electrochemical Etching CR-39 Foils for Personnel Fast Neutron Dosimetry — 424
L. Tommasino, G. Zapparoli, R. V. Griffith and J. C. Fisher

Classification of LiF- Dosimeters Using the Ratio of Peak Heights — 428
W. Wachter, N. J. Vana and H. Aiginger

Enquete sur la Qualite d'Exploitation des Detecteurs Gamma au Germanium — 431
J. C. Zerbib

Contents

APPLIED RADIATION PROTECTION

Radiation Protection for Industrial Radiography in the
  Aerospace Industry
  W. E. Morgan ... 437

Simulation and Computation in Health Physics Training
  J. R. A. Lakey, D. C. C. Gibbs and C. P. Marchant ... 443

Retention of Radioactive Substances, Aerosols and Poison
  Gas by Sandfilter
  L. Balcarczyk, V. Hess and H. Sorantin ... 447

A Study of Effluent Control Technologies Employed by
  Radiopharmaceutical Users and Suppliers
  L. Leventhal, J. Slider, E. Chakoff, J. I. Cehn and E. Savage ... 450

Surveying and Assessing the Hazards Associated with the
  Processing of Uranium
  J. Kruger ... 454

Rupture Accident in a BWR Offgas Charcoal Treatment System
  Michele Laraia ... 460

Tritium Control at the Tritium Systems Test Assembly
  R. A. Jalbert ... 464

Tritium Control Field Study at an 'Open-Concept' Candu (PHWR)
  Station
  L. J. Sennema, M. A. Maan and G. A. Vivian ... 468

The Control of Radioactivity in the Working Environment in the
  Factories for Production of Phosphate Fertilizers
  B. N. Ajdacic, S. S. Gnjatovic and P. V. Vujovic ... 472

Present State of the Monitoring for Internal Contamination at
  Tokai Research Establishment, Japan Atomic Energy Research
  Institute
  J. Akaishi, H. Fukuda and S. Mizushita ... 476

Uranium Compounds in Ceramic Enamels-Radioactivity Analysis and
  Use Hazards
  G. Cucchi and P. Amadesi ... 480

Emergency Plan for a Uranium and Plutonium Handling Laboratory
  A. Hefner, P. Mullner and H. Sorantin ... 483

Radiation Exposure of Personnel in a Reprocessing Plant
  G. Herrmann ... 487

Radiological Protection Aspects of ^{123}I Production
  C. J. Huyskens and R. L. P. van den Bosch ... 491

Experiences in Monitoring Airborne Radioactive Contamination
  in Jaeri
  Y. Ikezawa, T. Okamoto and A. Yabe ... 495

An Operational System of Radiation Protection for Occupationally
  Exposed Personnel
  R. Kramer and G. Drexler ... 499

Contents

Studies on the Radiation Burden Using ^{131}I for Thyroid Therapy J. W. Krzesniak and O. A. Chomicki	502
Evaluation of the Protection Factor of Half-Masks with Respirator Fitting Test Apparatus M. Murata, Y. Ikezawa, Y. Yoshida, H. Matsui and M. Kokubu	506
Mesures de Debit de Dose au Contact de Divers Materiels de Laboratoire Utilises en Radiochimie et Radiobiologie J. P. Perotin, A. Moreau, J. P. Guerre and J. Goubert	510
Contribution a l'Irradiation de l'Am 241 Present Dans le Plutonium. Risques Radiologiques Lies Aux Manipulations de l'Am 241 G. Pescayre and L. Piton	514
Radiation Protection Problems at Compact Cyclotrons for Medical and Other Use P. F. Sauermann, W. Friedrich, J. Knieper, K. Komnick and H. Printz	518
Development of Guidelines for Incorporation Monitoring Programs H. Sorantin, K. Irlweck and F. Steger	522
Characterization of Working Conditions for Handling Radionuclides Y. van der Feer	528

ETHICAL AND LEGAL ASPECTS

A Bioethical Perspective on Radiation Protection and "Safety" M. N. Maxey	535
Ethical Norms in the Use of Radiation and Nuclear Power P. Oftedal	543
The Irradiation of Human Volunteer Subjects in Research R. Rosen	549
Regulatory Limits in Radiation Injury Cases: Recent Developments G. Charnoff	553

ATMOSPHERIC DISPERSION AND AEROSOL BEHAVIOR

Long-Range Transport of Radioisotopes in the Atmosphere and the Calculation of Collective Dose H. M. ApSimon, A. J. H. Goddard and J. Wrigley	559
Comparison of the Atmospheric Dispersion Calculation with the Measurements of γ-Radiation Levels from Ar-41 Releases S. Chakraborty, E. Nagel and A. Zurkinden	563
Realistic Environmental Exposure Calculation for a Multisource Facility H. G. Ehrlich, K. Heinemann and K. J. Vogt	567
Reduction of the Environmental Concentration of Air Pollutants by Proper Geometrical Orientation of Industrial Line Sources J. Tadmor and A. Eisner	571
The Importance of Plume Rise in Risk Calculations L. S. Fryer and G. D. Kaiser	575

Contents

Statistical Studies on the Limitation of Short-Time Releases from
   Nuclear Facilities 579
   K. Heinemann and K. J. Vogt

Washout and Dry Deposition of Atmospheric Aerosols 583
   J. Porstendorfer, G. Robig and A. Ahmed

The Functional Dependence of the Total Hazard from an Air
   Pollution Incidence on the Environmental Parameters 588
   D. Skibin

Resuspension of Plutonium from Contaminated Soils 592
   V. D. Vashi, P. Kotrappa and S. D. Soman

A Revised Method to Calculate the Concentration Time Integral of
   Atmospheric Pollutants 596
   E. Voelz and H. Schultz

Comparison of Two- and Three-Dimensional Atmospheric Dispersion
   Models 600
   A. Zurkinden

CONTENTS

Volume 2

INTERNAL CONTAMINATION FROM Pu AND OTHER TRANSURANICS

A Case of Internal Contamination with Plutonium Oxide — 607
F. Breuer, G. F. Clemente, E. Strambi and C. Testa

Toxicity of Inhaled ^{238}PuO$_2$ I. Metabolism — 611
J. H. Diel and J. A. Mewhinney

Proposed Retention Model for Human Inhalation Exposure to ^{241}AmO$_2$ — 615
J. A. Mewhinney, W. C. Griffith and B. A. Muggenburg

Toxicity of Inhaled ^{238}PuO$_2$ II. Biological Effects in Beagle Dogs — 619
B. A. Muggenburg, J. A. Mewhinney, Barbara S. Merickel,
B. B. Boecker, F. F. Hahn, R. A. Guilmette, J. L. Mauderly and
R. O. McClellan

The Accuracy of a Routine Plutonium in Lung Assessment Programme — 623
D. Ramsden

Retention and Effects of ^{239}Pu in the Tree Shrew (*Tupaia Belangeri*) — 627
A. Seidel, R. M. Flügel, D. Komitowski and G. Darai

The Use of DTPA to Inhibit the Extrapulmonary Deposition of
Curium-244 in the Rat Following the Bronchial Intubation of
Oxide Suspensions — 631
J. R. Cooper, G. N. Stradling, H. Smith and Sandra E. Ham

The Effect of Oxidation State on the Absorption of Ingested or
Inhaled Plutonium — 635
M. F. Sullivan, Linda S. Gorham and J. L. Ryan

Pulmonary Carcinogenesis from Plutonium-Containing Particles — 638
R. G. Thomas, D. M. Smith and E. C. Anderson

Contents

Subcellular Distribution of ^{239}Pu in the Liver of Rat, Mouse,
  Syrian and Chinese Hamster 642
  R. Winter and A. Seidel

RADIOBIOLOGY

Biological Effects of Inhaled Radionuclides: Summary of ICRP
  Report 31 649
  W. J. Bair

An Approach to the Derivation of Radionuclide Intake Limits for
  Members of the Public 657
  R. C. Thompson

RBE of α-Particles vs β-Particles in Bone Sarcoma Induction 661
  C. W. Mays and Miriam P. Finkel

Are We at Risk from Low Level Radiation - DNA Repair Capacity
  As a Probe of Potential Damage and Recovery 666
  E. Riklis

Quatre Observations de Traitement Chirurgical de Blessures
  Contaminees par des Radionucleides 670
  F. Briot, P. Henry, P. Lalu, J. Mercier, J. Sarbach and J. Tourte

RADIOLOGICAL ASPECTS OF THE THREE MILE ISLAND ACCIDENT

Radiological Consequences of the Three Mile Island Accident 677
  L. Battist and H. T. Peterson, Jr.

Protective Action Guides: Theory and Application Lessons from
  the Three Mile Island Accident 685
  B. Shleien

Some Radiation Protection Implications of the Three Mile Island
  Incident 690
  D. Ilberg

NON-IONIZING RADIATION

Microwave/Radiofrequency Protection Standards: Concepts,
  Criteria and Applications 695
  S. M. Michaelson

Electromagnetic Pollution of the Environment 703
  L. Argiero and G. Rossi

The Evolution of Non-Ionizing Radiation Protection Standards
  in Norway 711
  H. Aamlid and G. Saxebol

A Recommended Permissible Environmental Standard for Microwave
  and Radiofrequency Radiation 715
  L. R. Solon

Insidious Ocular Effects of Laser Radiation   719
   D. H. Brennan

Effects of Microwave Radiation on Endocrine System of Mouse   723
   P. Deschaux, R. Santini, R. Fontanges and J. P. Pellissier

INSTRUMENTATION AND SHIELDING

Nuclear Shielding Analyses for an Intense Neutron Source
   Facility   731
   J. Celnik

Use of Informatic for Radiation Control Panels   736
   R. Cochinal, B. Grimont and V. Mai

Secondary Standard Dosimetry System with Automatic Dose/Rate
   Calculation   739
   K. E. Duftschmid, J. Bernhart, G. Stehno, W. Klosch, J. Hizo
   and K. Zsdanszky

Dose Rates During Experiments with Heavy Ions   743
   J. G. Festag

A Microprocessor Based Area Monitor System for Neutron and
   Gamma Radiation   747
   R. Wilhelm and G. Heusser

A Personal Tritium Monitor   751
   R. V. Osborne and A. S. Coveart

Measurement of Absorbed Dose-Rate in Skin for Low-Level
   Beta-Rays   755
   K. Shinohara, Y. Kishimoto, Y. Kitahara and S. Fukuda

Avantages Presentes par l'Introduction Industrielle de
   l'Informatique dans la Surveillance Centralisee des Niveaux
   de Rayonnements   759
   H. Vialettes and P. Leblanc

"Intelligent" Radiation Instruments   763
   A. Ward

Spectra, Differential Albedo and Shielding Data for Bremsstrahlung
   Scattered from Common Shielding Materials   766
   H.-P. Weise, P. Jost and W. Freundt

EMERGENCY PLANNING

Emergency Planning and Preparedness: Pre- and Post-Three Mile
   Island   773
   H. E. Collins

On the Extent of Emergency Actions for the Protection of the
   Public After Accidental Activity Releases from Nuclear Power
   Plants   781
   W. G. Hubschmann, A. Bayer, K. Burkart and S. Vogt

Contents

An Application of Cost-Effectiveness Analysis to Restrict the Damage Caused by an Accidental Release of Radioactive Material to the Environment L. Frittelli and A. Tamburrano	785
Discussion of an Environmental Dose Methodology to Obtain Compliance with Dose Limits in the Case of Postulated Accidents H. D. Brenk and K. J. Vogt	790
Protecting Front-Line Survey and Rescue Teams During Emergencies H. Tresise	794

BIOLOGICAL EFFECTS OF ENVIRONMENTAL RADIATION

Natural Background as an Indicator of Radiation-Induced Cancer J. J. Cohen	801
Two Decades of Research in the Brazilian Areas of High Natural Radioactivity T. L. Cullen, A. S. Paschoa, E. P. Franca, C. C. Ribeiro, M. Barcinski and M. Eisenbud	805
Evaluation of Cancer Incidence for Anglos in the Period 1969-1971 in Areas of Census Tracts with Measured Concentrations of Plutonium Soil Contamination Downwind from the Rocky Flats Plant in the Denver Standard Metropolitan Statistical Area C. J. Johnson	809
Hazards of Radon Daughters to the General Public D. K. Myers, J. R. Johnson and A. M. Marko	813
Lung Doses from Radon in Dwellings and Influencing Factors E. Stranden	817
Assessment of Biological Effects Resulting from Large Scale Applications of Coal Power Plant Wastes in Building Technology in Poland J. Pensko and J. Geisler	821

JUSTIFICATION AND OPTIMIZATION

Justification and Optimization in Radiation Protection D. Beninson	827
The Cost of Occupational Dose A. B. Fleishman and M. J. Clark	835
Optimisation et Controle des Rejets Radioactifs des Centrales a Eau Pressurisee du Programme Electronucleaire Francais J. Lochard, C. Maccia and P. Pagès	839
Optimisation de la Protection et Evaluation du Risque le Cas des Accidents de Transport de Matieres Dangereuses T. Meslin	843

Contents

Application of the Principles of Justification and Optimisation to Products Causing Public Exposure 847
A. B. Fleishman and A. D. Wrixon

RADIATION PROTECTION STANDARDS

Setting Standards for Trivial Concentrations of Radioactivity 853
D. van As

International Electrotechnical Commission (IEC) and Radiation Safety Requirements for Medical X-Equipment 858
E. Koivisto and H. Bertheau

The Laboratory Appraisal of Ionisation Chamber Smoke Detectors 862
B. T. Wilkins and D. W. Dixon

Occupational Dose Equivalent Limits 866
E. P. Goldfinch

The Development of the American National Standard, "Control of Radioactive Surface Contamination on Materials, Equipment and Facilities to be Released for Uncontrolled Use" 870
J. Shapiro

MEASUREMENT TECHNIQUES

A Simple Method for Collection of HTO from Air 877
A. N. Auf der Maur and T. Lauffenburger

A Procedure for Routine Radiation Protection Checking of Mammography Equipment 881
L. G. Bengtsson and I. Lundéhn

Health Physics Aspects of the *In Vivo* Analysis of Human Dental Enamel by Proton Activation 885
F. Bodart and L. Ghoos

A Relatively Fast Assay of Sr-90 by Measuring the Cherenkov Effect from the Ingrowing Y-90 889
B. Carmon and Y. Eliah

Decontamination of Tritiated Water Samples Prior to Tritium Assay 892
B. Carmon, S. Levinson and Y. Eliah

Evaluation of the Spectral Distribution of X-Ray Beams from Measurements on the Scattered Radiation 895
E. Casnati and C. Baraldi

Efficacite de Comptage de Gels Scintillants 899
M. Chauvet-Deroudilhe, M. Dell'Amico, M. Bourdeaux and C. Briand

Study and Measurement of the Atmospheric Pollution by ^{85}Kr 903
G. Eggermont, J. Buysse, A. Janssens and F. Raes

Neutron Spectra and Dose Equivalent Inside Reactor Containment 907
G. W. R. Endres, L. G. Faust, L. W. Brackenbush and R. V. Griffith

Contents

Dosimetry of Criticality Accidents Using Activations of the
Blood and Hair — 911
D. E. Hankins

A Passive Monitor for Radon Using Electrochemical Track Etch
Detector — 915
G. E. Massera, G. M. Hassib and E. Piesch

Beta Dosimetry with Surface Barrier Detectors — 919
M. F. M. Heinzelmann, H. Schuren and K. Dreesen

^{230}Th Assay by Epithermal Neutron Activation Analysis — 923
R. L. Kathren, A. E. Desrosiers and L. B. Church

Calibration of Radiation Protection Instruments at SSDL Level
in Israel — 927
M. E. Kuszpet, A. Donagi and T. Schlesinger

Atmospheric Dispersion Study with ^{85}Kr and SF_6 Gas — 931
O. Narita, Y. Kishimoto, Y. Kitahara and S. Fukuda

Methods of I-129 Analysis for Environmental Monitoring — 935
T. Nomura, H. Katagiri, Y. Kitahara and S. Fukuda

A New Technique for Neutron Monitoring in Stray Radiation Fields — 939
E. Piesch and B. Burgkhardt

Minidosimetry of Alpha-Radiation from 239-Pu in the Skeleton — 943
E. Polig

Decontamination and Modification of Liquid Scintillators — 947
S. R. Sachan and S. D. Soman

Tritium – Is It Underestimated? — 951
G. D. Whitlock

INTERNAL DOSIMETRY

Inhalation du Radon et de Ses Produits de Filiation : Doses et
Effets Pathologiques — 957
D. Mechali

Recent Past and Near Future Activities of ICRP Committee 2 — 965
R. C. Thompson

Les Effets Cancerigenes Combines des Radiations Ionisantes
et des Molecules Chimiques — 973
J. Lafuma

REGULATORY ASPECTS

Flexibility in Radiation Protection Legislation – The UK
Approach — 983
P. F. Beaver and J. R. Gill

Contents

A Review from the Regulatory Position of the Control of Occupational
Exposure Associated with the First 20 Years of the United Kingdom
Commercial Nuclear Power Programme — 987
B. W. Emmerson

Occupational Radiation Protection Legislation in Israel — 991
J. Tadmor, T. Schlesinger, A. Donagl and C. Lemesch

Legal Provisions Concerning the Handling and Disposal of
Radioactive Waste in International and National Law — 994
W. Bischof

WASTE DISPOSAL

Management of Radioactive Wastes in the United States of America — 1001
F. L. Parker

Environmental Monitoring and Deep Ocean Disposal of Packaged
Radioactive Waste — 1009
N. T. Mitchell and A. Preston

Absorption of Dose Rate in Great Bitumen Blocks - Experimental
Detection — 1014
P. R. M. Patek

A Review of the Disposal of Miscellaneous Radioactive Wastes in
the United Kingdom — 1018
B. Hookway

INTERNAL CONTAMINATION AND RADIOBIOLOGY

Chemical Protection and Sensitization to Ionizing Radiation:
Molecular Investigations — 1025
R. Badiello

Dynamics of Cs-137 Distribution in the Muscle Tissue of Swine
by Single and Repeated Contamination — 1029
J. Begovic, S. Stankovic and R. Mitrovic

Tumorigenic Responses from Single or Repeated Inhalation
Exposures to Relatively Insoluble Aerosols of ^{144}Ce — 1033
B. B. Boecker, F. F. Hahn, J. L. Mauderly and R. O. McClellan

Radiobiological and Radioecological Studies with the Unicellular
Marine Algae *Acetabularia*, *Batophora* and *Dunaliella* — 1037
S. Bonotto, A. Luttke, S. Strack, R. Kirchmann, D. Hoursiangou
and S. Puiseux-Dao

Para-Hydroxybenzoic Acid, a Hypoxic Radiosensitizer in Bacterial
Cells — 1041
N. Sade and G. P. Jacobs

A Dosimetric Model for Tissues of the Human Respiratory Tract at
Risk from Inhaled Radon and Thoron Daughters — 1045
A. C. James, J. Rosemary Greenhalgh and A. Birchall

Contents

Distribution of Plutonium and Americium in Human and Animal
Tissues after Chronic Exposures
J. K. Miettinen, H. Mussalo, M. Hakanen, T. Jaakkola,
M. Keinonen and P. Tähtinen ... 1049

Etude Experimentale des Cancers Induits Chez le Rat par des
Particules a Transfert Lineique d'Energie Elevee
Michèle E. Morin and J. E. Lafuma ... 1053

Present State of Radio-Strontium Decorporation Research with
Cryptand(222)
W. M. Müller ... 1056

Sustained Release of Radioprotective Agents *In Vitro*
J. Shani, S. Benita, A. Samuni and M. Donbrow ... 1060

Investigation of the Solubility of Yellowcake in the Lung of
Uranium Mill Yellowcake Workers by Assay for Uranium in Urine
and *In Vivo* Photon Measurements of Internally Deposited Uranium
Compounds
H. P. Spitz, B. Robinson, D. R. Fisher and K. R. Heid ... 1064

Excretion of Organic and Inorganic Tritiated Compounds in Cow's
Milk after Ingestion of Tritium Oxide
J. Van den Hoek, G. B. Gerber and R. Kirchmann ... 1068

Correlation Between Concentrations of ^{210}Pb in the Biologic
Samples from Miners and Individual Levels of Exposure to
Short Lived Radon-222 Daughter Products
H. I. M. Weissbuch, M. D. Grădinaru and G. T. Mihail ... 1072

Distribution and Clearance of Inhaled ^{63}NiCl$_2$ by Rats
P. L. Ziemer and S. M. Carvalho ... 1075

FATE OF RADIONUCLIDES IN THE ENVIRONMENT

Studies of Age-Dependent Strontium Metabolism with Application
to Fallout Data
S. R. Bernard and C. W. Nestor, Jr. ... 1083

Fate of Major Radionuclides in the Liquid Wastes Released to
Coastal Waters
I. S. Bhat, P. C. Verma, R. S. Iyer and S. Chandramouli ... 1087

Assessment of ^{210}Po Exposure for the Italian Population
G. F. Clemente, A. Renzetti, G. Santori and F. Breuer ... 1091

Representation Geographique de Diverses Donnees dans une Grille
Europeenne
A. Garnier, A. Sauve and C. Madelmont ... 1095

Studies of the Transfer Factors of Sr90 and Cs137 in the Food-Chain
Soil-Plant-Milk
K. Heine and A. Wiechen ... 1099

Studies of Concentration and Transfer Factors of Natural and
Artificial Actinide Elements in a Marine Environment
E. Holm, B. R. Persson and S. Mattsson ... 1103

Contents

^{226}Ra- and ^{222}Rn-Content of Drinking Water 1107
J. Kiefer, A. Wicke, F. Glaum and J. Porstendörfer

Mobility and Retention of ^{60}Co in Soils in Coastal Areas 1111
Y. Mahara and A. Kudo

Radioecological Studies of Activation Products Released from a Nuclear Power Plant into the Marine Environment 1115
S. Mattsson, M. Nilsson and E. Holm

Monitoring of Environmental Radon-222 in Selected Areas of Taiwan Province of the Republic of China 1119
T.-Y. C. Mei, C.-F. Wu and P.-S. Weng

Distribution of Uranium, ^{226}Ra, ^{210}Pb and ^{210}Po in the Ecological Cycle in Mountain Regions of Central Yugoslavia 1123
Z. Miloševic, E. Horšić, R. Kljajić and A. Bauman

Ra-226 Collective Dosimetry for Surface Waters in the Uranium Mining Region of Pocos De Caldas 1127
A. S. Paschoa, G. M. Sigaud, E. C. Montenegro and G. B. Baptista

The Radium Contamination in the Southern Black Forest 1131
H. Schüttelkopf and H. Kiefer

Evaluation of Small Scale Laboratory and Pot Experiments to Determine Realistic Transfer Factors for the Radionuclides ^{90}Sr, ^{137}Cs, ^{60}Co and ^{54}Mn 1135
W. Steffens, F. Führ and W. Mittelstaedt

The Transfer of Sr-90, Cs-137, Co-60 and Mn-54 from Soils to Plants - Results from Lysimeter Experiments 1139
W. Steffens, W. Mittelstaedt and F. Führ

The Assessment of Radon and its Daughters in North Sea Gas Used in the United Kingdom 1143
B. T. Wilkins

RADIOLOGICAL AND INDUSTRIAL HYGIENE

Epidemiology, Occupational Hygiene and Health Physics 1149
J. A. Bonnell

An Analytical Approach to the Comparison of Chemical and Radiation Hazards to Man 1155
H. P. Leenhouts, K. H. Chadwick and A. Cebulska-Wasilewska

Technology Transfer from Nuclear and Radiological to Industrial Safety 1159
Y. G. Gonen

Radiation Protection Principles Applied to Conventional Industries Producing Deleterious Environmental Effects 1163
J. Tadmor

A Health and Research Organization to Meet Complex Needs of Developing Energy Technologies 1167
R. V. Griffith

Contents

REVIEW OF ACTIVITIES OF INTERNATIONAL ORGANIZATIONS

Development and Trends in Radiological Protection and the NEA
  Programme in this Field ... 1173
  O. Ilari and E. Wallauschek

The Radiation Protection Programme Activities of the World Health
  Organization ... 1181
  E. Komarov and M. J. Suess

Review of Activities of International Organizations that Cooperate
  with IRPA ... 1185
  K. Z. Morgan

REPORTS OF INTERNATIONAL COMMITTEES

Summary Account on the Activities of the IRPA/International NIR
  Committee ... 1189
  H. Jammet

Activites du Comite International de l'IRPA et Problemes Poses
  par la Protection Contre les Rayonnements Non-Ionisants ... 1192
  H. Jammet

The Work of the ILO in the Field of Protection of Workers Against
  Ionising and Non-Ionising Radiations ... 1210
  G. H. Coppée

The BEIR-III Report and its Implications for Radiation Protection
  and Public Health Policy ... 1213
  J. I. Fabrikant

The Radiation Protection Research Programme of the Commission of
  European Communities ... 1233
  F. F. Van Hoeck

*Internal Contamination
From Pu and Other
Transuranics*

A CASE OF INTERNAL CONTAMINATION WITH PLUTONIUM OXIDE

F. Breuer, G. F. Clemente, E. Strambi and C. Testa

This paper describes a case of plutonium and americium internal contamination due to an accidental glove-box explosion occurred in 1974 at the Casaccia Plutonium Plant. The involved person showed a small contaminated wound (3x 0.5 cm) on his right cheek, a diffused contamination on the hair and a considerable activity in the nose which would indicate a possible incorporation by inhalation. On the basis of the information obtained at the Plutonium Plant the material contained in the exploded glove-box resulted to be a powder of PuO_2 calcinated at high temperature. In order to know just the isotopic and weight composition of the contaminating material the following radiometric and chemical measures were carried out on the nose-blow sample: gamma spectrometry (^{241}Am); gamma+X spectrometry with a thin NaI(Tl) crystal and Be window (^{241}Am and Pu); liquid scintillation (^{241}Pu and alpha emitters); alpha spectrometry ($^{238}Pu + {}^{241}Am$ and $^{239}Pu + {}^{240}Pu$); chemical separation of americium from plutonium. The following data were obtained: 97.56% in weight for alpha emitters (0.17% ^{238}Pu, 97.05% $^{239,240}Pu$, 0.34% ^{241}Am) and 2.44% ^{241}Pu; the activity distribution was 95.9% beta activity (^{241}Pu) and 4.1% alpha activity ($^{238,239,240}Pu$ and ^{241}Am); the distribution of alpha activity resulted to be 65% $^{239,240}Pu$, 25% ^{238}Pu and 10% ^{241}Am. The knowledge of the isotopic composition was necessary to correctly estimate the initial plutonium and americium activity in the wound, the lung burden calculated by W.B.C. and the dose committment to the different organs.

DIAGNOSTIC AND THERAPEUTIC ACTIONS

The following actions were taken to reduce the initial contamination and to get the maximum information on the residual contamination and on the dose commitment. a) The wound was washed with DTPA and the activity was removed by a surgical toilet; b) the hair and the nose were decontaminated; c) some direct lung countings were performed; d) many urine and fecal samples were analyzed for Pu and ^{241}Am; e) some blood samples were collected for the determination of plutonium and for the detection of possible chromosomial aberrations; f) at the second day a DTPA treatment was started consisting on three daily 0.5g DTPA intravenous injections followed by 3 others on alternate days and on a 0.5 g DTPA aerosol inhalation during 2 consecutive days.

RESULTS

The following results were obtained. a) The activity in the wound was determined (1) by using a special NaI(Tl) probe suitable to detect the weak X emission of plutonium (17 KeV) and the X+gamma emission of ^{241}Am (17 KeV and 60 KeV); the localization of the superficial alpha activity was obtained by using a probe with a 7 mm^2 solid state alpha detector. The initial activity resulted to be $\sim$ 30 nCi, and it was reduced to background levels by washing with DTPA and by carrying out a surgical toilet. b) The initial activity in the hair was $\sim$83 nCi and it was reduced to negligible values by using a shampoo containing DTPA. The activity of the nose blow, collected just after the incident, was 7.5 nCi. A direct lung counting (2,3) of the subject, based on the detection of both the 17 KeV X-rays emitted by the plutonium isotopes and the 60 KeV gamma-rays of the ^{241}Am, was performed at various times after the incident. A 12.5 cm diameter x 0.1 mm thick NaI(Tl) phosphwich crystal positioned on the right lung or over the sternum was employed. The calibration factor applied to lung counting of the plutonium isotopes was obtained on the basis of both phantom and "in vivo" calibration (3,4) taking into account both the chest size of the subject and the isotopic composition of the contaminating mixture. The calibration factor for the ^{241}Am in vivo counting was based on phantom calibration only. The ^{238,239,240}Pu lung contents $\pm$ 2σ as a function of time elapsed from the incident were the following: 56 $\pm$ 20 nCi (5 h.); 25 $\pm$ 15 nCi (22 h.); 13 $\pm$ 10 nCi (5 d.); <10 nCi (19 d.). The corresponding ^{241}Am lung contents $\pm$ 2σ resulted to be: 2.0 $\pm$ 0.5 nCi; 1.5 $\pm$ 0.5 nCi; 0.7 $\pm$ 0.3 nCi; 0.4 $\pm$ 0.3 nCi (40 d.); 0.15 nCi (70 d.). d) Taking into account the 55 urine analyses (37 of Pu and 18 of ^{241}Am) and the 34 feces analyses (26 of Pu and 8 of ^{241}Am) (5), the excretion curves shown in Fig. 1 and 2 have been obtained. e) No plutonium activity greater than the sensitivity limit (0.04 pCi) was detected in 10 ml of blood and no cromosomial aberration was found in 200 cells. f) No effect due to the DTPA treatment was shown in the urinary excretion curves.

DOSIMETRIC EVALUATION

Taking into account the data supplied by the lung counting and by the excretion curves, the following conclusions can be drown: a) the ratio ^{238,239,240}Pu/^{241}Am for fecal excretion is about 10, just as the ratio of the alpha activity present in the contaminating material: the similar metabolism observed for Pu and Am can be due to the fact

that both the elements were present as a very insoluble oxide; b) both the Pu and Am fecal excretion curves are very steep during the first few days (Peak activity/Plateau activity $\sim 10^5$) and this datum is in good agreement with the sharp decrease of the lung content in the same period: it appears threfore that the material granulometry was high (1 ÷ 10 μm), mainly deposited in the upper part of the respiratory tract and thus fastly removed by the ordinary clearance mechanisms. c) The high ratio E_f/E_u ($\sim 10^4$ in the first few days) and the ineffectiveness of DTPA confirm the biological non-trasportability of the contaminant. d) The plutonium activity excreted in the first few days with feces is ~ 130 nCi which may correspond to an initial lung burden comprised between 13 and 65 nCi; this value is in good agreement with that found by the W.B.C. at the first day (56 $\pm$ 20 nCi). e) Taking into account the fecal curve after the first ten days, a lung half-time of about 100 days can be deduced in accordance with the values reported in the literature (6) for insoluble compounds. f) The fecal excretion after 100 days (0.5 pCi) would indicate a plutonium residual lung burden of 0.25 ÷ 1 nCi (6). g) Taking into account the urinary excretion after 100 days (0.2 pCi) a plutonium systemic burden of 3 nCi can be obtained (7). h) The committed lung dose, calculated on the basis of reference (6) and a biological half-life of 500 days for Pu and Am, resulted in the range of 60 ÷ 240 mrad with a corresponding maximum dose rate of 30 ÷ 120 mrad/y.
i) The dose due to systemic contamination has been evaluated on the basis of reference (8) and considering the following percentage depositions and biological half-lives: Pu 42% in bone ($T_b = 5.5 \cdot 10^4$ d.), 56% in liver ($T_b = 5.5 \cdot 10^4$ d.); Am 25% in bone ($T_b = 7.3 \cdot 10^3$ d.), 35% in liver ($T_b = 3.5 \cdot 10^3$ d.) and 3% in kidneys ($T_b = 2.7 \cdot 10^4$ d.). For the contribution of lung contamination to systemic dose, the T_b in lung was considered 90 days. l) The calculated absorbed dose rate for bone was rather constant being in the range of 30 ÷ 40 mrad/y slowly increasing with time; for liver a rather constant dose rate of 200 mrad/y; for kidneys a rather constant dose rate of 3 mrad/y slowly decreasing with time. m) The committed dose equivalents, calculated on the basis of ICRP recent metabolic models (9,10) with Q = 20 for alpha particles, are: lung 1.2 ÷ 4.8 rem (12 ÷ 48 mSv); bone 40 rem (400 mSv); liver 100 rem (1 Sv); kidneys 3 rem (30 mSv). The effective total body committed dose equivalent is 7.5 ÷ 8 rem (75 ÷ 80 mSv). From a medical point of view, the operator was readmitted

to unlimited radiation work, but caution was taken not to involve him in high-risk contamination areas or operations.

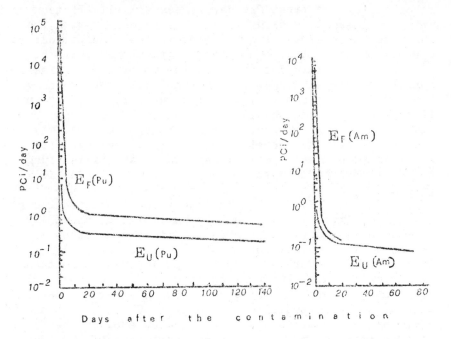

Fig.1 and 2. Urinary (E_u) and fecal (E_f) excretion of plutonium and americium.

REFERENCES

1) Testa C., Delle Site A., (1973), Proc. 2nd IRPA Regional Congress on Health Physics Problems on Internal Contamination, p. 593, Akademiai Kiado, Budapest.
2) Clemente G.F., ibid. p. 503.
3) Clemente G.F., (1973), Proc. Regional Cont. Radiation, Protection, vol. II, p. 805, IAEC, Yavne.
4) Newton D., Fry F.A., Taylor B.T., Eagle M.C., Sharma R. C., (1978), Health Physics, 35, 751.
5) Testa C., (1972), Proc. Conf. on Ass. of Rad. Cont. in Man, p. 405, IAEA, Vienna.
6) IAEA, (1973), Techn.Rep. Series n.142.
7) ICPR, (1968), Publ. n.10, Pergamon Press.
8) Fish B.R., (1972); Techn. Report ORNL-NFIC-74.
9) ICRP, (1977), Publ. n.26, Pergamon Press.
10) ICRP, (1978), Publ. n.30, Pergamon Press.

TOXICITY OF INHALED $^{238}PuO_2$
I. METABOLISM

J. H. Diel and J. A. Mewhinney

This paper reports the results of a study of the fate of inhaled $^{238}PuO_2$ in Beagle dogs. It complements a study on the relationship of biological response to radiation dose after inhalation of aerosols of $^{238}PuO_2$ (4). The aerosols used for the inhalation exposures were monodisperse (containing particles of only one size) or polydisperse. The use of monodisperse aerosols makes it possible to study the effect of particle size on the biological effectiveness of plutonium for various end points and to study the effect of particle size on the deposition and subsequent redistribution of plutonium in the body.

Other investigators (5) have shown that ^{238}Pu translocates more rapidly from the lung than does ^{239}Pu after inhalation of the dioxide form. This increased translocation may be due to a specific activity dependent breakup of the PuO_2 particles that results in more rapid dissolution or direct translocation of very small particles (1,7). In the present study, it was shown that ^{238}Pu was translocated relatively slowly up to about 100 days after inhalation, but the translocation rate increased more than twofold thereafter. A study of autoradiographs of the lungs of dogs that inhaled monodisperse aerosols revealed the presence of particle fragments in the lung. This indicated that the change in translocation was due to the breakup of particles as suggested.

MATERIALS AND METHODS

Young adult Beagle dogs received inhalation exposures to one of three sizes of a monodisperse aerosol or to a polydisperse aerosol of $^{238}PuO_2$ designed to produce an initial burden in the pulmonary region of 2.6 kBq per kg of body weight. Periodic excreta collections were made and analyzed radiochemically for Pu-238 content. Dogs were serially sacrificed after exposure. Their lungs were inflated, fixed and sampled for autoradiography. The remainder of the lung and other major organs taken at necropsy were analyzed radiochemically for Pu content. Samples taken for autoradiography were embedded, 5 μm thick sections obtained and autoradiographs were made.

Each dog's initial lung burden was calculated by summing the total excreta with the total activity in the tissues at sacrifice. The pulmonary retention of plutonium and its build-up in liver and skeleton were characterized by functions of exponentials. Radiation doses to lung, liver and skeleton were calculated by integration of the fitted curves. The number of particles, fragments and single

---

This research was performed under U.S. Department of Energy Contract No. EY-76-C-04-1013 in facilities fully accredited by the American Association for the Accreditation of Laboratory Animals.

tracks in a lung section autoradiograph were counted using an Olympus BHC microscope with darkfield illumination at 100X. Detection of fragments in lung depended on the uniform appearance of the autoradiographic images of the particles of a monodisperse aerosol. A set of concentric tracks (alpha star) which had an appearance similar to that of a alpha star in an animal sacrificed shortly after inhalation exposure was considered a particle. An alpha star with fewer tracks was called a fragment. Single tracks were either isolated alpha tracks or tracks in a group that did not originate from a common point. The diameter of a spherical $^{238}PuO_2$ particle which would be expected to have a given number of tracks was estimated (6).

RESULTS

The distribution of ^{238}Pu in lung, liver, skeleton and tracheobronchial lymph nodes of Beagle dogs sacrificed at various times after exposure is illustrated in Figure 1. Only the values for the

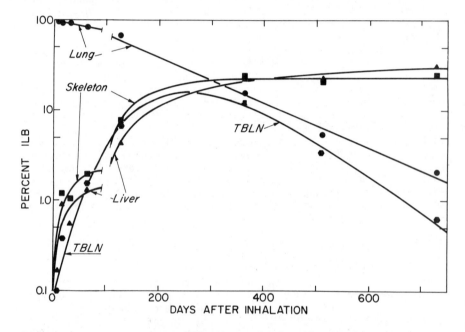

Figure 1. Distribution of ^{238}Pu in Beagle dogs following inhalation of a 1.4 μm AD monodisperse aerosol of $^{238}PuO_2$.

1.4 μm aerodynamic diameter monodisperse aerosol are shown in Figure 1. The retention and translocation of the other aerosols were similar except that the polydisperse aerosol translocated slightly more rapidly at early times after exposure. Up to 64 days after exposure, plutonium is cleared from the lung at a rate which would result in half of the material being cleared by 310 days after exposure. At later times this rate of clearance increased so that half of the material present at 100 days after exposure was cleared by 220 days after exposure resulting in a clearance half-time of 120 days.

As a result of the increased translocation from lung, the doses to liver and skeleton at 720 days after exposure were each about 10% of the dose to lung at that time and were increasing rapidly.

The fraction of the alpha activity in the autoradiographs which was considered particles, fragments or single tracks is given in Figure 2. These data are also for dogs that inhaled a 1.4 μm aerodynamic diameter monodisperse aerosol. Fragments ranged in size from

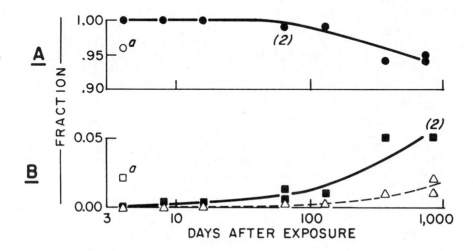

Figure 2. Fraction of activity from inhaled ^{238}PuO$_2$ particles in lung which in in (A) particles or in (B) fragments producing one (□———) or more (Δ - - - -) tracks.

1 track per fragment to about 100 tracks with an average size of about 7 tracks if single tracks were not included in the average and about 1.1 tracks with single tracks included.

DISCUSSION

The rate of clearance from the lung at times after 100 days was more than twice that at times before 100 days. Increased translocation of plutonium from the ^{238}PuO$_2$ particles in the tracheobronchial lymph nodes is illustrated by the decrease in the amount of Pu found in these nodes at times beyond 1 year after exposure. While urinary excretion of Pu began to increase at early times after inhalation, the peak level was not reached until after 100 days after inhalation. These data suggest that there was a definite change in the nature of the ^{238}PuO$_2$ particles at about 100 days after inhalation. Autoradiographic analysis of the Pu in the lung revealed an increased number of fragments beginning at about this time. These fragments were somewhat larger than those observed by Fleisher and Raabe (1) in the *in vitro* "dissolution" of $^{239}PuO_2$ particles stored dry for 3.75 years. The fragments observed in their study were all less than about 9 nm in diameter and hence could, in large part, be transferred directly from lung to blood (6). All or most of the fragments observed as single tracks in this study may be fragments larger than 10 nm in diameter, or those fragments which cannot be directly translocated to the systemic circulation (2,6). Thus, the fragmentation appears to have caused increased Pu translocation because of increased surface area (3) or direct translocation of extremely small particles into the systemic circulation (6).

CONCLUSIONS

There was breakup of $^{238}PuO_2$ particles deposited by inhalation in the lungs of Beagle dogs that resulted in less focal irradiation of the lung and in increased translocation of plutonium from the lung to other organs after about 100 days after exposure. This conclusion has several implications for the assessment of hazards following inhalation of $^{238}PuO_2$. First, in experiments using inhaled $^{238}PuO_2$, data must be obtained over a long time period, preferably at least 2 years after inhalation, to assess accurately the radiation dose to lung, liver and skeleton. Second, the increasing urinary excretion of ^{238}Pu with time following intake by inhalation must be considered when using urinary excreta data to assess the quantity of material present in lung following any inhalation incident. Third, because of the differences in translocation of Pu isotopes and the increased dispersion of ^{238}Pu in the lung, $^{238}PuO_2$ inhalation incidents must not be evaluated using factors derived from $^{239}PuO_2$ studies or incidents. Finally, the designation of organs at risk after inhalation of $^{238}PuO_2$ may require some reassessment.

REFERENCES

1. Fleischer, R. L. and Raabe, O. G. (1977): Health Phys., 32, 253.
2. Kanapilly, G. M. and Diel, J. H. (1980): Health Phys., (in press).
3. Mercer, T. T. (1967): Health Phys., 13, 1211.
4. Muggenburg, B. A., Mewhinney, J. A., Merickel, B. S., Rebar, A. H., Boecker, B. B., Guilmette, R. A. and McClellan, R. O. (1980): these proceedings.
5. Park, J. F., Case, A. C., Catt, D. L., Craig, D. K., Dagle, G. E., Madison, R. M., Powers, G. J., Ragan, H. A., Rowe, S. E., Stevens, D. L., Tadlock, J. R., Watson, C. R., Wierman, E. L. and Zwicker, G. M. (1978): In: Pacific Northwest Laboratory Annual Report, Part 1, p. 3.13.
6. Smith, H., Stradling, G. N., Loveless, B. W. and Ham, G. J. (1977): Health Phys., 33, 539.
7. Stradling, G. N., Ham, G. J., Smith, H., Cooper, J. and Breadmore, S. E. (1978): Int. J. Radiat. Biol., 34, 37.

PROPOSED RETENTION MODEL FOR HUMAN INHALATION EXPOSURE TO $^{241}AmO_2$

J. A. Mewhinney, W. C. Griffith and B. A. Muggenburg

Human exposures to ^{241}Am have been reported for four cases with measurements of lung retention near the exposure time (1-3), and five cases with long-term measurements of skeleton retention (4-6). These data were used to evaluate a model of ^{241}Am dissolution and retention developed using data from inhalation exposures of dogs to $^{241}AmO_2$. In several of the reports on human inhalation exposure, discrepancies have been shown with predictions of the Task Group Lung Model (1,3,4). The dissolution and retention model used in this paper takes into account the effects of particle size, distribution of particle sizes, and density of particles on lung retention. It is shown that the proposed dissolution and retention model is consistent with human inhalation exposures to ^{241}Am.

MATERIALS AND METHODS

The dissolution and retention model was developed using data from inhalation studies in Beagle dogs exposed to one of three sizes of monodisperse aerosols (0.75, 1.5, and 3.0 µm aerodynamic diameter) or a polydisperse aerosol (1.8 µm activity median aerodynamic diameter) of $^{241}AmO_2$ (7). Animals were sacrificed in pairs from 8 to 730 days after inhalation exposure to determine the organ retention and distribution patterns. Metabolic data from the studies using monodisperse aerosols were used to evaluate the effect of particle size on retention. Model parameters derived from the studies using monodisperse aerosols were used for modeling the study using polydisperse aerosols with adjustments only being made for the rate of mechanical clearance from lung to gastrointestinal tract. Model parameters from the study using polydisperse aerosols in Beagle dogs were compared with the human exposures in this paper.

The dissolution and retention model in Figure 1 described the lung as consisting of three regions — the nasopharynx, tracheobronchial and pulmonary — each cleared by competing pathways of mechanical clearance of particles to the gastrointestinal tract or dissolution and absorption into the general circulation. Dissolution and absorption was modeled as occurring through a dissolution pool and a compartment for the fraction of dissolved ^{241}Am bound locally to lung constituents. The locally bound ^{241}Am represents ^{241}Am seen as a diffuse distribution of single alpha tracts on autoradiographs of lung from the Beagle dog studies (7). Dissolution of the particles was described by equations developed by Mercer (8) who assumed the

---

This research was performed under U.S. Department of Energy Contract No. EY-76-C-04-1013 in facilities fully accredited by the American Association for the Accreditation of Laboratory Animals.

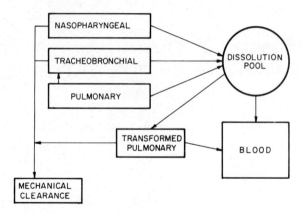

Figure 1. Dissolution and retention model for lung showing mechanical clearance pathways on the left to the gastrointestinal tract and dissolution pathways (described by equations of Mercer (8)) on the right from the 3 regions of lung to the dissolution pool. Other pathways for skeleton, liver, kidney, and soft tissue were modeled as exchanging with the blood compartment.

rate of dissolution is proportional to surface area of deposited particles. These equations take into account the distribution of particle sizes, density of particles, and shape of particles. The dissolution and retention model forced a total materials balance and used first order rate constants to describe transport, uptake and retention. In contrast, the Task Group Lung Model used by ICRP 30 (9) describes the clearance of lung as single exponential rates from subpools of the 3 regions of the lung. Also in contrast to ICRP 30, the model of retention in organs took into account redistribution of the absorbed ^{241}Am by representing the skeleton, liver, kidney, and soft tissue as two compartments in series exchanging ^{241}Am with the general circulation.

RESULTS

Predictions of lung clearance by the dissolution and retention model are shown for two particle sizes in Figure 2. Also shown are four cases of human inhalation exposures with early lung retention data (1-3) and the lung clearance of ^{241}Am as a Class W compound predicted by ICRP 30. Figure 3 shows predicted uptake by skeleton for three different particle sizes using the dissolution and retention model and for ^{241}Am as a Class W compound predicted by ICRP 30 (9). In applying the ICRP 30 lung model, it was assumed there was no absorption from the nasophyaryngeal and tracheobronchial regions since absorption from these regions appears to be low.

DISCUSSION

As can be seen from Figure 2, the dissolution and retention model can account for differences in early lung clearance observed in

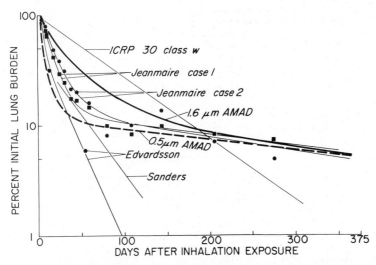

Figure 2. Comparison of early clearance from the lung for four human exposure cases (1-3), predictions of the dissolution and retention model, and predictions of ICRP 30 for ^{241}Am as a Class W compound (9). Lung retention is expressed as percent of the 4-day lung burden. Predictions of the dissolution and retention model are for a 0.5 μm activity median aerodynamic diameter (AMAD) aerosol (dashed line) and a 1.6 μm AMAD aerosol (heavy solid line) with a geometric standard deviation of 2 and density of 8 g/cm^3.

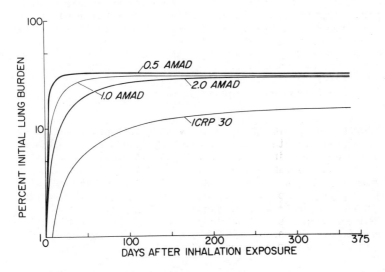

Figure 3. Predictions of skeletal uptake and retention from the dissolution and retention model for three particle sizes (geometric standard deviation of 2 and density of 8 g/cm^3) compared with the predictions of the ICRP 30 model of ^{241}Am as a Class W compound.

the human exposure cases by considering a range of activity median aerodynamic diameters from 0.5 μm to 1.6 μm. The surface area solubility rate constant was the same as that observed in dog studies (1.5 x 10^{-6} g/cm^2/day) and in *in vitro* solubility studies of AmO_2 (7). The particles were assumed to be spherical; however, this is not an essential assumption since changing the shape factor for irregular particles will have the same effect as changing the median particle size (8). Although characteristics of the aerosols were not reported for the human exposures, the parameters used to describe the aerosols in the calculations are typical of aerosols characterized in industrial facilities (10).

Comparison of the dissolution and retention model predictions with human inhalation exposure cases in which only long-term retention data were available for skeleton show reasonable agreement. Some discrepancies exist between the calculated half times of skeleton retention in humans of 17 years (5), 28 years (5), and 100 years (4,6), and the dissolution and retention model prediction of 10 years. These discrepancies probably occur because of a wide age range in human exposure cases (6 years to adults) and because the data used in developing the dissolution and retention model extends to only 2 years after inhalation exposure, making prediction of the long-term half-life in skeleton uncertain.

As shown in Figure 2, clearance of ^{241}Am from lung for the human exposure cases was more rapid than predicted in the ICRP 30 model with ^{241}Am as a Class W compound (9). The ICRP 30 model also fails to predict the presence of a long-term retained fraction as observed in several of the human exposures (1,4,5,6). Figure 3 shows the ICRP 30 model underpredicts the long term skeletal burden by a factor of 2 because the fraction of ^{241}Am absorbed from lung is underpredicted (4,7). Also the rate at which ^{241}Am accumulates in skeleton is underpredicted because ICRP 30 predicts a slower rate of absorption of ^{241}Am from the lung than was observed in the human exposures.

REFERENCES

1. Jeanmaire, L. and Ballada, J. (1971): In: Proceedings, Radiation Protection Problems Relating to Transuranium Elements, UR-4612, p 531.
2. Sanders, Jr. S. M. (1974): Health Phys., 27, 359.
3. Edvardsson, K. A. and Lindgren, L. (1976): In: Proceedings, Diagnosis and Treatment of Incorporated Radionuclides, IAEA, Vienna, p 497.
4. Fry, F. A. (1976): Health Phys., 31, 13.
5. Cohen, N., LoSasso, T. and Wrenn, M. E. (1978): In: New York University Institute of Environmental Medicine Radioactivity Studies Progress Report, COO-3382-17, p III-1.
6. Toohey, R. E. and Essling, M. A. (1980): Health Phys. (in press).
7. Mewhinney, J. A. and Griffith, W. C. (1978): In: Inhalation Toxicology Research Institute Annual Report, LF-60, p 43.
8. Mercer, T. T. (1967): Health Phys., 13, 1211.
9. ICRP Publication 30 (1979): Limits for Intakes of Radionuclides by Workers: Permagon Press: Oxford.
10. Raabe, O. G., Newton, G. J., Wilkinson, C. J., Teague, S. V. and Smith, R. C. (1978): Health Phys., 35, 649.

TOXICITY OF INHALED ^{238}PuO$_2$ II. BIOLOGICAL EFFECTS IN BEAGLE DOGS

B. A. Muggenburg, J. A. Mewhinney, Barbara S. Merickel,
B. B. Boecker, F. F. Hahn, R. A. Guilmette, J. L. Mauderly and
R. O. McClellan

Plutonium-238 is produced in nuclear reactors using ^{235}U fuel. It is used as a fuel for space nuclear auxillary power units and as a power source in cardiac pacemakers.

The most likely route of entry of ^{238}Pu into the body during many accidents is by inhalation. Because of its high specific activity, local dose around particles of ^{238}Pu can be high and the question of homogeneous versus non-homogeneous dose to lung and its influence on biological effects becomes important. To study that question, the use of particles all of the same size (monodisperse) is necessary.

Dogs serially sacrificed after inhalation of ^{238}PuO$_2$ had a significant amount of ^{238}Pu translocated to bone. Similar findings with significant numbers of bone tumors were found in another study (2).

MATERIALS AND METHODS

Seventy-two, 1 year old Beagle dogs, 36 males and 36 females, were given a single, nose-only exposure to an aerosol of 1.5 μm AD particles and an additional 72 dogs were given an exposure to 3.0 μm AD particles of ^{238}PuO$_2$. Each study had 6 desired activity levels: 0.56, 0.28, 0.14, 0.07, 0.03 and 0.01 μCi per kg body weight; 12 dogs per activity level (Table 1). An additional 24 control dogs were exposed only to the aerosol generation solution. Methods for the preparation of monodisperse aerosols and for inhalation exposure of dogs have been described (1,3). The ^{238}PuO$_2$ particles were tagged

TABLE 1. Experimental design.

Parameter	1.5 μm (AD)	3.0 μm (AD)
Physical size, μm	0.44	0.96
pCi per particle	4.9	51
Local dose rate, rads/day	280	3100
Number of particles, range	2×10^4 to 1×10^6	2×10^3 to 1×10^5
Fraction of lung irradiated	9×10^{-4} to 5×10^{-2}	8×10^{-5} to 5×10^{-3}
Initial lung burden, nCi	100 to 5600	100 to 5600
Avg. lung dose rate, rads/day	0.3 to 15	0.3 to 15

This research was performed under U.S. Department of Energy Contract No. EY-76-C-04-1013 in facilities fully accredited by the American Association for the Accreditation of Laboratory Animals.

with a gamma-emitting radionuclide, ^{169}Yb. Periodic whole-body counts of the ^{169}Yb tag were performed after exposure for the calculation of an initial lung burden (ILB). Medical examination of the dogs was daily observation, annual physical and radiographic examination, and semi-annual blood cell counts and serum chemistry tests. Sick dogs were examined and tested to establish a diagnosis. A few dogs died from their illness but most were euthanized. A necropsy examination was performed on all dogs and tissues were evaluated both histologically and radiometrically.

RESULTS

Initial lung burdens (ILB) ranged from 0.005 to 2.2 µCi/kg and 0.008 to 2.2 µCi/kg for dogs exposed to 1.5 µm AD and 3.0 µm AD particles, respectively.

The first biological effect observed was a lymphopenia. It was observed in all the dogs that died or were euthanized and occurred from 60 to 1200 days after exposure (90% of the dogs were diagnosed within 180 days). A 60% incidence of leucopenia was also noted.

Radiation pneumonitis with pulmonary fibrosis was found in dogs dying from 536 to 1213 days after exposure (Figures 1 and 2). The disease was characterized by a progressive and restrictive pulmonary disease. It was recognized clinically from 38 to 375 days before death, except 2 dogs died suddenly. About 80% of the dogs dying later with lung or bone tumors had histologic evidence of radiation pneumonitis and fibrosis.

Lung tumors were the primary disease at death in 4 dogs dying from 1107 to 1417 days after exposure (Figures 1 and 2). The tumors were in the peripheral portion of the lungs and were classified as adenocarcinomas or bronchioloalveolar carcinomas. They were distributed among all lung lobes and did not metastasize to organs outside of the thoracic cavity.

Bone tumors were the primary disease in 24 dogs euthanized from 1125 to 1918 days after exposure (Table 1). These osteosarcomas were located in the axial skeleton, pelvis or the proximal ends of the humerus or femur and one in the tibia. Some tumors (20%) metastasized to the lungs. Because these tumors caused paralysis or other serious locomotor problems, the dogs were euthanized from 4 to 156 days after the first observed clinical signs. Because the dogs were euthanized, survival time was slightly underestimated.

DISCUSSION

The initial lung burdens of ^{238}Pu in these two studies represent a continum of activity levels from very low to high levels. The dogs were exposed to ^{238}PuO$_2$ from 1200 to 2100 days ago and only dogs with high lung burdens have shown biological response.

The earliest response, lymphopenia, was probably due to the irradiation of lymphocytes as these passed through the lung. The leucopenia, which occurred later than the lymphopenia, was possibly related to the accumulation of plutonium in the endosteum and subsequent irradiation of the bone marrow.

Radiation pneumonitis was the earliest cause of death. Seven dogs died due to radiation pneumonitis and no additional deaths from

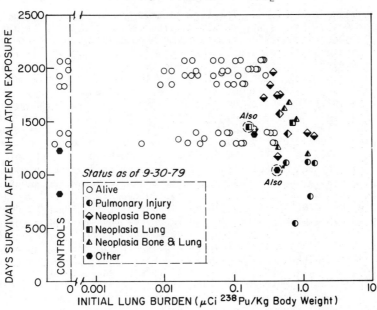

Figure 1. Survival time plotted vs. initial lung burden and major disease at death for dogs that inhaled 1.5 μm AD $^{238}PuO_2$ particles.

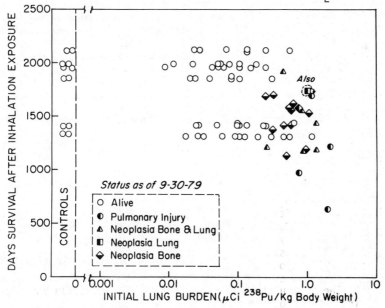

Figure 2. Survival time plotted vs. initial lung burden and major disease at death for dogs that inhaled 3.0 μm AD $^{238}PuO_2$ particles.

this cause are expected. Beagle dogs that inhaled polydisperse aerosols of $^{238}PuO_2$ had similar results with deaths from radiation pneumonitis occurring out to 3 years after exposure (2).

Lung tumors were observed beginning at 1107 days after exposure. This was earlier than the time of appearance of lung tumors with polydisperse aerosols of $^{238}PuO_2$ in Beagle dogs (2). A high incidence of lung tumors was observed in rats exposed to $^{238}PuO_2$ (4). The lobar distribution of primary lung tumors has been random.

The leading cause of death in the ^{238}Pu exposed dogs was osteosarcomas. These tumors occurred as early as 1161 days after exposure. In intravenous injection studies in Beagle dogs, osteosarcomas were found at about the same time in dogs injected with ~ 1.0 µCi of ^{239}Pu/kg body weight (5). In that study, tumors doubled in size about every 12 days. That suggested that tumors were initiated about 1.3 years before death. In this study, osteosarcomas appeared earlier for a given dose than in the injection studies. This may be due to the continuous dose to the bone surface from the continuous translocation of Pu from lung to bone. Bone tumors occurred somewhat later in studies in Beagle dogs exposed to polydisperse $^{238}PuO_2$ aerosols (2). Bone tumors were not observed in rats exposed to $^{238}PuO_2$ polydisperse aerosols. This may reflect differences in the bone metabolism of plutonium between dogs and rats. In injection studies, the sites of tumor formation (axial skeleton, pelvis and the proximal end of the humerus) agreed with those in this study. These were found to be areas with the higher trabecular bone turnover rates (5).

No clear biological response differences are evident to date between the dogs exposed to 1.5 µm and 3.0 µm particles of $^{238}PuO_2$. So far, the lung and bone have been equal targets for response in the dogs exposed to the 1.5 µm particles and bone the primary organ in the dogs exposed to 3.0 µm particles. This may be related to the more uniform radiation of the lung with the 10 times higher number of 1.5 µm particles compared to the 3.0 µm particles. The average dose to organs is comparable for the two particle sizes to 1500 days after exposure (absorbed alpha dose to 1500 days: lung, 700 rads, liver 230 rads, skeleton 100 rads).

The development of dose-response curves based on local dose as well as total organ dose is expected as this study continues. Observation of each surviving dog will continue with particular concern for late effects at low dose levels.

REFERENCES

1. Boecker, B. B., Aguilar, F. L., and Mercer, T. T. (1964): Health Phys., 10, 1077.
2. Park, J. F., Lund, J. E., Ragen, H. A., Hackett, P. L., and Frazier, M. E. (1976): Recent Results in Cancer Research, 54, 17.
3. Raabe, O. G., Boyd, H. A., Kanapilly, G. M., Wilkinson, C. J., and Newton, G. J. (1975): Health Phys., 28, 655.
4. Sanders, C. L., Dagle, G. E., Cannon, W. C., Powers, G. J., and Meier, D. M. (1977): Radiat. Res. 71, 528.
5. Thurman, G. B. (1971): Growth Dynamics of Osteosarcomas in Beagles, COO-119-243, University of Utah, 30.

THE ACCURACY OF A ROUTINE PLUTONIUM IN LUNG ASSESSMENT PROGRAMME

D. Ramsden

Systems for the detection of the radioisotopes of the transuranic elements in the human lung have been developed in several laboratories. Such systems have been in use for incident assessment since the early 1970s. There is justifiable criticism against the use of such systems for routine assessment programmes in that the minimum detectable activity (MDA) for pure plutonium 239 is such that any routine programme, at the best will produce ambiguous results and at the worst may be positively misleading. This short paper, by presenting a summary of our routine programmes over the past few years shows that meaningful conclusions can be made for the technologically important cases of reactor grade plutonium with MDAs of less than 6 nCi of alpha emitters in the lung and attempts to demonstrate that meaningful conclusions are still possible for most subjects in the case of pure plutonium 239.

THE SYSTEMS

Present systems for the assessment of insoluble particulates of plutonium in the lung rely on the detection of soft X-rays (@ 17 keV) external to the chest using phoswich detectors and proportional counters. Such systems have been described by several authors and will not be redetailed here.

SOURCES OF ERROR

For a measurement on a specific individual the statistical counting errors can be assessed together with the errors in predicting subject background and a calibration error. The sources of error and their approximate magnitudes are listed in Table 1.

Calibration error is not usually included in the overall assessed error and its relative importance ($\sim$30%) has been discussed elsewhere and will not be expanded here. For a 'standard man'* a typical standard deviation for pure plutonium 239 in the lung would be approximately 14.5 nCi giving an MDA of 29 nCi on a 2 σ criterion.

The use of single valued MDAs for such work is misleading as they depend critically on the subject's body size and a range of values between 4 nCi and 300 nCi is more illustrative.

An alternative approach is to study the distribution of 'observed' lung contents in several populations. This approach is followed here.

POPULATION OF "NORMALS"

A population of 305 male radiation workers with no history of plutonium work was studied to ensure that we would return nil lung contents for such a group. The distribution of excess counts (observed - predicted) was normal with a mean of -0.01 cpm and a standard deviation of 0.72 cpm. The 'standard' man calibration is 55 cpm/µCi plutonium 239 in the Winfrith system, ie the standard deviation of

TABLE 1. Sources of Error (approximate values for Standard Man*)

Source of Error	Pu239 equivalent
1. Counting statistics	11 nCi
2. Subject background	5 nCi
3. Body build correction	20%

*'Standard' man, approximates to ICRP standard men - a chest wall thickness of 25 mm is chosen.

this group would be approximately 13 nCi Pu239 equivalent. For high burn up plutoniums (see below) the calibration figure is of the order of 250 cpm/µCi alpha, giving a standard deviation for such a group of approximately 3 nCi.

ISOTOPIC COMPOSITION OF FUELS

Plutonium, as a potential nuclear fuel, is chemically extracted from high burn up uranium fuels when there are appreciable quantities (by weight) of the higher plutonium isotopes Pu240, Pu241, Pu242. When the mixture is expressed in terms of alpha activity, or X-ray activity, there are also appreciable amounts of Pu238 and Am241.

The X-ray/alpha emission of such fuels are usually of the order of 0.1 (cf Pu239 at 0.045) with Am241 alpha contents of between 3% and 8%. There are three main advantages in assessing such mixtures compared with the case of pure plutonium 239.
    (a) The increase of specific X-ray emission.
    (b) The independent assessment of Am241 via its 60 keV gamma ray.
    (c) The contribution of degraded 60 keV radiations in the 17 keV band.

Point (c) may need a little amplification. Fig 1 shows the absorption curves of an Am241 impregnated lung in a realistic chest phantom with muscle equivalent overlying tissue.

The 60 keV band shows the exponential absorption characteristic of that energy but the 17 keV band has a two component structure; that of true absorption (see the Pu238 line) and that of degraded 60 keV radiations. Thus for fuel with an appreciable Am241 content there is an additional component in the 17 keV band which is not as readily absorbed as true 17 keV X-rays. For obese subjects this component predominates when the Am241 contents are above 5% by alpha and hence reduces the large predicted errors on such subjects (see below). Table 2 lists typical values of system responses for three subjects of differing body size.

All the above argument presumes a knowledge of the isotopic composition and also assumes that Am241 will behave in the lung with the same characteristics as the plutonium particulate. The latter point is certainly not true if either the material is solubilized in the lung or if there has been a long time lapse since intake (years).

RESULTS

(a) A population of 448 plutonium workers is shown in Fig 2. The

Lung Assessment Programme

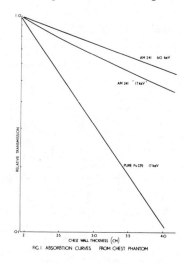

FIG. 1 ABSORBTION CURVES FROM CHEST PHANTOM

TABLE 2. Predicted system responses (from 17 keV energy band)

Composition of Mixture (% by Weight)		Sub-ject	Chest Wall Thickness (mm)	Response (cpm/µCi alpha)				Standard Deviation (nCi)	
				Pu239 Xrays	Mixture			Pu239	Mixture
					Xrays	γ	Total		
Pu238	0.1	A	18	110	237	157	394	7	2
Pu239	78.5								
Pu240	18.0	B	25	55	118	136	254	14	3
Pu241	3.0								
Pu242	0.4	C	45	8	17	81	98	96	8
Am241	0.2								

isotopic composition is unknown and is taken as pure plutonium 239. The results are expressed in nCi Pu239.

The standard deviation of the population at 15.7 nCi is comparable with that of a single measurement on a standard man. For any individual it would be obviously unwise to deduce the presence or absence of plutonium at levels below ½ MPLD (8 nCi) although repeat measurements and probability analysis can help. Conclusions can be drawn as to the status of the whole group. Fig 2 also splits this data in two groups - one of body build thinner than the standard man ($\sigma \sim 4.5$ nCi) and one group more obese than the standard man ($\sigma \sim 21$ nCi).

We can now draw conclusions as to the status of the individuals within Group A using 9 nCi as a crude investigation level. Group B still presents problems as to whether specific individuals do contain plutonium and these individuals must be studied by bioassay, via urine and faecal samples. Some guidance may be obtained from Am241 in lung contents which is always measured simultaneously with the plutonium in lung assessment. The mean Am241 lung contents for subjects in Group B was 0.04 nCi with a standard deviation of 0.10 nCi.

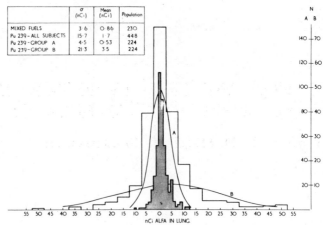

FIG 2 DISTRIBUTION OF OBSERVED RESULTS FROM TWO ROUTINE 'Pu IN LUNG' ASSESMENT PROGRAMMES.

(b) A population of 230 plutonium workers where the isotopic composition, although variable, was known and was of high burn up is also summarised by Fig 2. The standard deviation of the group is now only 3.6 nCi. For such a group one can deduce the presence of plutonium with better than 96% confidence at the 8 nCi level. Because of the relatively high Am241 contents there is no longer the large difference between body builds.

CONCLUSIONS

If the isotopic composition of the plutonium is known and if this is of high burn up, a routine plutonium lung assessment programme can be used with MDA below 8 nCi alpha activity in lung. Such a programme in conjunction with routine bioassay and a defined procedure of monitoring after suspected incidents (1) is the basis of the Winfrith internal dosimetry system.

If the isotopic composition is not known, and cannot be "bracketed", any routine programme is of more doubtful use. It can be used to confirm the status of a group as a whole and can be used for individuals of a subgroup thinner than standard man with an approximate MDA of 9 nCi Pu239. For the more obese individual, the limitations of direct lung monitoring with no prior knowledge of the isotopic composition of the contaminant, are sadly obvious. For such individuals, decisions as to their 'plutonium' status are still predominantly based on biological monitoring.

Any subject who, as a result of a routine lung count, gives indications that plutonium may be present in the chest region is remonitored. His whole history is re-examined together with personal air sampling and biological monitoring results. It is on a combination of all such methods that a conclusion as to his plutonium 'status' is made.

REFERENCES

(1) RAMSDEN D, (1976), 'Diagnosis and Treatment of incorporated radionuclides'. IAEA, Vienna, p139.

RETENTION AND EFFECTS OF ^{239}Pu IN THE TREE SHREW (TUPAIA BELANGERI)

A. Seidel, R. M. Flügel, D. Komitowski and G. Darai

For the improvement of the evaluation of risk to man condiser-sidably more information on the metabolism and effects of Pu-239 is needed in primates. The tree shrew (Tupaia belangeri) is considered to be the most primate-like non-primate or the most primitive of the living primates (e.g. 1). They are as small as rats, relatively easy to breed and maintain and can live from 8 to 14 years. One of the aims of this pilot study is to compare the retention of Pu-239 in tupaias with that in rodents and larger primates in order to evaluate the usefulness of tupaias for toxicological studies.

METHODS

The animals used were 19 female tree shrews, 8-18 months old, (150-200 g) from the breeding stock of the Institute for Animal Physiology, University of Bayreuth. Plutonium-239 citrate (The Radiochemical Centre, Amerham, UK), 0.5 µCi/kg, in essentially monomeric form was injected intramuscularly. The animals were sacrificed at the time intervals indicated in Fig. 1. The radioactivity remaining at the injection site and in the organs was determined by liquid scintillation counting (2). The distribution of the nuclide within the different parts of the skeleton was also measured. Thus, the data given in Fig. 1 for skeleton were not calculated from the activity in the femur but represent measured data for the whole skeleton. The data in Fig. 1 are expressed as a percentage of the dose absorbed from the injection site up to the time of sacrifice. The weight of the animals was followed continuously and the organs were carefully inspected at sacrifice. Sections for histological examination were prepared and hematological and chemical tests with blood were performed by routine methods.

Cells from different organs (muscle, kidney, spleen, thymus, and blood leukocytes) were established in tissue culture as described previously. The cell-free supernatants from these cultures were assayed for reverse transcriptase as indication for the expression of retroviruses as described previously (3).

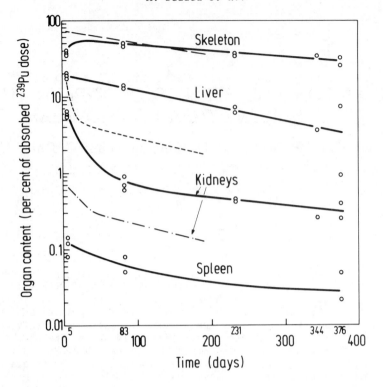

Figure 1. Retention of i.m. injected Pu-239 in tupaia belangeri (tree shrew). Each point represents one animal. Dotted lines are from rats (4).

RESULTS AND DISCUSSION

The retention of Pu-239 at the injection site decreased from ~ 15 % at day 5 to ~ 5 % at day 83, this value remained virtually constant thereafter. The retention in the organs is shown in Fig. 1, in which the corresponding retention functions for rats (4) are represented as dotted lines. As can be seen, the initial deposition in skeleton and liver is not very different from the well known picture in rats. The half life in tupaia skeleton is somewhat longer (one year as compared to 215 days) which may be due to the different biological age of the skeleton at the time of injection. During the first year, the elimination of Pu-239 from the liver can be described by a single exponential with a half life of 150 days, whereas ~ 75 % of the initially deposited nuclide are eliminated from the rat liver with a half life of < 10 days. Initial deposition in the kidneys is much higher than in rats but the elimination rates are similar.

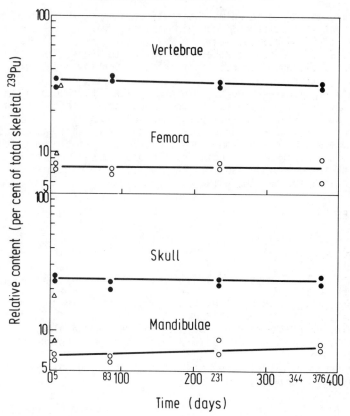

Figure 2. Retention of Pu-239 in different bones of the tupaia (tree shrew) skeleton expressed as a fraction of Pu-239 in total skeleton. Triangles are data from rats for Am-241 (5).

The retention in four bones is presented in Fig. 2 which also shows data for rats given Am-241 (5), (data for Pu-239 were not available). The fraction deposited in the different bones remained constant during the first year after injection, indicating a low rate of bone remodelling. The factor for calculating total Pu-239 in skeleton from activity in one femur was 27.

No statistically significant change in weight and no gross signs of toxicity were observed during the first year, except for one animal, which died for unknown reasons. No macroscopic or histological changes in the organs have been observed (till d 231). Hemogl., hematocrit and erythrocyte counts remained normal. There was a rise of glutamate-pyruvate- and oxalate-transaminase as well as of alkaline phosphatase in the three animals sacrificed at day 376. A total of 36 different cell cultures (from nine animals) were tested and found to be negative for retroviruses.

This study provides further information to suggest that the retention of plutonium in rat liver is exceptionally short when compared to other animal species, hamster, primate, dog. However, if the tupaia is also regarded as a lower primate, the rate of elimination of transuranium elements by sub-human primates tested so far is still much faster than that predicted for human liver (ref. in 6,7). These findings strenghten our view that the realistic assessment of the risks to man from deposited transuranium elements must be based on a clear understanding of the biochemical mechanisms underlying the deposition and elimination of these elements in various animal species including man.

REFERENCES

1. Simons, E.L. (1972): Primate Evolution. MacMillan Comp., New York.
2. Seidel, A., and Volf,V. (1972): Int. J. Appl. Radiat. Isotopes, 23, 1.
3. Flügel, R.M., Zentgraf, H., Munk, K. and Darai, G. (1978): Nature, 211, 543.
4. Gemenetzis, E. (1976): Thesis, University of Karlsruhe.
5. Seidel, A. (1977): Health Phys., 33, 83.
6. Durbin, P. (1973: Handbook of Experimental Pharmacology Vol.XXXVI, p. 739. Editors: H.C. Hodge, J.N. Stannard, J.B. Hursh, Springer, Heidelberg.
7. ICRP Publ. 19 (1972): The Metabolism of Compounds of Plutonium and Actinides. Pergamon, Oxford.

THE USE OF DTPA TO INHIBIT THE EXTRAPULMONARY DEPOSITION OF CURIUM-244 IN THE RAT FOLLOWING THE BRONCHIAL INTUBATION OF OXIDE SUSPENSIONS

J. R. Cooper, G. N. Stradling, H. Smith and Sandra E. Ham

INTRODUCTION

Much experimental evidence has accumulated on the use of the calcium or zinc salts of diethylenetriaminepenta-acetic acid (Na_3Ca DTPA and Na_3Zn DTPA) for removing actinides from animals (1). However, such studies are based on experiments in which the actinides have been administered as the citrate or nitrate complexes. The present work looks at the efficacy of Na_3Ca DTPA and Na_3Zn DTPA for enhancing the excretion of curium after pulmonary intubation of curium-244 dioxide (CmO_2).

MATERIALS AND METHODS

High fired CmO_2 was supplied by the Radiochemical Centre (Amersham, Bucks, U.K) and was fractionated by ultrafiltration as described previously (2).

In the animal experiments Na_3Zn DTPA or Na_3Ca DTPA were administered intravenously in isotonic saline. The diuretic Lasix, frusemide B.P., (0.2 ml, 8.9 mg kg^{-1}) was injected intravenously at intervals to promote a high urine flow to allow the collection of adequate volumes of urine for analysis by gel permeation chromatography.

The methods of pulmonary intubation, gel permeation chromatographic separation and radioactivity determinations are given by Stradling et al., (2).

RESULTS AND DISCUSSION

After intubation into the lung, 0.22-1.2 µm diameter curium dioxide particles rapidly form particles of 0.001 µm in diameter (2). These particles, believed to be of the hydroxide, then diffuse passively to the blood probably through pores in the alveolar epithelium (3). In the blood intact 0.001 µm particles of CmO_2 combine with serum proteins. The protein-bound CmO_2 rises from 45% of the circulating radioactivity at 35 minutes after pulmonary intubation to >90% at 24 hours; the remaining activity is 0.001 µm particles.

0.001 µm particles will also combine with serum proteins in vitro. For example, when serum labelled for 24 hours was chromatographed on Sephadex G-200, Cm eluted with the α and γ globulins, and the transferrin and albumin fractions in about equal amounts. Negligible activity (< 1%) was recovered in the low molecular weight fractions where unbound particles or curium would elute. However, if Na_3Ca DTPA or Na_3Zn DTPA is added at a concentration of 0.02 mg. ml^{-1} to the serum 6 minutes before the 0.001 μm particles the reaction between particles and proteins is inhibited and even after 24 hours 99% of the radioactivity eluted as intact particles. Similarly, intact 0.001 μm particles could be regenerated from protein-bound Cm by addition of Na_3Ca DTPA (2.5 $mg.ml^{-1}$). It is suggested that DTPA blocks receptor sites for the particles on the protein by a preferential binding process.

Previous work has shown that a major factor infuencing the urinary excretion of Cm following the intake CmO_2 into the lungs is the renal dialysis of 0.001 μm particles (2). The binding of particles to serum proteins may compete with this process. The above studies <u>in vitro</u> suggest that either Na_3Ca DTPA or Na_3Zn DTPA could maintain these 0.001 μm particles in the blood for long enough to permit the quantitative urinary excretion of Cm. The effect of administering Na_3Ca DTPA or Na_3Zn DTPA to rats exposed to CmO_2 suspensions is shown in Table 1. If the concentration of Na_3Ca DTPA or Na_3Zn DTPA in the blood is maintained above 0.002 $mg.ml^{-1}$ (Expt. 2), by administering 0.28 $mg.kg^{-1}$ body weight initially followed by injections of 0.14 $mg.kg^{-1}$ at 30 minute intervals, then deposition of Cm in the skeleton and liver is markedly reduced. The interval between successive injections corresponds to the half time of DTPA in the blood (4). At higher concentrations (Expt. 3) Na_3Ca DTPA is still effective in minimising tissue deposition even when administered 2 hours after small particle suspension. In all of the experiments where Na_3Ca DTPA or Na_3Zn DTPA and Lasix were administered before the oxide suspension the Cm was excreted as 0.001 μm particles. When the oxide suspension was administered before the DTPA and Lasix the Cm was present in the urine as 0.001 μm particles and Cm citrate. The Cm citrate is probably formed from particles and citrate in the renal tubular fluid (2).

The experiments outlined above demonstrate that (i) DTPA is not chelating Cm but inhibiting a reaction between 0.001 μm CmO_2 particles and serum proteins (ii) Na_3Ca DTPA and the less toxic Na_3Zn DTPA are equally effective and (iii) to obtain efficient urinary excretion of Cm the concentration of DTPA in the blood must be maintained above about 0.004 $mg.ml^{-1}$. Animal experiments indicate that following an accidental intake of $^{244}CmO_2$ by man, about 90% of that fraction destined to translocate to blood would do so during the next month (5). Therefore, for DTPA therapy to be most effective it should be administered continually over this period at a constant rate of 14 $mg.kg^{-1}.day^{-1}$. This is within the dose range normally used in clinical practice (6).

REFERENCES

1. Catsch, A., (1976) In: Diagnosis and Treatment of Incorporated Radionuclides, p 295. IAEA, Vienna.
2. Stradling, G. N., Cooper, J. R., Smith, H. and Ham, S. E. (1979): Int. J. Radiat. Biol., 36, 19.

Table 1. Injection schedules, tissue distribution and excretion of ^{244}Cm after administering ^{244}Cm oxide

Expt. No.	Injections	Injection schedule[a]		Tissue distribution and excretion[b] (%)				
		mg.kg^{-1} body wt	t min	Lungs	Liver	Skeleton	Urine	
1	Lasix	8.9	-5; 120, 210	11.1	15.6	33.5	32.4	
2	Na$_3$ZnDTPA	0.28;0.14[c]	-5;25,55,85..235	13.1	0.5	1.5	79.8	
3	Na$_3$CaDTPA	14;7[c]	120;140,180...360	12.2	0.3	1.5	83.5	

[a]Suspension of 0.001 µm diameter particles 100 µl, 500 Bq administered by tracheal intubation at zero time. The injection times shown for DTPA are relative to this labelling. The amount of DTPA administered in the first injection is twice that administered in subsequent injections. Lasix 0.2 ml, 8.9 mg.kg^{-1} administered intravenously at -5, 120 and 210 minutes except experiment 3 where these times are relative to the first injection of DTPA.

[b]Values expressed as a percentage of initial lung burden; animals killed 240 min after initial injection. Remainder of ^{244}Cm present in blood, kidneys and gastro-intestinal tract and contents. No faeces were passed during the course of the experiments.

[c]To convert to mg.ml^{-1} of blood divide by 70 (7).

The metabolic data were closely similar when Na$_3$CaDTPA or Na$_3$ZnDTPA were administered by the same injection schedule.

3. Cooper, J. R., Stradling, G. N., Smith, H. and Ham, S. E. (1979): Int. J. Radiat. Biol. (accepted for publication).
4. Stevens, W., Breuenger, F. W., Atherton, D. R., Buster, D. S. and Howerton, G. (1978): Radiat. Res. 75, 397.
5. Sanders, C. L. and Mahaffey, J. A. (1978): Radiat. Res. 76, 384.
6. Planas-Bohne, F. and Lohbreier, J. (1976) In: Diagnosis and Treatment of Incorporated Radionuclides, p 505. IAEA, Vienna.
7. Stradling, G. N., Cooper, J. R., Smith, H. and Ham, S. E. (1979): Nucl. Med. and Biol. 6, 183.

THE EFFECT OF OXIDATION STATE ON THE ABSORPTION OF INGESTED OR INHALED PLUTONIUM

M. F. Sullivan, Linda S. Gorham and J. L. Ryan

Larsen and Oldham found that chlorine, at the concentrations found in municiple water supplies, can oxidize quadrivalent plutonium to its hexavalent state (1). Since studies in this laboratory had shown one thousand times more $^{239}Pu(VI)$ was absorbed than $^{239}Pu(IV)$ (2), they suggested that the maximum permissible concentrations (MPC), apparently based on data from Pu(IV), should be lowered.

Our initial experiments performed with $^{238}Pu(VI)$ nitrate did not support those earlier results. This suggested that the conditions of fasting and oxidation used in those studies may have been responsible. Absorption of plutonium either by gavage or inhalation was compared in fasting and nonfasting rats to determine if the intestinal contents influence absorption of plutonium that was injected intragastrically or swallowed as a result of clearance from the lungs.

METHODS AND MATERIALS

Wistar female rats weighing about 200 g received plutonium nitrate (pH2) by gavage or by nose-only exposure from a nitric acid aerosol generated by a Lovelace nebulizer (3). Fasted rats were deprived of food 18 hours before Pu gavage and for 72 hours following it. Animals exposed by inhalation were fasted either before, or both before and after treatment with Pu. Excreta was collected daily from gavaged rats for four days.

All animals were killed five days after treatment. Femurs were removed from all animals and the total skeletons analyzed from many with which a femur factor was derived to determine total bone Pu content. The skin and GI tracts were discarded. Carcass values were determined by a summation of bone and soft tissue, excluding the liver and lung values. High lung values in gavaged animals indicated poor injections and the data from these animals was rejected.

The ^{239}Pu was purified by anion exchange on a Dowex MSA-1 resin and oxidized to its hexavalent state by passing a stream of O_3, O_2 and Ar through a 0.4M HNO_3 solution for six hours. It was shown to be 100% $^{239}Pu(VI)$ by spectrophotometric analysis. Solutions that included the holding oxidant $K_2Cr_2O_7$ were made 0.015 M by dilution. Gavaged animals received 1.0 ml of a solution containing 0.5 mg ^{239}Pu, 97% of which was filterable through a 0.01 μm filter at the time of treatment. The dose administered by gavage was 30 μCi and by inhalation 5 μCi.

Plutonium was analyzed by a modification of the Keough and Powers method (4). Carbon-free aliquots in mixture of 1.0% boric acid were mixed in a scintillation solution containing 1,4 bis-2-[5 phenyloxazolyl] benzene, 2,5,-diphenyloxazole (PPO), Triton X-100 toluene and water.

RESULTS

The data obtained by Weeks et al (2) after intragastric administration of ^{239}Pu(VI) are shown in Figure 1 along with data that we obtained simulating conditions used by those investigators. Other

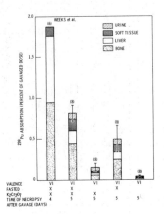

Figure 1. Absorption of ^{239}Pu(VI) by fasted and unfasted rats after intragastric administration of ^{239}Pu nitrate.

groups are included to show the effect that ad libitum feeding and an absence of the holding oxident, $K_2Cr_2O_7$ had on ^{239}Pu absorption from the GI tract. The fasting period lasted from 18 hours before gavage until 72 hours thereafter. Our absorption data under those conditions amounted to about half that of Weeks, possibly because their solutions were more acid, pH 1 versus pH 2, and their $K_2Cr_2O_7$ may have been more concentrated. Feeding reduced absorption 18-fold and the combination of feeding and elimination of dichromate from the solution reduced it about 26-fold.

Results obtained by exposing groups of rats to aerosols of either ^{239}Pu(IV) or ^{239}Pu(VI) (Figure 2) indicate that there was increased retention of plutonium after exposure to ^{239}Pu(VI) in comparison to ^{239}Pu(IV). The absence of food either before, or both before and after the inhalation exposure had no effect on the amount of ^{239}Pu retained by the liver and carcass.

DISCUSSION

Plutonium in its hexavalent state may under certain conditions be more readily absorbed from the GI tract. However, the experimental conditions in which increased uptake occurred, i.e. fasting for 90 hr, high acidity (pH 1-2) and a dose of 130 µCi/kg, are unlikely to occur in human exposure. The oxidation state may also influence absorption after inhalation because of the large fraction entering the bowel as a result of swallowing Pu cleared from the

The Effect of Oxidation State

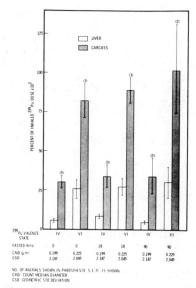

Figure 2. Retention of ^{239}Pu by fasted and unfasted rats after inhalation of either ^{239}Pu(IV) or ^{239}Pu(VI) nitrate.

lung. Although the Pu dose inhaled was lower than the dose gavaged (25 µCi/kg) some of it probably entered the stomach in its hexavalent state. Absorption from the GI tract was not higher than when ^{239}Pu(IV) was inhaled, even when the intestinal contents were depleted by fasting. The absence of an effect of food deprivation suggests that the increased absorption of plutonium was due to translocation of ^{239}Pu(VI) from the lung and not to an increase in transport across the GI tract.

This data support the observation made in the gavage studies that the MPC for drinking water, apparently based on the absorption of Pu(IV), is adequate for Pu(VI) ingestion in quantities that may be expected in the environment. Absorption of Pu(VI) from the lung, however, is higher than that of Pu(IV).

REFERENCES

1. Larsen, R. P., and Oldham, R. D., (1978): Science, 201, 1008.
2. Weeks, M. H., Katz, J., Oakley, W. D., Ballou, J. E., George, L. A., Bustad, L. K., Thompson, R. C., and Kornberg, H. A., (1956): Radiat. Res. 4, 339.
3. Mercer, T. T., Tillery, M. I., and Chow, H. Y. (1968): Am. Ind. Hyg. Ass. J., 29, 66.
4. Keough, R. F., and Powers, G. J., (1970): Anal. Chem., 42, 419.

ACKNOWLEDGEMENT

The assistance of E. F. Blanton, J. P. Herring and B. L. Southward with many of the technical problems involved in this research is gratefully acknowledged.

Work supported by the U. S. Department of Energy under contract EY-76-C-06-1830.

PULMONARY CARCINOGENESIS FROM PLUTONIUM-CONTAINING PARTICLES

R. G. Thomas, D. M. Smith and E. C. Anderson

The potential pulmonary effects of inhaled plutonium have been summarized in recent years (1,2) and the need for further animal research has been emphasized. The results presented here are the outgrowth of an effort to assess the tumorigenesis of focal plutonium sources in the Syrian hamster respiratory tract (3,4). Because localized radiation (hot spots) in bone was known to be more tumorigenic than diffuse radiation, it was thought that this same phenomenon may prevail in the lung tissue. Based primarily upon studies with beta irradiation of the skin (5) models were developed as guidelines for our "hot particle" research.

MATERIALS AND METHODS

To create well-controlled plutonium sources localized in specific numbers in the Syrian hamster lung it was decided to use the intravenous (IV) route of administration. Ceramic spherical particles of zirconium dioxide were manufactured such that fixed small amounts of plutonium could be homogeneously incorporated in them, to control the strength of each particle's radiation field (6). The particles were uniformly 10 μm in diameter (6) and were injected into the jugular vein (7), after which they lodge in the lung capillary bed. A ^{57}Co tag was also added so that retention characteristics could be determined through periodic whole-body counting.

The aerosol particles for inhalation (INH) studies were of two different chemistries. In the earlier work it was decided to nebulize a mixture of the ZrO_2 sol used in the hot particle microsphere manufacture; into the sol was incorporated the desired amounts of ^{238}Pu or ^{239}Pu and ^{57}Co. The sol was nebulized (8) and the droplets passed through a heating column at ∼ 900°C. The animals were exposed in a nose-only setup (9). The resulting aerosol was polydispersed with a aerodynamic mass median diameter of ∼ 1.8 μm and a geometric standard deviation of ∼ 1.8. The second type of aerosol was from plutonium dioxide generated in a similar fashion, but the starting material was a suspension of $^{239}PuO_2$ in distilled water. The aerosol samples for analysis were collected with cascade impactors (10) or electrostatic precipitators (11).

The Syrian hamsters were allowed to live out their lifespan and were sacrificed only when moribund. They were necropsied as soon as feasible, and routine gross and microscopic pathological examinations were performed (12, 13).

RESULTS

Data on the tumor incidences in all three (IV and INH) studies are presented in Tables 1-3. The histology slides for the PuO_2 studies have not been completely read so the available results only are presented in Table 3. The types of tumors are indicated in Tables 2 and 3, primarily to show the ratios of adenomas to adenocarcinomas.

TABLE 1. Pulmonary neoplasm incidence following intravenous injection of microspheres in Syrian hamsters

No. of Spheres Per Hamster	Lung Burden (nCi)	No. of Hamsters	Fraction of Lung Irradiated (%)	No. of Tumors	Lung Tumor Incidence (%)
CONTROL	0	521	0	3	0.6
2360	140	68	1	0	0
10 900	97	17	5	0	0
58 800	120	160	28	19	12
312 000	130	25	80	2	8

TABLE 2. Pulmonary neoplasm incidence following inhalation of $Pu-ZrO_2$ aerosol particles by Syrian hamsters

Initial Lung Burden (nCi)	No. Of Hamsters	Tumor Incidence (%)	Adenoma	Adeno- Carcinoma	Squamous Cell Carcinoma
0	144	0.7	1	0	0
6	40	5	2	0	0
8	43	12	5	0	0
76	50	28	11	6	0
87	50	40	12	8	0
101	44	50	10	9	3

DISCUSSION

It is obvious that plutonium alpha irradiation distributed focally is not tumorigenic (Table 1). When less than 5% of the pulmonary tissue receives the radiation dose, there are no observed tumors at death. More diffuse irradiation does lead to the formation of tumors.

TABLE 3. Pulmonary neoplasm incidence following inhalation of PuO_2 aerosol particles by Syrian hamsters

Initial Lung Burden (nCi)	No. of Hamsters	Tumor Incidence (%)	Fraction Slides Read*		
			Adenoma	Adeno-Carcinoma	Undifferentiated Tumors**
0	50	0	0	0	0
40	63	4	1/23	0	0
96	66	13	6/54	0	1/54
110	60	7	1/46	1/46	1/46
144	65	16	7/49	1/49	0

* All animals in this study have not been processed; hence, the incidences are based upon fewer than the total exposed in each group.

** Carcinomas and Sarcomas

Inhalation studies are much more productive in the induction of lung tumors, as shown in Table 2. A trend of increased tumor incidence with increasing initial lung burden is obvious. There is also an apparent trend from adenoma induction to more invasive types of tumor, with increasing initial lung burden. The same dosage-incidence trend may be forthcoming in the PuO_2 studies (Table 3), but more data await analysis. A currently unexplained effect is the apparently greater tumor yield produced by Pu-ZrO_2 compared to PuO_2. Averaging the last 3 dose groups of the latter gives 12 ± 3% tumors from a mean lung burden of 117 nCi while averaging the last 2 dose groups of the Pu-ZrO_2 aerosol gives 45 ± 6% from 94 nCi. Particle size and residence time do not account for any difference.

SUMMARY

Plutonium administered as an alpha radiation source to the respiratory tracts of Syrian hamsters has resulted in various incidences of neoplasia. Adenomas are the primary lung tumor observed, but adenocarcinomas are also prevalent.

ACKNOWLEDGEMENT

Many individuals in the Toxicology Group played an important role in this work, but G. A. Drake and J. E. London have carried on the experimentation and collection of data throughout the studies.

REFERENCES

1. Bair, W. J., Richmond, C. R., and Wacholz, B. W. (1974): United States Atomic Energy Commission Report, WASH-1320.
2. Biological Effects of Inhaled Radionuclides, International Commission on Radiological Protection, Report ICRP-31 (1979).
3. Dean, P. N., and Langham, W. H. (1969): Health Phys. 16, 79-84.
4. Richmond, C. R., Langham, W. H., and Stone, R. S. (1970): Health Phys. 18, 401-408.
5. Albert, R. E., Burns, F. J., and Heimbach, R. D. (1967): Radiat. Res. 39, 515-524.
6. Anderson, E. C. and Perrings, J. D. (1978): Health Phys. 34, 225-236.
7. Holland, L. M., Drake, G. A., London, J. E., and Wilson, J. S. (1971): Lab. Anim. Sci. 21, 913-915.
8. Mercer, T. T., Tillery, M. I., and Chow, H. Y. (1968): Amer. Ind. Hyg. Assoc. J. 29, 66-78.
9. Raabe, O. G., Bennick, J. E., Light, M. E., Hobbs, C. H., Thomas, R. L., and Tillery, M. I. (1973): Toxicol. Appl. Pharmacol. 26, 264-273.
10. Mercer, T. T., Tillery, M. I., and Ballew, C. W. (1962): AEC Research and Development Report, LF-5.
11. Mercer, T. T., Tillery, M. I., and Flores, M. A. (1963): AEC Research and Development Report, LF-7.
12. Anderson, E. C., Holland, L. M., Prine, J. R., and Smith, D. M. (1979): Radiat. Res. 78, 82-97.
13. Smith, D. M., Anderson, E. C., Prine, J. R., Holland, L. M., and Richmond, C. R. (1976): In Proceedings of Biological Effects of Low-Level Radiation, Vol. II., pp. 121-129, International Atomic Energy Agency.

SUBCELLULAR DISTRIBUTION OF ^{239}Pu IN THE LIVER OF RAT, MOUSE, SYRIAN AND CHINESE HAMSTER

R. Winter and A. Seidel

It is well known that the biological half life of transuranium elements in the liver of mammalian species varies from a few days to many years (ref. s.1). However, the reasons for these differences are unknown. The aim of our studies is to elucidate the biochemical mechanisms responsible for these species differences as a part of our attempt to improve risk estimation following incorporation of transuranium elements into man through a profound understanding of the biochemistry of these elements. We have chosen rats and mice as models for a short and two hamster species as models for a long biological half life of Pu-239 in liver.

METHODS

The animals used were adult females of the following strains: Rats (Heiligenberg strain), mice (NMRI), Syrian hamsters (commercial strain) and Chinese hamsters (Breeding stock of Institute for Zoology, Darmstadt). Radiochemically pure Pu-239- and Fe-59-citrate injection solutions were injected i.v. (5-10 µCi/kg Pu-239, 50 - several hundred µCi/kg Fe-59). On the sixth day 750 mg/kg of the non-ionic detergent Triton WR1339 were given i.p. and sacrifice took place on the tenth day. Triton WR1339 causes a shift of the density of the lysosomes from $\sim$ 1.2 to $\sim$ 1.1 g/cm^3 and any lysosomally-associated material can be recognized by a parallel shift. Control animals received 0.9 % NaCl. After differential centrifugation of liver homogenates, a fraction designated MLP was obtained, which contained most of the formed cell elements except the nuclei. Aliquots of this MLP fraction were centrifuged in a linear sucrose density gradient for 4 or 16 hours at 88 000 av. g. Radioactivity and marker enzymes (s. Figures) were determined in all fractions obtained after isopycnic centrifugation. Details of the methods are described elsewhere (2).

RESULTS AND DISCUSSION

The MLP fraction contained more then three quarters of the total cellular radioactivity. The results obtained by isopycnic centrifugation of that fraction are shown in Figs. 1 and 2. Certain results are common to all species: The extent to which the profile of the lysosomal marker, acid phosphatase, is shifted to light densities by injecting Triton WR1339 is very similar. The profile of glutamate dehydrogenase (mitochondria) is not influenced by Triton WR1339 and

Subcellular Distribution of ^{239}Pu

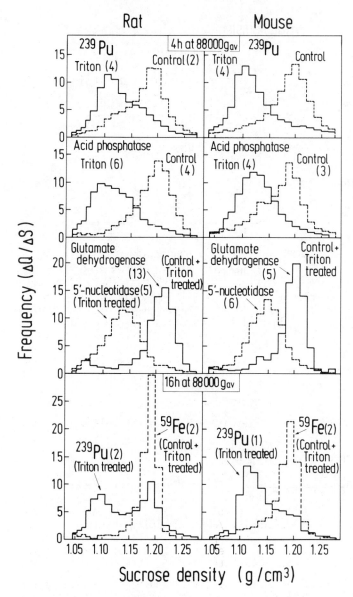

Figure 1. Distribution of radioactivity and marker enzymes (s.text) after centrifugation of the MLP fraction from liver cells in a sucrose density gradient. ΔQ: fractional amount of constituent found in that section; Δρ: density increment from one fraction to the other. The area under each section is proportional to the fractional amount and the total area is one. Values in brackets represent number of experiments.

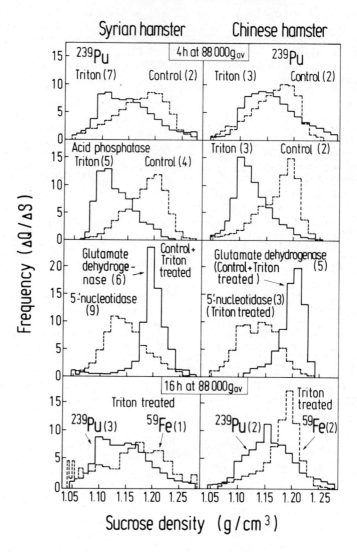

Figure 2. Distribution of radioactivity and marker enzymes after centrifugation of the MLP fraction from liver cells in sucrose density gradients. For presentation s. Fig. 1.

data for controls and treated animals could be combined. On the other hand, the Pu-profiles are considerably broader and the fraction which is unequivocally shifted is considerably smaller in both hamster species as compared to rats and mice (upper row of Figures). In Chinese hamsters, the peaks of Pu-239 and acid phosphatase after Triton treatment are not even congruent; a distinct shoulder at $\rho \sim 1.5$ g/cm^3 is visible in the Pu profile for Triton treated Syrian hamsters. With a centrifugation time of 16 instead of 4 hours (lower row of Figures) the profiles of the marker enzymes remain at virtually the same density. Again clear discrepancies between the Pu-profiles of the various species occur. In rats, and to a lesser degree also in mice, a second peak develops at the same density at which Fe-59 equilibrates. The profile for Pu in Chinese hamsters becomes unimodal with a peak at ~ 1.5, where a minimum exists in the corresponding profile for rats. The clear relationship to the Fe-profile seen in rats was not observed in Chinese hamsters. At this point it should be mentioned that we have seen only minor, or even no, differences between the Fe-profiles from control or treated animals, which, therefore, could be combined for presentation. There is no reason to assume any important association of Pu with plasma membranes, represented by 5'-nucleotidase, in rats, mice and Syrian hamsters. However, the role of the plasma membrane needs to be clarified by further experiments, especially in Chinese hamsters.

If the parallelism between the shift of Pu-239 and acid phosphatase is taken as a measure for the extent of lysosomal binding, there is clear evidence for association with these organelles for rats, mice and to some degree, also for Syrian hamsters. However, the data for Chinese hamsters are equivocal, at least at the tenth day, and it is possible that neither lysosomes nor iron-binding proteins are major sites of fixation in this species. It is interesting to note that the indications for lysosomal association are unequivocal in both rats and mice, the species with rapid Pu elimination, whereas these organelles appear to become less important in the species with longer retention, Syrian and especially Chinese hamster. However, investigations of the changes with time in the subcellular distribution and on the exact nature of the binding components for Pu-239 in the livers of all species, but especially of Chinese hamsters, are needed before further conclusions can be drawn.

REFERENCES

1. Seidel, A. (1978): Radiat. Res., 76, 60.
2. Seidel, A., Winter, R., Jentzsch, C., Gruner, R., Heumann, H.-G. and Hanke, S.: in: Proceedings of the International Symposium on Biological Implications of Radionuclides Released from Nuclear Industries, IAEA, Vienna, in press.

Radiobiology

BIOLOGICAL EFFECTS OF INHALED RADIONUCLIDES: SUMMARY OF ICRP REPORT 31

W. J. Bair

A report on the biological effects of inhaled radionuclides was prepared by an International Commission on Radiological Protection Task Group charged with evaluating the hazards associated with inhalation of plutonium and other radionuclides.[1] The Task Group enumerated the biological responses, identified tissues and cells at risk, derived risk coefficients for inhaled radionuclides from animal experiments for comparison with human data, and determined an Equal Effectiveness Ratio for alpha emitters relative to beta-gamma emitters.

BIOLOGICAL EFFECTS

Since there are no human populations (other than uranium, fluorspar, and other miners who worked in mines containing high concentrations of radon decay products) that have shown health effects that can be associated with radionuclide inhalation, it was necessary for the Task Group to use data from animal experiments to describe the biological effects. Radionuclides deposited in the respiratory tract are either "insoluble" (not readily translocated to other tissues or excreted) or "relatively soluble" (more readily translocated to other tissues or excreted). Biological effects resulting from the inhalation of radionuclides depend upon the distribution and retention of these radionuclides in the body and upon the doses to the tissues irradiated.

Life-Span Shortening

The Task Group compared the mean or median survival time of animals exposed to radionuclides with those of appropriate controls. Only data from alpha-emitter experiments were used because of the paucity of beta-gamma emitter data. Figure 1 shows the shortening of life span in animals that inhaled PuO_2. At deposition doses below about 0.01 µCi/g lung, the shortening of mean life span was less than about 10%. Several experimental groups had mean life spans greater than control groups. At doses above about 0.01 µCi/g lung, life-span shortening increased with dose. The dose causing a 50% reduction of mean life span was between 0.05 and 0.1 µCi/g lung. Similar results were obtained for soluble alpha emitters; the dose causing a 50% reduction of mean life span was about 0.1 µCi/g lung.

Pathologic and Clinical Responses

The shortening of life span following the inhalation of radionuclides was accompanied by and/or caused by certain pathologic changes, some of which were reflected in clinical signs and symptoms. Some of these effects appear to be nonstochastic, i.e., the degree of effect, rather than its occurrence, is a function of dose; examples are lymphocytopenia, respiratory insufficiency, pulmonary and lymph node fibrosis, and cellular metaplasia. Other effects, such as pulmonary and bone neoplasia, are stochastic since the probability of occurrence is related to dose. Since a full spectrum of doses has not been investigated for any inhaled radionuclide in any animal species, it is not possible

---

[a]This summary was prepared under contract EY-76-C-06-1830 with the U.S. Department of Energy.

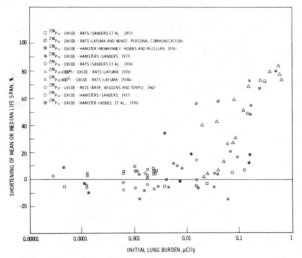

FIGURE 1. Shortening of Life Span by Inhaled PuO_2[a]

to know the lowest doses at which any biological effect *will occur* nor even the highest dose at which it *will not occur*. However, the available data were used to base judgments about the lowest dose range at which given nonstochastic effects are likely to be seen. For stochastic effects, such as death-causing neoplasia, estimates of the frequency of occurrence were developed.

Hematologic Effects. Hematologic effects occur after inhalation of radionuclides because of the irradiation of the hematopoietic tissue to which the radionuclides are translocated and deposited, and because of the direct irradiation of blood circulating in lungs, liver, and perhaps lymph nodes containing the radionuclide. Leukocytopenia was observed in animals after inhalation of soluble forms of alpha emitters or beta-gamma emitters. The most consistent hematologic response observed in dogs after inhalation of PuO_2 and insoluble beta-gamma emitters was lymphocytopenia. The Task Group concluded that lymphocytopenia is probably a nonstochastic effect, and that it probably can be detected after pulmonary deposition of ≥ 0.0005 µCi PuO_2/g lung. Much higher doses of inhaled insoluble beta-gamma emitters are required to cause a detectable lymphocytopenia, probably on the order of >0.01 µCi/g. Since both the magnitude and time of onset of lymphocytopenia are dose dependent, lymphocytopenia caused by these low doses would not occur until long after exposure to the radionuclide and might be marginally detectable.

Nonneoplastic Lesions. Respiratory insufficiency (characterized by increased respiratory rate, decreased arterial P_{O_2} and increased P_{CO_2} caused by diffuse fibrosis of the lungs) and alveolar edema may cause death within a month after inhalation of large amounts of alpha- or beta-gamma-emitting radionuclides. Death from respiratory insufficiency and fibrosis may also result after exposure to much lower doses and may occur in rodents after many months and in dogs after several years. Respiratory insufficiency is the major cause of death in those groups of animals that showed a greatly reduced mean life span, Figure 1.

The available data from several animal species indicate that respiratory insufficiency and possibly death resulting from extensive fibrosis in the lungs caused by radiation from inhaled radionuclides are nonstochastic processes and might be expected to occur after alveolar deposition of >0.01 µCi alpha emitters/g lung and of >0.5 µCi beta-

---

[a]See ICRP Publication 31 for references.

gamma emitters/g. If the retention times of the radionuclides in the lungs are short, larger quantities of the radionuclides would have to be deposited in the lungs to cause these lesions and early death.

Lymph nodes containing radionuclides may be severely damaged. The primary lesions are characterized by lymphadenitis and fibrosis with partial to complete depletion of germinal centers. Primary neoplasia does not appear to occur in lymph nodes of experimental animals that have inhaled alpha emitters or beta-gamma emitters. The dose-effect relationship for nonneoplastic lesions in the tracheobronchial lymph nodes is poorly known. Although lesions have been described in rodents, most of the dose-effect data from experiments with dogs suggest that fibrotic and/or atrophic lesions in the tracheobronchial lymph nodes are nonstochastic responses and that the dose at which these responses may be expected to begin to be observed is an alveolar deposition of about 0.001 µCi/g lung. This applies to insoluble alpha emitters such as $^{239}PuO_2$. In the case of more soluble alpha-emitting compounds, which are not as readily accumulated in the tracheobronchial lymph nodes as $^{239}PuO_2$, larger amount would be required to produce lymph node lesions. The dose of insoluble beta-gamma emitters required to produce such lesions is not known, but is probably greater than 0.05 µCi/g lung.

Neoplastic Lesions. Pulmonary neoplasia has been identified in miners of uranium and other substances as a consequence of inhaling alpha-emitting radon daughter radionuclides. Although pulmonary neoplasia has not been reported in human beings who have inhaled other radionuclides, such as the transuranic elements, it has been well demonstrated in experimental animals as a potential consequence of inhaling several different alpha-emitting and beta-gamma-emitting radionuclides.

In Figure 2 the incidences of lung cancer are plotted against the initial lung burdens of soluble alpha-emitting radionuclides. In rats, initial lung burdens above 0.001 µCi/g had an increasing probability of causing lung cancer. The low incidences at high doses reflected the shortened life spans due to deaths from causes other than neoplasia which prevented the full lung cancer potential from being expressed. At initial lung burdens below about 0.001 µCi/g, none of the experiments showed statistically significant increases in lung cancer, although several lung cancers occurred.

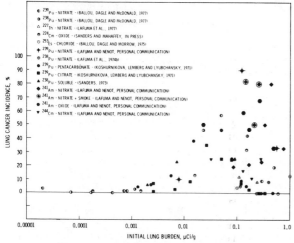

FIGURE 2. Relationship Between Incidence of Lung Cancer in Experimental Animals and Initial Lung Burdens of Inhaled, Relatively Soluble, Alpha-Emitting Radionuclides[a]

[a] See ICRP Publication 31 for references.

Like the data for relatively soluble alpha emitters, the lung cancer incidence for insoluble PuO$_2$ increased markedly at initial lung burdens above 0.001 µCi/g. Low-dose experiments are still in progress in which groups of dogs were exposed to initial lung burdens as low as 0.00003 µCi ^{239}PuO$_2$/g lung and 0.000016 µCi ^{238}PuO$_2$/g lung. In several experiments at different laboratories it was observed that hamsters were relatively insensitive to the induction of lung cancer. Thus, comparing species susceptibility to lung cancer caused by inhaled PuO$_2$ at the dose range where data exist (~0.01 µCi/g initial lung burden), beagle dogs were more sensitive than rats, which were more sensitive than mice, which were more sensitive than hamsters.

In studies of beta-gamma-emitting radionuclides deposited by inhalation or by intratracheal injection, initial lung burdens above 0.1 µCi/g led to increasing incidences of lung cancers. There are no data at lower doses.

While recognizing the possible shortcomings of using experimental animal cancer incidence data, the Task Group believed that, in the absence of a human data base, a quantitative descriptive model for radionuclide carcinogenesis in the lungs based on the animal data could be useful for risk assessment. Since a model could not be devised from hypotheses of the mechanism(s) of induction of cancer by inhaled radionuclides, the Task Group chose the logarithmic probit model usually employed in dose-response analysis and contrasted it with the linear model usually used to reflect conservatism. The incidence values for the linear regression model were weighted on a basis of the number of observations.

The original data for alpha emitters, uncorrected for control mortality, and the fitted functions (heavy solid lines) in Figure 3. suggest that insoluble alpha emitters were slightly more effective than soluble alpha emitters in causing pulmonary cancer in experimental animals.

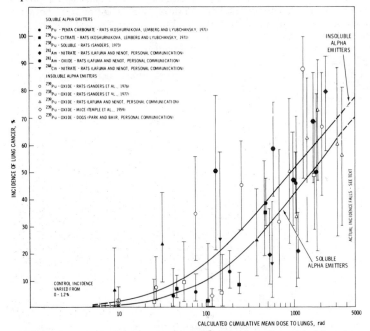

FIGURE 3. Relationship Between Incidence of Lung Cancer and Alpha Dose to Lungs from Inhaled Soluble and Insoluble Alpha Emitters; Probit Analysis[a]

---

[a] See ICRP Publication 31 for references.

Both the linear and probit models gave an adequate description of the alpha-emitter incidence results over the range of observed doses, Figure 4. However, the linear model used to describe the beta-gamma-emitter incidence data gave unrealistically high projected results at low or zero doses, Figure 5. Although the probit model gave more realistic values at low doses, the slope of the line was very shallow.

The improved Mantel-Bryan procedure also was used to obtain lung cancer risk estimates based on data from animal experiments. The Task Group compared these estimates with published estimates obtained by other methods (usually by linear extrapolations) from both experimental animal data and from limited human data, the latter mostly from external radiation exposures, Table 1.

The risk coefficients obtained by the Mantel-Bryan procedure and by extrapolating the fitted probit models are lower than those obtained from the fitted linear model and other published linear models using animal or human data. This means that for these experimental data, linear models yield a more conservative estimate of lung cancer risk than the other models. The applicability of these lung cancer risk coefficients to human beings who have inhaled or may inhale radionuclides is not known. However, the Task Group believed the risk estimates calculated from available animal data, summarized in Table 1, are supportive of the ICRP decision in Publication 26 (1978) to use a risk factor for the lungs of 2×10^{-3} Sv^{-1} (20×10^{-6} rem^{-1}).[2]

Comparison of the risk estimates obtained by analysis of all the alpha-emitter data, 25 and 36×10^{-6} rad^{-1}, with the risk estimate of 0.84×10^{-6} rad^{-1} for beta-gamma emitters gave an Equal Effectiveness Ratio of about 30 for inhaled alpha-emitting radionuclides. Thus, the experimental animal data tend to support the decision by the ICRP to change the recommended quality factor from 10 to 20 for alpha radiation (ICRP, 1977).[2]

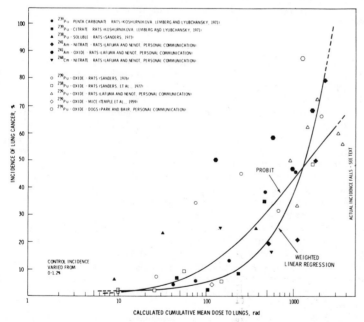

FIGURE 4. Comparison of Weighted Linear Regression and Probit Models as Descriptors of Alpha-Induced Animal Lung Cancer Data[a]

---

[a] See ICRP Publication 31 for references.

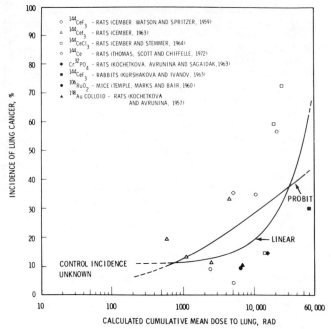

FIGURE 5. Comparison of Weighted Linear Regression and Probit Models as Descriptors of Beta-Gamma-Induced Animal Lung Cancer Data

Extrapulmonary Lesions. Various tumors in tissues other than those of the respiratory tract have been observed in animals after inhalation of alpha-emitting or beta-gamma-emitting radionuclides. These have occurred mostly in animals that have inhaled relatively soluble radionuclides which translocated from lungs to other tissues in the body. It is well documented that alpha-emitting radionuclides such as radium deposited in bone tissue can cause osteosarcomas in human beings as well as in experimental animals. Thus, bone neoplasia can be expected to be a potential consequence of inhaling alpha-emitting radionuclides if sufficient quantities are translocated to bone. In experimental animals, bone neoplasia has occurred at dose levels of about 0.01 µCi/g lung of inhaled soluble plutonium and other transuranic elements. Inhaled beta-gamma-emitting radionuclides such as ^{90}Sr and ^{144}Ce, which deposited in skeleton, also have been shown to cause skeletal neoplasia in experimental animals.

With the more highly transportable transuranic elements such as einsteinium, curium, and americium, the incidences of extrapulmonary and extraskeletal cancers were increased in experimental animals. These included kidney and bladder carcinomas and thoracic and abdominal lymphoreticulosarcomas. Although many of the inhaled radionuclides studied in experimental animals translocated to liver, relatively few liver cancers were reported. Since these usually occurred at long times after exposure it was concluded that liver cancer could be one of the predominate late consequences of inhaling radionuclides.

Numerous neoplasias were observed in the nasal cavities of animals that inhaled alpha emitters such as radon and uranium and beta-gamma emitters such as ^{91}Y, ^{144}Ce, and ^{90}Sr, probably as a result of continuous irradiation of the nasal epithelium

All of these neoplasias are associated with tissues and organs in which inhaled radionuclides are deposited or accumulated following translocation from the respiratory tract. However, the fact that a tissue accumulates radionuclides does not necessarily

TABLE 1. Summary of Risk Coefficients for Radiation-Induced Lung Cancer

Animal Species	Model	Risk Coefficients (cases of lung cancer per million animals or persons per rad)				Reference[a]
		Alpha Radiation			Beta-Gamma Radiation	
		Insoluble	Soluble	All		
Rodents and dogs	Improved Mantel-Bryan	70	20	25	0.84	ICRP-31
Rodents and dogs	Probit	65	20	36	-	ICRP-31
Rodents and dogs	Linear	-	-	360	-	ICRP-31
Rats	Linear	1600	800	1250	-	Bair and Thomas, 1976
Dogs	Linear	-	-	600[b]	-	Bair and Thomas, 1976
Man	Linear	-	-	400[b]	20	Thorne and Vennart, 1976
Man	Linear	-	-	500[b]	25	MRC, 1975
Man	Linear	-	-	200	-	Mays, 1976
Man	Linear	-	-	400[b]	20	BIER, 1972
Man	Linear	-	-	200-800[b]	10-40	UNSCEAR, 1972
Man	Linear	-	-	1000[b]	25-50	UNSCEAR, 1977
Man	Linear	-	-	400[b]	20	ICRP-26, 1977

[a] See ICRP Publication 31 for references.
[b] Values converted to rad from rem on basis of a Q factor of 20 for alpha radiation.

indicate a susceptibility to cancer induction. For example, the thoracic and, to a lesser extent, hepatic lymph nodes have been shown to accumulate concentrations of radionuclides that exceed by many times the concentrations retained in lungs or that occur in other tissues. This is especially true for insoluble compounds. Primary neoplasia of thoracic and hepatic lymphatic tissue has not been reported in any of the experiments with inhaled alpha- or beta-gamma-emitting radionuclides. In life-span studies with beagle dogs that inhaled ^{239}PuO$_2$ or insoluble ^{144}Ce particles, metastases of primary lung cancers were found in thoracic lymph nodes, as were occasional lymphangiosarcomas and several hemangiosarcomas. There were no other lymph node cancers. Thus, lymph nodes in experimental animals appear to be much less susceptible to cancer induction than other tissues in which inhaled radionuclides are deposited or accumulated. Further, inhalation of radionuclides is not known to be related to the induction of lymph node tumors in any human being. These observations influenced the ICRP's decision to consider the lymph nodes with the lungs as one composite organ for radiation protection purposes.[3]

CELLS AND TISSUES AT RISK

High lung burdens of inhaled radionuclides result in profound structural and functional changes in which the pulmonary capillary endothelial cells are the most prominent cells at risk. Pulmonary carcinogenesis is the most serious effect of low doses of inhaled radionuclides. The cells at risk are the precursor cells and basal cell layers of the respiratory tract epithelia. The Task Group considered the possibility that certain types of neoplasia induced by inhalation of radioactive material could be related to the presumed cell(s) of origin, and the cell lines at risk, but recognized the types of neoplasias produced may also depend on the pattern of spontaneous tumor development in a given species and strain. For instance in uranium miners small cell carcinomas, associated with lower radiation exposures, and epidermoid carcinomas, associated with higher initial exposures, appeared to originate in the larger proximal bronchi. In animals after inhalation of radionuclides, adenocarcinomas and epidermoid carcinomas appeared to originate in peripheral regions of the lungs. If the types of neoplasia observed after inhalation of radionuclides do suggest the cell line at risk, basal cell layers and Kulchitsky cells of bronchial epithelia may be at risk in uranium miners who develop undifferentiated small cell and epidermoid carcinomas. For adenocarcinoma in experimental animals, the cells at risk appear to be bronchiolar and, possibly, type II pneumocytes or Clara cells.

The occurrence of hemangiomas in lungs of animals that have inhaled radioactive particles suggests that the endothelial cells of pulmonary capillaries are also at risk. At risk also are the cellular constituents of bronchial and tracheobronchial lymph nodes, which may accumulate large doses or radiation. These include the T and B lymphocytes, germinal center cells, plasma cell precursors, endothelial cells and possibly reticulum cells, histiocytes, fibroblasts, and other mesenchymal elements. Endosteal bone tissue, hematopoietic marrow, and liver and spleen tissue may also be at risk from radionuclides translocated from the lungs. Lymphocytes appear to be at high risk from inhaled radioactive particles because blood lymphocytopenia is among the earliest and most sensitive changes observed in experimental animals. Since inhaled radionuclide-induced lymph node tumors as well as lymphomas and leukemias have been very rare in animal experiments, nothing can be said about tissues and cells at risk for these types of neoplasia. The possibility of genetic damage to germ cells after inhalation of radionuclides was not excluded by the Task Group but was not addressed because of the lack of data.

HOT PARTICLES

The Task Group believed that knowledge about the behavior of inhaled alpha-emitter particles in the lungs and the interaction of alpha irradiation with cells is inadequate either to support or completely deny the hot particle theory concerning the induction of lung cancer. Animal experiments indicate that the lung cancer risk associated with inhaled plutonium particles in quantities that could be distributed in hot spots may be slightly greater than for more soluble and, therefore, more diffusely distributed alpha emitters. Other experiments with plutonium microspheres clearly showed that "diffuse" radiation sources in the lungs of hamsters were much more likely to cause both malignant and benign lung tumors than highly localized sources. The Task Group concluded that the risk of lung cancer from inhaled radioactive particles will be greatly overestimated if based on hot particle concepts.

OTHER FACTORS

The Task Group considered possible modification of the effects of radionuclides by inhalation of other potentially damaging agents. Only a few animal experiments have addressed the question of combined effects of inhaled radionuclides and air pollutants or smoking. Results of these few studies are inconclusive. Therefore, such factors had to be ignored in addressing the objectives of the report. However, it was stressed that the possibility that the effects of inhaled radionuclides could be greatly influenced by smoking, and air pollutants should not be ignored in protecting human beings from airborne radionuclides.

ACKNOWLEDGEMENT

The members of the Task Group were: Dr. W. J. Bair, Chairman, Dr. B. B. Boecker, Dr. H. Cottier, Dr. P. E. Morrow, Dr. J. C. Nenot, Dr. J. F. Park, Dr. J. M. Thomas, and Dr. R. G. Thomas.

REFERENCES

1. International Commission on Radiological Protection (1979): *Report of ICRP Task Group on Biological Effects of Inhaled Radionuclides.* Committee 2. Oxford: Pergamon Press.
2. International Commission on Radiological Protection (1977): *Recommendations of the International Commission on Radiological Protection.* Publication 26. Oxford: Pergamon Press.
3. International Commission on Radiological Protection (1979): *Limits for Intakes of Radionuclides by Workers.* Committee 2. Oxford: Pergamon Press.

AN APPROACH TO THE DERIVATION OF RADIONUCLIDE INTAKE LIMITS FOR MEMBERS OF THE PUBLIC

R. C. Thompson

This paper describes a systematic approach to the development of radionuclide exposure limits for members of the public--an approach which starts with the occupational ALI (Annual Limit on Intake) and applies an adjustment factor, separately derived for each radionuclide. It seems expedient to utilize the extensive body of data and calculations relevant to occupational radionuclide limits, as collected in ICRP Publication 30 (1), rather than to attempt the derivation of public exposure limits from first principles. This also follows the past practice of deriving limits for external whole-body exposure of members of the public by applying a factor (usually 1/10) to the occupational limit for external exposure.

Lower exposure limits for members of the public, as compared to radiation workers, have been justified for such reasons as the following: (a) the radiation worker receives a specific benefit in the form of wages and other career satisfactions that is not received by the involuntarily exposed member of the public; (b) radiation workers constitute a population less susceptible to damaging effects, because of their age and general state of health; (c) radiation workers constitute a much smaller population, which is important from genetic considerations; (d) radiation workers are a more carefully monitored and medically treated population; (e) the timespan of occupational exposure is only a fraction of the lifespan--a minor fraction for the case of genetically significant exposure. Whatever numerical factor may be justified by these arguments, the external whole-body exposure limit for a member of the public may be defined by the following equation:

$$H_{L_p} = F_p \cdot H_{L_w}$$

where H_L is the annual dose equivalent limit for the member of the public (p) or the worker (w) and F_p is the *general population exposure reduction factor* justified by the cited arguments.

The considerations that lead to F_p for external exposure also apply to the derivation of intake limits for radionuclides. An additional factor is required, however, relating to specific interactions between radionuclide and individual, which differ for the worker and the member of the public. Radionuclide intake limits for members of the public may then be defined as:

$$(ALI)_p = F_p \cdot F_j \cdot (ALI)_w$$

---

*Work supported by U.S. Department of Energy Contract EY-76-C-06-1830.

where F_j is the *radionuclide-specific exposure adjustment factor*. The derivation of radionuclide intake limits for the member of the public then becomes a problem of developing appropriate values for F_j.

I am assuming that a single limit for a member of the public is desirable. One might set limits applicable to any number of special categories of exposed persons, but for practical purposes it would seem necessary to control to a single limit, chosen to provide an acceptable apportionment of dose among all these special categories. In practice, of course, this $(ALI)_p$ will not be employed directly, but will be used as a basis for derived limits, which will control the amount of radionuclide to which the member of the public has access.

Some of the more important factors, varying with age, state-of-health, etc., that might be involved in the development of values for F_j, include the following: (a) radiosensitivity factors altering risk; (b) morphological factors altering dose; (c) metabolic factors altering retention, distribution, and absorption from gastrointestinal tract or lung; (d) environmental factors altering biological availability; and (e) a physical half-life factor that might alter all of the above. I would like to illustrate this approach by applying it to the derivation of exposure limits for plutonium.

TABLE 1. Derivation of Member-of-the-Public Radionuclide-Specific Exposure Adjustment Factor for ^{239}Pu Ingestion

Parameter	Factor for: Infancy	Lifespan
Organ size	0.1	1
Food consumption	10	1
Food selection	5	1
Enhanced gastrointestinal absorption	0.001	0.7
Distribution and retention in: Bone (2x for infant) Liver (0.5x for infant) G.I. tract (100x for infant)	0.6	1
Environmental availability	1	0.2
Fraction of committed dose received	40	1
Overall value of F_j	0.12	0.14

Table 1 summarizes adjustments required for the ALI for ingestion, considered on a lifespan basis. Also shown are the adjustments required if only the first year of life is considered, as a check on the adequacy of protection during this critical period. There is no unusual factor of radiosensitivity relating to plutonium; any general effect of age on radiosensitivity is presumably included in the general population exposure reduction factor, F_p. Because of the smaller size of infant organs, the same level of intake would result in a higher dose rate in the infant than in the adult. For purposes of this illustration, a factor of 0.1 is assumed to equalize the infant and adult dose. Over a lifespan, this factor is insignificant. Food consumption by the infant is smaller than that assumed in deriving the ALI_w, by an assumed factor of 10. Because the infant's food is likely to be of lower than average plutonium content, e.g., milk and canned foods rather than leafy and root vegetables and seafoods, a factor of 5 is introduced for food selection.

The Derivation of Radionuclide Intake Limits

A major factor of 0.001 is required to correct for the enhanced gastrointestinal absorption of plutonium by the infant, as suggested by studies in miniature swine (2). Applying this factor of 0.001 to 1/50 of the lifespan and correcting for the lower rate of intake during the period of infancy, the overall lifespan effect requires adjustment by only a factor of 0.7.

From animal data it appears that the very young deposit more plutonium in bone (by a factor of 2 we will assume) and less plutonium in liver (0.5 assumed) (3), and retain plutonium for a much longer time in the gastrointestinal tract (a probably overly conservative factor of 100 assumed) (3). When converted to weighted dose equivalent according to ICRP practices (1), these altered distribution and retention parameters result in about a 60% increased effective dose equivalent, adjusted for by the factor of 0.6. A 5-fold enhanced environmental availability of plutonium is assumed throughout the lifespan; this factor is not applied to the infant because it seems unreasonable to add this to the already conservative assumption of a 1000-fold increased absorption from the infant gut.

Because the ALI limits dose commitment rather than current dose, and since only about 1/40 of the committed dose is actually delivered during the first year of exposure, adjustment by a factor of 40 is required if one wishes to evaluate only the period of infancy.

Multiplied together, these factors lead to a value for F_j of 0.12 for infancy and 0.14 for lifespan. A factor of 0.1 should therefore afford protection for the first year of life as well as for the adult. If a general population dose reduction factor of 0.1 is assumed, we have a final ALI for a member of the public, which is one-hundredth of the occupational ALI for ingestion.

TABLE 2. Derivation of Member-of-the-Public Radionuclide-Specific Exposure Adjustment Factor for ^{239}Pu Inhalation

Parameter	Factor for: Infancy	Lifespan
Organ size	0.1	1
Volume inhaled	10	1
Fraction deposited and/or retained	2	1
Enhanced gastrointestinal and/or pulmonary absorption	0.2	1
Distribution and retention in: Bone (2x for infant) Liver (0.5x for infant) G.I. tract (100x for infant)	0.6	1
Fraction of committed dose received	40 (W)* 4 (Y)*	1
Overall value of F_j	10 (W)* 1 (Y)*	1

* Compound solubility class (1)

Table 2 summarizes a similar derivation for inhaled plutonium, differentiating, where necessary, between Class W and Y compounds (1). Again, there is no radiosensitivity adjustment. It is assumed that smaller lung size is exactly compensated by a reduced volume inhaled. Based on limited experimental animal data (4,5), alveolar deposition and/or retention is assumed to be lower by a factor of 2 in the infant. Enhanced absorption from the gastrointestinal tract during infancy is a less significant factor for inhaled plutonium, since a substantial fraction is directly absorbed from the lung. The situation is also complicated by the lack of data on possibly enhanced infant absorption from the lung. These interactive factors were lumped and conservative adjustment factors assumed. Distribution parameters for the infant are the same as those assumed for the ingestion case and result in a similar adjustment. There is no correction for environmental availability, since this is primarily a food-chain phenomenon.

The correction for the fraction of committed dose received during the first year is about a factor of 40 for Class W compounds. It is only about a factor of 4 for Class Y compounds, because something approaching 25% of the weighted committed dose equivalent will be delivered to the lung during the first year following exposure.

Multiplied together, these factors lead to F_j values of 1 for lifespan exposure and for Class Y infant exposure, indicating that no further adjustment is required beyond that provided by the general population dose reduction factor, F_p. For Class W compounds the infant is relatively overprotected by limits based on lifespan exposure.

Let me emphasize that my purpose in this presentation has been to illustrate an approach to systematic consideration of radionuclide exposure limits for members of the public, and not to suggest specific limits for plutonium. A more definitive analysis might well lead to different numbers, but I hope the exercise has shown how such analyses could be conducted for all radionuclides.

As a member of ICRP Committee 2 on Secondary Limits, and of the U.S. NCRP Committee 57 on Internal Emitter Standards, I have benefited from much discussion in these committees on the subject matter of this paper. Views expressed in this paper should not, however, be interpreted as reflecting official ICRP or NCRP opinions.

REFERENCES

1. ICRP Publication 30 (1979): Limits for Intake of Radionuclides by Workers. Pergamon Press, Oxford.
2. Sullivan, M.F. (1979): In: Pacific Northwest Laboratory Annual Report for 1978 to the DOE Assistant Secretary for Environment, Part I Biomedical Sciences (Doc. PNL-2850), p. 3.89. Pacific Northwest Laboratory, Richland, Washington.
3. Sikov, M.R. and Mahlum, D.D. (1972): Health Phys., 22, 707.
4. Sullivan, M.F. (1979): In: Pacific Northwest Laboratory Annual Report for 1978 to the DOE Assistant Secretary for Environment, Part I Biomedical Sciences (Doc. PNL-2850), p. 3.95. Pacific Northwest Laboratory, Richland, Washington.
5. Sikov, M.R., Cannon, W.C., Buschbom, R.L. and Mahlum, D.D. (1977). In: Pacific Northwest Laboratory Annual Report for 1976 to the ERDA Assistant Administrator for Environment and Safety, Part I Biomedical Sciences (Doc. PNL-2100 Pt 1), p. 112. Battelle Pacific Northwest Laboratories, Richland, Washington.

RBE OF α-PARTICLES vs β-PARTICLES IN BONE SARCOMA INDUCTION

C. W. Mays and M. P. Finkel

^{226}Ra and ^{90}Sr were injected intravenously into 17-month-old beagles at the University of Utah (13) and 70-day-old CF1 female mice at the Argonne National Laboratory (2,3,4). Bone sarcomas, mostly osteosarcomas, were the main radiation-induced cancers (Tables 1 and 2).

^{226}Ra is an α-emitter and ^{90}Sr is a β-emitter. Both are bone volume seekers, so that the mean endosteal dose is roughly equal to the skeletal dose averaged over both bone and marrow (1,8). The average skeletal dose in rads was computed for ^{226}Ra including its retained α-emitting daughters (7,9), and for ^{90}Sr including its β-emitting daughter, ^{90}Y (7,10). The average skeletal dose was calculated at the assumed start of tumor growth, which was taken as 1 year before death in the beagles (10), 140 days before death with bone sarcoma in the mice treated with ^{90}Sr (10), and 100 days before radiographic appearance of the tumors in the mice treated with ^{226}Ra (9). Since the shapes of the retention curves for ^{226}Ra and ^{90}Sr are similar in beagles (and in mice), assumptions on the time span of the "wasted" radiation have little influence on the calculated RBE (11).

The relative biological effectiveness (RBE) of α-particles vs. β-particles in producing bone sarcomas was taken as the ratio of ^{90}Sr dose/^{226}Ra dose at a given level of incidence. The RBE progressively increased as the incidence decreased, reaching RBE = 26 at 8.7% incidence in beagles, and RBE = 25 at 7.7% incidence in mice (Table 3). The increase in RBE was largely due to the decreased effectiveness per rad of ^{90}Sr β-radiation at low doses and low dose-rates (10). Because of statistical fluctuations, the RBE's are not shown below an incidence of 7.5%, but the trends are compatible with even higher RBE's. In these experiments all of the mice have died. None of the beagles treated with ^{90}Sr remain alive. However, if future bone sarcomas appear among the 9 surviving beagles that received low levels of ^{226}Ra, the α-particle RBE at low doses will be increased above the values indicated in this paper. Of special relevance is the RBE at the low doses and low risks that are considered permissible for man. The ICRP recently increased the quality factor for α-particles up to 20 (5). But is that enough? Additional information should come in a few years from beagles treated with ^{226}Ra and ^{90}Sr at Davis, California.

The increase of RBE with decreasing dose seems a general property of densely-ionizing radiation. It also applies to the fast-neutron-induction of leukemia in man, chromosome aberrations in human lymphocytes, skin damage, in man, rat, mouse, pig; mammary tumors in rats, cataracts in mice, inactivation of intestinal crypt cells in mice, mutations in Tradescantia, and growth reduction in Vicia Faba (6,12).

---
*Research supported by the U.S. Department of Energy.

TABLE 1. Bone sarcomas in *beagles* treated with ^{226}Ra or ^{90}Sr

Nuclide	Inj. µCi/kg	Av. yr inj. to death	Treated dogs	Sar. dogs	Incidence (%)	Av. skel. rads 1 yr before death
^{226}Ra	10.4	2.86	10	9	90.0	13400
	3.21	4.13	13	12	92.3	5700
	1.07	6.12	12	11	91.7	2500
	0.339	10.05	13	5	38.5	1100
	0.166	9.40	14	1	7.1	447
	0.062	---	23(3)*	2	8.7	~210
	0.022	---	25(4)*	1	4.0	~74
	0.0074	---	10(2)*	0	0	~25
	0	---	44(14)*	0	0	0
^{90}Sr	97.9	3.40	14	8	57.1	10100
	63.6	5.82	12	8	66.7	9360
	32.7	9.98	12	2	16.7	7940
	10.8	12.27	12	0	0	2870
	3.46	10.79	12	0	0	798
	1.72	11.31	13	0	0	445
	0.57	12.93	12	0	0	143
	0	11.49	13	0	0	0

*Living dogs, as of 1 January 1980, shown in parentheses.

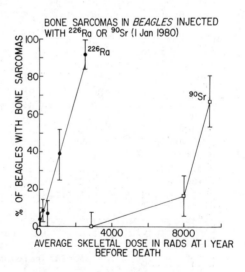

FIGURE 1. Bone sarcoma incidence in beagles. The response seems approximately linear up to 2500 rads from ^{226}Ra, but is strongly concave upwards for ^{90}Sr. Standard deviations are shown here and on Fig. 2.

RBE of α-Particles vs β-Particles

TABLE 2. Bone sarcomas in *female CF1 mice* treated with ^{226}Ra or ^{90}Sr.

	Inj. µCi/kg	Mice at 150 days Post Inj.	Sar. Mice	Inc. (%)	Bone Sarcoma Mice Av. days, inj. to appear. (Ra) or death (Sr)	Av. skeletal rads 100 d before appear. or 140 d before death
^{226}Ra	120	45	14	31.1	328	28900
	80	44	31	70.5	359	21300
	40	45	33	73.3	394	11800
	20	44	38	86.4	428	6420
	10	43	34	79.1	484	3640
	5	45	28	62.2	544	2040
	2.5	104	45	43.3	639	1190
	1.25	104	22	21.2	657	614
	1.00	239	56	23.4	643	480
	0.75	504	94	18.7	686	383
	0.50	683	80	11.7	655	244
	0.25	247	19	7.7	580	109
	0.10	252	5	2.0	853	62
	0.05	254	11	4.3	710	26
	0	521	6	1.2	730	0
^{90}Sr	2200	26	19	73.1	216	12000
	880	45	41	91.1	260	6630
	440	42	34	81.0	440	6300
	200	59	8	13.6	510	3310
	88	74	2	2.7	760	2090
	44	83	3	3.6	640	900
	8.9	104	0	0	---	172*
	4.5	119	2	1.7	600	87
	1.3	148	2	1.4	630	26
	0	149	2	1.3	550	0

*Dose for 8.9 µCi/kg level calculated at (600-140) = 460 days.

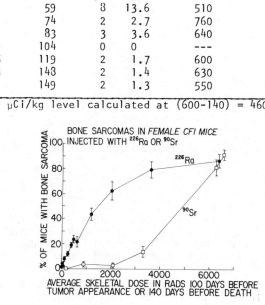

FIGURE 2. Bone sarcoma incidence in mice. ^{226}Ra is much more effective than ^{90}Sr at low doses, but at very high doses the effectivenesses converge.

TABLE 3. Bone sarcoma RBE of ^{226}Ra vs. ^{90}Sr.

Species	Incidence (%)	^{90}Sr β-particles (rads)	^{226}Ra α-particles (rads)	RBE (α vs. β)
Beagles	66.7	9360	1900*	5
	38.5	8600*	1100	8
	16.7	7940	480*	17
	8.7	5500*	210	26
Mice	86.4	6500*	6420	1
	81.0	6300	4400*	1.4
	79.1	6200*	3640	2
	62.2	5500*	2040	3
	43.3	4600*	1190	4
	21.2	3700*	614	6
	23.4	3800*	480	8
	18.7	3500*	383	9
	13.6	3310	280*	12
	11.7	3100*	244	13
	7.7	2700*	109	25

*Interpolated from curves on Figures 1 and 2.

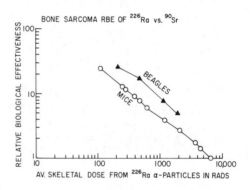

FIGURE 3. Relative biological effectiveness of α-particles from ^{226}Ra and retained daughters, relative to β-particles from ^{90}Sr and its daughter, ^{90}Y. The RBE increases as dose decreases, both in beagles and in mice. The effect is mainly due to decreased effectiveness per rad in bone sarcoma induction by β-particles at low doses and low dose-rates.

REFERENCES

1. Beddoe A.H. and Spiers F.W. (1979): A comparative study of the dosimetry of bone-seeking radionuclides in man, Rhesus monkey, beagle, and miniature pig, Radiat. Res. 80, 423-439.
2. Finkel M.P., Biskis B.O., and Jinkins P.B. (1969): Toxicity of radium-226 in mice, in Radiation-Induced Cancer, A. Ericson (Ed.), IAEA, Vienna, pp. 369-391, updated by Miriam Finkel in 1971.
3. Finkel M.P. and Biskis B.O. (1959): The induction of malignant bone tumors in mice by radioisotopes, Acta Unio International Contra Cancrum 15, 99-106, updated by Miriam Finkel in 1971.
4. Finkel M.P., Biskis B.O. and Scribner (1959): The influence of strontium-90 upon lifespan and neoplasms of mice, in Progress in Nuclear Energy, Series IV, vol. 2 - Biological Sciences, Pergamon Press, London, pp. 199-209, updated by Miriam Finkel in 1971.
5. ICRP Publication 26 (1977): Recommendations of the International Commission on Radiological Protection, Pergamon Press, see p. 5.
6. Kellerer A.M. and Rossi H.H.(1972): The theory of dual radiation action, in Current Topics in Radiation Research 8, 85-158.
7. Lloyd, R.D., Mays C.W., Atherton D.R., Taylor G.N., and Van Dilla M.A. (1976): Retention and skeletal dosimetry of injected ^{226}Ra, ^{228}Ra, and ^{90}Sr in beagles, Radiat. Res. 66, 274-287.
8. Mays C.W. and Lloyd R.D.(1972): Predicted toxicity of ^{90}Sr in humans, in Second International Conference on Strontium Metabolism, J.M.A. Lenihan, chairman, USAEC CONF-720818, pp. 181-205.
9. Mays C.W. and Lloyd R.D.(1972): Bone sarcoma incidence vs. alpha particle dose, in Radiobiology of Plutonium, B.J. Stover and W.S.S. Jee (Eds.), J.W. Press, Salt Lake City, Utah, pp. 409-430.
10. Mays C.W. and Lloyd R.D. (1972): Bone sarcoma risk from ^{90}Sr, in Biomedical Implications of Radiostrontium Exposure, M. Goldman and L.K. Bustad (Eds.), USAEC Symposium Series 25, CONF-710201, pp. 352-375.
11. Mays C.W., Dougherty T.F., Taylor G.N., Lloyd R.D., Stover B.J., Jee W.S.S., Christensen W.R., Dougherty J.H., and Atherton D.R. (1969): Radiation-induced bone cancer in beagles, in Delayed Effects of Bone-Seeking Radionuclides, Univ. of Utah Press, Salt Lake City, pp. 387-408; see p. 394.
12. Rossi H.H. and Mays C.W. (1978): Leukemia risk from neutrons, Health Physics 34, 353-360.
13. Wrenn M.E. (1979): Tabular data on the experimental dogs, in Research in Radiobiology, University of Utah Report COO-119-254, pp. A1 to A57; updated to 1 January 1980.

ARE WE AT RISK FROM LOW LEVEL RADIATION. DNA REPAIR CAPACITY AS A PROBE OF POTENTIAL DAMAGE AND RECOVERY

E. Riklis

Can the question of risk from low level radiation be answered by an experimental molecular approach? Can we assess accurately biological damage caused by low level radiation? Is there a threshold ? Is there a way to predict future genetic or late somatic expression of damage? What are the processes that occur in a cell and lead, following assaults by radiation or chemicals, towards mutagenicity and possibly carcinogenicity? What is the main reason for differences in radiosensitivity between cells which are otherwise equal?
This last question was answered in the case of bacteria first, when differences in survival from ultraviolet radiation was shown to be dependent on the existence of enzymatic repair systems which are lacking in radiation sensitive mutants (1,2,3). This discovery, soon followed by finding the existence of similar enzymatic repair mechanisms for mammalian cells and for assaults also by ionizing radiations as well as chemicals, opened new prospects for approaching by similar ways the problems of risk from radiation and the raging debate on the question of threshold.

Biological dosimetry and risk estimation may be approached from two different angles: one is to look at damage already inflicted by radiation, and expressed in a biologically important molecule, such as the chromosome, at such low doses that the damage is not apparent in the normal functions of the cell. The other is more directly related to risk evaluation, that is to look for a cellular system which will indicate the potential risk of a radiation assault. Cytological techniques have been developed to enable observing chromosome aberrations in lymphocytes, the most sensitive human cell, and determine in this way the dose of radiation. Aberrations are however the end product of a series of events occuring in the irradiated cell, furthermore they can be easily and accurately seen only following a dose of about 20 rads or more. The biochemical systems responsible for repair of damage to DNA are controlling the ultimate fate of radiation damage, be it specific products (4) or strand breaks (5), and thus are determining both the number of aberrations as well as the radiosensitivity of the cell. Indeed, excision repair and/or postreplication repair function in most normal human cells, and can repair most types of damage to DNA. An impaired activity of any of the enzymes of repair results in lack or reduced capacity of repair. This may result in extreme sensitivity of such cells to radiation, and such a situation has been found to exist in humans carrying genetic auto-immune diseases. Cells from patients with the disease Xeroderma pigmentosum (XP) show extreme sensitivity to UV light, and patients with the disease

Are We at Risk from Low Level Radiation

Ataxia telangiectasia (AT) show extreme sensitivity to X-rays. These cells may be considered as the human mutants, identical in response to the bacterial mutants which show radiation sensitivity, in comparison to their wild type strains. Impairment in excision repair of UV-irradiated DNA is clearly implicated in XP, and similar impairment is suspected in Ataxia (6,7) and in several other diseases (although the existence of variants with normal DNA excision repair activity sheds doubt whether this is the only factor involved in the appearance of the disease), such as Fanconi's anemia, chronic lymphocytic leukemia and more. An interesting feature of these diseases is characterized in their "chromosome breakage syndrome" and their predisposition to cancer. Several recent reports indicate also an impairment in mitogen-induced transformation of lymphocytes of patients with these diseases. The possibility of a link between chromosome aberrations lymphocyte transformation and DNA repair was recognised by us (8) and led to a decision to develop a biochemical method, based on measurement of DNA repair capability, which would enable pre-determination of radiation sensitivity, and a more important feature — an indication of inherent sensitivity which might be expressed ultimately in the future — when the cell or organism might be faced with a situation in which its repair ability will have to function to its full capacity.

EXPERIMENTAL APPROACH AND RESULTS

The aims and steps of these studies have been previously formulated (9) and described (10). We have already shown that induced transformation of human lymphocytes is somewhat impaired following acute radiation dose, in cases where the person from whom the lymphocytes have been drawn was chronically exposed previously to low level beta (tritium) radiation (8). Induced transformation is, however, a black box and not a natural event in the life cycle of the lymphocyte. A clear and exact biochemical reaction was needed and none is better than the study of DNA repair itself. DNA repair may be followed by employing the technique of repair synthesis, whereby the amount of incorporation of labelled thymidine into cellular DNA is measured under conditions which do not permit the normal semiconservative DNA synthesis to occur. The commonly used method of treatment with hydroxyurea results in still too high a background and is subject to increasing criticism as affecting repair. A novel new method was therefore developed in our laboratory in which the cells are treated with trioxalen (trimethylpsoralen, TMP) and near UV light (NUV), bringing about an almost complete cessation of semiconservative DNA synthesis, 99.5 to 99.8% inhibition (Heimer, Kol and Riklis, in preparation). This treatment was previously successfully used by us to study gene expression and the control and regulation of inducible enzymes (11). Now it enabled the accurate measurement of incorporation of labelled thymidine (^{3}H-TdR) into DNA, following assaults by radiation or chemicals, indicating that repair synthesis is occuring.

Details of the experimental procedures are described elsewhere
(11, and, Heimer, Kol and Riklis, in preparation; Kol, Heimer
and Riklis, in preparation for detailed results). Briefly, it
involves irradiation of cells with near UV light for 3-5 minutes
in the presence of 5×10^{-6} M TMP, incubation for 3 hrs, assault by
far UV, gamma radiation or a chemical (MMS), incubation with
^{3}H-TdR for 90 min., washing, TCA precipitation and counting in
liquid scintillation counter. The method has been found suitable
for human fibroblasts, human breast cancer cells, Chinese hamster
cells and human lymphocytes. The number of counts incorporated
is many thousands above background, indicating repair synthesis
following the acute assault. A linear increase of counts is observed up to a certain radiation dose, this being designated as the
repair capacity of the cell, given by absolute numbers when the
number of counts incorporated while repairing a high dose damage
is divided by the counts incorporated following the lowest acute
dose or no irradiation-control (Fig.1). The counts were taken up
into double stranded DNA, as shown by the S_1-nuclease method (13)
indicating true repair replication. Thus, the 'repair capacity"
is a true measure of the capability of the cells to repair damage.

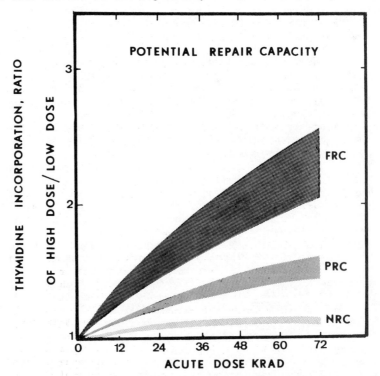

Figure 1. Potential Repair Capacity. Number of counts incorporated
per 10^6 cells following each dose is divided by cpm/10^6 of control
cells. FRC=full repair capacity; PRC=partial repair capacity;
NRC=no repair capacity(12). A hypothetical graph.

Are We at Risk from Low Level Radiation

A most significant result was obtained in showing that chronically irradiated Chinese hamster cells, with 20 rads per day for 3 months showed the same repair capacity as controls. Their survival was also not affected although their growth rate was slower.

DISCUSSION

Repair synthesis as measured by thymidine incorporation does not indicate the type of repair which has taken place, the exact nature of it is yet to be determined. A more important question to be asked is whether this repair is error-free or error-prone, namely, can we determine whether the cell is fully functional and not mutated. This is clearly a difficult problem, but attempts are being made to answer it by following the performance of an inducible enzyme, Ornithine decarboxylase (14), which if induced only after a radiation assault and if it functions to full capacity, indicates the wholesomeness of the DNA template.

Since the psoralen + NUV method of repair capacity measurement is applicable for lymphocytes, it can easily be used for a world-wide interlaboratory comparative study, in which the range of "repair capacity" of normal healthy humans will be established, and average figures will be determined for fully repairing cells (FRC), partially repairing cells (PRC) and non-repairing cells (NRC). A threshold for radiation damage will be between the PRC and the NRC regions. Individuals with no repair capacity (NRC), are 'at risk' from any level of radiation above background, while the general public, showing normal repair capacity, may be considered safe from the effects of low level radiations.

REFERENCES

1. Setlow R.B. and Carrier W.L. (1964): Proc. Natl. Acad. Sci. USA, 51, 226.
2. Boyce R.B. and Howard-Flanders P. (1964): ibid 51, 239.
3. Riklis E. (1965): Canad. J. Biochem., 43, 1207.
4. Riklis E. (1973): in: New Trends in Photobiology, Intl. Symp. Rio de Janeiro, Anais Acad. Bras. Sci. suppl. 45, 221.
5. Elkind M.M. (1979): Intl. J. Rad. Oncol. Biol. Phys. 5, 1089.
6. Cleaver J.E. (1970): J. Invest. Dermatol. 54, 181.
7. Paterson M.C., Smith B.P., Lohman P.H.M., Anderson A.K. and Fishman L. (1976): Nature 260, 444.
8. Riklis E. and Kol R. (1978): in: Late Biological Effects of Ionizing Radiation, IAEA,SM 224, vol I, 429, Vienna.
9. Riklis E. (1978): in: DNA Repair and Late Effects, IGEGM Intl. Symp., Editors: E. Riklis, H. Slor, H. Altmann, NRCN.
10. Riklis E. (1979): in: Cell Biology and Immunology of Leukocyte Function, p. 925, Editor: M.R. Quastel, Acad. Press.
11. Heimer Y.M. and Riklis E. (1979): Biochem. J. 183, 179.
12. Riklis E., Kol R. and Heimer Y.M. (1979): in: Transactions, Joint Ann. Meet. Nucl. Soc. Israel, vol.7, p.IV-14.
13. Ben-Hur E., Prager A. and Riklis E. (1979): Photochem. Photobiol. 29, 921.
14. Ben-Hur E., Heimer Y.M. and Riklis E. (1980): Intl. J. Radiat. Biol., submitted.

QUATRE OBSERVATIONS DE TRAITEMENT CHIRURGICAL DE BLESSURES CONTAMINEES PAR DES RADIONUCLEIDES

F. Briot, P. Henry, P. Lalu, J. Mercier, J. Sarbach and J. Tourte

INTRODUCTION

A travers quatre cas concrets survenus dans des Etablissements Nucléaires Industriels différents, il nous est apparu utile d'évoquer les problèmes médicaux liés au traitement de blessures contaminées.

Les enseignements tirés de ces cas permettent de dégager les points essentiels concernant l'organisation de la sécurité, les moyens thérapeutiques et les méthodes de mesure.

I - DESCRIPTION DES CAS OBSERVES

Observation n° 1 -

Au cours d'un nettoyage de l'intérieur d'une boîte à gants fortement contaminée en 239 Pu, un agent s'est légèrement coupé l'auriculaire droit à travers les gants de protection.

Après application immédiate d'une solution de DTPA, le blessé est dirigé vers le Service Médical de l'établissement. Il existe une plaie linéaire de 3 mm de longueur et de 3 à 5 mm de profondeur. Une mesure avec la sonde α donne 10 chocs/seconde. On pratique une première excision superficielle et on fait saigner la plaie sous DTPA. Les compresses sont comptées et indiquent une contamination profonde. On décide alors de pratiquer une injection intra-veineuse de DTPA (1 gr), suivie d'un comptage à la sonde X. Le résultat du comptage permet d'évaluer la quantité de Pu incluse à 200 nCi. On décide alors de faire appel au chirurgien pour pratiquer une excision. Les comptages X effectués extemporanément pendant l'intervention permettent de déterminer l'importance des excisions à réaliser : quatre excisions successives sont ainsi effectuées, jusqu'à une activité résiduelle acceptable, évaluée à 14 nCi.

Compte tenu du traitement par le DTPA, l'étude de l'excrétion urinaire et fécale de la première semaine conduit aux conclusions suivantes :

- la quantité de Pu qui a diffusé de la plaie vers le sang pendant la période comprise entre l'incident et l'intervention chirurgicale est restée très faible.
- dans l'hypothèse où tout le Pu résiduel finisse par diffuser dans les liquides extracellulaires, la charge corporelle qui serait acquise après l'élimination complète du Pu de la blessure sera au maximum celle restant dans le doigt après l'excision des tissus, soit 14 nCi dont environ 6 nCi dans l'os.

Quatre Observations de Traitement Chirurgical

Observation n° 2 - JL.M.

Un ouvrier travaillant en vêtement pressurisé au démantèlement d'une boîte à gants se blesse sur une arête de matière plastique à 10 h. Il s'agit d'une plaie linéaire de 6 mm, profonde de 1 mm, de l'éminence thénar gauche.

Conformément aux consignes en usage, le blessé, après déshabillage et isolement de la blessure par l'agent de radioprotection de l'atelier est conduit à l'infirmier du bâtiment. Ce dernier constatant une contamination (sonde α) procède à une décontamination. A la fin du traitement le comptage superficiel est négatif, mais le blessé est adressé au service médical de l'établissement pour mesure de contrôle, faisant appel à une sonde X. Cet examen met en évidence au niveau de la blessure une contamination profonde (Pu 239 : 3,86 nCi). Un traitement chélatant (DTPA local et général) est immédiatement institué. La décision est prise d'exciser la plaie. Le comptage effectué en fin d'intervention montre que 1,7 nCi (44 % de l'activité initiale) a été enlevé. (2 h 30 après l'accident).

Compte tenu de la nature chimique du contaminant (Nitrate de Pu 239) et du traitement déjà institué, il est décidé de ne pas augmenter l'excision. Un traitement chélatant itératif est mis en oeuvre et la victime placée sous surveillance radiotoxicologique.

En moins de deux mois, les mesures urinaires deviennent négatives. La partie transférable du radionucléide peut être considérée comme éliminée de la blessure.

Observation n° 3 - M.C. J. Pierre

Piqûre des 2° phalanges des index droit et gauche par deux brins d'un cable métallique effiloché contaminé par des produits de fission.

Contrôle : 150 chocs/seconde en β sur chaque doigt. On pratique sans succès une décontamination cutanée. Un examen au microscope ne permet pas de repérer les points de pénétration des brins de cable. On gratte superficiellement l'épiderme sans obtenir de modification de l'activité.

L'examen anthroporadiamétrique du corps entier (mains exclues maintenues derrière la tête) met en évidence 5,8 nCi de 106 Ru. Une mesure simultanée des 2 doigts donne 19 nCi.

24 heures après : Contrôle des doigts à la sonde β : **même résultat** que la veille. L'existence ce jour d'une légère réaction inflammatoire localisée, permet de visualiser sous microscope les points de pénétration des brins de cable.

Un grattage et une excision très superficielle amènent une diminution de l'activité : 120 chocs/seconde à gauche, 60 chocs/seconde à droite.

48 heures après : Pas de modification par rapport à la veille. Etant donné le radionucléide en cause et l'activité mesurée, la contamination interne de cet agent pouvait être considérée comme négligeable et l'irradiation localisée en βγ d'un niveau très bas.

Cependant, la persistance de cette contamination localisée rendant difficile les contrôles systématiques lors du travail en zone contrôlée, la décision d'excision cutanée a été prise.

Dans un premier temps, des comptages successifs à travers des écrans en aluminium percés de trous de diamètres décroissants (de 20 mm à 2 mm) ont permis de localiser très précisément les zones à exciser.

L'intervention (six jours après l'incident) permet :
- l'ablation de fragments dermoépidermiques en quartiers d'oranges dans les zones repérées,
- l'examen spectrométrique γ a montré que l'activité mesurée antérieurement "in vivo" au niveau des doigts se retrouvait au niveau des fragments excisés :
        5 nCi provenant du doigt droit,
        9,9 nCi provenant du doigt gauche.

Lors de la reprise du travail, l'anthroporadiamétrie de contrôle montrait une courbe normale. Les prélèvements d'urines et de selles effectués après l'incident avaient par ailleurs été négatifs.

Observation n° 4 - J.R.

J.R. 45 ans, ouvrier chimiste, en changeant en boîte à gants un élément de canalisation ayant contenu du 239 Pu irradié se blesse au pouce droit à 11 h.

Il existe une plaie de 1,5 cm de longueur au niveau de la face palmaire, 2ème phalange, du pouce droit, avec une contamination superficielle importante décelée à la sonde α.

Après examen par l'infirmier du bâtiment, la victime est transférée au service médical. Après un essai de décontamination suivi de lavage de plaie au DTPA et injection de DTPA on décide de pratiquer une excision, compte tenu de la première évaluation de la contamination sous jacente (239 Pu et 241 A ; 18 nCi au niveau de la plaie). Après l'excision, terminée 4 h 30 après l'incident, le comptage de la région contaminée est négatif, et le comptage du lambeau excisé fait apparaître 36 nCi de 239 Pu et 31 nCi de 241 Am.

Les examens radiotoxicologiques des excretas de surveillance sont toujours restés négatifs.

Cette observation a fait l'objet du film qui va être projeté.

Quatre Observations de Traitement Chirurgical

II - ENSEIGNEMENTS TIRES DE CES OBSERVATIONS

II-1 Organisation de la sécurité

II-1.1 Personnel et installations spécialisés

La présence d'un agent de radioprotection dans l'atelier et éventuellement d'une infirmerie de bâtiment pour les premières mesures et les premiers soins est bénéfique.

En outre, le service médical de l'Etablissement doit posséder les installations nécessaires pour pouvoir effectuer des interventions chirurgicales avec contrôle pré, per et post opératoires des lésions contaminées. Ces interventions nécessitent la disponibilité d'une équipe multidisciplinaire associant chirurgiens et radiotoxicologues.

II-1.2 Consignes sur les lieux de travail

Dans tous les lieux de travail où sont manipulés des radionucléides doivent exister des consignes indiquant la conduite à suivre à différents niveaux :

- appel immédiat de l'agent de radioprotection par l'agent blessé ou par ses compagnons de travail, quelle que soit l'importance de la blessure,
- lavage immédiat de la plaie à l'eau ou par des chélateurs si nécessaire (terres rares, transuraniens),
- mesure rapide α ou β de la contamination superficielle et évacuation vers le Service Médical.

II-1.3 Formation du personnel

Les personnels d'exécution ainsi que ceux des services de sécurité doivent être informés sur la nature des risques et sur la conduite à tenir. Des exercices périodiques permettent de s'assurer que les consignes sont correctement comprises et exécutées.

II-2 Moyens thérapeutiques

Lorsqu'existe une présomption de contamination par des terres rares ou des transuraniens il est indispensable de procéder rapidement à une injection de chélateur (par exemple DTPA). Ce traitement général permet d'éviter la fixation dans les organes du radionucléide qui diffuse dans le sang et permet de procéder sans précipitation aux mesures et interventions nécessaires.

L'exploration des plaies, la recherche éventuelle de petits frangments radioactifs sont facilitées par l'utilisation de microscope et de micro instruments chirurgicaux (micro instrumentation type chirurgie de l'oreille).

Toute intervention chirurgicale nécessitée par une contamination doit être aussi limitée que possible. Il serait aberrant d'aboutir à des interventions mutilantes pour éliminer complètement des contaminations résiduelles peu importantes.

II-3 Méthodes de mesure

II-3.1 Mesure de l'activité dans la plaie

La mesure de l'activité β et surtout α est insuffisante elle doit être doublée d'une mesure systématique des rayonnements γ ou X associés permettant de déceler la contamination profonde. Si le radionucléide contaminant est inconnu il est nécessaire de faire appel à des mesures spectrométriques.

II-3.2 Mesure de la contamination interne

Il est indispensable de surveiller par des examens radiotoxicologiques des excrétas et/ou anthroporadiamétriques l'efficacité de la thérapeutique et la contamination résiduelle.

Radiological Aspects of the Three Mile Island Accident

RADIOLOGICAL CONSEQUENCES OF THE THREE MILE ISLAND ACCIDENT

L. Battist and H. T. Peterson, Jr.

SITE AND ENVIRONS

The Three Mile Island Nuclear Station (TMI) is located on an island in the Susquehanna River approximately 14 km southeast of Harrisburg, Pennsylvania. The station is operated by a private utility, the Metropolitan Edison Company, and consists of two reactors, Unit 1, a 2535 megawatt (thermal) pressurized water reactor (PWR), and Unit 2, a 2772 megawatt (thermal) PWR. Unit 1 went into commercial operation on September 2, 1974 and Unit 2 went into commercial operation on December 30, 1978, approximately 3 months prior to the accident.

Three Mile Island is one of a number of islands in the Susquehanna River. It is located approximately 275 m from the east bank of the river and approximately 2 km from the west bank. Several private residences are located along the east shore within 0.8-1.2 km of the reactor buildings. Approximately 200 summer cottages are located on the nearby islands. Goldsboro, a community of approximately 900 people, is situated approximately 1.9 km west of the site and Middletown (approximately 10,000 people) is located 4.0 km to the north. Major population centers in the area are Harrisburg (∼70,000 people) which is 14 km NW and York (∼50,000 people) which is 21 km South. There are approximately 2,000,000 people residing within 80 km of the TMI site.

THE ACCIDENT

Three Mile Island Unit 2 was operating at 97 percent (916 MWe) of its licensed power level on the morning of March 28, 1979. At 0400 a series of events resulted in a substantial loss of primary coolant and the reactor's core being partially uncovered for several periods during the next 16 hours. High cladding temperatures resulted in metal-water reactions between the zirconium fuel cladding and the water (or steam). Oxidation and failure of the cladding resulted, releasing substantial quantities of fission products into the coolant and production of hydrogen. A primary coolant sample collected on March 29 shows the degree of the fission product contamination (Table I).

RADIOACTIVE MATERIALS RELEASED: PATHWAY AND QUANTITY

The fission products released to the coolant were transported to the auxiliary building in the primary coolant through the normal coolant purification system. The noble gas radionuclides and a fraction of the radioiodines were stripped into the gas phase and leaked into the supporting equipment buildings. Ventilation air transported these gases to the auxiliary building stack (10 feet below the top of the containment building) through high efficiency particulate filters (HEPA) and a charcoal absorber.

Although substantial noble gas activity was released, estimates range from 2.4-14 MCi, the water-to-air partition process and the filters reduced radioiodine release, estimated to be 15 Ci. The distribution of the noble gases released is shown in Table II, along with the core inventory of these radionuclides at the time of the accident.

TABLE I. The Major Radionuclides in a Sample of Reactor Coolant Taken on March 29, 1979.*

Nuclide	Half Life	Coolant Concentration µCi/cc
Iodine-131	8 d	1.3×10^4
Iodine-133	20.8 h	4.6×10^4
Cesium-134	2 y	6.3×10^1
Cesium-136	13 d	1.8×10^2
Cesium-137	30 y	2.8×10^2
Barium-140	12.8 d	21.0×10^1
Strontium-89/90	50 d/29y	5.3

*Reactor coolant sample taken at approximately 1700 on March 29. Sample was analyzed by the Bettis Atomic Power Laboratory, Pittsburgh, Pa.

TABLE II

Radionuclides Released to the Environment as a Result of TMI-2 Accident.

RADIONUCLIDE	HALF-LIFE	QUANTITY IN CORE AT TIME OF SHUTDOWN (Curies)	QUANTITY RELEASED ESTIMATED (Curies)	ESTIMATED FRACTION OF TOTAL RELEASED
Kr-88	2.8 hours	6.92×10^7	3.75×10^5	0.15
Xe-133	5.2 days	1.42×10^8	1.58×10^6	0.63
Xe-133m	2.2 days	2.11×10^7	2.25×10^5	0.09
Xe-135	9.1	3.31×10^7	3.0×10^5	0.12
Xe-135m	15.3 min.	2.60×10^7	2.5×10^4	0.01
I-131	8.0 days	6.55×10^7	15	*

* On an estimated fractional basis of total nuclides released, iodine-131 was very small.

The Three Mile Island Accident

Almost all (99%) of the noble gas emissions occurred in the period from March 28 until April 1 and 70% of these releases occurred within the first 36 hours. Radioiodine releases persisted until the end of April due to evaporation of liquids in the auxiliary building and degeneration of the charcoal filter performance.

Releases of fission products in liquid effluents were very small and consisted primarily of radioiodine and cesium-137. The total activity released in liquid effluents during the first three months following the accident was 0.23 Ci of iodine-131 and 0.24 Ci of other radionuclides.

RADIOLOGICAL MONITORING RESULTS

The U.S. Nuclear Regulatory Commission requires all reactor licensees in the United States to have an environmental monitoring program. In addition, each reactor is to have an emergency plan. Once it was realized that significant radiological releases might occur, the licensee dispatched teams to determine radiation levels offsite, particularly in the anticipated plume direction. Several State and Federal agencies also responded to the emergency and established environmental monitoring programs, sampling air, milk, water, vegetation, foodstuff and deploying additional thermoluminescent dosimeters (TLD's). The U.S. Department of Energy (DOE) used helicopters for tracking and measuring the activity in the plume. Metropolitan Edison used helicopters for monitoring on the site. During the 3 months following the accident, several thousand sample analyses were performed by the Commonwealth of Pennsylvania and several U.S. Federal agencies including the Department of Health, Education and Welfare, Bureau of Radiological Health and the Environmental Protection Agency.

Environmental

As a result of increasing in-plant radiation levels, beginning around 0700 on March 28, monitoring teams were dispatched to make radiation measurements outside the plant both onsite and offsite. Initial measurements made onsite starting at 0748 and offsite starting at 0832 were less than the minimal detectable level of the instruments (1 mR/hr). Radiation levels first began to increase at 1020 on March 28 when onsite monitoring teams detected exposure levels of 3 mR/hr. The instruments used for the offsite survey were Geiger-Muller detectors and ion chamber (RO-2) survey type instruments. Many of the reported readings were open window measurements and reported as β,γ-mR/hr, which is an undefined exposure rate. Where "β,γ" readings are known, they are so indicated. The instruments were not calibrated against a beta source, nor were they calibrated for an immersion situation. What the influence is on the total reading of the beta component is not known. These levels generally increased over the next 12 to 13 hours. Peak onsite radiation exposure rates of 300-365 (β,γ) mR/hr were reached between 2130 and 2330. Offsite radiation exposure rates were generally very low (maximum of 3 mR/hr). A reading of 50 mR/hr measured along the east river bank at 1548 was the highest reported offsite exposure rate. Noble gas emissions continued to be high until late in the morning of March 29. A reading of 30 mR/hr was recorded in Goldsboro (1.9 km WSW) at 0600 on March 29.

The maximum onsite dose rate on March 29 was 150 mR/hr (β,γ) at 0532. During the remainder of March 29, onsite levels were generally less than 10 mR/hr and offsite levels less than 1 mR/hr and did not exceed 2 mR/hr. Wind direction throughout the night of March 28-29 was generally in a northwesterly direction (toward Harrisburg). During the afternoon of March 29, a helicopter above the stack measured 3 R/hr (β,γ), 400 mR/hr gamma.

A second period of noble gas emissions occurred on March 30-March 31. This release resulted from intentional venting of the waste tanks in the auxiliary building required to reduce excessive pressure buildup in the tanks. Onsite exposure rates associated with this release reached a peak of 110 mR/hr. The highest offsite levels were 5-15 mR/hr at a point approximately 1.6 km to the south. However, a helicopter reading taken ~40 meters above the stack was 1.2 R/hr (β,γ) at 0800 hours. The reading could not be repeated, indicating a probable puff release.

Radioiodine Analyses

Offsite radioiodine was detected in analyses of milk samples collected for the first seven days following the accident with 68 positive iodine-131 results out of 264 samples collected. The concentrations ranged from 1 to 41 pCi/l (the 41 pCi/l was in a sample of goat's milk, which was not used for human consumption). In the subsequent 2 weeks, only 8 out of 80 samples taken by the U.S. Food and Drug Administration yielded positive results. Concentrations ranged from 15-36 pCi/l.

Initial measurements of airborne radioiodine concentrations made using portable air sampling equipment having charcoal adsorption cartridges were reported as 10^{-9}-10^{-10} μCi/ml at Goldsboro at 0900 and 0940 on March 28. However, laboratory gamma spectrometric analysis of the second cartridge by the Pennsylvania Bureau of Radiation Protection showed that this activity was primarily due to xenon-133 and xenon-135 and that actual radioiodine concentrations were less than 10^{-11} μCi/ml. The highest reported offsite radioiodine concentrations and measurable deposition occurred in mid-April in conjunction with replacement of the effluent filters in the auxiliary building, onsite 4 x 10^{-10} μCi/ml, offsite 1 x 10^{-10} μCi/ml.

In Plant

The highest radiation levels encountered by Met Ed personnel were in the auxiliary and fuel handling buildings. Radiation levels in excess of 1000 R/hr were measured during the first days of the accident at entrances to the cubicles containing tanks of primary coolant. General area radiation levels in these buildings ranged from 5 R/hr to 100 R/hr. Radiation levels in the reactor control room and other areas were generally low, less than 0.5 mR/hr. Due to the airborne activity (noble gases) in the Health Physics Control Station, counting and gamma spectrometry facilities had to be evacuated.

Population Exposure

The ground surveys that were performed and the analyses of local foods indicated that there was no measurable deposition of radioactive materials released from TMI. Of primary concern, however, was the need to assess the dose to the population and evaluation of the potential long-term consequences.

The Three Mile Island Accident

As part of Metropolitan Edison's environmental monitoring program, 20 TLD stations both on and offsite were located around the site at the time of the accident at distances up to 22 km. In addition, ten stations had a quality control TLD of a different type. Commencing on March 31, the Nuclear Regulatory Commission (NRC) placed an additional 37 TLD's around the TMI site. These were analyzed daily for a period of one week and at longer intervals thereafter.

The first evaluation of the population dose was performed by Battist, et. al[2], using the TLD's in place at the time of the accident and those subsequently placed by the NRC. This was accomplished by an interpolation equivalent to plotting the measured doses for each sector on logarithmic coordinate graph paper and joining the measured values by straight line segments. The intersection of each line segment with a standard distance for the grid was taken as the dose at that distance. In instances where the net dose calculated for a location was not greater than zero, this method could not be used. In such cases, linear interpolation was used to estimate the doses at standard distances.

Doses at distances beyond the outermost dosimeter or within the innermost dosimeter were estimated by extrapolation using the assumption that the dispersion in a sector is proportional to distance to the (-1.5) power. A DOE analysis concludes that their airborne measurements and the TLD data suggest a more rapid decrease of exposure with distance, more consistent with an exponential function or a power function with an exponent of (-2). The (-1.5) power assumption is therefore conservative, yielding a higher collective dose.

Doses for the standard distances in sectors in which no measurements were made were estimated by interpolating linearly between the dose values of the adjacent sectors for which measured data were available.

The mean dose within each sector segment was estimated by weighting the dose, H(r), by the area within the sector

$$\bar{H} = \frac{\int_{r_1}^{r_2} H(r) r \, dr}{\int_{r_1}^{r_2} r \, dr}$$

where $\bar{H}$ is the mean dose, $H(r)$ is the dose as a function of distance, r, and r_1 and r_2 are the inner and outer radii of the sector segment, respectively.

The collective dose for each sector segment is the product of the corresponding mean dose and the population in that sector. The sum of the collective doses for all sector segments and periods is the total collective dose for the entire assessment area for the total period under consideration.

Utilizing the available TLD data, a range of collective dose equivalent estimates were determined. These values ranged from 1600 to 5300 person-rem, with the most probable value being 3300 person-rem. This range resulted from different sets of dosimeter data used in individual determinations.

The highest value, 5300 person-rem, was the result of including all of the NRC and Metropolitan Edison dosimeters. However, the first day's set of NRC TLD data contained several inconsistencies. Later analyses[3] indicated that these dosimeters were most likely exposed prior to deployment and that no controls were included to evaluate these effects. Use of the Metropolitan Edison dosimeters, including the quality control badges, resulted in a collective dose equivalent of 3300 person-rem. The other two values, 2800 and 1600 person-rem, were obtained by using the TLD data within 10 km of the plant; first including all of the TLD data, and later by excluding the NRC TLD data.

A second evaluation of the TLD data was performed by Auxier, et. al.[3] Their estimate of the collective dose equivalent was 2800 person-rem. Taking into account the shelter factor for the low photon energy of xenon-133 reduced the collective dose equivalent estimate to 2000 person-rem.

The agreement between these independent analyses is quite good, and the collective dose equivalent is in the range of 1600-3300 person-rem. Attempts were also made to determine the collective dose equivalent using meteorological dispersion calculations.

In-Plant Exposures

Although high radiation fields existed in the auxiliary building, and several entries were made, only three individuals exceeded NRC's quarterly whole-body exposure limits of 3 Rem. The exposures were 4.1, 3.6, and 4.2 Rem, respectively. In total, during the seven-month period following the accident, only seven individuals received doses in excess of 3 Rem. The total collective occupational exposure through September 30 was approximately 1200 person-rem.

In August 1979 several workers were contaminated by beta activity when working in contaminated areas. Extremity exposures were high, approximately 40-50 Rem, due to residual contamination. No whole-body exposures in excess of regulatory limits were reported.

DISCUSSION

Although the accident at TMI Unit 2 was the most severe reactor accident to date, the release of several megacuries of radioactive noble gases resulted in a relatively small population exposure estimated to be in the range of 1600 to 3300 person-rems, as determined from TLD measurements. The sparseness of the data and the extrapolation of individual dosimeter results to assess the dose to the population in a large sector contribute to the uncertainty. However, the continual low offsite exposure readings, lack of residual ground activity and other dosimeters placed in the environs of the site by Federal agencies all tend to confirm that the population dose could not have been significantly different than that defined above. The maximum individual offsite dose was stated to be less than 100 mrem in the Ad Hoc group study,[2] and about 50 mrem by Auxier, et.al.[3]

In-plant personnel exposures have been maintained at reasonable levels. The fact that only three overexposures were recorded on the first two days of the accident is remarkable in view of the high radiation fields that existed. However, the cleanup operations could result in a significant collective worker dose unless significant health physics control is exercised.

The defense-in-depth concept under which nuclear plants are designed worked well in practice. Radiological releases were quite small in view of the magnitude of the fuel damage. The containment building, requirement of filtered pathways, and backup systems all functioned to minimize the potential radiological consequences. However, the accident indicated that better health physics instrumentation and personnel training is required to obtain more meaningful survey results and to control in-plant exposures.

HEALTH EFFECTS

As a result of the radiation exposure to the offsite population within 50 miles of the TMI site, the projected incidence of fatal cancer is less than 1; and fatal plus non-fatal cancers is less than 1.5, with zero not excluded. This is to be contrasted to the nearly 541,000 cancers (325,000 fatal and 216,000 non-fatal) expected in this population over its remaining lifetime, not related to the TMI accident.

The additional lifetime fatal cancer risk to the individual receiving the maximum probable dose offsite (less than 100 mrem) is about 1 in 100,000. The additional risk of fatal cancer to an individual receiving the average individual offsite dose (1.4 mrem) is about 1 in 5,000,000. The risks of non-fatal cancer induction are the same as those for fatal cancers.

The additional cancer risks due to internal irradiation and skin irradiation are very small compared to the above values and can be regarded as being included in the values presented above for whole-body gamma irradiation. Even if the cancer risks defined above were to be expressed, the resultant cancers would not be detectable among the population in the vicinity of TMI. (Note that zero additional incidence is not excluded.)

The whole-body external occupational exposure of 1,000 person-rem has potential total cancer risk of less than 0.5 (zero not excluded). The risk to the maximally occupationally exposed individual (4.1 rems) is about 1.2 in 1,000 for both fatal and non-fatal cancers.

The potential incidence of genetically related ill-health is considerably smaller than that of producing a fatal or non-fatal cancer. This risk is estimated to be about 0.002 cases per year, and about one case per million live births for all future human existence. This contrasts with an estimated 3,000 cases pwer year of genetically related ill health among the offsprings of the population in the vicinity of TMI based on present birth rate (28,000 births per year), and not related to the TMI 2 accident.[5]

REFERENCES

1. U.S. Nuclear Regulatory Commission (1979) Investigation Into the March 28, 1979 Three Mile Island Accident by the Office of Inspection and Enforcement, U.S. NRC Report NUREG-0600.

2. Battist, L., Buchanan, J., Congel, F., Nelson, C., Nelson, M., Peterson, H., and Rosenstein, M. (1979) Population Dose and Health Impact of the Accident at the Three Mile Island Nuclear Station, U.S. Nuclear Regulatory Commission Report NUREG-0558.

3. Auxier, S. et al, Report of the Task Group on Health Physics and Dosimetry to the President's Commission on the Accident at Three Mile Island, October 31, 1979.

4. Woodard, K., Assessment of Offsite Radiation Doses from the Three Mile Island Unit 2 Accident, TDR-TMI-116, Rev. 0, July 31, 1979, Pickard Lowe and Garrick, Washington, D.C.

5. President's Commission on the Accident at Three Mile Island, 1979, Report of the Radiation Health Effects Task Group, October 31, 1979, Washington, D.C.

PROTECTIVE ACTION GUIDES THEORY AND APPLICATION LESSONS FROM THE THREE MILE ISLAND ACCIDENT

B. Shleien

During the Three Mile Island Nuclear accident, the Food and Drug Administration (FDA), Department of Health, Education and Welfare (HEW), fulfilled its traditional role in insuring a safe food supply, and together with other elements of HEW, provided advice and assistance in the response to health concerns. This paper will present two aspects of FDA's role: 1) actions relative to the use and availability of potassium iodide (KI) as a thyroid-blocking agent; and 2) protective action guidance (PAG's) relative to human food and animal feed.

KI AS A THYROID BLOCKING-AGENT

By a notice in the <u>Federal Register</u> of December 15, 1978[1], entitled "Potassium Iodide as a Thyroid-Blocking Agent in a Radiation Emergency," the FDA did the following:
1) In the interest of public safety it requested submission of New Drug Applications (NDA's) for potassium iodide in oral form for use as a thyroid-blocking agent in a radiation emergency.
2) It announced the availability of labeling guidelines for potassium iodide for such use.

Many organic and inorganic drugs were considered for emergency use as thyroid-blocking agents. Potassium iodide was chosen over the other drugs because of: 1) the high degree of blocking achieved; 2) the rapidity of onset of action; 3) a long high duration of its blocking effect; and 4) its relative safety Although potassium iodide acts on the thyroid in several ways, its usefulness for this purpose is primarily predicated on its ability to saturate the iodide transport system, effectively abolishing entry of radioiodine. Complete blocking is achieved, that is, over 90 percent of radioiodine is blocked by the recommended oral administation of 130 milligrams of KI. This is equivalent to 100 milligrams of iodide. The recommended dose for infants under one year of age is 65 milligrams. Onset of blocking occurs within 30 minutes. Therefore, administration before or immediately after initial exposure yields the best results. However, substantial benefit can still be achieved if potassium iodide is given within three or four hours after exposure. The duration of time that a blocking agent is required is not likely to exceed 10 days.

As with any drug, certain cautions have to be recognized. The number of reports of adverse reactions for the use of potassium iodide when it is given in greater doses than for blocking and over a longer period of time has been very low. The risk is judged to be low for the short-term use of potassium iodide in a radiation emergency.

The <u>Federal Register</u> notice offers no guidance as to when to use potassium iodide as a blocking agent. Other sources of guidance are available. The National Commission on Radiation Protection and Measurements Report No. 55[2] suggests that potassium iodide should be considered for blocking at a projected or anticipated thyroid dose from radioiodine of 10 - 30 rads to the thyroid.

The Three Mile Island episode graphically demonstrated the need for effective planning for the use of potassium iodide as a component of public health emergency reponse capability. The drug was not available for mass distribution in the proper dosage forms at the time of the Three Mile Island accident. FDA had not as yet received nor approved an NDA for potassium iodide as a thyroid-blocking agent. To meet the emergency, FDA arranged for the manufacture of a supply of potassium iodide before any decision to employ the drug had been made. Pennsylvania Officials had a plan for distribution of the supplies of KI and patient information. The Three Mile Island emergency was unique in that FDA supplied the drug. The Agency does not stockpile nor will it be a source of the drug in the event of other accident situations.

Among significant factors which influence the decision as to whether or not KI should be employed as a thyroid-blocking agent in a radiation accident are:
1. The efficacy of potassium iodide as a thyroid-blocking agent depends on the pathway of radiation exposure.
2. There is an absence of Federal guidelines for the use of potassium iodide as a thyroid-blocking agent.
3. The area and population potentially affected by the release of radioactivity are important to the decision as to when and how to use potassium iodide.
4. There are logistical problems of storage and distribution because of the need for prompt administration in the event of an accident.
5. Alternative protection actions must be considered instead of, or in concert with, the use of potassium iodide.

Discussion of these problem areas requires more space and time than are available here, and is the subject of a paper in preparation[3].

PAG's FOR HUMAN FOOD AND ANIMAL FEED

"Proposed" Protective Action Guidance for the food pathway also appears in the <u>Federal Register</u> of December 15, 1978[1]. These recommendations are for use by appropriate State and local agencies in response planning for radiation emergencies in the event of an accident resulting in the contamination of food or animal feed by radioactive substances. Events which may result in radiation emergencies include, but are not limited to, accidents at nuclear facilities, transportation accidents, and fallout from nuclear devices. Although such incidents could lead to a general contamination of the environment, the FDA's recommendations are limited to the food pathway.

The proposed protective actions are intended for implementation within hours or days from the time an emergency is recognized, and the duration should not exceed a month or two. Other Federal agencies are responsible for guidance in the event of exposure to the population from pathways other than for food, and for action of a longer duration.

The FDA protection action guidance does not imply an acceptable radiation dose from food containing radioactivity during normal peacetime conditions. Rather, their purpose is to reduce or avoid further radiation dose to the population via the food chain in the event of a radiation accident. These recommendations are not a license to needlesssly permit environmental levels of radiation to rise.

Protective action guides or PAG's define the projected dose commitment to individuals in the general population that warrants protective actions following the release of radioactive material. Projected dose commitment, is defined as the dose commitment that would be received in the future by individuals in a population group from a contaminating event if no protective action were taken. In other words, the protective action guidance is based on anticipated or projected doses. The purpose of the PAG's is to provide guidance in order to prevent additional radioactive contamination from entering the human food chain and to reduce or avoid future radiation doses to the population after an accidental contaminating event.

Two protective action guidance levels have been proposed. They are:
1. The <u>Preventive PAG</u> is applicable to situations where protective actions causing minimal impact are justified. These protective actions prevent or reduce concentrations of radioactivity in food. The preventive PAG is 1.5 rem projected dose commitment to the thyroid, or 0.5 rem projected dose commitment to the whole body, bone marrow, or any other organ.
2. The <u>Emergency PAG</u> is applicable to incidents where protective actions of high impact are justified because of the greater projected health hazards. Levels at which food should be isolated from commerce are appropriate at the emergency PAG level. The emergency PAG is 15 rem projected dose commitment to the thyroid, or 5 rem projected dose commitment to the whole body, bone marrow, or any other organ.

A practical means of employing the PAG's is through the use of derived response levels. Derived response levels refer to the activity of a specific radioactive substance per unit weight or volume of food, or animal feed, which corresponds to a particular numerical PAG limit previously mentioned. Response levels have been calculated based on recent metabolic and agricultural models. Specific derived response levels are given in the FDA recommendations for radioactive substances which are thought to be relatively abundant under emergency conditions, easily enter the food chain, and are taken up and retained by the human body. These radionuclides are iodine-131, cesium-137, strontium-90, and strontium-89. Derived response levels are given specifically for initial deposition on pasture, concentration in forage and in milk, and total intake. Variations in the basic model permit calculation of derived response levels for different food products and mixtures of radionuclides.

A large number of milk samples were collected following the Three Mile Island accident from farms and dairies in the vicinity of the accident site. Based on the peak concentration of iodine-131 detected in these samples, the dose to the thyroid of an infant drinking one liter of milk daily for the entire time during which positive milk samples were found, would be 0.005 rem over the lifetime of the individual[4]. The maximum iodine-131 levels detected at Three Mile Island were thus 300 times smaller than the preventive action level.

Two problems which arose during the Three Mile Island accident relative to the PAG's were their application over an extended period of time and their effect on the "marketability" of food. The PAG's are intended for use up to one or two months. They include a derived response limit for total radionuclide intake which may be used as a basis on which to evaluate chronic intake during this period.

Producer's general experience with food contaminated with other materials indicates the general public's reluctance to accept "contaminated" products. This problem involves public perception of radiation risks and governmental credibility. The issuance of PAG recommendations should influence public perceptions of radiation risks in a manner to encourage rational action relative to maketing and acceptance of foods.

REFERENCES
1. Accidental Radioactive Contamination of Human Food and Animal Feeds and Potassium Iodide as a Thyroid-Blocking Agent, **Federal Register**, December 15, 1978.
2. National Commission on Radiation Protection and Measurments, Protection of the Thyroid Gland in the Event of Releases of Radioiodine, NCRP Report No. 55, August 1, 1977.
3. Halperin, J.A., Shleien, B. and Kahana, S.E., Thyroid-Blocking with Potassium Iodide: Useful Radiation Protection Guidance, in preparation.
4. Ad Hoc Population Dose Assessment Group, Population Dose and Health Impact of the Accident at the Three Mile Island Nuclear Station, May 10, 1979.

Therefore, the main Reactor Coolant Pumps were used in their place even though they were not intended for cold shutdown operation. Thus, further contamination of the Auxiliary Building and radioactive fission product release to the environment were prevented.

During the accident it was required several times to enter the Auxiliary Building to align valves, start pumps or acquire samples from the containment atmosphere. Several of these operations were delayed or were not completed. The radiation fields within that building only allowed for a short stay of several minutes, which were insufficient to complete the required assignments.

(b) Undue exposures of operating personnel:

The review of the recovery operations reveal that some operations are needed more frequently than others. Among these are:
- Reactor Coolant sampling for boron analysis
- Operation of equipment from radwaste panels
- Change of filters
- Surveillance of equipment, monitors and instrumentations.

These operations have resulted in personnel exposures approaching the quarterly dose limit at each entry, and resulted in high extremity doses. Such high doses may be warranted in nonroutine one-time assignments and should be avoided in operations required on a frequent basis.

(c) Failure of radiation-protection instrumentation and monitors to provide correct information.

Radiation-protection instrumentation and monitors have been designed mainly to control normal operation including conditions of anticipated operational occurrences. Postulated accidents were also considered, but all protective systems were assumed to perform successfully and to reduce the fission product release. The TMI experience reveals many cases of monitors which were driven out of range. It also points out cases of monitors which measured the background radiation created by large amounts of noble gases rather than iodine or particulates being released through them. In the last case the monitors were exaggerating the actual release.

It should be pointed out that the measuring equipment operated successfully from the electro-mechanical point of view in most cases. i.e., the equipment was available and redundant equipment reached their set points within several minutes.

(d) Failure of evaluating radiation protection measurements, alarms and other information to determine the actual reactor situation:

In spite of much instrumentation going out of range, there were quite a number of high radiation alarms, high level measurements by the containment dome monitor (which did not go out of range) or by operators surveilling the Auxiliary Building. In addition the gaseous effluent monitoring system was indicating high effluent discharge (exaggerated by radiation from the noble gases). These measurements could be related to core conditions and fuel failures in the core. During the TMI incident the above information was not correctly interpreted to indicate that significant fuel failure was taking place in core. It was rather explained as steam generator leakage to containment atmosphere combined with some steam generator tube failures. This indicates a need for improved training of personnel to distinguish radiation fields indicating abnormal occurrences from normally encountered fields.

SOME RADIATION PROTECTION IMPLICATIONS OF THE THREE MILE ISLAND INCIDENT

D. Ilberg

In many cases in the past, the design of Nuclear Power Plants (NPP) in the area of system safety and accident analysis has considered the radiation protection aspects mainly under conditions of normal operation including anticipated operational occurences. The Regulatory Guide 8.19 which was published recently by the USNRC (7) provides an example.

The Three Mile Island (TMI) incident and the recovery operations which followed it, focus attention on additional aspects which require careful considerations in the design of NPP:
(a) Safety related systems should be available for operation without any delay, even though a significant amount of fuel failures have already occurred in the reactor core.
(b) The operating personnel should be able to conduct recovery operations without undue exposures even when an accident has resulted in some fuel damage.
(c) Radiation-protection instrumentations should provide useful information even in the case of accidents which produce high radiation fields in the area of their reading.

High radiation fields that can be expected during severe accidents have been considered in the past mainly with respect to radiation measuring instrumentation (2,8) and radiation qualification of components in the containment (3). Only conditions of large-break LOCA (l_{oss} of Coolant Accident) were used to estimate the radiation fields (2). The desgin of shielding to allow access to safety systems and other vital equipment was in many cases based only on normal operating radiation levels (4).

In the next sections we discuss these safety problems, the reccommendations (5,6) made by the Nuclear Regulatory Commission (NRC) and by the Kemeny Commission (1) to treat these problems and arrive at some additional conclusions and suggestions.

SOME SAFETY PROBLEMS REVEALED IN THE TMI RECOVERY OPERATIONS

Our review of the recovery operations which followed the accident (1,5,6) revealed some safety problems. These safety problems may be divided into four general groups:

(a) Delay or prevention of the use of safety systems important to the recovery operations:
The Decay Heat Removal System is the main system planned for operation to maintain the reactor in the cold-shutdown mode; however, it was realized during the accident that this system is not sufficient leak-tight for use with highly radioactive primary water.

Some Radiation Protection Implications

RECOMMENDATIONS MADE BY INVESTIGATING COMMITTEES

Several investigations into the TMI incident were performed. The NRC investigation (5,6) and the Kemeny Commission investigation (1) resulted in some recommendations related to radiation protection from the system safety point of view.

With respect to the safety problem (a) above, it is recommended by the NRC (5,6) to improve the integrity of systems outside containment likely to contain radioactive materials. No design improvements are required for, at least, the short term. The recommendations call for implementation of all practical leak reduction measures for the systems and performance of leakage rate tests on a periodic basis to keep the leakage rate at a constant level.

With respect to safety problem (b) above, it is recommended by the NRC (5) to perform a design review of the radiation fields and the shielding in the spaces arround systems that may contain highly radioactive materials. The design review should identify vital areas and equipment required during post-accident recovery operations. Measures to be taken to provide adequate access to vital areas should include post-accident procedural controls, permanent or temporary shielding and when required also redesign of facilities, components or systems. A quantitative source-term is suggested for the design review, i.e., the Regulatory Guide 1.3 source term (9).

With respect to safety problem (c) above it is recommended by the NRC (5) that no high range radiation monitors for noble gases in plant effluent lines and in the reactor containment be installed. In addition instrumentations for the monitoring of radioiodine and particulate effluents under accident conditions would also be provided.

With respect to safety problem (d) above it is recommended by the NRC (5,6) to improve post-accident sampling capability. A design and operational review of the reactor coolant and containment atmosphere sampling systems should be performed to determine the capability of personnel to obtain a sample within an hour under accident conditions, without incurring a radiation exposure exceeding the quaterly dose limit to whole body or extremities for radiation workers. Timely information from such samples can be important for an early understanding of core conditions.

The Kemeny Commission recommendations (1) are more qualitative in nature. They call on the NRC to include as part of its licensing requirements plans for the mitigation of consequences of accidents, including the cleanup and the recovery of a contaminated plant. The Kemeny Commission urges correcting inadequacies in equipment required for the mitigation of the accident (i.e., safety problem (a) above). It recommends that consideration should be given to overall gas-tight enclosure of systems processing highly radioactive water during the accident and the recovery phases. It urges improvements of radiation monitors and the provision of the capability to take and quickly analyze samples of containment atmosphere and reactor coolant.

D. Ilberg

SUMMARY

This paper presents four safety problems and the recommendations made by the NRC and the Kemeny Commission to treat them. It can be seen that the recommendations respond adequately to the safety problem mainly, in the above mentioned problems (b) and (c).

In its recommendations the NRC provides a source term to allow for more quantitative design review and for determination of cases which require some improvements. It is suggested here to add two supplemental steps to this quantitative approach:

(a) The use of specific scenarios of postulated accidents to determine the required recovery operations, which are the vital plant areas for post accident access and the required systems and equipment during the recovery phase. In particular, such scenarios may include, in additions to a large break LOCA, the small break LOCA, an ATWS event, a steam line break case and a control rod ejection followed by a small break LOCA.

(b) The use of a dollar value per man-rem as a criterion for determination when design improvements are required rather than procedural controls.

The design review of the NPP will therfore include the assumption of a Reg. Guide 1.4 source term and a specific postulated accident scenario. Design improvements which have the potential for dose reduction both to personnel or to the population, would be judged by their dollar value per man-rem reduced. The criterion may be a lower dollar value than used for ALARA purposes today.

REFERENCES

1. Kemeny, J.G. (1979): Report of the President's Commission On The Three Mile Island: The President's Comm., Washington D.C.
2. Koler, J.M., McCarty, J.R. and Olson, N.C., (1977): Nucl. Tech., 36, 74.
3. Naber, J.A. and Lurie, N.A. (1977): Nucl. Tech., 36, 40.
4. Sejvar, J. (1977): Nucl. Tech. 36, 48.
5. USNRC, (1979): TMI-2 Lessons Learned Task Force Status Report And Short Term Recommendations: NUREG-0578, USNRC, Wash. D.C.
6. USNRC, (1979): Investigation Into The March 28, 1979 Three Mile Island Accident: NUREG-0600, USNRC, Wash. D.C.
7. USNRC, (1979): Regulatory Guide 8.19 Rev.1. USNRC, Wash. D.C.
8. USNRC, (1975): Regulatory Guide 1.89 Rev. 1. USNRC, Wash. D.C.
9. USNRC, (1974): Regulatory Guide 1.4 Rev.2 USNRC, Wash. D.C.

Non-Ionizing Radiation

MICROWAVE/RADIOFREQUENCY PROTECTION STANDARDS CONCEPTS, CRITERIA AND APPLICATIONS*

S. M. Michaelson

Electromagnetic energies i.e. 300 KHz to 300 MHz (Radio-frequency) and 300 MHz to 300 GHz (Microwaves) can produce biological effects or injury depending on power levels and exposure durations. There is thus a need to set limits on the amount of exposure to radiant energies individuals can accept with safety. Protection standards should be based on scientific evidence but quite often are the result of empirical approaches to various problems reflecting current qualitative and quantitative knowledge.

In considering standards, it is necessary to keep in mind the essential differences between a "personnel exposure" standard and a "performance" standard for a piece of equipment. An exposure standard refers to the maximum safe (incorporating a safety factor) level of power density and exposure time for the whole body or any of its parts. This standard is a guide to people on how to limit exposure for safety. An emission standard (or performance standard) refers not to people but to equipment and specifies the maximum limit of emission close to a device which ensures that likely human exposure will be at levels considerably below personnel exposure limits.

Basic Considerations Illumination of biological systems with MW/RF energy leads to temperature elevation when the rate of energy absorption exceeds the rate of energy dissipation. Whether the resultant temperature elevation is diffuse or confined to specific anatomical sites, depends on: the electromagnetic field characteristics and distributions within the body as well as the passive and active thermoregulatory mechanisms available to the particular biological entity.

Analysis of Literature Elucidation of the biologic effects of microwave exposure requires a careful review and critical analysis of the available literature. Such review requires differentiation of the established effects and mechanisms from speculative and unsubstantiated reports.

Although there is considerable agreement among scientists concerning the biological effects and potential hazards of microwaves, there are areas of disagreement. There also is a philosophical question about the definition of hazard. All effects are not necessarily hazards. In fact, some effects may have beneficial applications under appropriately controlled conditions. MW/RF induced changes must be understood sufficiently so that their clinical significance can be determined, their hazard potential assessed and the appropriate benefit/risk analyses applied. It is important to determine whether an observed effect is irreparable, transient or reversible, disappearing when the electromagnetic field is removed or after some interval of time. Of course, even reversible effects are unacceptable if they transiently impair the ability of the individual to function properly or to perform a required task.

In an analysis of scientific literature to determine the probability of a biological response from exposure to a noxious agent, we must consider the consistency of experimental results claimed, both the nature of the response and the biological system involved, the ability to replicate the results of studies with consistency and whether the results claimed and observations reported can be explained by accepted biological principles.

Experiments with small animals, such as mice and rats, to evaluate the potential effects of MW/RF energy, must be carefully designed and performed. The responses may be the result of another, unrelated agent inadvertently introduced into the experimental design rather than the factor intended to be studied. The fact that a living organism responds to many stimuli is a part of the process of living; such responses are examples of biological "effects". Since biological organisms have considerable tolerance to change, these "effects" may be well within the capability of the organism to maintain a normal equilibrium or condition of homeostasis. If, on the other hand, an effect is of such an intense nature that it compromises the individual's ability to function properly or overcomes the recovery capability of the individual, then the "effect" should be considered a "hazard". In any discussion of the potential for biological "effects" from exposure to electromagnetic energies we must first determine whether any "effect" can be demonstrated; and then determine whether such an observed "effect" is "hazardous".

When assessing the results of research on biological effects of MW/RF exposure it is important to note whether the techniques used are such that possible effects of intervening factors i.e. noise, vibration, chemicals, variation in temperature, humidity, air flow are avoided and care is taken to avoid population densities that perturb the field to the extent that measurements become meaningless. The sensitivity of the experiment should be adequate to ensure a reasonable probability that an effect would be detected if indeed any exists. The experiment and observational techniques should be objective. Data should be subjected to acceptable analytical methods with no relevant data deleted from consideration. If an effect is claimed, it should be demonstrated at an acceptable level of statistical significance by application of appropriate tests. A given experiment, should be internally consistent with respect to the effect of interest. Finally, the results should be quantifiable and susceptible to confirmation by other investigators.

<u>Principles of Biologic Experiments and Interpretation</u> Proper investigation of the biologic effects of electromagnetic fields requires an understanding and appreciation of biophysical principles and "comparative medicine". Such studies require interspecies "scaling", the selection of biomedical parameters which consider basic

physiological functions, identification of specific and nonspecific reactions, and differentiation of adaptational or compensatory changes from pathological manifestations.

Scaling Much of the research on biological effects of microwaves has been done with small rodents that have coefficients of heat absorption, field concentration effects, body surface areas, and thermal regulatory mechanisms significantly different from man. Adverse reaction in animals does not prove adverse effect in man, and lack of reaction in animals does not prove that man will not be affected. Even closely related species can differ widely in their response. The literature is replete with "anomalous" reactions. Thus, results of exposure of common laboratory animals cannot be readily extrapolated to man unless some form of "scaling" among different animal species, and from animal to man, can be invoked in an accurate way to obtain a quantitatively valid extrapolation from the actual data observed.

The physical factors that must be considered include: frequency of radiation, intensity, animal orientation with respect to the source, size of animal with respect to the wavelength, portion of the body irradiated, exposure time-tensity factors, environmental conditions (temperature, humidity, air flow), and absorbed heat distribution in the body. In addition, variables such as restraint, metabolic rate, body ratio of volume/surface area, and thermoregulatory mechanisms will affect the biological response to microwaves.

The need for proper dosimetry in experimental procedures and the importance of realistic scaling factors required for extrapolation of data obtained with small laboratory animals to man are clearly required. Maximum absorption in man occurs at 80 MHz and falls off at higher frequencies. Formulas for scaling factors among species are available (1). Five milliwatts per centimeter square, 2450MHz, 5 mW/cm^2 exposure of a small animal such as a mouse or a rat, can result in a thermal effect that could influence the central nervous system and elicit behavioral and other physiologic responses in that animal, but not necessarily in a larger animal or man.

Epidemiology A number of retrospective studies have been done on human populations exposed or believed to have been exposed to MW/RF energies. Those performed in the U.S. (2,3) and Poland (4), have not revealed any relationship of altered morbidity or mortality to MW/RF exposure. Nervous system and cardiovascular alterations in humans exposed to microwaves has been reported in Eastern European literature (5,6,7). Most of the reported effects are subjective, consisting of fatigability, headache, sleepiness, irritability, loss of appetite, and memory difficulties. Psychic changes that include unstable mood, hypochondriasis, and anxiety have been reported. The symptoms are reversible, and pathological damage to neural structures is insignificant. There is considerable difficulty in establishing the presence of, and quantifying the frequency and severity of "subjective" complaints. Individuals suffering from a variety of chronic diseases may exhibit the same dysfunctions of the central nervous and cardiovascular systems as those reported to be a result of exposure to microwaves; thus, it is extremely difficult, if not impossible, to rule out other factors in attempting to relate microwave exposure to clinical conditions.

Protection Guides and Standards: Exposure Standards   The first standards for controlling exposure to MW/RF were introduced in the 1950's, in the USA and the USSR. The maximum permissible exposure levels proposed then have remained substantially unchanged, i.e., for continuous exposure these are respectively 10 mW/cm^2 and 10 µW/cm^2. Most countries that developed national standards based them on either the US (8) or the Soviet (9) values. Subsequently, however, some countries have proposed standards intermediate between these extremes.

Current U.S. Government Standards Include: Occupational Standard (general) - OSHA Standard, adopted in 1972, applies to employees in the private sector. An addendum, adopted in 1975, applies to work conditions particularly in the telecommunications industry. OSHA Standards are mandatory for federal employees including the military. Maximum permissible exposure limit is 10 mW/cm^2, for durations greater than 6 min, over the frequency range 10 MHz-100 GHz.

Product Emission Standard - "Radiation Control for Health and Safety Act of 1968" (PL 90-602), administered by HEW/FDA (BRH), provides authority for controlling radiation from electronic devices. BRH microwave oven standard, effective October, 1971: Ovens may not emit (leak) more than 1 mW/cm^2 at time of manufacture and 5 mW/cm^2 subsequently, for the life of the product--measured at a distance of 5 cm and under conditions specified in the standard.

Additionally, there are non-goverment organizations which develop recommended standards and safety criteria, e.g.: American National Standards Institute (ANSI) - A voluntary body with members from government, industry, various associations and the academic community which develops consensus standards (guides) in various areas. ANSI issued a nonionizing radiation safety standard in 1966 with maximum permissible exposures of 10 mW/cm^2, as averaged over any 6 minute period, for frequencies from 10 MHz to 100 GHz which was essentially adopted by OSHA. This standard was reviewed and reissued with minor modifications in 1975. ANSI must review and withdraw, revise or reissue ANSI Standards every five years. Presently Subcommittee 4 of ANSI Committee C-95, which deals with hazards to personnel is reevaluating ANSI's radiofrequency exposure standard for adoption in 1980. The recommendations, based on frequency dependence and specific absorption rates (SAR) state: For human exposure to electromagnetic energy of radiofrequencies from 300 KHz to 100 GHz, the radiofrequency protection guides, in terms of equivalent plane wave free space power density, and in terms of the mean squared electric ($\underline{E}^2$) and magnetic ($\underline{H}^2$) field strengths as a function of frequency, are:

Microwave/Radiofrequency Protection Standards

Frequency (MHz)	Power Density (mW/cm^2)	E^2 (V^2/m^2)	H^2 (A^2/m^2)
0.3 - 3	100	400,000	2.5
3 - 30	900/f^2	4,000 (900/f^2)	0.025 (900/f^2)
30 - 300	1.0	4,000	0.025
300 - 1500	f/300	4,000 (f/300)	0.025 (f/300)
1500 - 100,000	5	20,000	0.125

Note: f is the frequency, in <u>megahertz</u> (MHz)

For near field exposure, the only applicable radiofrequency protection guides are the mean squared electric and magnetic field strengths given in columns 3 and 4. For convenience, these guides may be expressed in equivalent plane wave power density.

For both pulsed and non-pulsed fields, the power density and the mean squares of the field strengths, as applicable, are averaged over any 0.1 hour period and should not exceed the values given in the Table. For situations involving exposure of the whole body, the radiofrequency protection guide is believed to result in energy deposition averaged over the entire body mass for any 0.1 hour period of about 144 joules per kilogram (J/kg) or less. This is equivalent to a specific absorption rate (SAR) of about 0.40 watts per kilogram (W/kg) spatially and temporally averaged over the entire body mass. This recommendation will no doubt be adopted with only minor modifications if any.

The National Institute for Occupational Safety and Health (NIOSH) is developing a criteria document with recommended standards for occupational MW/RF exposures, which is, except for certain modifications, comparable to that recommended by ANSI.

The Radiation Protection Bureau of Health and Welfare Canada is considering "Emission and Exposure Standards for Microwave Radiation". The maximum permissible levels (MPL'S) are 1 mW-hr/cm^2 average energy flux for whole body exposure as averaged over an hour and a maximum exposure during any one minute of 25 mW/cm^2 for occupational settings. The MPL's would apply for the frequency range of 10 MHz-300 GHz. No distinction is made between CW and pulsed waveforms. There is no lower MPL for the general population.

The State Committee on Standards of the Council of Ministers of the USSR has promulgated "Occupational Safety Standards for Electromagnetic Fields of Radiofrequency (GOST 12.1.006-76)," effective January 1, 1977. It specifies the maximum permissible magnitudes of voltage and current density of an EM field in the workplace. It does not apply to personnel of the Ministry of Defense. Maximum permissible RF fields in the workplace must not, during the course of the workday, exceed;

		P (mW/cm^2)	E (V/m)	H (A/m)	Frequency Range
Stationary Source (See Note 1)	Rotating, Scanning			5	60 KHz-1.5 MHz
			50		1.5 MHz-3.0 MHz
			20		3.0 MHz-30 MHz
			10	0.3	30 MHz-50 MHz
			5		50 MHz-300 MHz
0.01	0.1		(entire workday)		300 MHz-300 MHz
0.10	1.0	(2 hr. period during workday)			
1.00		(20 min. period during workday)			

Note 1: Also applies in environments with ambient temperatures above 28°C and/or in the presence of X-ray radiation, except, under these conditions, the maximum during a 20 minute period is restricted to 0.1 mW/cm^2.

There is some indication that the USSR Ministry of Health has endorsed guidelines for maximum exposure limits for the general population which stipulates the maximum allowable levels of electromagnetic energy in human dwellings or in areas of human dwellings, as follows:

P (μW/cm^2)	E (V/m)	Frequency Range
	20	30-300 KHz
	10	300 KHz-3.0 MHz
	4	3.0-30 MHz
	2	30-300 MHz
5		300 MHz-300 GHz

In 1977, the Polish Ministries of Work, Wages and Social Affairs and of Health and Social Welfare promulgated a change in the Polish Standard for occupational exposure. The change extends the frequency range down from 300 to 0.1 MHz;

Hazardous Zone II		Hazardous Zone I		Intermediate Zone		Safe Zone		Frequency Range
T_p 0	250 A/m 1000 V/m	T_p	40/H 150/E	T_p	10 A/m 70 V/m Entire 8 hr Workday	T_p	2 A/m 20 V/m	No limit	0.1-10 MHz
			300 V/m 3200/E2		20 V/m		7 V/m	10-300 MHz

T_p = Permissible time of exposure/workday (minutes).
E = Electric field (volts/meter).
H = Magnetic field (amps/meter).

Microwave/Radiofrequency Protection Standards

In 1976, the Swedish National Board for Industrial Safety, promulgated a nonionizing radiofrequency standard (Worker Protection Authority Instruction No. 111) effective January 1, 1977. This regulation applies to all work which may involve exposure to radiofrequencies between 10 MHz and 300 GHz. The instruction specifically excludes applications involving the treatment of patients. Maximum permissible exposures (as averaged over a six minute period) are:

Power Density	Frequency Range		
5 mW/cm^2	10 MHz	to	300 MHz
1 mW/cm^2	300 MHz	to	300 GHz

The maximum permissible momentary exposure is 25 mW/cm^2.

Emission Standards The best known emission standards concern the maximum permissible leakage from microwave ovens. The Canadian standard (10,11) restricts the maximum leakage to 1 mW/cm^2 at 5 cm from the oven (consumer, commercial and industrial). The U.S. standard (12) specifies a maximum emission level at 5 cm of 1 mW/cm^2 before purchase and 5 mW/cm^2 thereafter which is consistent with standards for the general population in the USSR and Poland (13). The standard applies to domestic and commercial ovens, but not to industrial equipment. This has been adopted in Japan and most of Western Europe.

Conclusion

International cooperation in the development of compatible standards should be encouraged. Towards this end the International Radiation Protection Association (IRPA) charter was broadened in April 1977, to include nonionizing radiation. IRPA has cooperated with the World Health Organization (WHO) in the preparation of a criteria document which is scheduled for 1980. The European Regional Office (ERO) of the World Health Organization is presently preparing a manual on health aspects of exposure to nonionizing radiation. The manual is intended to provide guidance in nonionizing radiation protection and to summarize international experience in the field. Among the topic areas to be included are health aspects of ultraviolet, optical, infrared and laser; microwave RF and ELF fields; ultrasound; licensing, legislation and regulations.

ACKNOWLEDGEMENT

*This paper is based on work performed under contract with The U.S. Department of Energy at the University of Rochester Department of Radiation Biology and Biophysics and has been assigned Report No. UR-3490-1622.

REFERENCES

1. Durney, C.H., C.C. Jacobson, P.W. Barber, H. Massoudi, M.F. Iskander, J.L. Lords, D.K. Ryser, S.J. Allen and J.C. Mitchell (1978). Radiofrequency Radiation Dosimetry Handbook (2nd edition). USAF Report SAM-TR-78-22, Brooks Air Force Base, TX.

2. Robinette, C.D. and C. Silverman (1977). Cause of Death Following Occupational Exposure to Microwave Radiation (Radar) 1950-1974. Pages 337-344, in: Symposium on Biological Effects and Measurement of Radio Frequency/Microwaves. HEW Publication (FDA) 77-8026.

3. Lilienfeld, A.M., J. Tonascia, S. Tonascia, C.H. Libauer, G.M. Canthen, J.A. Markowitz and S. Weida (1978). Foreign Service Health Status Study: Evaluation of Health Status of Foreign Service and other Employees from Selected Eastern European Posts-Final Report July 31, 1978 Contract No. 6025-619073 Dept. of Epidemiol. Johns Hopkins Univ. Baltimore, NTIS PB-288, 1963.

4. Czerski, P. and M. Piotrowski (1972). Proposals for specification of allowable levels of microwave radiation, Medycyna Lotnicza (Polish), No. 39, 127-139.

5. Marha, K., J. Musil and H. Tuha (1968). Electromagnetic Fields and the Living Environment, State Health Publishing House Prague. San Francisco Press, 1971).

6. Petrov, I.R., (ed.) (1970). Influence of Microwave Radiation on the Organism of Man and Animals, Meditsina Press Leningrad.

7. Gordon, Z.V. (1970). Occupational health aspects of radio-frequency electromagnetic radiation. In: Ergonomics and Physical Environmental Factors. Occupational Safety and Health Series, No. 21, International Labour Office, p. 159, Geneva.

8. ANSI (1966). Safety Level of Electromagnetic Radiation with Respect to Personnel, American National Standards Institute, C95.1-1966, C95.1 - 1974, New York.

9. USSR (1958). Temporary Sanitary Rules for Working with Centimeter Waves. Ministry of Health Protection of the USSR.

10. DHEW Canada (1974). Radiation Emitting Devices Regulations. SOR/74-601 23 October 1974. Part III Microwave Ovens Canada Gazette Part V 108, pp. 2822-2825.

11. Repacholi, M.H. (1978). Proposed exposure limits for microwave and radiofrequency radiations in Canada. J. Microwave Power 13, 199-277.

12. USDHEW (1974). Regulations for Administration and Enforcement of the Radiation Control of Health and Safety Act of 1968 paragraph 1030.10 Microwave Ovens DHEW Publ. No. (FDA) 75-8003 pp. 36-37, July, 1974.

13. Baranski, S. and P. Czerski (1976). Biological Effects of Microwaves. Dowden, Hutchinson and Ross, Stroudsburg, PA, 234 p.

ELECTROMAGNETIC POLLUTION OF THE ENVIRONMENT

L. Argiero and G. Rossi

The recent development of the radar and telecomunication via satellite devices have imposed to the attention of the researches the necessity to solve the problem regarding the protection of radiofrequency electromagnetic and microwave radiations.

The use of high power devices have besides stimulated many interests in the research of the possible biological effects from human exposition to the electromagnetic radiations in acute or chronic form.

The present research deals with national situation concerning the ambient pollution by the radiofrequency electromagnetic waves and the microwaves (1)(2)(3).

The experimental part is limited from 0.5 to 500 MHz. This one is divided into section:
- devices numbering: broadcasting, telecomunication, television, air traffic control, air and shipborne navigation, meteorology, metallurgy, food processing, etc.;
- environmental tests: rural zones, towns, industries.

In this report are shown the results of measures that have been effected in Livorno (Fig. 1), to make out the real entity of electromagnetic pollution, in order to intervene if it will be necessary.

METHODOLOGY AND MEASUREMENT

Livorno has a surface of 104.6 km^2, a population of 177523 persons, a density of population of 1697 persons/km^2, medium altitude on the sea-level is 3 m, but some places have a highter position, as Valle Benedetta (338 m), Montenero (293 m), Quercianella (91 m), Castellaccio (300 m), Collinaia (52 m), Monterotondo (149 m).

Livorno is endowed with several factories and an industrial part that is characterized by an uninterrupted flux of ships.

According to the distribution of the installations we have divided the territory in three areas (Fig. 2 and Table 1):
- area A, including the port in which are situated about thirthy stations for marine support and aid, working in HF, UHF and VHF band; their maximum power is 150-400 W and they have an inconstant time of work.
  You can find them both on earth and on boats;
- area B, including Castellaccio and the places near there, where you can find 6 relay stations in VHF band for the comunications of power 10 W and 6 relay stations with parabolic antennas in UHF band of power is 1 kW and they work almost uninterrupting;
- area C: in this area all the broadcasting and television stations of

the town, public and private (transmitters and relay stations) are grouped and they have a power with fixed time of work.

About 12 km far from the town in Coltano there is another station with remarkable power, in Monte Serra; about 35 km from Livorno, there are antennas with great power for broadcasting, television and telephony.

The stations of O.M. and C.B. are not individualized.

We have choosen 50 test points (Fig. 3 and Table 2):
- area of the port: 6 stations;
- central area: 6 stations;
- peripherical area: 8 stations;
- area of Montenero: 5 stations;
- area of Castellaccio: 10 stations;
- the surroundings: 15 stations.

We have used the Electric and Magnetic Field Sensor TE 307, by Aeritalia (Torino), maximum range 10 V/m, the measure the electric field. The RAHAM, by General Microwave, has been used to measure the density of power.

RESULTS AND DISCUSSION

The results of our measurements are shown in Table 3.

The following areas can be individualized (Fig. 4):
- area 1 – including the area of the port; obtained results: $0.1 \div 0.2$ V/m and $0.05 \div 0.3$ mW/cm^2;
- area 2 – Castellaccio – obtained results: $1 \div 5$ V/m and $0.2 \div 0.5$ mW/cm^2;
- area 3 (coinciding with area C): obtained results: $0.2 \div 1$ V/m and $0.05 \div 1$ mW/cm^2;
- area 4 – it is the central area, excepted area C: obtained results: $0.05 \div 0.2$ V/m and $0.03 \div 0.1$ mW/cm^2;
- area 5 – it is the peripherical area: obtained results: $0.05 \div 0.2$ V/m and $0.05 \div 0.1$ mW/cm^2;
- area 6 – including the surroundings: obtained results: 0.1 V/m and $0.05 \div 0.1$ mW/cm^2;

CONCLUSIONS

- The stations of area around the port have an influence that can't be measured on the electric field and the density of power in the central area.
- The stations of area C have a greater influence on the central area.
- According to our expectations, the stations of area B give the greatest contributions to values of the electric field and of the density of power.
- In the peripherical area and in the surroundings the electric field and the density of power have the same values of the central area (area 4).

Electromagnetic Pollution of the Environment

Making a comparation between the obtained results and the maximum values that have been fixed in the national rules we can believe that this are near these maximum values expecially regarding to the density of power.

TABLE 1 - Distribution of the installations

Area A : the port
Area B : Castellaccio
Area C : Montenero, Castellaccio, Cavour and Roma square

TABLE 2 - Measurement stations

AREA	STATION N°
The port	1-2-3-4-5-6
Central	7-8-9-10-11-12
Peripherical	13-14-15-16-34-35-36-37
Montenero	17-18-19-20-21
Castellaccio	22-23-24-25-26-27-28-29-30-31
Surroundings	32-33-38-39-40-41-42-43-44-45-46--47-48-49-50

TABLE 3 - Results of the measures

STATION n°	ELECTRIC FIELD V/m	E.M. DENSITY mW/cm^2	STATION n°	ELECTRIC FIELD V/m	E.M. DENSITY mW/cm^2	STATION n°	ELECTRIC FIELD V/m	E.M. DENSITY mW/cm^2
1	0.2	0.05	18	0.1	0.1	35	0.1	0.05
2	0.2	0.1	19	0.2	0.15	36	0.1	0.05
3	0.2	0.1	20	0.1	0.05	37	0.2	0.05
4	0.2	0.3	21	0.2	1	38	0.1	0.1
5	0.2	0.3	22	2	0.2	39	0.1	0.1
6	0.1	0.1	23	0.2	0.15	40	0.1	0.05
7	0.1	0.1	24	0.8	0.21	41	0.1	0.05
8	0.05	0.05	25	1	0.2	42	0.1	0.05
9	1	0.1	26	2.4	0.2	43	0.1	0.05
10	0.5	0.05	27	3.4	0.25	44	0.1	0.05
11	0.2	0.03	28	4	0.4	45	0.1	0.05
12	0.2	0.15	29	4.4	0.5	46	0.1	0.05
13	0.1	0.15	30	5	0.5	47	0.1	0.05
14	0.2	0.1	31	1	0.1	48	0.1	0.05
15	0.05	0.05	32	0.2	0.1	49	0.1	0.05
16	0.1	0.05	33	0.1	0.05	50	0.1	0.0⁵
17	0.8	0.1	34	0.1	0.1			

The measurements are been executed in collaboration with the Italian Navy.

REFERENCES

1. Argiero L. (1976): L'Elettrotecnica LXIII, 11.
2. Argiero L. (1979): La Medicina del Lavoro 70,3.
3. Argiero L. (1979): Radio Rivista 9.

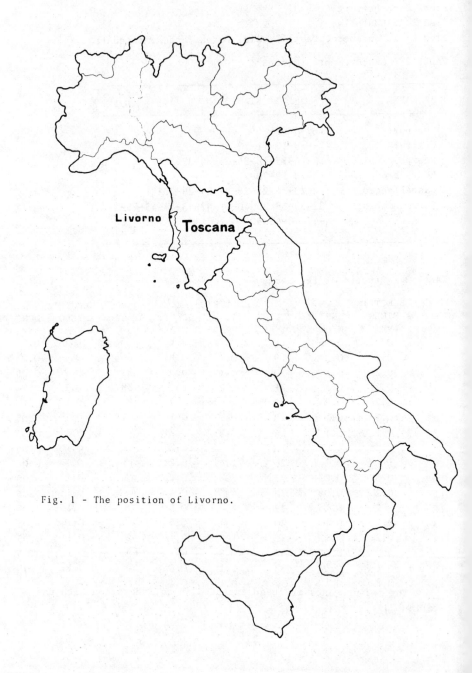

Fig. 1 - The position of Livorno.

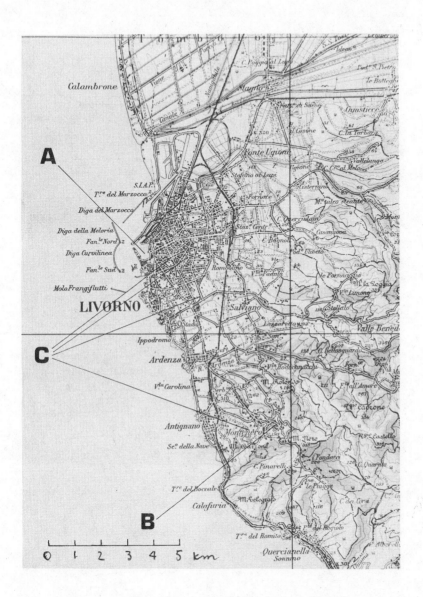

Fig. 2 - Distribution of the installations.

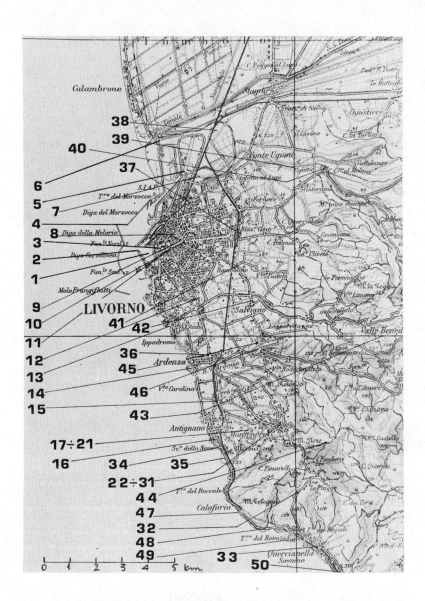

Fig. 3 - Measurement stations.

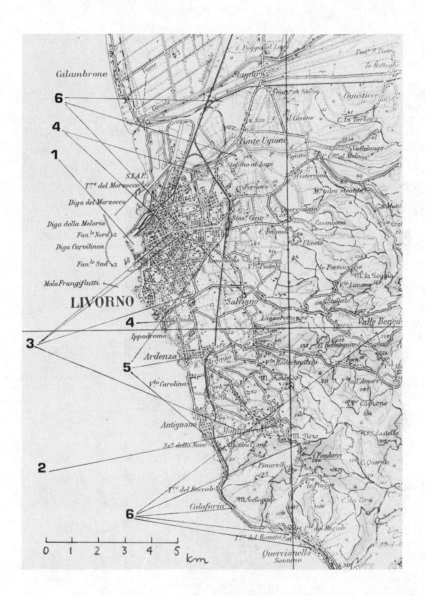

Fig. 4 - Irradiation areas.

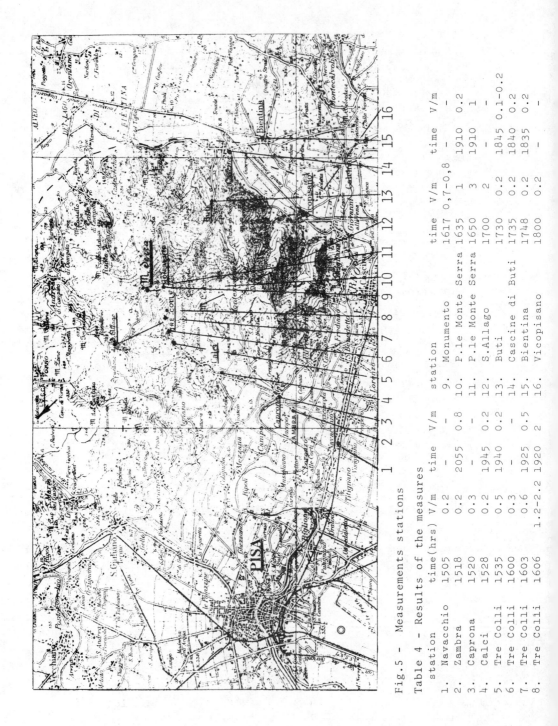

Fig.5 - Measurements stations

Table 4 - Results of the measures

station	time(hrs)	V/m	time	V/m	station	time	V/m	time	V/m
1. Navacchio	1505	0.2	–	–	9. Monumento	1617	0,7-0,8	–	–
2. Zambra	1518	0.2	2055	0.8	10. P.le Monte Serra	1635	1	1910	0.2
3. Caprona	1520	0.3	–	–	11. P.le Monte Serra	1650	3	1910	1
4. Calci	1528	0.2	1945	0.2	12. S.Allago	1700	2	–	–
5. Tre Colli	1535	0.5	1940	0.2	13. Buti	1730	0.2	1845	0.1-0.2
6. Tre Colli	1600	0.3	–	–	14. Cascine di Buti	1735	0.2	1840	0.2
7. Tre Colli	1603	0.6	1925	0.5	15. Bientina	1748	0.2	1835	0.2
8. Tre Colli	1606	1.2-2.2	1920	2	16. Vicopisano	1800	0.2	–	–

THE EVOLUTION OF NON-IONIZING RADIATION PROTECTION STANDARDS IN NORWAY

H. Aamlid and G. Saxeböl

In Norway a Committee[x] was set up in 1975 to review the risks associated with the use of non-ionizing radiation in Norway. The Committee's mandate is also to recommend amendments to the existing standards on non-ionizing radiation protection. For a small country with limited resources for research on biological effects of radiation the Committee soon decided that its recommendations had to be based on research mainly performed in other countries.

However, such a decision does not solve all the problems. As long as disagreement exists among competent international sources any national standard may be open to question. Some approaches to non-ionizing radiation protection standards are discussed in this paper.

COMPARISON IONIZING-NON IONIZING RADIATION

Concern for potential non-ionizing radiation hazards appears to be universal. But, to date, there seems to be a universal lack of agreement as to the nature and the severity of the hazards involved.

For ionizing radiation, however, the situation is different. Most countries have long regulated the use of ionizing radiation through legislation, and only minor differences exist from country to country in this legislation. The main reason for this is the fact that the fundamental concepts of risk are internationally accepted. Several international bodies are involved in radiation

---

[x]
The Norwegian Non-Ionizing Radiation Hazard Committee comprises the following members: Dr. F. Devik (Chairman, pathology), Dr. T. Hvinden (physics), Dr. H.H. Tjønn (occupational medicine), Dr. P. Syrdalen (ophthalmology), Dr. F. Storm Davidsen (military medicine), T. Liholt (telecommunications and broadcasting), H. Aamlid (radiation physics), G. Saxeböl (radiation physics) and M. Brady (microwave biological effects).

protection and they all take advantage in the basic recommendations issued by the International Commission on Radiological Protection (ICRP). The early identification of ionizing radiation hazards and the resultant early establishment of ICRP has for many countries, especially smaller countries, been of fundamental value in legislative work. The resultant situation today is that most groups in society - workers, politicians, employers and legal authorities - respect and accept the recommendations of the ICRP. Therefore the evaluation of ionizing radiation hazards and the relevant questions of risk are identified and treated similarly in different countries.

For non-ionizing radiation hazards, the non-existence of an international body capable of working out and recommending standards such as the ICRP for ionizing radiation causes both international and national difficulties. No one country, or groups within a country, can set or use standards with confidence when the international picture of the subject seems confused.

Additionally, an important difference exists in the nature of risk associated with non-ionizing radiation when compared to ionizing radiation. The presence of a stochastic effect of ionizing radiation which has given rise to the concept of the linear dose-effect relationship for protection purposes is well established. With respect to non-ionizing radiation most effects are considered non-stochastic i.e. there is a threshold above which an effect will occur and below which the effect will not occur. It should be expected that this nature of the dose (or irradiation)- effect relationship would make the setting of standards more simple and make the struggle for international agreement more easy. However, this is not always the case. While the risk and hazards and guides for protection standards are relatively uniform for the optical region of the spectrum, the existing protection standards for microwaves and radio-frequency show discrepancies both in principles and figures.

NORWEGIAN POSITION

In Norway the State Institute of Radiation Hygiene under the Ministry of Health and Social Affairs is the competent authority for protection against ionizing radiation as well as non-ionizing radiation. Many countries have organized the radiation protection system in a different way. This difference in administration may cause problems as far as international cooperation is concerned. For instance the IRPA will have problems in collecting the international expertise in non-ionizing radiation simply because these experts are not members of the national radiation protection societies.

The general radiation protection Law in Norway which was passed by the Norwegian Parliament in 1938 opens possibility for the Government to implement this Law also for non-ionizing radiation such as light, ultraviolet radiation, diathermy, short waves, etc. This widening of the 1938-Act was made by regulations issued by the Ministry of Health and Social Affairs in 1976 about the same time as the multidisiplinary Committee on NIR was set up. Working under the auspices of the State Institute of Radiation Hygiene the Committee decided to evaluate an approach to protection against non-ionizing radiation based upon the same philisophy as for ionizing radiation.

RISK PHILOSOPHY

Any standard concerning a hazard must be built on a philosophy of acceptable risk. Because of difference in physical properties and biological effects the spectrum is divided into two parts: the optical and the radiofrequency.

The risk and hazard questions associated with the optical spectrum, i.e. wavelengths between 100 nm and 1 mm, seem to be relatively uniform. For instance the standards for exposure to lasers have reached a high degree of international agreement . Standards and recommendations have been proposed by international bodies such as IEC and WHO. The ultraviolet part of the optical spectrum is a region of growing interest. One reason for this is that the number of skin cancers including malignant melanomas in many countries has shown a steady increase over the last 10-20 years. Medical experts feel confident that the increased exposure to sunshine for tanning purposes have influence upon the observed increasing number of cases. For UV-radiation the erythema curve is the main basis for safety standards. So far the evaluation of standards has not been influenced by the acceptance of the carcinogenic risk. In Norway the NIR-Committee has recommended that all sunlamps intended for total body exposure should be equipped with a warning label reading: "By increased UV-exposure the cancer risk may increase". The argument for the NIR-Committee to propose this recommendation was the principle of optimization used by the ICRP: "All exposures shall be kept as low as reasonably achievable, economic and social factors being taken into account". The Committee found that this part of the basic philosophy of ICRP should also be valid for non-ionizing radiation especially when such radiation can be associated with risk for stochastic effects.

APPROACH TO SAFETY STANDARDS FOR MICROWAVE AND RF

The NIR-Committee started its work on microwaves and radiofrequency by stating that the general exposure limit of 100 W/m^2 being inofficially used in Norway for the frequency range 300 MHz - 300 GHz should not be lowered unless clear biological evidence was found. However, by taking the ICRP principle of optimization into account, the Committee has found that 100 W/m² is an <u>unnecessary</u> high limit for whole body all day occupational exposure. On the other hand the Committee of course recognizes the need for time limited or partial body exposure to higher levels. Accordingly the Committee now discusses the implementation of a new limitation concept which will start off with a general and lower exposure limit for the general public. Exposure to higher levels may take place for radiation workers under controlled conditions. This includes principles such as time limitation and measurements of exposure levels. The exposure limit for the general public being discussed at the moment is 10 W/m² for the frequency range 10 MHz - 300 GHz. This limit is also thought to be implemented for full day occupational exposure. The Committee recognizes the need for additional exposure limits for partial body exposure. However, the Committee strongly feels that a set of regulations should be as simple as possible. Unnecessary frequency dependent specifications should be avoided. One item that will be seriously discussed in the Committee is the need for differentiation in standards between radiation workers and the general public. The following way of thinking will support the idea of such a differentiation even if only non-stochastic effects of radiation are considered As the number of irradiated individuals increases the mean value of the threshold-effect will be found more exact. However, the scattering of the results will increase and the slope of the envelope curve (or ED_o-curve) will decrease and the ordinary S-curve will be streched out and show a more linear form. We may come to the conclusion that this concept should be realized in the regulations. Following this way of thinking the exposure limits for large population groups may be set lower than exposure limits for full time radiation workers. The idea is to take care of the varying biological sensitivity to radiation in the population.

REFERENCES

1. Aamlid, Saxeböl and Brady: "Towards prudent definitions of non-ionizing radiation hazard". IMPI-Microwave Power Symposium 1979.

2. ICRP, Publication 26.

A RECOMMENDED PERMISSIBLE ENVIRONMENTAL STANDARD FOR MICROWAVE AND RADIOFREQUENCY RADIATION

L. R. Solon

Non-ionizing radiation in the radiofrequency and microwave region of the electromagnetic spectrum has received comparatively insubstantial investigation and environmental regulatory attention. With the increasing utilization of the radiofrequency spectrum in recent decades, including radio and television broadcasting, radar for commercial transportation control and military observation, consumer products such as microwave ovens, diathermy equipment employed in medical therapy, and other applications, there has been a surge of warranted concern and interest in public health control of possible harmful biological effects of non-ionizing radiation in humans.

In June 1978, in my capacity as Director of the Bureau for Radiation Control of the New York City Health Department, I recommended to the Board of Health of the City, adoption of a regulation to the Health Code which would set maximum permissible levels for potential exposure to microwave and radiofrequency radiation to members of the general public in uncontrolled or unregulated areas.

The overall framework of the recommended regulation was derived from these main considerations.
(1) Biological and clinical effects have been exhibited on a sufficiently broad scale in laboratory experiments and clinical observation to demonstrate the potential for physiological impairment in humans from microwave/radiofrequency radiation from various power or energy density levels. Physiological effects conclusively or almost certainly demonstrated in animal experiments include cataractogenesis, hormonal alterations, chromosomal anomalies, and hematological changes. Less certain but probable effects include central nervous system impairment and mutagenesis. In the realm of conjectural but possible effects, warranting however the most careful public health scrutiny, are mutagenic, oncogenic and teratogenic influences.
(2) The very approximate threshold for some observable effects of varying public health implications appears to be between one and 10 milliwatts per cm^2 for noncontinuous exposures. Of course, a microwave or radiofrequency induced physiological perturbation may not be clinically significant in the sense of hazard or impairment.
(3) Also legally licensed mobile units employed by police, fire and emergency response departments have a potential for exposing individuals in the public environment to hundreds of milliwatts/cm^2 in the microwave and ultra high frequency range. Presumably these transmissions are of an intermittent and non-sustaining nature as far as individual radiation exposures are concerned. Their aggregate public health impact deserves careful future scrutiny.

Biological Effects and Clinical Observations for Microwave and Radiofrequency Radiation

It should be emphasized at the outset that the frequency region of interest i.e. 30 KHz-300 GHz is in the non-ionizing portion of the electromagnetic spectrum. Unlike x-or gamma radiation typical of the ionizing region of the electromagnetic spectrum, microwave/radiofrequency radiations are relatively low energy in the quantum sense and incapable of separating bound electrons from atoms and molecules.

The biological modes of action may be characterized in three very general ways:
(1) Macroscopic heating or hyperthermia of a whole living organism or substantial part thereof resulting in the sustained elevation of temperature producing reversible, irreversible or partly reversible biological changes.
(2) Macroscopic heating of individual cells or very small sections of an organ producing biological changes of various persistence without perceptible temperature rise in the macroscopic sense.
(3) Non-thermal or only partly thermal effects relating to interaction of impinging electromagnetic radiation with the electric or magnetic fields of living tissues or cells.

Very approximately, mode one, microscopic heating or hyperthermia, would be associated with power densities in excess of 10 milliwatts/cm^2. The second mode, microscopic heating, would identify with power densities roughly in the range of one to 10 milliwatts/cm^2. The third or athermal mode, would be related to biological effects produced below one milliwatt/cm^2 or in the microwatt/cm^2 power density range.

In Table 1 are listed the principal representative biological effects purported to have been observed in various biota by investigators as the result of microwave or radiofrequency exposure. Also included are representative effects alleged to have been seen clinically or by epidemiological inference in human beings. Most such reports have addressed observations in workers occupationally connected with microwaves or radiofrequency radiation and information on the frequency range or power density is not available at all or incomplete.

Measured Background Environmental Levels for Radiofrequency Radiation

In one survey performed by the United States Environmental Protection Agency (EPA) covering the radiofrequency ambient environmental bands between 46 MHz and 900 MHz, the power densities encountered generally fell into the range of between 0.001 and 1 microwatt per square centimeter ($\mu W/cm^2$) with a median value of between 0.02 - 0.03 $\mu W/cm^2$. Forty locations were surveyed in the metropolitan New York area. The range of ambient values encountered in this region from a minimum .000068 $\mu W/cm^2$ (Tottenville, Staten Island) to a maximum of 4.6 $\mu W/cm^2$ (Mount Pleasant Street, West Orange, New Jersey). The average for these forty observed locations was 0.22 uW/cm^2 including the two maximum values. If the two maxiumum values (i.e. 4.6 $\mu W/cm^2$ and 1.9 $\mu W/cm^2$) are omitted, the average for the remaining 38 sites is 0.069 $\mu W/cm^2$.

Table 1. Biological and Clinical Effects Observed from Exposure to
Microwaves or Radiofrequency Radiation.

In Non-humans (biological effects)

Chromosomal anomalies
in Chinese hamster and drosophila melanogaster; also breaks in human
lymphocytes in culture; frequency 5 to 40 megahertz (pulsed).

Mutagenesis
in Swiss male mice; DNA changes; high mutagenicity index; in sperm;
exposure ranges 50 milliwatts per square centimeter at 17 gigahertz.
Also, basic changes in cell structure and density of bacteria between 10 to 50 milliwatts per square centimeter at 50 to 90 gigahertz.

Teratogenesis
in mealworms; preceded presumed thermal lethality at 9 to 10 gigahertz with power density not reported but total power of 20 to 80
milliwatts.

Behavioral impairment
in rats; adverse motor coordination and balance; induced "docility";
exposure ranges between 0.4 and 2.8 milliwatts per square centimeter
at 1.3 to 1.5 gigahertz.

Neuroendocrine and hormonal alterations
in rats and dogs; transient changes ascribed to temperature increase.
Exposure 20 to 60 milliwatts per square centimeter for 30 to 60
minutes at 2,450 megahertz.

Prenatal impairment of body and brain weight
in rats; 10 milliwatts per square centimeter at 2,450 megahertz for
5 hours daily for 17-day gestation period.

Blood-brain barrier alterations
in hamsters; 10 milliwatts per square centimeter at 2,450 megahertz.

Central nervous system influence
in chicks; 1 to 2 milliwatts per square centimeter at 147 megahertz
increase in calcium ion release.

Mortality
in rats, rabbits, dogs; effects dominantly thermal; typical exposure
ranges 40 milliwatts per square centimeter for minutes to hours, 2800
megahertz; various wavelengths between 1 millimeter to 10 centimeters.

In Humans (biological or clinical effects)

Cataractogenesis and other ocular effects
Epidemiological studies among microwave workers show increase in lens
opacities and retinal lesions; frequency range, power density incompletely reported.

Central nervous system influence
Auditory nerve response not necessarily hazardous.

Oncogenesis
Speculative; epidemiological suggestion of carcinogenesis in North
Karelia region of Finland.

Biochemical imbalances
Many studies but exposure parameters not reported.

Subjective physchological complaints
Wide variety of alleged clinical effects between 0.01 to 10 milliwatts
per square centimeter.

Hematological changes
Many blood changes cited; main instability in leukocyte indices.

These measurements were all done at street level. Total field strength measurements reported by the EPA yielded maximum levels for New York City as follows:

	Total Field Strength ($\mu W/cm^2$)
World Trade Center	
Outdoor Observation Deck	6.8
Indoor Observation Deck	1.2
Empire State Building	32.50
Pan Am Building	10.3

Two locations, one in Miami, Florida (2 Biscayne Boulevard) and one in Chicago (Sears Tower) yielded higher field strengths, 96.85 $\mu W/cm^2$ and 65.73 $\mu W/cm^2$ respectively.

Conclusion

The thrust of the evidence would indicate an occupational permissible level for sustained working exposures (viz. in excess of 0.1 hours) of 500 $\mu W/cm^2$ i.e. about ten percent of the approximate midpoint of the 1-10 mW/cm^2 potentially hazardous interval.

Public health prudence, in the absence of more definitive research to the contrary, would dictate that unregulated or uncontrolled areas which are available to men, women and children by virtue of residence, recreation or general public access maintain a microwave/radiofrequency power density environment not exceeding ten percent of the indicated occupational permissible level, or 50 $\mu W/cm^2$. This reduction is in sound recognition that the public environment is not subject to the same presumed level of biomedical surveillance and detailed health and safety awareness as the occupational workplace.

(1) Solon, L.R.(1979) Bulletin of the New York Academy of Medicine, p. 1251-1266 V. 55 No. 11 December 1979
(with references).

(2) Solon, L.R.(1979) The Bulletin of the Atomic Scientists, p. 51 V. 35, No. 8 October, 1979 (with references).

(3) Athey, T.W. et al (1978) Nonionizing Radiation Levels and Population Exposure in Urban Areas of the Eastern United States, EPA-520/2-77-008 (Washington, D.C. U.S. Environmental Protection Agency, 1978).

(4) Environmental Protection Agency Staff, Nonionizing Radiation in the New York Metropolitan Area, EPA-902/4-78 (Washington, D.C.; U.S. Environmental Protection Agency, 1978).

INSIDIOUS OCULAR EFFECTS OF LASER RADIATION

D. H. Brennan

Most safety codes for lasers emitting in the wavelength band 400-1400 nm are based on threshold studies for a single event thermal lesion, where ocular damage is related to wavelength, retinal image size, pulse duration and energy density. Current codes of practice are derived from a 50% damage probability for different laser systems producing diffraction limited image sizes.

Noell in 1965 (1) discovered that the rat retina could be damaged by exposure to moderate light sources. Marshall (2) showed that pigeons exposed to moderate white light luminances suffered cone loss; he continued his work with fish (3) where he was able to selectively damage specific cones responding to one primary colour, by illuminating the aquarium with monochromatic light. Harwerth & Sperling (4) in their behavioural studies produced temporary and permanent colour blindness in monkeys following exposure to intense spectral sources. Ham (5) in his studies showed that retinal damage thresholds decreased for short wavelengths. Zwick (6) exposed 2 monkeys to very low luminances of argon laser irradiation on a hemisphere and was able to show that photopic visual function was substantially depressed and that recovery was minimal over a 12 month period.

Concern is now felt that current codes of practice may be inadequate to protect individuals exposed, for long periods, to sub-threshold levels of laser irradiation, particularly so with lasers emitting at the blue end of the spectrum.

Many workers involved in activities such as research, holography, data processing and laser light shows are exposed over long periods to subthreshold laser irradiation. This has caused us to investigate visual function in one such group of workers who have been exposed to argon laser irradiation over a 2-year period in the development of a laser scan visual flight simulator.

MATERIALS AND METHODS

The laser scan, at present, employs two argon lasers of 5 watts nominal output working multimode. The laser camera illuminates a terrain model Fig I and is normally operated at 500 milliwatts. The information from the model modulates a laser projector which has operated for 90% of the time at 1 watt output and for remaining 10% of the time at 4 watts output. The display is presented on a hemisphere, the luminance of which has varied between 3.5-7.0 cd/m^2. The field of view to the observer is 180° in the horizontal and 60° in the vertical. The resolution of the display is 5,280 television lines and the colour is blue-green. Production systems will be in full colour, this will be achieved by separately modulating the blue and green lines of the argon laser and by adding red from a krypton laser.

Eight workers have been monitored and their exposure times for viewing the display and for being in the vicinity of the lasers are given in Table I.

As evidence from the literature suggests that cones are primarily at risk, the measures of visual performance have mainly concentrated on photopic function. The tests were:-

D. H. Brennan

1. Clinical eye examination.
2. Liminal brightness increment for white and blue light at photopic and scotopic luminances.
3. Colour vision testing with Farnsworth-Munsell 100-Hue Test. Illumination of 10 Lux.
4. Perimetric field of view measurements for 2 mm white and blue stimuli.
5. Macula thresholds for white and blue light stimuli.
6. Central visual fields for white and blue light stimuli
7. Dark adaptation curves for white and blue light stimuli, following light adaptation.

The tests 5, 6 & 7 were made on a calibrated Friedmann Visual Field analyser. The unattenuated luminances of the white and blue light sources were 3.0 and 0.1 photopic nit seconds respectively.

The blue filter used in these tests was an Ilford 623, a spectrophotometric trace of which is shown in Fig 2.

TABLE I. Estimated exposure times, of workers exposed to argon laser light.

SUBJECT	AGE	HOURS LOOKING AT HEMISPHERE DISPLAY	HOURS WORKING IN LASER ENVIRONMENT
A	47	0	0
B	44	2	100
C	40	3	1
D	32	3	0
E	45	1	1
F	25	300	120
G	40	150	10
H	32	20	100

Fig I. View of laser camera and terrain model.

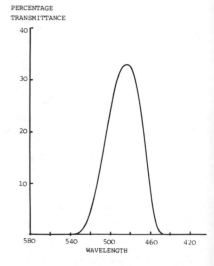

Fig 2. Spectrophotometric trace of Ilford 623 filter.

RESULTS

The clinical examination included an ocular history, tests of pupillary function, visual acuity estimates on a Snellen chart under white and blue illumination with a refraction where necessary, macula function tests using an Amsler grid, a fundoscopic examination and an examination of the anterior segment of the eye with a slit lamp. These tests did not reveal any pathology which could be attributed to work in a laser environment. The expected addition of a -0.50 sphere was necessary to restore visual acuity under blue illumination.

Colour vision testing with the Ishihara pseudo-isochromatic plates did not show any deficiencies. Colour vision testing with the Farnsworth-Munsell 100-Hue test under low illumination did reveal some error scores, particularly in the blue-green and purple hue regions.

TABLE 2. Farnsworth-Munsell 100-Hue test

SUB-JECT	HUE ERROR SCORES				TOTAL ERROR SCORES
	610-570 nm	570-500 nm	500-470 nm	470-630 nm	
	RED-YELLOW	YELLOW-GREEN	PURPLE GREEN-BLUE	PURPLE BLUE — RED PURPLE	
A	4	8	4	0	16
B	0	4	12	0	16
C	4	15	32	8	59
D	4	0	16	4	24
E	4	8	24	20	56
F	0	0	0	0	0
G	0	4	4	0	8
H	0	8	4	4	16

Perimetric assay did not show any loss of peripheral field for white or blue stimuli.

The log densities for the white and the blue macula thresholds are the maximum densities at which no error was made for ten consecutive stimuli. The log densities for the central fields are the maximum densities at which the central field was full for the white and the blue stimuli.

TABLE 3. Central fields and macula thresholds.

SUB-JECT	AGE	NORMAL LOG DENSITY FOR AGE (WHITE)	LOG DENSITY FOR FULL FIELD				MACULA THRESHOLDS. LOG DENSITIES.			
			WHITE		BLUE		WHITE		BLUE	
			Rt.	Lt.	Rt.	Lt.	Rt.	Lt.	Rt.	Lt.
A	47	1.8	1.8	1.8	1.2	1.2	2.0	2.2	1.0	1.2
B	44	1.8	1.8	1.8	1.0	1.2	2.2	2.2	1.0	1.0
C	40	2.0	2.0	2.0	0.8	1.0	2.4	2.4	0.8	0.8
D	32	2.0	2.0	1.8	1.0	1.0	2.6	2.4	1.2	1.2
E	45	1.8	1.8	1.8	1.0	0.8	2.6	2.4	1.0	1.0
F	25	2.0	2.0	2.0	1.2	1.2	2.4	2.4	1.2	1.2
G	40	2.0	2.0	2.0	1.2	1.2	2.4	2.4	0.8	0.8
H	32	2.0	2.0	2.0	1.0	1.0	2.8	2.8	1.2	1.0

The results of the liminal brightness increment measures showed the expected increase in contrast threshold $(\frac{\Delta I_s}{I})$ at the scotopic luminances for both white and blue light.

TABLE 4. Liminal brightness increment

BACKGROUND LUMINANCE cd/m²		ΔI+I cd/m² mean	I cd/m² mean	$\Delta I/I\%$ RANGE	MEAN
3.2×10^{-3} SCOTOPIC	BLUE	3.46	3.18	2.8-14.0	8.76
	WHITE	3.49	3.19	3.7-14.0	9.49
5.0×10^{-2} PHOTOPIC	BLUE	5.15	4.96	2.0-7.6	3.92
	WHITE	5.17	5.00	1.2-5.6	3.42

Dark adaptation curves after white light adaptation did not show any significant departures from normal for white and blue stimuli. The scotopic portion of the curve was not continued beyond 20 minutes.

DISCUSSION

This is a preliminary survey into the visual function of workers exposed to long term argon laser irradiation and as yet it has not been possible to demonstrate any visual decrement which could be, directly, attributed to work with lasers. Paradoxically the workers with the longest exposure times performed as well or better than those with minimal exposure. The decrements found were those which would be expected in any random group of varying age. The survey is limited, it involves eight workers of whom only half have been exposed for a significant period. It is intended to repeat the tests of visual performance at six monthly intervals as exposure times increase.

It will be understood that by the nature of the work exposure times are approximate, particularly with the variable exposure incurred when working in the laser environment. The exposure time for looking at the hemisphere display is more precise, as is the luminance. Should visual decrements develop it would be valuable to correlate these with display times over the six monthly intervals.

The dosimetry of laser exposure in man can never be as precise as that in experimental animals. It is considered, however, that the subtle changes in vision which may occur with long term low level laser irradiation may only be detected in man.

REFERENCES

1. Noell, W.K., Walker, V.S., Kang, B.S. and Berman, S. (1966): Retinal damage by light in rats. Invest. Ophthal., 5, 450-473.
2. Marshall, J., Mellerio, J. and Palmer, D.A.P. (1972): Damage to pigeon retinae by moderate illumination from fluorescent lamps. Exp. Eye Res., 14, 164-169.
3. Marshall, J. (1978): Retinal injury from chronic exposure to light and the delayed effects from retinal exposure to intense light. Current concepts in erg ophthalmology. Ed. Tengroth, B., Stockholm, 81-105.
4. Harwerth, R.S. and Sperling, H.G. (1971): Prolonged colour blindness induced by intense spectral lights in Rhesus monkeys. Science, 174, 520-523.
5. Ham, W.T., Mueller, H.A. and Sliney, D.H. (1976): Retinal sensitivity to damage from short wavelength light. Nature, 260, 153-155.
6. Zwick, H. and Beatrice, E.S. (1978): Long term changes in spectral sensitivity after low level laser (514 nm) exposure. Mod. Prob. in Ophthalmol., 19, 319-325.

EFFECTS OF MICROWAVE RADIATION ON ENDOCRINE SYSTEM OF MOUSE

P. Deschaux, R. Santini, R. Fontanges and J. P. Pellissier

Swiss male mouse, 1 or 2 months old were irradiated in an exposure chamber (32,5x32x54) cm : microwave frequency 2450 MHz ; time of irradiation : 1, 2, 3 or 4 hours during 5 days. The sacrifice of animals is performed by decapitation immediatly or 24 hours or 5 days after the last irradiation. The density power was calculated from the relationship

$$P = N \times S \times 10^{-2}$$

P : density power (watt)
N : number of animals in the exposure chamber
S : surface of an animal (cm^2)

Ambient temperature was maintained between 23-25°C. In any case a total of irradiated 40 mice and an equal number of non irradiated animals (controls) were studied.

When the mice were decapited immediatly after removal from the exposure cage, colonic temperature was measured within 60 s : we had never observed a variation.

After decapitation trunk blood was collected in iced tubes containing ethylenediamine-tetraacetic acid (EDTA), centrifuged at 2,500 rpm for 10-15 min in a refrigerated centrifuge, and the plasma was frozen at -23°C until it was assayed for hormones.

Testosterone assay. A modification of the radioimmunoassay method of Nieschlag and Loriaux (3) was used for the measurement of testosterone.

The intrassay coefficient of variation was less than 10%, as determined by assaying 10 ml samples of plasma. The base values obtained from water (charcoal treated) were between 0 and 8 pg ; 10 pg was considered significantly different from 0.

Corticosterone assay. The plasma corticosterone level was determined by the competitive binding radioassay of Murphy (2). Rat transcortin was used as the binding protein. Test plasma was extracted with carbon tetrachloride followed by celite chromatography to remove other steroids which might interfere with the assay.

LH assay. Plasma luteinizing hormone assay was performed exactly as described in the radioimmunoassay kit supplied by the Rat Pituitary Distribution Program (NIAMDD). The second antibody used was antirabbit gamma-globulin purchased from Welcome laboratories (England). For the assays to be acceptable, the coefficient of variation for hormone concentration at the 50% level of bound radioactivity intercept was not greated than 6.0% in this LH assay.

ACTH assay. A rabbit antiserum against porcine ACTH (ACTH retard de poc Choay) obtained from Dr Depieds (1) was used. This antibody is produced chiefly against the biological fragment 1-24 ACTH. Buffer used in this procedure consisted of 1.5 ml 20% human albumin, 200 ·µl 1-2 mercaptoethanol, 1.0 ml zymofren at 10 000 U/ml (Specia), 0.02 M veronal (pH 8.6), and water to a final volume of 100 ml. This buffer had no inhibitory effect on the antigen-antibody reaction. The standard curve was done in triplicate using ACTH-free plasma from hypophysectomized animals. Assays of plasma were done in duplicate. After a 48 h incubation at room temperature, free ACTH was adsorbed with 50 mg of talc. The radioactivity was determined in the precipitate after centrifugation.

The statistical significance of differences among data from the different groups of animals was determined using Student's t-test.

The testosterone and LH levels in mice after various microwave exposures are shown in table I, corticosterone and ACTH levels in table II.

The results of the experiments suggest that the mice adrenal axis is transitory stimulated during microwave exposure without regulation by a possible feed-back on ACTH secretion.

This endocrine perturbation seems to be a concomitant to increased testosterone plasma level. We have never observed perturbations concerning spermatogenesis.

REFERENCES

1. Cros, G, Vague, P, Oliver, C and Depieds, R. (1970) : C.R. Soc Biol. 164, 1289.
2. Murphy, B.E.P. (1967) : J. Endocrinol. Metab. 27, 973.
3. Nieschlag, E. and Loriaux, D.L. (1972) : Z. Klin. Biochem. 10. 164.

TABLE I

Time irradiation (hour/day)	Number of irradiation per day	Time between last irradiation and sacrifice	age of animals (day)	Plasma testosterone level (μg/100 ml) control		irradiated	Plasma LH level (ρg/100 ml) control		irradiated
1	1	0	30	87,2 ± 2,3	↗	106 ± 2,0	94 ± 0,6	↗	108 ± 0,8
			60	313 ± 18,3	↗	341 ± 14,0	82 ± 4,4	↗	112 ± 1,1
		24 h	30	55 ± 2,2	NS	48,7 ± 7,1	92,8 ± 6,1	NS	100 ± 4
			60	297,8 ± 12,2	NS	281 ± 29	83,4 ± 0,8	NS	82 ± 2
		5 d	30	64,2 ± 4,3	NS	65 ± 3,1	101,4 ± 10,2	NS	96,6 ± 7,3
			60	316 ± 18,0	NS	308,8 ± 5,2	79 ± 1	NS	87 ± 22
2	2	0	30	76,4 ± 4,2	↗	100,6 ± 5,6	91,9 ± 1	↗	126 ± 2,2
			60	318,4 ± 8,4	↗	410 ± 20,7	92,4 ± 0,8	↗	101 ± 1
		24 h	30	66,4 ± 3,0	↗	76 ± 2,4	100 ± 9,3	↗	118 ± 2
			60	268,2 ± 11,1	↗	294 ± 6,9	79 ± 4,5	↗	94 ± 0,8
		5 d	30	53,8 ± 5,0	NS	61,6 ± 0,9	88,2 ± 0,9	NS	88 ± 2
			60	243,4 ± 11,0	NS	262 ± 10,7	86 ± 2,4	NS	86 ± 2,4
3	3	0	30	90,8 ± 3,9	↗	105 ± 4,5	96 ± 2,4	↗	124 ± 2,4
			60	314,8 ± 1,3	↗	429 ± 85,0	96 ± 5,3	↗	120 ± 3,1
		24 h	30	88,4 ± 4,1	↗	121,4 ± 4,6	80 ± 2,7	↗	120 ± 1,1
			60	312 ± 22,8	↗	344 ± 5,0	87 ± 4,3	↗	120 ± 4,4
		5 d	30	51,4 ± 3,1	NS	55 ± 5,7	106 ± 5,0	NS	112 ± 4,8
			60	344 ± 8,7	NS	351 ± 11,2	79 ± 3,3	NS	80 ± 2
4	4	0	30	74,2 ± 2,4	↗	152 ± 10,3	91 ± 9	↗	104 ± 2,4
			60	376 ± 2,9	↗	475 ± 30,9	98 ± 8	↗	128 ± 2
		24 h	30	88,8 ± 3,0	↗	162 ± 4,8	105 ± 2,2	↗	122 ± 5,8
			60	392 ± 15,9	↗	423 ± 3,7	99 ± 3,3	↗	120 ± 7,0
		5 d	30	93,4 ± 5,1	NS	107,2 ± 1,7	96 ± 9,79	NS	101 ± 2,69

Effects of Microwave Radiation

Time irradiation	Number of irradiation day	Time between last irradiation and sacrifice	age of animals (day)	Plasma corticosterone level (µg/100 ml) control		irradiated	Plasma ACTH level (pg/ml) control		irradiated
1	1	0	30	14,8 ± 0,5	↑	18,6 ± 2,1	108 ± 3,7	↑	124 ± 2,4
			60	15,2 ± 0,8	↑	25,4 ± 2,0	76 ± 6	↑	124 ± 4
		24 h	30	14,6 ± 0,8	↑	18,6 ± 0,4	102 ± 2	↑	122 ± 2
			60	12 ± 0,4	↑	14,6 ± 0,9	85 ± 5,3	↑	107 ± 8
		5 d	30	14,6 ± 1,4	NS	14,4 ± 1,1	118 ± 4,8	NS	119 ± 5,5
			60	14,6 ± 0,7	NS	13,2 ± 0,3	97 ± 2,5	NS	104 ± 6
2	2	0	30	14,2 ± 0,5	↑	18,4 ± 0,4	101 ± 4	↑	117 ± 2
			60	12,2 ± 0,8	↑	16,4 ± 1,3	80 ± 3,5	↑	122 ± 3,3
		24 h	30	14,8 ± 0,3	↑	18 ± 1,6	117 ± 4,8	↑	124 ± 4
			60	15,6 ± 0,8	↑	17,4 ± 1,4	58 ± 1,6	↑	90 ± 3,6
		5 d	30	12 ± 0,3	NS	12,8 ± 1,2	83 ± 7	NS	84 ± 14,6
			60	12,5 ± 0,6	NS	12 ± 0,3	92 ± 8	NS	105 ± 6,3
3	3	0	30	13,2 ± 0,8	↑	16,8 ± 0,7	120 ± 8,9	↑	134 ± 4
			60	14,6 ± 1,0	↑	18,6 ± 0,3	90 ± 4,1	↑	122 ± 2
		24 h	30	13 ± 0,6	↑	18,6 ± 1,6	118 ± 5,8	↑	132 ± 3,7
			60	14 ± 0,7	↑	18 ± 1,5	81 ± 4,8	↑	122 ± 2
		5 d	30	13,2 ± 0,7	NS	12,4 ± 1,8	110 ± 3,4	NS	102 ± 11,1
			60	13 ± 1,6	NS	12 ± 0,5	96 ± 8,1	NS	98 ± 9,6
4	4	0	30	10,8 ± 0,1	↑	20 ± 1	81 ± 3,3	↑	108 ± 7,3
			60	12,4 ± 0,3	↑	19,2 ± 0,5	66 ± 9,1	↑	112 ± 9,6
		24 h	30	18,2 ± 0,7	↑	21 ± 0,3	38 ± 5,8	↑	56 ± 3,6
			60	17,6 ± 1,4	↑	20 ± 2,0	45 ± 6,3	↑	73 ± 9,5
		5 d	30	12 ± 0,3	NS	12,2 ± 0,3	76 ± 2,4	NS	78 ± 9,5
			60	17,2 ± 0,2	NS	17,6 ± 1,4	74 ± 7,4	NS	78 ± 5,8

Instrumentation and Shielding

NUCLEAR SHIELDING ANALYSES FOR AN INTENSE NEUTRON SOURCE FACILITY

J. Celnik

1. INTRODUCTION

This paper summarizes nuclear shielding analyses applicable to fusion, and fusion related, facilities. The analyses were performed during the design of an Intense Neutron Source Facility, to provide an experimental neutron irradiation facility yielding a neutronic environment similar to that encountered in a fusion power reactor.

The analyses included:
- bulk shield
- skyshine
- various generic and specific penetrations, single and multi-legged
- source cell door design.

All results are based on a neutronic environment generated by a D-T source yielding 3×10^{15} -14 MeV neutrons/second. Analyses included the effect of secondary gamma rays produced by the interaction of primary and scattered neutrons with air and the proposed shielding materials.

2. BULK SHIELD ANALYSIS

The adequacy of an eleven-foot concrete wall, composed of an inner layer of one-foot borated gypsum and ten feet of ordinary concrete to meet the design dose rate criteria was calculated using ANISN. The ANISN computer program calculates radiation transport in a one-dimensional geometry via the discrete ordinates method. This computational technique is commonly used for the solution of deep penetration problems.

To evaluate the computational uncertainty the results were compared using:
- different cross section data sets
- increasing the order of the Legendre expansion of the scattering cross section with a comparable increase in the angular quadrature representation.

Some highlights of the analyses are:
a) An outer wall of one-foot borated gypsum (to minimize activation) followed by ten feet of concrete yields a dose rate of 135 mRem/occupational year.
b) The total dose rate drops rapidly when the detector is placed off the source axis.
c) Use of the CASK 22 neutron-18 gamma ray coupled cross section data set underestimates the neutron leakage by a factor of about three and the total dose rate by more than a factor of two, as shown in Table 1.

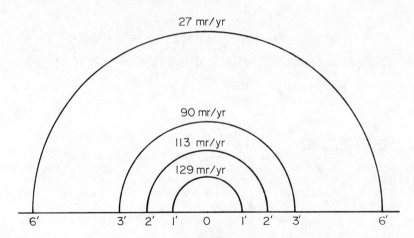

FIGURE 1: Total Dose Rate on Exterior of Source Cell Wall as a Function of Off-Axis Position

TABLE 1: Dose Rates Outside 1' Borated Gypsum and 10' Concrete Wall

Cross Section Data Set	Dose Rate (mRem/hr)		Total
	Neutron	Gamma Ray	
CASK (22N + 18G)	5.25-3	3.63-2	4.16-2
DLC-31 (37N + 21G)	1.55-2	7.30-2	8.85-2

d) Table 2 shows the effect of using a more detailed description of the anisotropic scattering, for an eleven foot concrete shield wall.

TABLE 2: Dose Rate Outside 11' Concrete Wall

Cross Section Data Set	Dose Rate (mRem/hr)		Total
	Neutron	Gamma Ray	
P_3S_8 - DLC 31	1.31-2	7.80-2	9.11-2
P_5S_{12} - DLC 27	6.44-3	6.11-2	6.76-2

3. SKYSHINE ANALYSIS

Some highlights of the analysis include:

a) A source cell roof of eight feet concrete is required to reduce the skyshine dose rate contribution to acceptable levels.
b) For proper skyshine evaluation, the leakage spectrum on top of the source cell roof should be calculated using a two-dimensional discrete-ordinates code. A one-dimensional code will under-estimate the leakage and hence the skyshine contribution by about a factor of ten.
c) The 2-D leakage results can be used as the source for the full 3-D Monte Carlo analysis of the skyshine contribution.

4. PENETRATION ANALYSES

An investigation was done to estimate radiation streaming effects for a variety of penetration configurations. The results may be used to design general penetration layouts, which could then be analyzed in greater detail and accuracy. All analyses were performed with the Monte Carlo method. Preliminary analyses were done with MORSE CG program using the CASK cross section data set, with more detailed results obtained with the LASL MCNP program using the LASL recommended cross section data.

All penetrations analyzed were for a ten foot ordinary concrete wall. Some of the general conclusions are:

a) The line-of-sight uncollided dose rate dominates for penetrations directly facing the source;
b) The scattered dose rate is large (> 1 Rem/hour for a source of 10^{15}-14 MeV neutrons/second) even for a 12 cm penetration angled $45°$ through a 10 foot wall;
c) Placement of large (two foot diameter) penetrations in an extreme corner of a 12 foot X 12 foot cell is not adequate to reduce streaming to acceptable levels;
d) Use of a multi-legged, non-coplanar design can reduce the exit dose rate, for a large diameter penetration, by several orders of magnitude;
e) Decreasing the penetration size from 2' X 2' to a 1' X 1' opening will decrease the dose rate at the end of the first leg by about a factor of 2, and decreases it by a factor of about 100 at the exit of the four-legged non-coplanar penetration;
f) The above comparison is also valid when an 18 inch (45.7 cm) radiation flux trap is included at the end of the first leg;
g) For some configurations, use of a flux trap will increase the total attenuation by about a factor of two;
h) When using a flux trap, the dose rate at a point on the outer wall surface opposite the first leg may be significantly higher than the dose rate at the exit face of the last leg. A magnetite concrete plug may be used to match the dose rate through the first leg with the dose rate at the penetration exit.
i) The dose rate at the penetration exit, for the multi-legged penetrations analyzed in this study, is due primary to secondary gamma ray leakage. Inclusion of an 8% borated polyethylene liner will decrease gamma ray leakage by a factor of about 100 and the total dose rate by a factor of about 15.

5. SOURCE CELL DOOR DESIGN

Access to the source cell is provided by a system of hydraulic doors. The inner door, adjacent to the cell, is composed of a one-foot liner of borated gypsum followed by 3'-3" of magnetite concrete. The outer door, consisting of 3' of magnetite concrete, contains a lead glass viewing window. The doors are stepped to minimize streaming between the doors and the wall in which they are located. In addition, a steel plate is placed in the cavity beneath the door to eliminate the potential for radiation scattering into the cavity and then re-emerging in front of the door.

TABLE 3: Results of Source Cell Door Analysis

Shield Configuration*	Total Dose Rate (mRem/hr)
Bulk wall (6'-6" mag. concrete)	0.8
Through both doors (6'-3" mag. concrete)	1.4
Through window (3'-3" mag. concrete + 3' Pb-glass)	9.0
Through alternate window design	5.5
Through window lining (3'-3" mag. concrete + 3' Fe)	18.5
Window lining shielded by oil layer	3.6

*All shield configurations included a liner of one foot of borated gypsum.

The results indicate that the dose rate behind the wall and behind the doors meet the dose rate criterion of 10 Rem/occupational year (equivalent to 5 mRem/hr). However, use of a "standard" lead glass configuration will lead to a significant hot spot. Use of an alternate lead glass window composition can reduce the radiation streaming to acceptable levels.

In addition, design of the window frame should include provision for inclusion of a one-foot oil layer in front of the frame to reduce neutron streaming.

6. CONCLUSIONS

This paper presents highlights of nuclear shielding analyses performed during the design of an intense neutron source facility. It is expected that the results, though preliminary in some areas, would be useful in the design of similar fusion-related facilities as well as in the conceptual design of a fusion power reactor complex.

7. REFERENCES
   1. J. L. Liverman (1976): "Intense Neutron Source Facility", ERDA-1548.
   2. M. B. Emmett (1975): "The MORSE Monte Carlo Radiation Transport Code System", ORNL-4972.
   3. J. MacDonald (1977): "MCNP Users Manual", LASL.
   4. W. W. Engle, Jr. (1973): "ANISN-Multigroup 1-D Discrete Ordinates Transport Code with Anisotropic Scattering", K-1693, ORNL.
   5. "CASK-40 Group Cross Section Data", (1975) DLC-23, ORNL/RSIC.

Nuclear Shielding Analyses

6. D. E. Bartine et al. (1975): "DLC-31; 37 Neutron, 21 Gamma Ray Coupled, P_3, Multigroup Library", ORNL/RSIC.
7. "DLC-27; Coupled 104 Neutron, 22 Gamma Ray Groups, P_5" ORNL/RSIC.

USE OF INFORMATIC FOR RADIATION CONTROL PANELS

R. Cochinal, B. Grimont and V. Mai

Radiation control panels (R.C.P.) are systems which enable irradiation and contamination risks to be quantitatively determined and monitored.

Such systems automatically control the immediate activation of warning devices (audible and visible in the zones concerned so as to alert personnel at their working stations).

For a few years now, the CEA has been developing a programmed system generation of radiation control panels.

R.C.P. can be divided in to three main elements :

1°) a series of monitoring stations
   Each station monotoring a zone consists of :
. a detector adapted to the characteristics of the radiation to be detected together with part of all of the associated electronics (power supply, amplifier...)
. an audible and visible alarm unit alerting personnel of the risks to which they are exposed.

Measured exposure levels fall into four scales as seen in the table of figure 1, which also indicates the corresponding visible and audible signals.

2° a central station
   All the radioprotection data recorded converge into this station In general, an operator is posted here, whose responsibility is to monitor, and when necessary, record, the risks encountered at each individual station.

3° more or less sophisticated information processing facilities (between 1° and 2°)

A recent orientation in the design of the R.C.P. programmed system generation is to locate totally autonomous units in the various different zones. These units provide signals when given thresholds expressed in LMA or CMA are exceeded. In this way safety and availability are improved.

This unit, which together with it s detector constitutes an autonomously operating monitoring station, can be connected to a centralizing unit (e.g. minicomputer).

DESCRIPTION OF THE MICROPROCESSOR PROCESSING AND SIGNALLING UNIT (PSU)

The monitoring station of a zone figure 2 consits of :
- a detector and its associated electronics which delivers a standard pulse for all types of detector.
- a processing and signalling unit assuring the following functions :
  . acquisition of information detected by the detector
  . processing of this information to determine LMA dose rates, while taking into account parameters such as the radiotoxicity of the radioelement, which can be memorized in the unit

- generation of different output signals to be transmitted to the
  central station (100 to 200 meters)
- warning outputs for synoptic
- analogic dose rate output for recordings
- asynchronous line output for centralizer
- upon cyclic interrogation by minicomputer, the unit transmits :
  - dose rates
  - the threshold exceeded by the unit
  - the state of the unit
  - the memorized values of the thresholds and coefficients

    The correct operation of each station is verified :
- the complete system, by measurements with a permanent low activity
  control source, which triggers correct functioning threshold
- the quasi-totality of the system with periodic tests (generator
  simulating levels)

An MC 6800 microprocessor is used.
A maximum of 8K REPROM and 2K RAM memories is available.

DESCRIPTION OF CENTRALIZER

At the central station, all or part of the following facilities are available :
- minicomputer which acquires (via asynchronous lines) informations from units for determining cumulative doses and different logs
- a detailed visual synoptic providing, for each station, the number of the threshold exceeded
- recorders

The last two devices <u>are independent from the computer</u> and can constitute :
- either, by themselves, the centralizer
- or, a back up system for the computer in the case of failure.

A first realisation of this system is being used to control an effluent treatment plant. 30 units are connected to a MULTI 6 minicomputer.

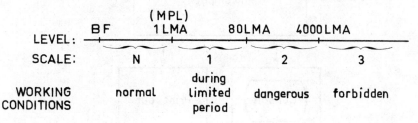

Fig. 1

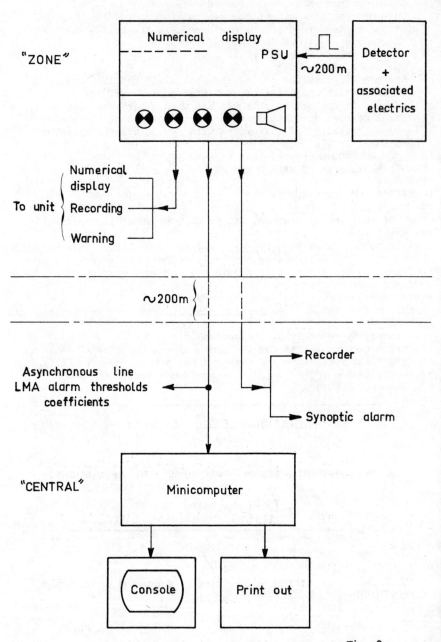

Fig. 2

SECONDARY STANDARD DOSIMETRY* SYSTEM WITH AUTOMATIC DOSE/RATE CALCULATION

K. E. Duftschmid, J. Bernhart, G. Stehno, W. Klösch, J. Hizo and K. Zsdanszky

INTRODUCTION

In view of the increasing requirements for standardization measurements in radiation dosimetry a versatile and automated secondary standard instrument*) has been designed for quick and accurate dose/rate measurement in a wide range of radiation intensity and quality for protection- and therapy level dosimetry. The system is based on a series of secondary standard ionization chambers /1/ connected to a precision digital current integrator /2/ with microprocessor circuitry for data evaluation and control. Input of measurement parameters and calibration factors stored in an exchangeable memory chip provide computation of dose/rate values in the desired units.

IONIZATION CHAMBERS

Secondary standard ionization chambers require excellent reproducibility and long-term stability of the sensitive volume. Therefore graphite is generally used as the wall material in combination with Al collecting electrodes to achieve a flat energy response. This compensation however is only valid for free-air measurement and may introduce significant errors when used in-phantom due to the inhomogeneous construction materials /3/.

In contrast the described chamber design uses walls and electrodes made from Polyacetal resins $(CH_2O)_x$. This material provides superior mechanical properties assuring the necessary long-term stability of dimensions and a most suitable chemical composition. By choosing the proper mixture of Polyacetal with Polytetrafluoroethylene (PTFE) and small additions of higher Z-material such as CaO, the chambers can be made virtually tissue-, water-, or air-equivalent as desired.

In order to achieve electronic equilibrium for photon energies above 1 MeV the wall thickness has to be at least 2 mm. For soft X-rays the absorption in the wall is compensated due to a thin vacuum-deposited layer of Al on the inner wall surface. In this way the energy response is within $\pm$ 2% between 0,02 - 1,2 MeV without any additional build-up caps etc.

---

*) International patents.

The ionization chambers (see fig. 1) are tailored to the different applications. For radiation protection measurements at low doserates a large spherical air-equivalent chamber of 10 l volume can be used down to environmental levels. An internal check-source of ^{241}Am can be introduced into the center of the chamber through a hollow axial tube protruding from the stem to the other pole of the sphere. For high doserates in therapy level dosimetry a small water- or tissue-equivalent thimble chamber of 1 cm^3 volume has been designed which can directly be put into a water phantom. For the intermediate doserate range a 100 cm^3 spherical chamber can be used. In addition a backscatter chamber for soft X-ray therapy measurements is in preparation.

ELECTRONIC CIRCUITRY

The basic components of the electronic system are shown in the simplified block diagram fig. 2.

The ionization chambers are connected to a MOS/FET electrometer amplifier through a series of reed switches (R1-R3). The exchangeable measuring capacitor (100 pF - 100 nF) which determines the range of measurements is normally shorted and the input grounded. During a measurement cycle R2 is closed and R1/R3 opened. The ionization current generates an increasing voltage signal at C, which is measured by an automatic TOWNSEND-balance circuitry consisting of a 5 digit dual slope DVM integrator with compensation by a feedback amplifier system. With an offset current of less than 10^{-15}A ionization currents in the range of 10^{-12}A to 10^{-7}A can be measured within $\pm$ 1% error.

The system is controlled by a microprocessor central processing unit (CPU) chip containing a 1 K x 8 bit EPROM and 64 byte RAM with 6 MHz quartz clock and internal timer/counter. An additional arithmetic processing unit (APU) performes all calculations. A 320 byte RAM is used as a data buffer controlled by the CPU. Up to 100 chamber calibration factors (for 10 chambers at 10 qualities) and 10 capacitance values (C) are stored in an exchangeable memory chip (2 K x 8 bit EPROM). This reusable chip is loaded after calibration and exchanged with each new chamber.

The measurement parameters (atmospheric pressure, temperature, radiation quality and number of cycles) are manually set on BCD-thumb wheel switches on the front panel. Preset dose values and additional calibration factors not contained in the memory can be manually selected if required.

The LED-display contains the voltage signal (5 digits),

integrating time (4 digits), dose/rate (4 digits) with 5 prefix-symbols and the unit of measurement (Gy, R,/h, /min,/s).

With a built-in miniaturized alphanumeric printer (16 characters/line) dose/rate, meanvalue, standard deviation, time, calibration factor and capacitance value are recorded.

The ionization chambers and measurement capacitors are identified by encoding resistors contained in the connectors. The programmable high voltage supply (0 - 2 kV,2 mA) is automatically set to the correct chamber high voltage by the CPU as a function of the decoded chamber number.

DISCUSSION

The system described is designed for secondary standard measurements in protection- and therapy level dosimetry. It covers a wide range of measurement between 1 µR and 100 kR (0,2 nC/kg - 20C/kg) with proper chamber and capacitance and automatically calculates dose/rate due to its microprocessor circuitry. The ionization chambers provide excellent long-term stability and energy response and can be used with internal check sources to test validity of calibration. The system is a useful tool particularely for daily measurements in a secondary standard dosimetry laboratory or radiation therapy center.

REFERENCES

/1/ ZSDANSZKY,K., Primary and secondary standards of dosimetry. Proc. IAEA Symp. Atlanta 1977, SM-222/63, Vol. 1 (1978) 107.

/2/ ZSDANSZKY,K., Precise measurement of small currents. Nucl.Instrum. and Methods 112 (1973) 299.

/3/ WILL,W., RAKOW,A., Untersuchungen zur Luftäquivalenz von Ionisationskammermaterialien. Strahlentherapie 141, 2 (1971) 166.

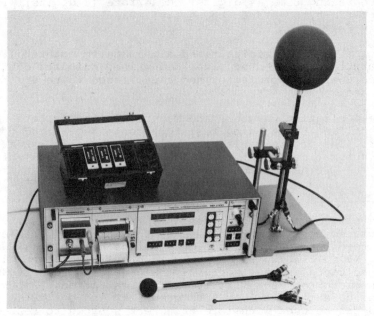

Fig.1 PHOTOGRAPH OF THE SECONDARY STANDARD SYSTEM

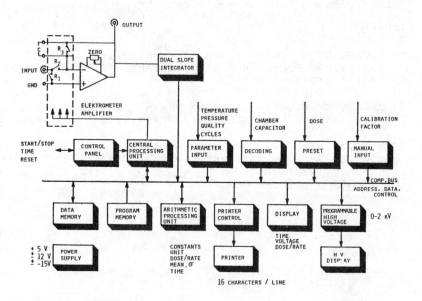

Fig.2 SIMPLIFIED BLOCK-DIAGRAM OF THE ELECTRONIC SYSTEM

DOSE RATES DURING EXPERIMENTS WITH HEAVY IONS

J. G. Festag

The field of nuclear physics with heavy ions is enlarging. The UNILAC (<u>uni</u>versal <u>lin</u>ear <u>ac</u>celerator) at Darmstadt has accelerated ions from argon to uranium up to energies of 1o MeV/u during the last few years.

As, equivalent dose rates originating from reactions with heavy ions ($A \geq 4o$) are not well known, we have undertaken efforts to measure them.

The share of the equivalent dose-rate belonging to neutrons is measured by rem-counters, and that due to gammas by ionization-chambers or G.-M. tubes with energy filters.

The measurements are made around thick targets during experiments of nuclear chemists (Fig. 1). The accelerated ions (5 to 1o MeV/u) are completely stopped in the target. It is reasonable to measure the neutron equivalent dose rates by rem-counters.

Slow neutrons registered by a bare BF_3-counters contribute less than 5 % to the equivalent dose rate.

The results obtained with activation detectors show that the share of neutrons with energies higher than 2o MeV $[^{12}C\,(n,2n)\,^{11}C]$ is smaller than one tenth of that with energies above 11 MeV $[^{19}F\,(n,2n)\,^{18}F]$.

In almost all cases (except for experiments with Xe) the portion of the equivalent dose rate due to gammas is smaller than 1o % and often smaller than 1 % (Fig. 2).

A typical angular distribution of equivalent dose rates due to neutrons is given in figure 3.

Possible sources of measurement errors:

1. Measurements were not always possible in the same geometry (distance from the target and angular position with respect to the beam).
2. The accuracy of the beam current measurements and the beam currents fluctuated.
3. The accelerated ions may have hit not only the target but also the beam pipe, slits, or other material around the target.
4. The results may be influenced by stray-neutrons from the walls and the ground. However, this source of experimental uncertainty seems to be less significant than those listed above.

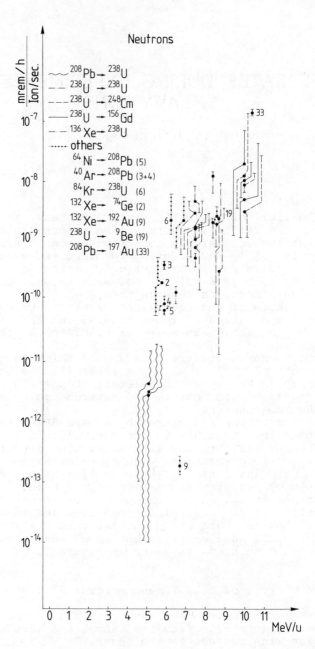

Fig. 1 Equivalent dose rate at 1 m from target in forward direction (± 3o°) caused by neutrons. (The average is given; the error bars show the biggest and the smallest value).

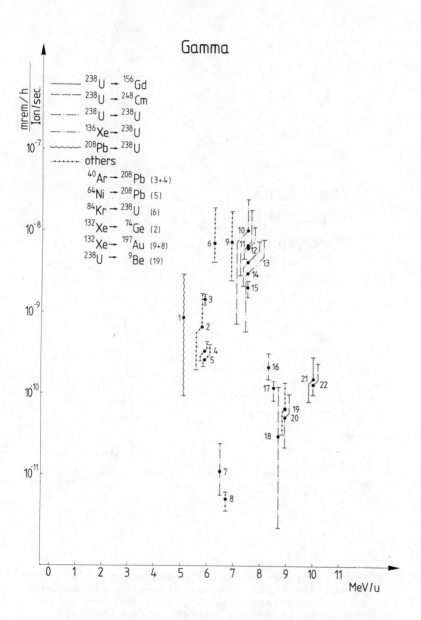

Fig.2 Equivalent dose rate at 1 m from target in forward direction (± 30°) caused by gammas. (The average is given; the error bars show the biggest and the smallest value).

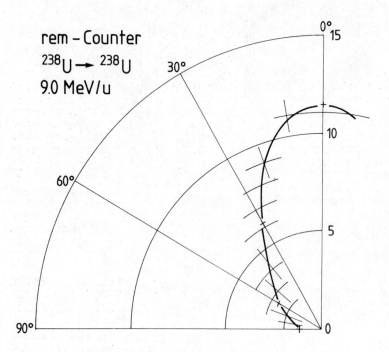

Fig. 3 Angular distribution of neutron equivalent dose rates. (Arbitrary units; direction of the beam at 0°; measurement at 0°, 3o°, 6o° and 9o°. The curve is only a guideline for the eye).

A MICROPROCESSOR BASED AREA MONITOR SYSTEM FOR NEUTRON AND GAMMA RADIATION

R. Wilhelm and G. Heusser

At the MPI-Heidelberg two Tandem van de Graaff (6 and 12 MV) and one postaccelerator are in operation. The 10 MV postaccelerator which consists of independently phased spiral resonators is coupled to the 12 MV Tandem. Depending on the specific ions accelerated, their energy and current, the dose rates range from nondetectable values up to some 10 rem/h. High-energy neutrons contribute 80 to 90% of the stray radiation total rem dose and gamma rays make up the balance. The extreme dose rate situations request flexible radiation protection regulations. They have to ensure that the absorbed dose of the personnel is kept as low as achievable without restricting the experimental operations more than necessary and that at beamtimes with high dose rates accidental exposures are strictly avoided. Since the radiation regulations are based on dose rate measurements of the area monitor system it has to operate in a wide dynamic range at a high reliability level.

The old system was no longer able to fulfil this requirement. There have been too many failures of the electronics. The choice for its replacement is a system consisting of individually microprocessor controlled area monitors which are connected to a microcomputer. A system without the monitor processors which would have been possible, too, was not chosen because of insufficient reliability. Failures of the central microcomputer would have blocked the whole monitor system. A conventional hardware solution would have been less secure and at least two times more expensive.

In Fig. 1 a block diagram of the complete new system is shown. The area monitors as well as the microcomputer have been developed on the basis of the 16 bit TMS 9900 microprocessor. The pulses of the neutron rem counter and of the gamma counter are each fed into 8 bit binary counters of the area monitor microprocessor system. Fig. 2 shows how the pulses are processed by the developed software. All operations are executed in the work space register of the TMS 9900. The overflow rate of the counters is stored in incremental registers. After 1 second the contents of the counters are read into one register and then added to the product of the overflow rate by 8 bit in another register. Now, the γ-count rate is normalized to give the same relative dose equivalent unit as the neutron count rate and finally converted into mrem/h. The γ- and neutron dose rates and their sum are stored in the memory. A comparison of the sum with two thresholds identifies three different dose levels. For failure surveillance the results of the γ- and the neutron measurement are integrated for a long enough time to judge from background statistics if there are too few counts. A subprogram converts the data from the hexadecimal into the decimal notation.

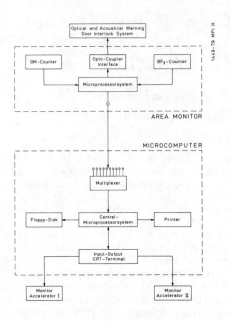

Fig. 1: Block diagram of the area monitor system.

Dose rate and failure indications are transmitted to different warning installations and to the door interlock system as shown in Fig. 1. All in- and output signals of the microprocessors are transmitted through opto-couplers. Each channel can process count rates up to 65000 per sec, which corresponds to about 30 rem/h for γ-rays and neutrons, respectively . The whole area monitor as encircled in Fig. 1 is housed in a box with a transparent front cover.

The dose equivalents of the individual area monitors are read by the central microcomputer via a multiplexer. Here, the data are processed for printing, display on TV monitors and recording on floppy disk. The printed data give information on the near past, whereas long time records are taken from the floppy disks. The TV monitors, with a turnover time of 10 sec, display the actual dose rate situation at the radiation protection office and the control rooms of the Tandems. Here the high dose level is also indicated on inspection panels of the door interlock systems. Fig. 3 shows the inspection panel of the 12 MV Tandem and the postaccelerator. 12 monitors survey this area. If a high dose threshold is surpassed, the red LED within the radiation sign lights up.The doors adjacent to that area can be locked either by switches on the console or automatically. Their state (open, locked and closed, locked but open) are also signaled by different LED. In the automatic mode the beam is interrupted if a door that should be

A Microprocessor Based Area Monitor System

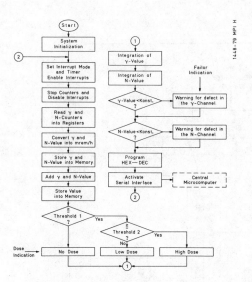

Fig. 2: Flow chart of the area monitor microprocessor program.

closed stays open for more than 2 minutes. In the experimental area, the low and high radiation levels are indicated by acoustic and light signals inside the individual target places and by a kind of traffic lights at the entrance doors to these places. Green light "no dose", yellow light "low dose", red light "high dose".

The microprocessor-based area monitor system provides high flexibility, which is desirable in view of the rather frequent changes in accelerator experiments.

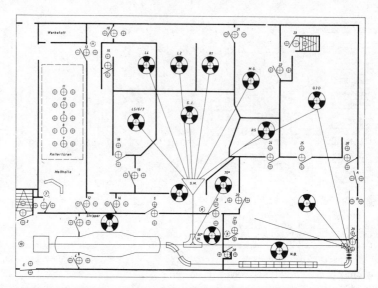

Fig. 3: Inspection panel of the MP Tandem.

ACKNOWLEDGEMENT

We are very much indebted to W. Heinecke and W. Schreiner for their help and fruitful discussions.

A PERSONAL TRITIUM MONITOR
R. V. Osborne and A. S. Coveart

A tritium monitor, similar in size to a normal gamma survey meter, is being developed as part of the systematic approach to providing an improved capability for measuring tritiated water vapour (HTO) near workers in CANDU nuclear power plants. The prototype is shown in Fig. 1.

Tritium, formed by neutron capture in deuterium, is present in the reactor heavy water so that small vapour leakages may result in unexpected airborne contamination in some working areas. The requirement is for a quick indication of concentrations of tritiated water vapour

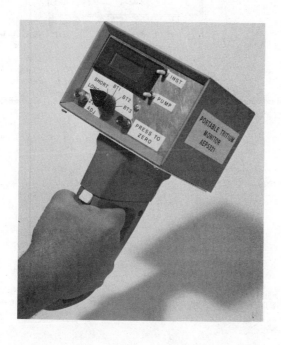

FIGURE 1. Prototype Personal Tritium Monitor

in air that are the order of one maximum permissible concentration and above, wherever a worker happens to be. Sensitive methods have already been developed for sampling [1] and for monitoring with a centrally-located instrument [2]. However, size, mass and cost are more important than sensitivity in an instrument that is designed to meet this requirement as opposed to a central monitor. Tritium monitors, even portable ones developed and described to date, have not been such that they can be available in large quantities. The instrument described here is intended to complement other monitoring methods by allowing on-the-spot assessments of tritium hazards and is designed to meet the severe constraints posed by the needs for portability and availability.

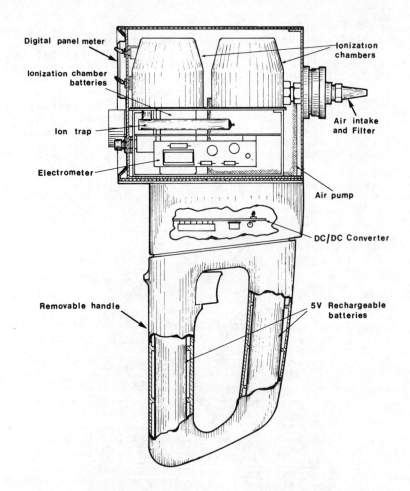

FIGURE 2. Mechanical assembly of the tritium monitor

As illustrated, the instrument is small enough to be easily handled and weighs only 2 kg. Only inexpensive, readily-obtainable mechanical and electrical components have been used in the instrument. Components need little machining and the assembly is uncomplicated.

The mechanical assembly is shown in Fig. 2. The tritium detector is an 80 cm^3 ionization chamber made from a nickel crucible (as used in chemical analyses) soldered to a fibre-glass printed circuit board on which guard rings and electrical connections have been etched and the copper nickel plated. No O-ring seals are used in the entire assembly which may be thrown away if it should be contaminated. Sample air is sucked through the molded plastic filter assembly and the cylindrical ion trap to the tritium ionization chamber by the loudspeaker coil-driven air pump [3]. A second ionization chamber, also 80 cm^3, is adjacent to the tritium chamber and, in the prototype shown, is sealed. The two ionization chambers are oppositely polarized so that the current from the sealed chamber partially cancels the background current in the tritium chamber caused by ambient gamma radiation. The electronics, on a 3 cm x 7 cm printed circuit board, are mounted under the ionization chamber assembly. The chamber and electronics can be removed as a single package for maintenance or replacement. The power supply is an inexpensive power handle, as sold for portable tools [4], that includes a rechargeable nickel-cadmium cell. The required voltages are derived in the DC/DC converter mounted in the mating assembly to the power handle to which the instrument is attached. The power handle may be detached for recharging whilst other handles are used. The switch on the power handle may be used to control intermittent operation of the pump.

The electrical schematic is outlined in block form in Fig. 3. The 4.2 V battery in the power handle drives the air pump directly and is converted to 5.6 V and 10 V for the electrometer and digital display meter. As shown, the major current drain from the power handle is the air pump. The battery lifetime with the current drain shown is about 5 hours. If an external (hand-operated) air pump is used then the battery lifetime is an order of magnitude longer.

The MOSFET input stage (3N155As, connected differentially as source followers) and one amplifier of a quad-amplifier DIP, LM324, comprise an operational amplifier with high input impedance and an effective feedback resistance of 4 TΩ. A smoothing circuit may be switched into the output stage to lengthen the electronic response time constant from less than one second to 3.3 seconds. The controls shown are on the monitor front panel; three additional positions on the slow/fast switch allow the battery voltages to be monitored.

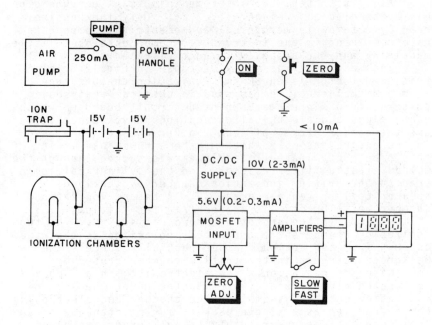

FIGURE 3. Block diagram of the electrical schematic of the tritium monitor

The sensitivity of the ionization chamber/amplifier combination is 3.2 mV/(MPC)$_a$. (The (MPC)$_a$ used is 10 μCi/m^3 [0.37 MBq/m^3]). The input to the 3½ digit meter is scaled to give an output range from 1 to 1999 (MPC)$_a$.

REFERENCES

1. Osborne, R.V., "Sampling for tritiated water vapour", Proceedings, Third International Congress of IRPA (Washington, D.C., 1973) 2 1428.
2. Osborne, R.V., Central tritium monitor for CANDU nuclear power stations, IEEE Trans, Nucl. Sci, NS 22 (1975) 676.
3. Spectrex Corp., Redwood City, California 94063, USA.
4. Black and Decker Inc.

MEASUREMENT OF ABSORBED DOSE-RATE IN SKIN FOR LOW-LEVEL BETA-RAYS

K. Shinohara, Y. Kishimoto, Y. Kitahara and S. Fukuda

Health and Safety Division, Power Reactor and Nuclear Fuel Development Corporation, Tokai-works, Tokai-mura, Ibaraki-ken, Japan

A new type of beta-ray absorbed dose-rate meter has been manufactured applying the detection method developed by K. Bingo et. al. (1) to evaluate the absorbed dose in skin at a depth of 7 mg/cm^2 lying above a contaminated sandy beach.

The instrument uses a plastic scintillator with 2.5 mm thick and a single channel pulse height analyzer(SCA) to obtain the best correlation between the instrument response and the absorbed dose-rate.

The absorbed dose-rates for beach sands were measured by the instrument and were compared with the calculated values by the computer code "BETA-SAND" developed by PNC and JAERI (2). With this instrument the absorbed dose-rate of about 2 μrad/hr can be measured by 60 minutes counting.

METHOD AND RESULTS

The beach sands were sampled at eight points around PNC Tokai-works and were dried at about 80°C by a drying oven. The concentration of potassium-40 in the samples were determined by a Ge(Li) spectrometer. The concentration of K-40 in the samples showed a range from 6.3 to 14.9 pCi/g-dry.

The schematic block diagram of the beta-ray absorbed dose-rate meter used in this experiment is shown in Fig. 1.

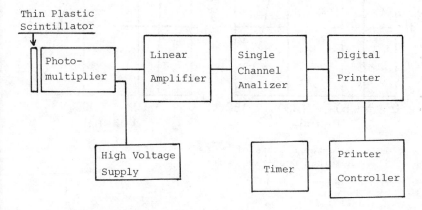

Fig. 1 Schematic Block Diagram of the Beta-ray Absorbed Dose-Rate Meter

The reference beta-ray sources shown in Fig. 2 such as Sr-90/ Y-90, Cs-137, Pm-147 and Tl-204 were used to determine the discrimination level and the window width of the SCA. The absorbed dose-rates in skin above the sources were determined by the theoretical value calculated by W. G. Cross (3). The conversion factor obtained from the calibration was 0.4 μrad/hr per cpm for the maximum beta-ray energy ranging from 0.5 to 2.27 MeV. The beta-ray spectra of Cs-137 and Tl-204 measured by the instrument are shown in Fig. 3.

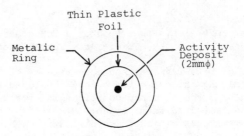

Fig. 2 Reference Beta-ray Source (LMRI made)

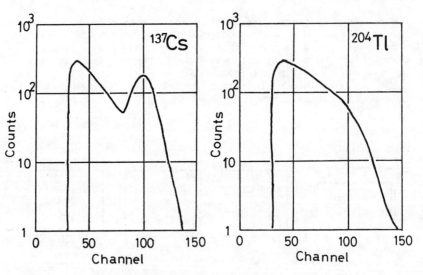

Fig. 3 Beta-ray Spectra of ^{137}Cs and ^{204}Tl measured by the Absorbed Dose-rate Meter

Measurement of Absorbed Dose-Rate

The absorbed dose-rates of the samples were measured by following steps;

(1). Counting the background (C_B) for 60 minutes,

(2). Counting the sample including the background (C_{BS}) for 60 minutes, and

(3). Calculation of the net counts (C_S)
$$C_S = C_{BS} - C_B$$
and the absorbed dose-rates (D)
$$D = (C_S/60) * (\text{conversion factor})$$

An experiment was done to know the relation between the absorbed dose-rate and the depth of sands. The sample prepared for the experiment was adsorbed K-40 on the sand particles. As shown in Fig. 4, the result presents that the absorbed dose-rate is independent with the depth of sands when it is greater than 0.4 g/cm^2.

The dominant natural radionuclide contributing to the absorbed dose is considered to be K-40 (4), so that we compared the measured values and the calculated K-40 dose-rates in the sands. The results of this experiment is shown in Fig. 5 and presents a good correlation.

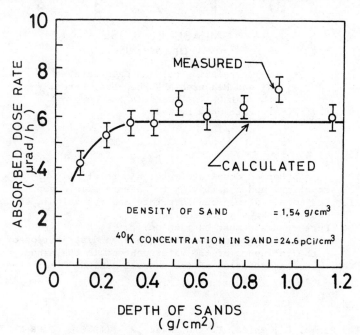

Fig. 4 Change of Absorbed Dose-Rate with Depth of Sands. "CALCULATED" is the Dose Rate by the Beta-ray from Potassium-40 Adsorbed on Particles of Sand.

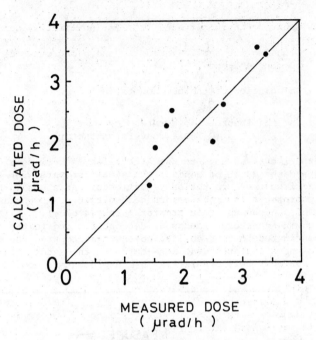

Fig. 5 Correlation between Calculated and Measured Absorbed Dose Rates. "CALCULATED DOSE" is the Dose Rate only by K-40.

SUMMARY

(1). This absorbed dose-rate meter has relatively high stability and sensitivity to measure skin doses.
(2). The measured absorbed dose-rate and the calculated dose-rate has shown a good correlation.
(3). Further study must be made on;
- the influence of the particle size distribution of sands,
- the influence of the water content in the sands, and
- the contribution of the natural radionuclides other than K-40.

REFERENCES

(1) Bingo, K., Chida, T. and Kawai, K. (1976): JAER-M-6753
(2) Itoh, N. et. al. (1977): JAERI-memo-7389
(3) Cross, W. G. (1967): AECL-2793
(4) Miyanaga, I. et. al. (1976): JAERI-memo-6842

AVANTAGES PRESENTES PAR L'INTRODUCTION INDUSTRIELLE DE L'INFORMATIQUE DANS LA SURVEILLANCE CENTRALISEE DES NIVEAUX DE RAYONNEMENTS

H. Vialettes and P. Leblanc

L'expérience acquise au CEN/SACLAY depuis plus de vingt ans montre que dans les installations nucléaires importantes, la surveillance des risques d'irradiation et de contamination radioactives doit être permanente et centralisée. La technologie a, évidemment, considérablement évolué au cours du temps, et récemment, le Service de Protection contre les Rayonnements du CEN-SACLAY a défini un type de matériel entièrement nouveau, faisant notamment appel aux techniques numériques. Une réalisation pilote de 20 voies de mesure a été décrite lors de l'exposition NUCLEX 75 /1/ et les résultats expérimentaux obtenus par la méthode de traitement numérique pour le contrôle de la radioactivité de l'air ont été exposés lors d'un colloque de l'AIEA /2/. La présente communication concerne une réalisation industrielle de 120 voies de mesure assurant la surveillance effective de l'usine de préparation de radioéléments du CEN-SACLAY.

PRESENTATION DU MATERIEL.

La figure jointe décrit le fonctionnement général de l'ensemble regroupant le "Tableau de Contrôle des Rayonnements (T.C.R.) et le Tableau des Contrôles Techniques (T.C.T.) de l'installation. Cet ensemble comprend :
- 62 voies de mesure pour le contrôle de la contamination de l'air par les aérosols.
- 54 voies de mesure pour le contrôle de l'irradiation externe.
- 4 voies de mesure pour le contrôle de la contamination de l'air par les gaz.
- 279 entrées de contrôles techniques (portes, niveaux de cuves à effluents, ventilations).

Le dispositif de mesure est bâti autour de deux calculateurs identiques, l'un d'eux ayant le rôle de calculateur de mesure. Tous les capteurs qui ont leur électronique propre, délivrent le même signal impulsionnel standardisé de telle sorte qu'une banalisation des voies de mesure est obtenue sur toute la chaîne, capteur exclu bien entendu. Au centralisateur, les informations d'entrée sont distribuées, en parallèle, aux deux calculateurs qui les traitent simultanément et de manière indépendante. Seul un système de surveillance mutuelle des calculateurs par "chien de garde" permet une commutation automatique à tout moment du calculateur maître (celui qui dispose des périphéries) vers le calculateur esclave.

/1/ H. JOFFRE - NUCLEX - Bâle, octobre 1975

/2/ H. VIALETTES and al. - IAEA-SM-217/13

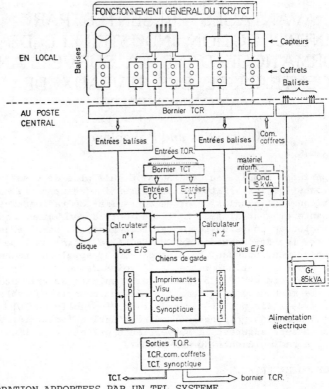

AMELIORATION APPORTEES PAR UN TEL SYSTEME.

Qualité de l'information : Les indications brutes des détecteurs sont converties, par le calcul, en niveaux d'exposition exprimé en nombre de LMA et en expositions cumulées exprimées en nombre de LMAh. Cette conversion tient compte de paramètres d'exploitation tels que facteur de correction de position du capteur, facteurs de rendement de filtration et de détection et facteur de radiotoxicité ; elle nécessite des algorithmes de traitements spécifiques pour chacun des deux types de voies de mesure ; elle aboutit dans tous les cas à des résultats homogènes et plus complets puisqu'elle permet d'exprimer d'une part le niveau instantané en rem/h et en CMA, et, d'autre part, l'exposition cumulée en rem et en CMAh. Grâce à ces résultats, la signalisation locale est relative soit à un débit de dose soit à une activité volumique dans l'air de sorte qu'elle est directement reliée, dans les deux cas, à une limite dérivée des recommandations de la CIPR, et qu'elle a donc une même signification vis-à-vis des risques pour les travailleurs.

Dans le cas de la contamination atmosphérique, l'algorithme de calcul permet, à partir de l'activité déposée sur un filtre fixe, d'évaluer au mieux - c'est-à-dire en temps réel et compte tenu des fluctuations statistiques - les concentrations de ces radionucléides dans l'air. Par conséquent, le système décrit permet d'assurer le suivi de l'évolution d'une contamination accidentelle, en particulier d'observer la phase de décroissance, sans qu'aucune intervention sur le dispositif de prélèvement ne soit nécessaire (figure ci-contre). Parallèlement, il est clair que le système informatisé est rapidement disponible et permettrait de mettre en évidence une deuxième bouffée de contamination, même à un niveau de quelques CMA, intervenant peu de temps après la première ; avec le système classique, au contraire, le comptage correspondant à cette deuxième bouffée risquerait de ne pas pouvoir être distingué des fluctuations statistiques du comptage correspondant à l'activité déjà déposée sur le filtre de prélèvement.

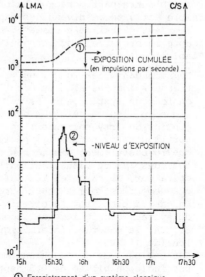

① Enregistrement d'un système classique
② Enregistrement du système informatisé

GESTION DES RESULTATS.

Outre les avantages bien connus qu'apporte l'utilisation de l'informatique dans la gestion en temps réel des résultats et dans leur archivage, la conception du système permet de ne pas encombrer, d'informations sans intérêt, les supports d'exploitation directe ou différée, comme l'étaient les rouleaux d'enregistrement des anciens systèmes. Pour autant, l'exploitant n'est privé d'aucune information grâce à des possibilités très riches de dialogue avec le système informatique mais surtout grâce aux trois dispositions complémentaires suivantes :

- la visualisation permanente des niveaux d'exposition (en clair et sous forme d'histogramme) et des états de toutes les voies de mesure.
- l'enregistrement permanent sur disque magnétique de toutes les informations des voies de mesure ; cet enregistrement permet de reconstituer intégralement l'historique de l'une quelconque des voies de mesure sur simple commande opérateur.
- le tracé en temps réel des courbes de variation des niveaux d'exposition par commande manuelle ou par commande automatique à la suite du dépassement d'un seuil de consigne.

SOUPLESSE D'UTILISATION ET FACILITES D'EXPLOITATION.

La standardisation des voies de mesure jusqu'au niveau du capteur permet de brancher, sans aucune modification mécanique ni électronique, n'importe quel type de capteur sur n'importe quelle voie. La seule intervention consiste à taper au clavier de commande la nouvelle identification de la balise de façon que ses informations soient traitées par le programme ad hoc.

D'autre part, grâce à la présence d'une source de contrôle de faible activité au niveau des capteurs, le contrôle de bon fonctionnement est global et permanent. Le traitement numérique prend en compte le bruit de fond introduit par cette source.

Fiabilité.

- La redondance des fonctions vitales (acquisitions, traitement et transmission des alarmes) et l'autocontrôle du système informatique en font un système sûr. De ce point de vue, l'arrêt d'un périphérique ou d'un calculateur ne perturbe en rien le système de contrôle continu des rayonnements ; en effet, le basculement d'un calculateur sur l'autre avec tous les périphériques est automatique, de même que le basculement des tâches d'une imprimante sur l'autre ; pour les consoles de visualisation, il est possible, sur simple commande opérateur, de basculer les informations de l'une sur l'autre, le changement de format étant automatique.
- En cas de panne sur le circuit d'alimentation générale, l'ensemble du système décrit ici est secouru par un groupe électrogène de 85 kVA, le matériel informatique étant lui-même secouru par un moduleur de 15 kVA.
- Sur une période de fonctionnement de 10 000 heures, on a observé une indisponibilité de 0,5 heures, ce qui confirme les calculs de fiabilité effectués par le fournisseur qui aboutissaient à 0,72 heures d'indisponibilité pour 10 000 heures de fonctionnement.

Prix de revient.

En janvier 1977, le système complet décrit ci-dessus a coûté, pour la partie TCR, 5 millions de francs se répartissant en 37 % pour les capteurs, 16 % pour les coffrets de signalisation, 31 % pour le centralisateur (matériel + logiciel), 16 % pour l'installation et la maîtrise d'oeuvre. Il est intéressant de noter que :
- ce coût est inférieur d'environ 30 % à une installation identique en version classique, sans traitement de l'information.
- par la réunion du TCR et du TCT, qui permet notamment une gestion commune, le prix de revient du TCT a été notablement diminué.

CONCLUSIONS.

Le contrôle centralisé des rayonnements par des méthodes de traitement numérique apporte des améliorations fondamentales pour la surveillance d'une installation nucléaire d'une certaine importance. Ces améliorations sont sensibles, tant sur le plan de la qualité de l'information, qui joue directement sur la radioprotection des travailleurs, que sur le plan de l'exploitation centralisée qui intervient directement sur la qualité des prestations de radioprotection. Etant donné qu'il est démontré que la fiabilité d'un tel système est excellente et que son prix de revient est sensiblement inférieur à celui d'un système classique donnant des informations beaucoup plus simplistes, il nous paraît clair que la généralisation s'impose pour toutes les installations.

Une variante de ce système existe en version décentralisée pour ce qui est du traitement de l'information et de la signalisation. Il est encore trop tôt pour la comparer valablement au système décrit ici, car elle n'a pas encore été appliquée à une réalisation vraiment industrielle et sa fiabilité n'a pas fait, non plus, l'objet des études mentionnées ici.

"INTELLIGENT" RADIATION INSTRUMENTS

A. Ward

Any activity involving the exchange, storage and analysis of information is likely to be strongly affected by the development of microprocessor technology. In the field of radiation protection we are concerned with information in the form of medical records, dose records, approved schemes of work and lists of classified workers. We are also concerned with the collection and examination of information from a wide range of radiation monitoring instruments. We are even concerned to an increasing extent with information being exchanged from one machine to another with only the most casual human interference. Further development of microprocessors is occurring and it seems possible that quite intelligent machines may be available in the next decade for such major tasks as the complete automation of nuclear waste management and perhaps also to assist in the eventual dismantlement of the present generation of nuclear power reactors as they come to the end of their useful life. In this paper a description is given of three straightforward applications of current microprocessor technology. These applications are characterised by the use of the microprocessor to impart a degree of intelligence to what would otherwise be conventional radiation detection techniques.

GEIGER COUNTER

We are all very well aware that the output from a geiger counter consists of a series of identical pulses. Furthermore, considerable background information with regard to the type of radiation sources involved is necessary to infer a value for the exposure dose rate from the observed counting rate. In the hands of an unskilled person a serious error can occur in estimating the radiation exposure. The arrangement described here turns over to the micro computer the task of computing the radiation dose from the observed counting rate.
Slide 1 is a schematic diagram of the system. Two identical geiger tubes are employed but in front of one of the tubes an absorber is placed which has the effect of causing the ratio of the counting rates on the two tubes to depend significantly on the effective gamma ray energy in the incident radiation. Each output pulse from either tube causes an "interrupt" to the main program on the micro computer which promptly postpones its main program and operates an auxiliary program which "services" the interrupt. The micro computer is programmed to keep a record of the number of pulses from each of the two geiger tubes and to form a ratio of these two numbers. The observed ratio in a particular situation is compared with stored data and in this way the incident gamma ray energy is estimated. The program then continues using additional stored data on the sensitivity of the geiger tube as a function of gamma energy to correct the counting rate (on the tube with no absorber) and thus obtain the required

dose rate. The dose rate thus obtained is then printed out in a
convenient form together with auxiliary information which includes
the maximum permissible exposure level, any special precautions which
are relevant and the results of a check made by the micro computer
into whether the entire system is functioning correctly.

SCINTILLATION COUNTER

The spectrum of pulses from a scintillation counter contains
information concerning individual gamma energies. However, the cost
of conventional multi channel pulse height analysers can be considerably more than the cost of the scintillation counter itself. It is
possible to use a micro computer not only to carry out the function of
obtaining the pulse height distribution but also to go a stage further
and employ a pattern recognition program to identify the incident
gamma ray energies and thus the radioisotope or isotopes involved.

Slide 2 is a schematic diagram of a typical arrangement. As
each pulse emerges from the counter it is held for a period at its
peak height and during this period an analogue to digital converter
produces a binary code number proportional to the pulse height.
This binary number is then latched on to an input port of the micro
computer. At the same time the converter produces a control pulse
which operates the interrupt request line on the micro computer.
This interrupt causes the main program to stop temporarily and a
service program comes into action. This service program reads the
binary number at the input port and compares its value with a list
of all possible values (channel numbers). When the matching stored
value is found the content of a linked memory location is incremen-
ted by one. The interrupt terminates and the micro computer returns
to the main program where it is ready for a further pulse. The contents of the linked memory locations form a conventional pulse height
distribution. The period of counting is controlled by a timer in
the micro computer. At the end of the counting period the micro computer changes to a program which is essentially a pattern recognition
search using stored data on the pulse height distribution observed
for different gamma ray energies. There are fairly strict limits on
pattern recognition programs before the problem of identifying the
gamma rays simply becomes too complex. However, in radiation protection situations it is usually the case that a limited number of radioisotopes are likely to be involved. For example, the laboratory concerned may use Cobalt-60, Caesium-137 and Iodine-125. Pattern recognition programs to identify these radioisotopes are quite straightforward.

The system described here can be left to itself to operate as a
long-term monitor so arranged that it will generate, from within the
micro computer, audible and visual warnings in a hazard situation.
Thus, for example, it is possible to set the system to give priority
to the detection of the radiation from Iodine-125 with its wellknown
associated thyroid hazard.

MONITOR CALIBRATION BENCH

The calibration of radiation monitors is an important part of
a radiation protection service. However, it is rather time consuming
when carried out rigorously and there is a certain risk of radiation

"Intelligent" Radiation Instruments

exposure particularly in the calibration of monitors intended to measure high dose radiation. These practical difficulties are such that many radiation protection laboratories make only the most casual check on whether the monitor is performing satisfactorily.

The new legislation emerging in the European Community following the Euratom Directive and the response to the Directive from national governments emphasises the need for rigorous calibration of radiation monitors.

Slide 3 shows in schematic form an arrangement for monitor calibration based on micro computer control which is currently under construction in the authors laboratory. The arrangement is simple and straightforward. A large number of standardised radiation sources are mounted on a slim rod which passes between two heavily absorbing rods which have a small gap between them. The arrangement is such that only one source can be exposed at any one time within the gap. The radiation monitor to be calibrated is carried forward through a series of halt positions. At each halt the monitor output is read and stored together with a record of the monitor halt location and the source involved. After measurements have been made at each halt the source is changed by moving the rod and the set of measurements repeated. This procedure is continued until the output of the monitor has been recorded (for a group of sources relevant to that monitor) at all halt locations. The entire procedure is controlled by the micro computer which also stores the data and prints out the results in a very convenient form. Initially all of the sources are shielded by one of the thick absorbers so that the operator is not significantly exposed when placing the monitor to be tested in position. The procedure once started is fully automatic under the control of the micro computer which also provides an audible and visual warning while calibration is in progress.

A very important point is that since the procedure is under software control by a program stored in the micro computer it is quite easy to arrange to store a number of programs one for each type of monitor brought to the system for calibration. It is this flexibility coupled with the convenient facilities of the micro computer which renders it much more suitable than any other technique for this type of control. It is also the case that the cost of micro computer electronics is extremely low. The micro computer employed in all three applications described here cost only a few hundred dollars.

SPECTRA, DIFFERENTIAL ALBEDO AND SHIELDING DATA FOR BREMSSTRAHLUNG SCATTERED FROM COMMON SHIELDING MATERIALS

H.-P. Weise, P. Jost and W. Freundt

INTRODUCTION

In the design of radiation shields for X-ray machines and electron accelerators not only the primary radiation but also the radiation scattered from irradiated material must be taken into account. The thickness of shielding barriers against secondary radiation can be calculated if the intensity and the quality of the radiation are known. For radiation protection purposes the scattered intensity is most conveniently described in terms of the differential albedo:

$$d\dot{J}_s = A_{JX}(E, \Theta_o, \Theta, \varphi) \, \dot{J}_o \, \frac{df \cos\Theta_o}{r^2}$$

A_{JX} differential exposure albedo

$d\dot{J}_s$ scattered exposure rate at a distance r from the surface area df of the irradiated material

E energy of incident photons

Θ_o angle of incidence; Θ polar angle of exit

φ azimuthal angle of exit

Θ_s scattering angle $\Theta_s = \pi - \Theta_o - \Theta$ for $\varphi = 0$

In the case of very high energy photons it is more appropriate to replace the exposure by the absorbed dose in a tissue equivalent material.

Albedo data were measured for bremsstrahlung with maximum photon energies in the ranges from 100 keV to 400 keV and from 10 MeV up to 35 MeV. In the low energy region spectra of scattered radiation were measured with a semiconductor detector. From these spectra attenuation curves were calculated using known broad beam transmission factors for monoenergetic photons. For high energy bremsstrahlung (10 MeV to 35 MeV) the quality of scattered radiation was determined by direct attenuation measurements.

EXPERIMENTAL

Bremsstrahlung with maximum photon energies between 100 keV and 400 keV was produced by means of a conventional industrial 400 kV X-ray machine (inherent filtration: 4 mm Al). The beam profile of the primary radiation was defined by a series of matched lead collimators.

The size of the irradiated area df was determined by exposing an X-ray film on the surface of the scattering material. Ion chambers with a flat energy response were used for measuring the exposure rate of primary and scattered radiation. The differential exposure albedo of ordinary concrete, barytes concrete, iron and lead was determined for normally incident radiation ($\theta_o = 0$) as a function of the polar angle of exit θ as well as for the special geometry $\theta_o = 45°$, $\theta = 45°$, $\varphi = 0°$.

High energy bremsstrahlung was produced using a 35 MeV travelling wave linear accelerator. The electron beam is deflected by an achromatic magnet system and focused on a tantalum bremsstrahlung target (7 mm thick) where it is completely absorbed. The cross section of the photon beam is defined by several steel collimators which are coaxially mounted in a channel through the concrete wall (3 m thick) between the accelerator hall and the experimental area. The dose rate of the incident and the scattered radiation is measured with ion chambers surrounded by a sufficient amount of tissue equivalent buildup material.

RESULTS AND DISCUSSION

Some selected albedo values are compiled in Fig. 1 to Fig. 3 and Table I for normally incident radiation.
In the low energy region there is a complicated relationship between the exposure albedo, the radiation energy and the scattering material (1), but for $\theta_o = 0$ the scattered dose rate always increases with increasing scattering angle θ_s (Fig. 1). Between 10 MeV and 35 MeV the differential albedo of materials with low mean atomic number (ordinary concrete, brick) decreases with increasing energy (Fig. 2). In the case of lead however (Fig. 3) higher scattering intensity is observed at higher energies because at high primary photon energies the main contribution to the albedo of high-Z material is due to the production of secondary bremsstrahlung which rapidly increases with energy (5).
Fig. 4 and Fig. 5 show photon spectra of low energy bremsstrahlung scattered from ordinary concrete. Whereas the quality of radiation scattered from low-Z material is determined by compton scattering, the spectra from high-Z materials (barytes concrete, lead) show a large contribution of characteristic X-radiation to the total scattered intensity. As an example calculated attenuation curves for low energy photons scattered from ordinary concrete are plotted in Fig. 6. They reasonably agree with measured values (3). Similar shielding data were calculated for several combinations of scattering and shielding materials.

TABLE I. Differential exposure albedo for low energy bremsstrahlung normally incident on common shielding materials ($\Theta_o = 0$)

tube voltage U [kV]	Θ_s[deg.]	barytes concrete	steel	lead
100	110	$1,5 \cdot 10^{-2}$	$2,6 \cdot 10^{-3}$	$7,7 \cdot 10^{-3}$
	135	$2,5 \cdot 10^{-2}$	$5,2 \cdot 10^{-3}$	$1,5 \cdot 10^{-2}$
	160	$3,1 \cdot 10^{-2}$	$7,5 \cdot 10^{-3}$	$2,0 \cdot 10^{-2}$
200	110	$1,3 \cdot 10^{-2}$	$3,7 \cdot 10^{-3}$	$9,5 \cdot 10^{-3}$
	135	$2,2 \cdot 10^{-2}$	$6,8 \cdot 10^{-3}$	$1,7 \cdot 10^{-2}$
	160	$2,9 \cdot 10^{-2}$	$1,0 \cdot 10^{-2}$	$2,1 \cdot 10^{-2}$
400	110	$9,3 \cdot 10^{-3}$	$6,1 \cdot 10^{-3}$	$7,1 \cdot 10^{-3}$
	135	$1,6 \cdot 10^{-2}$	$1,1 \cdot 10^{-2}$	$1,2 \cdot 10^{-2}$
	160	$2,1 \cdot 10^{-2}$	$1,5 \cdot 10^{-2}$	$1,6 \cdot 10^{-2}$

REFERENCES

1. Wachsmann, F., Tiefel, H. und Berger, E. (1964): Fortschritte auf dem Gebiete der Röntgenstrahlen und der Nuklearmedizin, 101, 308.
2. Vogt, H. G. (1972): Nucl.Eng.Design 22, 138.
3. Papke, W. H., unpublished data.
4. Karzmark, C. J. and Capone, T. (1968): Br.J.Radiol. 41, 222.
5. Maruyama, T. et al. (1975): Health Phys. 28, 777.

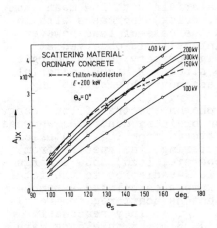

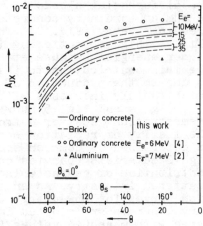

Fig. 1 Differential exposure albedo of ordinary concrete for low energy bremsstrahlung

Fig. 2 Differential dose albedo of concrete and brick for high energy bremsstrahlung

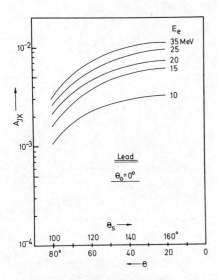

Fig. 3 Differential dose albedo of lead for high energy bremsstrahlung

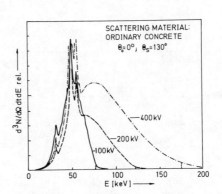

Fig. 4 Spectra of low energy bremsstrahlung scattered from ordinary concrete as a function of tube voltage

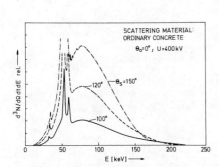

Fig. 5 Spectra of 400 kV bremsstrahlung scattered from ordinary concrete as a function of scattering angle

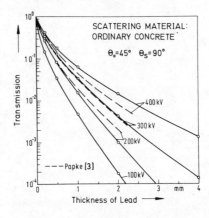

Fig. 6 Attenuation curves for scattered radiation. Scattering material: Concrete; shielding material: lead

Emergency Planning

EMERGENCY PLANNING AND PREPAREDNESS: PRE- AND POST-THREE MILE ISLAND

H. E. Collins

Prior to the accident at the Three Mile Island nuclear generating station, radiological emergency response planning and attendant preparedness as it relates to nuclear facilities, was never in a position of high visibility within the nuclear industry or within the Federal, State and local governments in the U.S. Further, very few resources, in terms of personnel and funds, were devoted to it. There were a variety of reasons for this state of affairs.

First and foremost, were the two long cherished notions: (1), that nuclear facilities were designed, constructed and operated with such integrity, <u>the chances of a serious accident occurring were extremely remote</u>; and (2), that even if an accident were to happen, because of the integrity of design, construction and operation, <u>any accident would have little effect in terms of offsite radiological consequences</u>. Although the record of nuclear power safety is excellent in general terms, it is not flawless and <u>we have been given some serious warnings</u>.

The first of these two notions, that is "chances" or "probabilities" of accidents happening, has, in my view and the views of others, been essentially "knocked into a cocked-hat." Two relatively serious events, in terms of "chance", have occurred in large power reactor facilities in this country within the last four years: the serious fire at the Browns Ferry nuclear power facility and the accident at the Three Mile Island nuclear power facility.

The corrolary or second of these two notions, that is that little would happen in terms of offsite consequences, is to some measure still supported by the integrity of the facilities themselves. One cannot say too much with respect to the role and actions of operators and nuclear facility management during both of these events, except to say that tardy notification of offsite organizations occurred, some correct moves were made, but at the same time, many incorrect moves were also made. The point to be made here is that we were all very fortunate in both of these accidents in that offsite radiological consequences were either non-existent or relatively minimal. However, we came uncomfortably close in both of these accidents to potential consequences that could have caused grievous harm to individuals, our society, our environment, and our national energy program.

The warning has clearly manifested itself. Dr. Stephen Hanauer, of the NRC, who was the Chairman of the NRC Special Review Group (of which I was a member), which prepared the report (NUREG-0050) (1) concerning the fire at the Browns Ferry nuclear power facility, remarked at one point during that investigation, with words to the effect -- "Maybe it was like a mild heart attack -- it woke us up." We have had a second "mild heart attack" at Three Mile Island. So, it behooves all of us, industry, government and everyone else involved, to learn from this experience because we may not get another chance to improve matters in the interim, should another accident occur -- especially a fast-breaking accident, as opposed to the drawn-out Three Mile Island event.

Other reasons for a relatively weak radiological emergency response planning and preparedness program with respect to the operation of nuclear facilities, are rooted in long-seated deficiencies in general emergency planning and preparedness programs at the Federal, State and local government levels in the U.S. Notwithstanding the massive Federal emergency operational response and industry response at Three Mile Island, advance emergency planning and coordination leaves much to be desired. Initially at Three Mile Island, coordination between Federal, State and local authorities, was a problem.

General emergency planning and preparedness at the governmental levels has suffered a period which can be best characterized as relative "benign neglect," ever since the end of World War II. Civil Defense or Emergency Services programs at the Federal, State and local government level have fallen into disarray and mediocrity due to fragmentation of efforts, lack of motivation, lack of effective leadership, inadequate attention, and inadequate funding. This is partially the reason why the new U.S. Federal Emergency Management Agency (FEMA) was established on April 1, 1979. FEMA brings together the major Federal agencies who have had responsibilities in civil preparedness, continuity of government during a national emergency, and disaster control and mitigation.

Any radiological emergency response planning and preparedness program that is mounted, must depend ultimately on an adequate general emergency planning base, at Federal, State, and local government levels. Efforts to build a proper radiological emergency response posture in support of these nuclear facilities, has suffered because one cannot build a "golden idol" on "feet of clay." If the base is defective, which it is, the idol will not stand for very long, if at all.

Adequate, well-conceived general emergency planning and preparedness at all levels of government, to cover the wide range of hazards in our technological society, is the key to an improved radiological emergency response planning and preparedness program. The NRC and other technical agencies must and will work with the new FEMA to improve this program.

POST-THREE MILE ISLAND - PROBLEMS AND PROGRESS

I have presented the overriding problem in my foregoing remarks. But, there are a number of specific problems related to radiological emergency response planning and preparedness. All of these problems existed before the accident at Three Mile Island, but the accident has speeded-up progress in these areas. There are many problems, but let me discuss five of the more salient ones:

1. An Adequate Planning Basis

What is an adequate planning basis for radiological emergencies at fixed nuclear facilities? This question, (rephrased as -- "What kind of an accident at a nuclear facility should we plan and prepare for handling?") was essentially asked by many of the U.S. States and local governments, and their national organizations some years ago. This resulted in two Federal agencies, NRC and EPA, launching an effort to examine this question.

In August of 1976, a joint U.S. Nuclear Regulatory Commission/ U.S. Environmental Protection Agency Task Force on Emergency Planning was formally appointed to look into this matter. In December of 1978, after over two years of work, the joint NRC/EPA eleven-member Task Force unanimously concurred in and published its report, "Planning Basis for the Development of State and Local Government Radiological Emergency Response Plans In Support of Light Water Nuclear Power Plants" NUREG-0396/EPA-520/1-78-016. (2)

The "bottom line" on this Task Force report is, that there is no specific nuclear power plant accident that one can identify as being the accident for which plans and preparedness programs should be in place. Rather, the Task Force came down on the side of planning for consequences, with only minimal concern for the uncertainties of probabilities. And, to define an adequate, improved planning basis, the Task Force recommended that essentially generic Emergency Planning Zones (EPZs) be established around all nuclear power facilities in the U.S. The Task Force further determined and recognized that the U.S. Low Population Zone (LPZ) concept used for siting purposes had little real meaning in terms of offsite emergency planning and preparedness. The Task Force, in essence, rejected the concept of the "LPZ" for definitive and comprehensive emergency planning offsite. Further, the Task Force recognized the need to develop an emergency planning basis to address the so-called "Class 9" accidents, or accidents resulting in extensive damage to, or melting of, the nuclear fuel core.

This need for a capability to accommodate emergency situations beyond the so-called "design basis accidents" used in plant and site evaluation, makes generic rather than site specific areas appropriate. The Task Force decided that the establishment of Emergency Planning Zones (EPZs) of about 10 miles for the airborne "plume" radiological exposure pathway, and about 50 miles for the ingestion or food radiological exposure pathway would be sufficient to define the areas in which planning for the initiation of predetermined protective measures is warranted for any given nuclear power plant. The Emergency Planning Zone concept is illustrated in Figure 1.

As a side note and independent of the work of the NRC/EPA Task Force, the Swiss Federal Office of Energy, Nuclear Safety Division, was developing an Emergency Planning Zone concept very similar to the zones recommended by the NRC/EPA Task Force. The Swiss have 3 zones; an inner "Fast Alarm Zone" of 2 to 6 kilometers, a second zone of 20 kilometers (12.5 miles), and a third zone (for the ingestion pathway) with no radius prescribed.

Although not without some initial controversy and resistance from many quarters, the Task Force report is a major milestone along the way toward defining an adequate radiological emergency response planning basis. The report, and the recommendations contained in the report have been formally endorsed by the Commissioners of the U.S. NRC as of October 5, 1979, and were endorsed by the EPA Administrator on January 15, 1980. Plans are to establish these Emergency Planning Zones in the U.S.

2. Accident Assessment

Accident assessment has been, and continues to be, a problem area. Although defined as an essential emergency planning element in 1970 in the AEC (now NRC) emergency planning regulations 10 CFR 50 Appendix 'E' (3) for nuclear facility NRC licensees, and later in the former AEC's emergency planning guidance document for States and local governments, "WASH-1293" (now NRC publication "NUREG-75/111"), (4) much needs to be done to improve accident assessment, both onsite and offsite.

Steps are underway to improve this accident assessment capability. On the nuclear facility side, improved in-plant instrumentation specifically designed for assessing accident situations has been indicated and will now be required. On the Federal, State and local side, standardized offsite accident assessment techniques and systems need to be developed and improved, especially in the areas of coordination between agencies at all levels of government and in the evaluative/decisionmaking process. The coordination of accident assessment information must also be improved between the nuclear facility operator and the offsite agencies. Guidance concerning the types of emergency instrumentation which might be useful, and the acquisition of instruments and systems themselves, are needed in many localities.

Several programs are now moving to address these problems. Nuclear facility operators will be required to upgrade their emergency plans. Further, they will be required to implement the related recommendations of the NRC "Lessons Learned Task Force" (5) involving instrumentation to follow the course of an accident, and relate the information provided by this instrumentation to <u>emergency action level guidelines</u> (6) promulgated by the NRC. This will include instrumentation for post-accident sampling, high range radioactivity monitors, and improved in-plant radioiodine instrumentation since radioiodine can be a dominant radioisotope of concern in airborne radiological releases. The implementation of the "Lessons Learned" recommendation on instrumentation for detection of inadequate nuclear core cooling will also be factored into the emergency plan action level criteria.

Guidance in the area of radiological instrumentation and offsite accident assessment techniques for States and local governments, are being prepared by the Idaho National Engineering Laboratory under contract to the NRC. Plans are also afoot to test an inexpensive airborne radioiodine sampling and collection device, which together with an existing modified Civil Defense radiological instrument, has the potential to help provide quick, rough "go" - "no go" information to authorities responding to an accident in offsite areas where a radioiodine release may be the dominant radioisotope of concern in certain accidents. This portable device, invented and recently patented by researchers at the Brookhaven National Laboratory (7) under contract to NRC, is being independently evaluated by the Idaho National Engineering Laboratory. If the device passes muster, NRC has plans to put it into the existing inventory of civil defense radiological monitoring instruments currently available to State and local government personnel.

Recently, the Commission has approved relatively modest budget resources to allow us to proceed with a few "pilot-demonstrations" of the Lawrence Livermore Laboratories (LLL) Atmospheric Release Advisory Capability (ARAC) system. The system, in its ultimate form, is capable of providing rapid atmospheric and radiological consequence assessment offsite, thus freeing nuclear facility operators and State and local organizations from laborious "1890"-type operations, with maps, plastic sheets, overlays, and grease pencils, which is the "State-of-the-art" in many nuclear power plants today.

ARAC was employed by the U.S. DOE response team, on an ad hoc basis at Three Mile Island. NRC intends to establish the first pilot-demonstration of ARAC in the State of New York by installing ARAC computer terminals and other hardware in the New York State Emergency Operating Center, and a local government Emergency Operating Center located near Consolidated Edison's Indian Point Nuclear Power Facility.

3. Training

Since March 1, 1975, the NRC with the assistance of other Federal agencies, has conducted formal training programs for Federal, State and local government personnel in both radiological emergency response planning and operations. The training programs have been well received and are of excellent quality, thanks to competent and dedicated faculty members. Much remains to be done in terms of retraining because of the high turn-over (roughly 10% per year) among State and local government personnel and also to keep pace with new developments in the emergency planning and preparedness area. NRC's plans are to continue to improve these training programs and to develop new ones where necessary. Nuclear facility personnel training must also be accelerated and improved as well.

Related to training, is the matter of standardized exercise-scenarios to test emergency plans. The NRC is developing exercise-scenarios to realistically test onsite and offsite emergency plans which should result in improving the emergency response capability at all levels of government.

4. Funding

Adequate funding for general and radiological emergency response planning and preparedness has been a problem at all levels of government; Federal, State and local. The funding problem is particularly acute at the local government level, where often many of the involved personnel are low-paid employees, part-time employees or volunteers with meager resources available to them. The funding situation needs to be improved. The amount of money required for a substantial improvement in the radiological emergency planning and preparedness effort, (as a sub-set of general emergency planning and preparedness), does not appear to be staggering. As a matter of fact, it is very small when compared to the investment made in a single nuclear power unit, of say, 1000 Megawatts-Electric, the gross cost of which today is well over the one billion dollar mark, in today's dollars, and we have some 70 nuclear power facilities licensed to operate in this nation today, and many more under construction.

Where can these funds come from? -- and more importantly -- where should they come from?

Dr. Stephen Salomon, an Environmental Economist of the NRC's

Office of State Programs, has recently completed a year-long study of this matter. His report, which was released in draft form as "NUREG-0553" (8) in the spring of this year, one day before the Three Mile Island accident, examines this question of emergency planning funding in significant detail. His findings depict a wide range of funding situations, from relative "affluence" -- to "abject poverty," -- concerning personnel and resources to do a proper job in this area, particularly at local government levels. Even where funding was adequate, in some cases there was no motivation or encouragement to spend funds on radiological emergency response planning and preparedness. These problems have at their roots, the individual, political, social, governmental and industrial perceptions of the relative safety of a high technology facility. Three Mile Island has changed a lot of heretofore complacent views.

But in those communities with little available to them to improve matters, the recognition of a need to do more does not always translate to, or result in, improvement. Help is needed. And, although the Federal government can and should provide some assistance, the nuclear industry has an obligation to provide financial assistance as well. Dr. Salomon's report, "Beyond Defense-in-Depth", NUREG-0553, was published as a final NRC staff report in October, 1979. The report should be useful to not only those of us involved in the regulation and management of the nuclear industry, but to the new U.S. Federal Emergency Management Agency (FEMA), and the Congress of the United States.

5. Emergency Planning Guidance

The accident at Three Mile Island, has in great measure, validated existing emergency planning guidance. Existing guidance on Protective Action Guides (PAGs) (9) (10) for radiological exposure needs to be completed by the U.S. Environmental Protection Agency and the U.S. Department of Health, Education, and Welfare, agencies charged with this responsibility. A Federal policy on the administration of radioprotective drugs, such as the use of potassium iodide as a thyroid blocking agent in some circumstances, needs to be developed by DHEW who is also charged with this responsibility. (11) (12) The NRC/EPA Task Force recommendations on the establishment of Emergency Planning Zones, must and should be quickly adopted. Specific technical guidance, such as emergency instrumentation and accident assessment guidance, needs to be developed. Guidance on interdicting or controlling the accidental radiological exposure to humans via domestic animals and agricultural products in the food chain, needs to be developed as well.

SUMMARY

The last bastion of the often quoted "Defense-in-Depth" concept against consequences of accidents at nuclear facilities, which has governed the development of commercial nuclear power for two-and-one-half decades, is a proper and effective emergency planning and preparedness program with respect to these facilities. This bastion, has not received the support which it deserves. Proper and adequate emergency planning can help alleviate many of the fears surrounding the safe operation of nuclear power facilities. This accident has given us a golden opportunity to improve things and we must not fail, collectively, to take advantage of it and to learn from it; to act on it. We are unlikely to have another chance to do so.

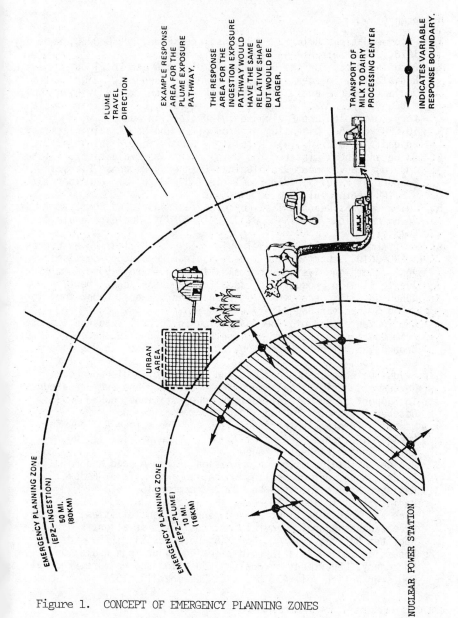

Figure 1. CONCEPT OF EMERGENCY PLANNING ZONES

REFERENCES

1. Recommendations Related to Browns Ferry Fire, NUREG-0050, February 1976, U.S. Nuclear Regulatory Commission.
2. Planning Basis for the Development of State and Local Government Radiological Emergency Response Plans in Support of Light Water Nuclear Power Plants, (NRC/EPA Task Force on Emergency Planning), NUREG-0396/EPA 520/1-78-016, December 1978, U.S. Nuclear Regulatory Commission and Environmental Protection Agency.
3. 10 CFR Part 50, Licensing of Production and Utilization Facility, Appendix E, U.S. Nuclear Regulatory Commission, Washington, D.C.
4. Guide and Check List for the Development and Evaluation of State and Local Government Radiological Emergency Response Plans in Support of Fixed Nuclear Facilities, NUREG-75/111, December 1974, U.S. Nuclear Regulatory Commission.
5. TMI-2 Lessons Learned Task Force Status Report and Short-term Recommendations, NUREG-0578, July 1979, U.S. Nuclear Regulatory Commission
6. Draft Emergency Action Level Guidelines for Nuclear Power Plants, NUREG-0610, September 1979, U.S. Nuclear Regulatory Commission.
7. An Air Sampling System for Evaluating the Thyroid Dose Commitment Due to Fission Products Released from Reactor Containment, NUREG-CR-0314, December 1978, Brookhaven National Laboratory for the U.S. Nuclear Regulatory Commission.
   Environmental Radioiodine Monitoring to Control Exposure Expected From Containment Release Accidents, NUREG-CR-0315, December 1978, Brookhaven National Laboratory for the U.S. Nuclear Regulatory Commission.
8. Beyond Defense-in-Depth, Cost and Funding of State and Local Government Radiological Emergency Response Plans and Preparedness in Support of Commercial Nuclear Power Stations, NUREG-0553, October 1979, U.S. Nuclear Regulatory Commission.
9. Manual of Protective Action Guides and Protective Actions for Nuclear Incidents, EPA-520/1-75-001, September 1975, U.S. Environmental Protection Agency.
10. Accidental Radioactive Contamination of Human Food and Animal Feeds, Department of Health, Education and Welfare, Federal Register Notice, Vol. 43, No. 242, December 15, 1978.
11. Radiological Incident Emergency Response Planning, Fixed Facilities and Transportation: Interagency Responsibilities, Federal Preparedness Agency, General Services Administration, Federal Register Notice, Vol. 40, No. 248, December 24, 1975.
12. Potassium Iodide as a Thyroid-Blocking Agent in a Radiation Emergency, Department of Health, Education and Welfare, Federal Register Notice Vol. 43, No. 242, December 15, 1978.

ON THE EXTENT OF EMERGENCY ACTIONS FOR THE PROTECTION OF THE PUBLIC AFTER ACCIDENTAL ACTIVITY RELEASES FROM NUCLEAR POWER PLANTS

W. G. Hübschmann, A. Bayer, K. Burkart and S. Vogt

Emergency actions for the protection of the public are necessary only after those nuclear accidents which are termed "hypothetical" and which involve core meltdown and failure of the leak tightness of the containment. Such accidents have been analysed in the German Reactor Risk Study (1). The analysis showed, however, that only a small part of the core meltdown accidents analysed have the potential of releasing radioactive material into the atmosphere to an extent that - if they occur in combination with unfavourable meteorological situations - they require extended and immediate protective actions. Such unfavourable meteorological conditions mostly involve precipitation which is an effective means to wash out the radioactive material and to deposit it on the ground. For this reason the accident consequence calculation model applied in (1) has been analysed for the accuracy of the contaminated areas calculated.

AREAS COVERED BY PROTECTIVE ACTIONS

The protective action model applied in (1) has been developed in consistency with official German recommendations (2). It takes into account the specific problems of nuclear accidents (ground contamination, time scale of radioactive decay, efficiency of decontamination, etc.) as well as the high density of population in the F.R.G.. The areas are defined as follows, see fig. 1:
- area A, fixed size, evacuation in any case of core meltdown,
- areas B_1 and B_2, potential early fatalities (bone marrow dose due to external irradiation during 7 days exceeds 100 rad),
- area C, no early fatalities, but ground contamination too high for early decontamination (whole body dose due to external irradiation during 30 years exceeds 250 rad),
- area D, ground decontamination necessary and sufficient (whole body dose due to external irradiation during 30 years exceeds 25 rad).

Consecutive steps of the actions are:
- sheltering in buildings or basements in areas A and B_1,
- evacuation of area A after 8 hours,
- subsequent relocation of the population in areas B_1 and B_2, later relocation of the population in area C,
- immediate decontamination of area D,
- later decontamination of area A, B_1, B_2 and C,
- crop and milk interdiction.

ACTIVITY DEPOSITION MODEL

The external radiation from the ground is the main exposure pathway considered in the following investigation. The activity depo-

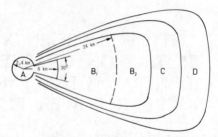

Fig. 1. Areas of protective actions.

sited on the ground is calculated according to conventional models (1). Dry deposition is characterised by the deposition velocity, wet deposition by the washout coefficient. The values of these parameters are choosen in accordance with (3), but the washout coefficient is assumed to depend on the precipitation intensity. The plume depletion is described by the "source depletion model:" the plume inventory is decreased by the deposited amount of activity.

Fig. 2 is a three-dimensional graph of the dose equivalent by external irradiation from the ground. The two pronounced peaks are mainly due to washout by rain during two separate periods of time. The graph is characteristic of the local dose rate distribution in cases where ground contamination is caused by washout.

The following feature of the consequence model tends to overpredict the area contaminated by washout: Activity concentration and ground contamination are calculated at discrete mesh points. The number of mesh points is limited for reasons of computer costs. At distances greater than about 20 km from the source the plume travel time from one grid point to the next exceeds one hour. If washout takes place during only part of that time, the activity being washed out is distributed nevertheless over the total interval. The resulting overestimate of the area and underestimate of the contamination would tend to sometimes overestimate and sometimes underestimate the areas B, C and D. It will be investigated below whether a serious error is involved by this feature of the model.

RESULTS

The area size distribution function of the B_1, $B_1 + B_2$ and C areas is shown in fig. 3. For this figure, the occurence of one of the accidents considered is assumed. The areas of most concern are B_1 and B_2, as here the external radiation is high enough to cause early fatalities, and therefore, the time available for relocation is limited. Such areas are mainly a consequence of a release characterised by the release category 1 (core meltdown and "steam explosion") or 2 (core meltdown and large containment leak). It should be added here that it is still questionable whether the release category 1 involving "steam explosion" is to be postulated in reactor risk assessments or not.

The calculation is based on the reference sample and on the test sample of weather sequences. A weather sequence contains the atmospheric transport and diffusion data (wind speed, diffusion category,

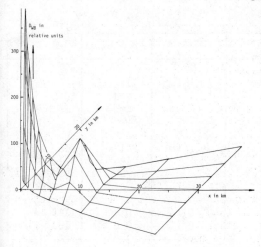

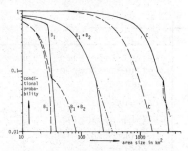

Fig. 2. Whole body dose D_{WB} due to external radiation from the ground. Release category 2

Fig. 3. Protective action area size distribution

——— release category 1 fine meshed grid
—·—· coarse "   "
————— release category 2 coarse meshed grid

precipitation) of subsequent hours, starting at the presumed time of emission. These data are inputs to the plume dispersion calculation. The weather sequences chosen for calculation should adequately cover the variety of meteorological situations, but their number is limited. In (1) 115 reference weather sequences have been taken into account. To investigate the accuracy of the model, a test sample of 98 weather sequences has been chosen alternatively, each involving precipitation during at least one of the first hours. The area sizes have been calculated using two different spacings of the grid mesh-points:
a) fine-mesh grid (the travel time from one mesh pont to the next does not exceed one hour);
b) coarse-mesh grid (the reference spacing of 18 grid points up to a distance of 500 km).

The result is shown in tab. 1 and fig. 3. The area B_1 is not affected, as up to a distance of 24 km (size limit of area B_1) the reference grid is fine enough. The maximum of the areas $B_1 + B_2$ is overestimated by 45 % when using the coarse-mesh grid, whereas the average area $B_1 + B_2$ is slightly underestimated. The same is found for area C, but the overestimate is not as drastic. For area D the effect is negligible, since this area is mainly caused by dry deposition which is not affected by the grid spacing. In fig. 3 the distribution functions of the reference and the test sample (coarse-mesh grid) do not differ markedly.

Tab. 1. Emergency Action Areas

	A	B_1	$B_1 + B_2$	C

average area size
   fine-mesh grid
   coarse " "
	33 km^2
	 "
		19.8 km^2
		 "
			71.2 km^2
			 66,4 "
				1101 km^2
				 1090 "

error by coarse grid	-	-	-7 %	-1 %

maximum area size
   fine-mesh grid
   coarse " "
	"
	 "
		39.3 km^2
		 "
			241 km^2
			 349 "
				3180 km^2
				 3710 "

error by coarse grid	-	-	+45 %	+17 %

minimum area size
   fine-mesh grid
   coarse " "
	"
	 "
		0
		 0
			0
			 0
				117
				 133

CONCLUSION

The models applied in reactor risk calculations are designed to give satisfactory answers in terms of the overall risk which is related rather to average than to maximum values of single parameters. Maximum values, however, are likely to be used as a guideline for emergency planning. It has been shown that the average areas covered by protective actions are only negligibly affected by model refinements, that the maximum of the area of fast relocation (area $B_1 + B_2$), however, is overestimated by a factor of about 1.5 unless a refined analysis is applied. Such methods will be studied in more detail in the second phase of the German Reactor Risk Study.

REFERENCES

1. Der Bundesminister für Forschung und Technologie (1979):
   Die deutsche Risikostudie, GRS-A-329.
2. Der Bundesminister des Innern (1977):
   Rahmenempfehlungen für den Katastrophenschutz in der Umgebung kerntechnischer Anlagen.
   Gemeins. Ministerialblatt 31, 638-718.
3. Reactor Safety Study (1975).
   WASH-1400. Nuclear Regulatory Commission.

AN APPLICATION OF COST-EFFECTIVENESS ANALYSIS TO RESTRICT THE DAMAGE CAUSED BY AN ACCIDENTAL RELEASE OF RADIOACTIVE MATERIAL TO THE ENVIRONMENT

L. Frittelli and A. Tamburrano

When an accidental release of radioactive material occurs, the mitigation of health effects in the exposed population can be achieved only by Remedial Actions (RA) applied to individuals or their environment. RA adoption should be based on a balance of the damage they carry and the reduction in the health effects they can achieve.

In this paper a "cost-effectiveness" analysis is performed by comparing the costs of RA with the monetary value of the collective dose avoided by them. The extent of the resulting damage is partly determined by the Intervention Level (IL) chosen for defining RA time and space features. In a general fashion, the higher the value of IL is, the smaller is the economic damage DCRA caused by RA, but smaller is the health damage DARA avoided by them. If DWRA is the health damage in absence of RA and α is the monetary value of "health damage" ($ per men - Sv), the "total social damage" DT will be equal to αDWRA + DCRA - αDARA.

METHODOLOGY

1) The Effective Dose Equivalent (EDE) received or committed by the Reference Member of the Public (RMP) has been evaluated by accounting a) external exposure from the cloud and from the contaminated ground; b) internal exposure from material inhaled from the passing cloud. Collective EDE for different groups of RMPs has been evaluated on the basis of a commitment time of 30 years for EDE from inhaled material and up to ∞ for exposure to contaminated ground. The RMP is characterized by organ dose conversion factors as in WASH 1400 (1) and organ weighting factors as in ICRP 26 (2).

2) At every distance from the source RA begin at the same time TE after the outset of the accident and all the members of the public in the "Non-Stochastic Effects Area" (NSEA) and in the "Stochastic Effects Area" (SEA) are evacuated. The boundaries of NSEA and SEA are defined by a maximum value RLIM of the distance from the source and the following criteria: a) within NSEA the doses received during or committed for a time span of 30 days by the RMP exceed the following values: Total Body = 5 Sv; GI tract

= 35 Sv; Red Marrow = 5 Sv; Lung = 45 Sv; Thyroid = 250 Sv; b) within SEA the Evacuation Dose (ED) exceeds IL; ED has been defined as the EDE which will be received by RMP from TE to TE + 70 years if RA are lacking; in other words ED is the EDE which could be reduced by RA. After the passage of the cloud a location within NSEA or SEA is interdicted for the Interdiction Time (IT) required for reducing to IL the EDE which will be received by RMP in the 70 years following IT, owing to the exposure to contaminated ground. In SEA, IT cannot be less than the time ITO, required for decontaminating (if needed) and for surveying the area; in NSEA, a larger minimum value IT1 for IT is adopted, because we assume that the evacuees (which suffered NS effects) or other people in their stead cannot return back before IT1.

3) By summarizing, the economic consequences of RA have been evaluated as follows: a) property within NSEA and SEA are expropriated at TE; the economic charge has been computed as the difference between the market values at TE and TE + IT converted at the present worth at time TE; b) the evacuation of SEA is definitive if IT exceeds TEV, the mean time required for evacuees from SEA to undertake a new work out of the interdicted area. During IT (or up to TEV) the earnings of the evacuees are secured by the community by means of subsides or services; c) decontamination is equally effective but its cost is a function of land use; d) only a fraction of the goods in the contaminated areas can be removed out of the area for being again utilized after decontamination. A "RA-tree" (Fig. 1) can be built by means of the options; b) - Interdiction, c) - Decontamination and d) - Goods Removal. The options which minimize the total costs of RA are supposed to be undertaken at every location.

RESULTS

The values of parameters for Reference Cases (RC) are in Table 1. The WASH 1400 accidents PWR2, PWR6 and PWR 7 have been selected as large-extent, mean-extent and contained accidents, which last one approximates a Design Basis Accident. Figg. 2, 3 and 4 show the values of DCRA, ∂ DARA and DCRA-∂DARA as function of IL, expressed in a normalized manner as the ratio of total damage to the unitary value of the land around the plant. We could draw the following conclusions:

a) - There is a value ILO of IL which optimizes the balance between DCRA and ∂ DARA: for PWR2 and PWR6 the value of ILO is about 0.1 Sv.

b) For PWR6 there is a value of IL1 of about 0.01 Sv below which DCRA is larger than ∂ DARA. In other words, if IL is less than IL1, the intervention enlarges, in a broad sense, the social cost of the accident.

c) For a contained accident like PWR7, no value of IL optimizes the balance between health and economic consequences, because DCRA is always larger than α DARA.

The optimization process is scarcely influenced by the values for the parameters of the model. In Fig. 5 the meteorological and the RA scenarios have been modified for PWR2 by changing, one at time, the wind speed, the dispersion coefficients, the deposition velocities and by putting TE = 0.

Also shown are the results for a value of α as low as 10^4 \$/man-Sv and for a "Developed area", whose features are between the RC and an area highly industrializated and with high density of tertiary activity. The strong influence of α on ILO and IL_1 is obvious, but we must notice how DARA is small compared to DWRA.

CONCLUSIONS

1 - The damage related to an accident in a NPP can be classified as follows: a) - <u>Early Damage</u> (t < TE); related to health effects from exposures before RA; it could be reduced only by ties on land usage around the plant, and cannot be avoided by RA; b) - <u>Damage which can be controlled by evacuation and interdiction of the land</u> (t > TE), related to health effects and to RA costs; it could be minimized by a value of ILO around 0.1 Sv (for PWR2 and PWR6), but the reduction of damage obtained by RA is very small (a few percent); c) - <u>Damage which can be controlled by impoundment of agricultural products,</u> not considered in this paper.

2 - If RA were managed only on the basis of the "cost-effectiveness" analysis as carried out in this paper (but we are aware that this cannot be true, mainly owing to the psychological factors which could enhance the reduction of the health effects), an Intervention Plan for reducing the "social damage" would be justified only for large and mean extent accidents. For large accidents the time and space features of RA could be so large that also the probability of the accident should be accounted for in planning the intervention.

REFERENCES

1 - WASH 1400 - Reactor Safety Study - 1975

2 - ICRP Publication 26 - 1977

Table 1 Values of the parameters for the Reference Case

TE (evacuation time) = 6 hours	= 6 hours	RLIM	= 250 km
ITO (Interdiction Time for SEA)	= 60 days	Pasquill categorie	D
IT1 (Interdiction Time foe SEA)	= 1 year	Wind speed	u = 5 m/sec
TEV (for SEA)	= 1 year	Noble gases	= 0
Decontamination Factor DF	= 99 %	Deposition Velocity Particulates	= 0.1 cm/sec
Accidents categories PWR2, PWR6, PWR7 (as in WASH 1400)		Halogenes	= 1 cm/sec
		$\alpha = 10^5$ \$/men-Sv	

	Reference Case	Developed area
Value of property (M\$/km^2)	0.75	50
Population Density (km^{-2})	10	500
Perishable goods	10%	10%
Real Estate	70%	70%
Chattel	20%	20%

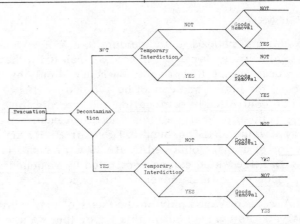

Figure 1 - Remedial Actions Tree. At every location are adopted the RA which minimize the total costs.

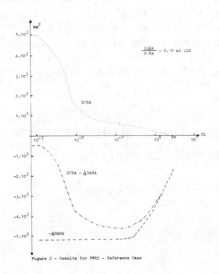

Figure 2 - Results for PWR2 - Reference Case

An Application of Cost-Effectiveness Analysis

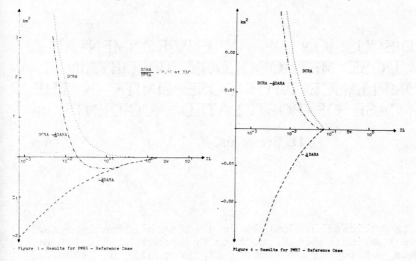

Figure 3 - Results for PWR6 - Reference Case

Figure 4 - Results for PWR7 - Reference Case

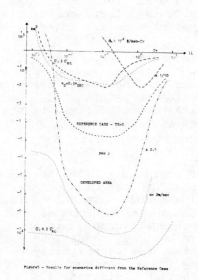

Figure 5 - Results for scenarios different from the Reference Case

DISCUSSION OF AN ENVIRONMENTAL DOSE METHODOLOGY TO OBTAIN COMPLIANCE WITH DOSE LIMITS IN THE CASE OF POSTULATED ACCIDENTS

H. D. Brenk and K. J. Vogt

INTRODUCTION

In the case of postulated accidents in nuclear facilities the Radiation Protection Ordinance of the F.R.G. demands sufficient protection of the public by technical design of the plant. Thus the worst case individual annual dose shall be kept within the given accident dose limits of 50, 150 and 300 mSv for total body, thyroid and bone respectively. These limits comprise all exposure pathways including the exposure via ingestion of contaminated food.

In order to achieve compliance with these demands, a practical methodology for the assessment of radiological impacts caused by postulated accidents has been developed. Thereby special attention is paid to the conservatism of its assumptions and its compatibility with the methodology of routine releases according to the existing regulations (1).

DOSE CONCEPT

The judgement of accidental releases concentrates on the radiological consequences of short-term emissions. Consequently the accident dose methodology is based on the concept of the committed environmental dose. Due to the equivalence between equilibrium dose and dose commitment based on the same boundary conditions (2), (3), (4), the calculated committed dose can directly be compared with the dose limits given as maximum annual doses in the German Radiation Protection Ordinance.

For the purpose of dose assessment it is significant to base the calculations on the three-compartment scheme of the ecosphere shown in fig. 1. Then one transfer function for each compartment can be defined which is determined by its output to input ratio. Due to the time dependence of the release rate in the case of an accident, the radionuclides undergo a time dependent transfer in the environment. Thus each of the transfer functions is also principally determined by the time dependence of the release rate. However, as our investigations concerning the nuclide dynamic in the environment have proved

---

This paper was sponsored by the Federal Ministry of the Interior of the F.R.G. under contract number St.Sch. 687

(5), the time dependence of the different ecosystems "atmosphere", "biosphere" and "man" can be temporally decoupled. Hence, each of the corresponding transfer functions A, B und D can now be determined irrespective of the time dependence of the output of the donor compartment as a peak response.

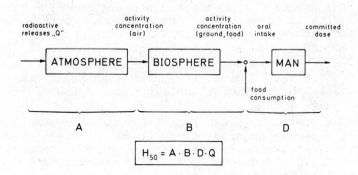

fig. 1

Due to this procedure the committed environmental dose for all exposure pathways "i" is now generally given by the following simplified formula

$$H_{50} = A_i \cdot B_{50,i} \cdot D_{50,i} \cdot Q \qquad [1]$$

where A is the atmospheric dispersion or deposition factor, B the transfer factor for the biosphere, D the dose conversion factor and Q the accidental source strength.

The dosimetric models on which the determination of the different transfer factors in equation [1] is based, are explained in more detail in (5).

DISCUSSION OF THE MODELS

The <u>meteorological dispersion and deposition factors (A)</u> represent the most unfavourable weather conditions, e.g. pessimistic wind velocities, high precipitation rates, etc., and thus lead to an upper limit of air and ground contamination.

According to the conclusions in (5) all <u>dose conversion factors (D)</u> derived for continuous releases can be applied to short-term releases as well.

In the case of external irradiation from the cloud or from the contaminated ground, the dose conversion factors are in fact "dose rate conversion factors" being defined as dose rate to air concentration ratio, and thus do not depend on the type of releases.

For internal irradiation (inhalation and ingestion) the applicability of the dose conversion factors for continuous releases to short-term releases is due to the equivalence between the committed dose and the equilibrium dose concept.

The use of the proposed <u>biosphere transfer factors</u> (B) results in conservative doses for the corresponding exposure pathways.

This is valid for the inhalation exposure, particularly caused by the conservative assumptions for the respiration rates. In case of resuspension being considered, this causes an additional contribution to the conservatism of the dose.

For γ-ground irradiation the model represents the dose of a person exposed for 50 years on contaminated paved surface. Possible shielding or wash-off effects are neglected. This neglection causes an overestimation of the dose up to a factor of 7 (5).

In the case of ingestion there are various simplifying assumptions which give rise to different systematic errors in dose calculation. So the temporal decoupling of the compartments in fig. 1 may cause systematic underestimation of the dose up to 90 % (factor 10). Moreover the soil contamination for the milk and beef pathway may be underestimated up to 50 % (factor 2). The remaining assumptions, i.e. using soil-plant transfer factors for continuous instead of short-term releases with no distinction between vegetation period and the rest of the year, and the application of pessimistic consumption rates for the different food pathways result in reasonable overestimations of the dose between a factor of about 3 to 11. Hence the overall overestimation is a factor of 2 to 10.

This systematic error is valid only for activity uptake via root transfer. For the consumption of food directly contaminated by the deposition of radionuclides on the above surface parts of the vegetation, the systematic error is considerably smaller. Thus, due to the dominating importance of the latter type of contamination, the total overestimation of the ingestion dose does not exceed a factor of 3.

If one assumes administrative preclusion of food consumption after an accident, the calculable ingestion dose would then be reduced to values between 60 % for Sr 90 and 0.03 % for I 131 compared to the case without any restrictions.

CONCLUSIONS

The proposed methodology for dose assessments in the case of postulated accidents shows compliance with the

F.R.G. Radiation Protection Ordinance and has the following main characteristics

> 1) Due to the equivalence between the committed dose concept and the concept of equilibrium dose, the methodology shows sufficient compatibility with that for routine releases. This is valid not only for the dose calculation, i.e. the total ecosystem but also for each subsystem in the ecosphere. Therefore both the routine and the accident calculations can be based on the same ecological data and dose conversion factors.
>
> 2) The dose calculations for postulated accidents based on the ecological parameters which are determined for the equilibrium case result in conservative assessments.
>
> 3) The determination of one transfer function for each compartment ("atmosphere", "biosphere" and "man") supplies theoretical measures, which partly offer the opportunity of direct validation of the corresponding model by measurements following accidental releases.
>
> 4) For a planning engineer these measures, calculated and fixed beforehand, enhance the practicability of the methodology as a tool in decision-making. Using these measures dose assessments become simple and feasible without dealing with the radiological problems being involved.
>
> 5) The methodology results in considerable reduction of the computational efforts.

REFERENCES
1. Der Bundesminister des Inneren (1977): Allgemeine Berechnungsgrundlagen für die Bestimmung der Strahlenexposition durch Emission radioaktiver Stoffe mit der Abluft, Bonn
2. Lindell, B. (1973): Assessment of Population Exp., IAEA-SM-179/B, Aix-en-Provence
3. Brenk, H.D. (1979): Konzept zur Berechnung der Umweltbelastung durch kerntechnische Anlagen in der BR Deutschland, Fachtagung Radioökologie des Deutschen Atomforum, Bonn
4. Brenk, H.D. (1978): Ein anwendungsbezogenes Konzept zur Berechnung der Umweltbelastung durch Abluftemissionen kerntechnischer Anlagen für Standorte in der BR Deutschland, Jül-1485
5. Brenk, H.D., Vogt, K.J. (1979): Radiation Exposure Caused by Postulated Accidents in Nuclear Installations (Jül-report to be printed)
6. ICRP-Publ. 29 (1979): Radionuclide Release into the Environment: Assessment of Doses to Man, Pergamon Press New York

PROTECTING FRONT-LINE SURVEY AND RESCUE TEAMS DURING EMERGENCIES

H. Tresise

Teams trained in First-Aid, Fire Fighting and Rescue are available from the shift personnel at British civil nuclear power stations to provide assistance during an emergency. Before the Emergency Controller can deploy these resources safely, he needs information about the situation on the plant. A reconnaissance team with monitoring instruments, protective clothing and self-contained breathing apparatus may first have to enter the accident zone before successful remedial action can be planned. Procedures are described which try to ensure the safety of personnel who have to approach unknown dangers.

The hazards which the reconnaissance team may encounter include:
1. Fire or high temperatures
2. Smoke or poor visibility
3. Concentrations of irrespirable gas (CO_2 at Magnox power stations)
4. Unsafe access routes due to building or plant damage
5. High levels of radiation or contamination.

In most of these accident situations the team may need to wear breathing apparatus, so the personnel involved must train regularly.

TEAM COMPOSITION

The minimum number for a team is 5 trained men, each equipped with breathing apparatus. One member should be a plant engineer with detailed knowledge of the conventional hazards of the site; a second should be used to making radiological measurements; a third should be trained in first aid.

In addition to breathing apparatus, protective helmet, ear defenders and a personal dosimetry pack, the team have found it advisable to take a radiation dose-rate meter, a gas detector, a walkie-talkie radio or jack-in telephone, first aid kit, torches, master keys, a guide line and a B.A. Control Board with clock and pen.

SELECTING THE INCIDENT CONTROL POINT

The Team Leader must use considerable discretion in approaching the unknown situation. When the accident has been located, the team establish an Incident Control Point in a safe place near to the boundary of the hazardous area, taking account of any forseeable short-term changes. Areas in the reactor buildings with easy access to stairs or lifts and equipped with telephones have been selected as potential control points. The Emergency Controller is informed of the situation and can send forward reinforcements together with a mobile trolley described later.

If the situation demands more than one access route to the area, a separate Incident Control Point is set up at each entry.

If the Control Point becomes untenable, teams already in the

Protecting Front-Line Survey and Rescue Teams

hazardous area are withdrawn at once. When all are out, the Control Point can be resited at a safe location in clean air.

INCIDENT CONTROL POINT PROCEDURES

A trained man is nominated by the Team Leader to take charge of entry procedures. It helps if he puts on a distinguishing armband or helmet. He is referred to as the Control Officer and all personnel who enter the hazardous area must report to him first.

No man is allowed to enter or remain alone in an area of hazard and normally men work in teams of 2 to 4. The opportunity to save life may motivate a man to go beyond these procedures, but he should recognise the risk involved both to himself and to his colleagues.

Usually two men go forward to investigate the situation while two more are immediately available to help if required. A U.H.F. walkie-talkie radio is carried by one member of each team and a "talk through" facility at the Control Point enables the Emergency Controller and other teams to know what information is available.

If self-contained breathing apparatus is required, the tally from each set in use is marked with the wearer's name, the air cylinder pressure and the time of entry to the hazardous area. This tally is pushed into a slot in a Control Board faced with transparent plastic, so that the information cannot be obliterated inadvertently. Men are instructed to return at once to the Control Point if the warning whistle on the B.A. set sounds, indicating 10 minutes air reserve left. The safe working time is entered on the Control Board next to the tally, using initially a standard table and the clock provided. It can be updated if the team members can inform the Control Officer of the reading on the pressure gauges of their air cylinders, because the duration is very dependent on the physical exertion required. The destination or function may also be written next to the tallies. The Control Officer will arrange for each team to be replaced by fresh men at the appropriate time, until the task is completed.

On returning to the Control Point, each B.A. set wearer collects his tally and replaces it on the set when the air cylinder has been changed. If the team do not reappear within the expected time, the Control Officer will send in the reserve pair to locate and assist those missing. In addition, he will ask the Emergency Controller for another pair of men to be available to him before the reserve team are due out, in case further reinforcements are needed.

The Control Officer can cope with 10 to 12 B.A. set wearers at one time. If more need to enter simultaneously, a second Control Point should be considered.

The written record of personnel entries reduces the risk to those in a hazardous area because their safety does not depend on the memory of a man who could himself become a casualty.

USE OF GUIDE LINES

Guide lines are used at the discretion of the Team Leader. They ensure that a team can retrace their entry route to the Control Point even when visibility is nil. Each wearer of a B.A. set has a "personal line" secured at one end to the set harness. The free end has a snap hook which can be clipped over the guide line.

A main guide line consists of 60 meters of nylon cord carried in a cylindrical canvas container. One end of the cord is attached to the container and the line is laid up so that it pays out through a hole in the lid as the user carries it away from the fixed end.

The container is carried by the team or attached to the harness of a B.A. set. The last man in the team ties the line to convenient objects at sufficient intervals to keep the line off the ground. The knot must be easily untied and a slip knot is normally used.

Two main guide lines may start from one Control Point and these are then marked with an 'A' or 'B' circular tally.

Every $2\frac{1}{2}$ meters along the guide line, two tabs are attached 15cm. apart. One tab has two separate knots while the other is unknotted and longer. The knotted tab is on the "way out" side of the plain tab.

The team leader may decide to start laying a main guide line after he has left the Control Point. In this case, the way back to the Control Point must be well defined and have good visibility.

Up to four branch lines may be attached to the main guide lines from one Control Point. These lines are marked where they start from the main guide lines by a rectangular tally containing 1, 2, 3 or 4 finger-sized holes, so that they can be identified by touch alone.

The Incident Control Point trolley described later carries six similar lines each in a separate container. When used, they function as a main guide line if attached at one end to the Control Point and as a branch guide line if attached as a spur to a main guide line. A person working at the far end of a branch line can be 120 meters away from his Control Point.

A personal line consists of 6 meters of lighter nylon cord secured to a pouch attached to the B.A. set. A 'D' ring is attached to the line $1\frac{1}{4}$ meters from the other end which has a snap hook to clip over a guide line. Normally this 'D' ring is secured inside the pouch so that the wearer is within $1\frac{1}{4}$ meters of the guide line. However, the personal line can be increased to 6 meters to allow the B.A. set wearer to extend his area of movement away from the guide lines.

After a guide line has been laid, all teams should attach the snap hook of their personal line to it or to the B.A. set harness of the man in front, with only the leading man coupled to the line.

Away from the Control Point, the personal line should only be unclipped from the guide line on two occasions. The first is when the team is transferring from a Main Guide Line to a Branch Guide Line, or returning. The second is to allow another team to pass. Out-going teams have precedence in the use of the guide line because their air supply may be low. The in-going team uncouple their personal lines and stand aside to allow the other team to pass.

If personnel do not return to the Control Point, a Rescue Team can find them by following the guide line and the personal line.

If a casualty is found but cannot be brought out, the line can be terminated there so that rescuers can go straight to the location.

Subsequent teams can move rapidly along a guide line knowing that hazards have been avoided.

CONTROL POINT TROLLEY AND EQUIPMENT

A 4-wheel trolley has provided a convenient mobile store for

Protecting Front-Line Survey and Rescue Teams

Control Point equipment. It is small enough to be wheeled through doorways and into lifts, so that it can be brought to the location chosen for the Control Point. Its presence helps to remind personnel of the need to avoid cross-contamination by segregating clean and used equipment, active team members and reserves etc. The Oldbury trolley is about 115 cm x 65 cm x 100 cm tall and the stock of equipment shown in the table below has proved useful in a variety of exercises and simulated accidents.

Radiation Dose-Rate Meter	Contamination Monitor and Probe
Gas-in-Air Detector	Clean Protective Clothing (6 sets)
Filter Packs (6 off)	Battery Operated Air Sampler
B.A. Set & Spare Cylinder	Notice for "Empty Cylinders"
Wax Crayons (2 off)	B.A. Control Boards (2 off)
Telephone Extension Lead	Guide Line & Container (6 off)
Jack-in Telephone	Guide Line Tallies (1 set)
Communications Amplifier	Communication Headsets (2 off)
Communications Lead	U.H.F. Walkie-Talkie Radio
QFE Charging Unit	Dosimeter Issue & Record Forms
First Aid Kit	Hazard Warning Rope (2 reels)
Adhesive Tape (1 roll)	100 cm. wide Plastic Sheet (1 roll)
Plastic Bags (3 off)	Facemask Disinfectant (1 bottle)
Paper Sacks (3 off)	Paper Tissues (1 box)
Torches (6 off)	Chemi-Luminescent Light (6 sticks)
Floor Plans of Plant	Personal Line & Pouch (6 off)

DISCUSSION

    For brevity, the preceeding sections have described equipment at Oldbury as examples, but the control principles can be used without such aids. The purpose-built B.A. control board is convenient to use, but a pencil or chalk on a wall or door could provide the essential information of who went where and at what time. The U.H.F. radio system by which each walkie-talkie receives all the messages transmitted reduces delays because trained men can anticipate developments, but the safety of team personnel is not reduced if communications are by telephone. The trolley has provided a useful focal point for the control procedures but other methods of personal discipline can be used to contain the hazard.

    Plant employees who are also part-time firemen use similar techniques for non-radiological hazards, so they are not subjected to conflicting training methods on and off site. Reinforcements from full-time services can be deployed on site quickly because the control system in operation will be familiar to them.

    These methods have been criticised for slowing down the first stages of rescue or remedial work in an emergency. While it takes time to lay out a guide line initially, subsequent movements in and out of the area are eased and the whole operation may be completed more quickly. During exercises, time scales tend to be shortened because participants do not expect to come to any harm. When the risks are real or the consequences of the accident are not yet known, personnel do not make snap decisions to send teams into areas of danger.

*Biological Effects
of Environmental Radiation*

NATURAL BACKGROUND AS AN INDICATOR* OF RADIATION-INDUCED CANCER

J. J. Cohen

A recurrent theme in discussions about nuclear energy is the uncertainty regarding the effects of exposure to low levels of radiation. Statements on this subject are often misleading by implying a general level of knowledge approaching total ignorance. This paper attempts to place the problem in a probablistic perspective.

Recent theoretical predictions on the effects of low-level ionizing radiation indicate induction rates as high as 8×10^{-3} cancers per man-rem.(1) Assuming this estimate is correct, then roughly half of all cancer incidence in the U.S.A. could be attributed to natural background radiation. Previous studies (2,3), however, indicate no correlation between cancer incidence and levels of background radiation.

Table 1 reviews some past estimates of radiation-induced cancer rates. These estimates pertain to effects of low-level radiation (sub MPC) whole body doses to large populations. The estimates shown were not all originally given in terms of cancers per man-rem. Some were calculated from doubling dose estimates, or were inferred from estimated effects of given population exposures. No judgment is made here as to the quality of these estimates or how they were derived. They are given to demonstrate their wide diversity.

Table 2 presents a summary of data on external background radiation and cancer incidence for the U.S.A. by state. The radiation values were taken from the USEPA (15), and were corrected for structural and body shielding according to the method suggested by the NCRP. (16) They are presented as corrected external background dose (CXBD) in mrem/yr. The cancer incidence values were taken from U.S. statistical abstracts. (17) For purposes of this analysis, a simplified but reasonable method was used to correct the raw data for age variation and other factors affecting the general state of health. The raw incidence data were normalized according to overall death rate, which itself is a function of the population age distribution and general state of health. The normalized cancer rate (NCR) was calculated by

$$\text{NCR} = \frac{\text{national death rate}}{\text{state death rate}} \times \text{state cancer rate}.$$

TABLE 1. Estimated radiation-induced cancer rates

Source	Date	Risk estimate (cancers/man-rem)	Source	Date	Risk estimate (cancers/man-rem)
ICRP: 8 (4)	1966	5.0×10^{-4}	Rasmussen (10)	1975	1.2×10^{-4}
Robison & Anspaugh (5)	1969	1.0×10^{-3}	NUREG-0216 (11)	1977	1.4×10^{-4}
Gofman & Tamplin (6)	1969	1.7×10^{-3}	Mancuso (12)	1977	6.8×10^{-3}
Otway (7)	1971	7.0×10^{-4}	Gofman (13)	1977	7.3×10^{-3}
Hull (8)	1971	1.0×10^{-4}	Tamplin & Cochran (1)	1977	8.0×10^{-3}
BEIR (9) Committee Low	1972	8.8×10^{-5}	Morgan (14)	1978	6.0×10^{-4}
BEIR (9) Committee "most likely"	1972	1.8×10^{-4}			
BEIR (9) Committee High	1972	4.4×10^{-4}			

*Work performed under the auspices of the U.S. D.O.E., contract No. W-7405-Eng-48.

TABLE 2. Background radiation and cancer incidence

STATE	NCR[a]	CXBD[b]	STATE	NCR	CXBD	STATE	NCR	CXBD
Alabama	150.9	81	Louisiana	161.3	57	Ohio	171.4	87
Alaska	112.2	79	Maine	175.1	93	Oklahoma	159.3	83
Arizona	164.5	92	Maryland	186.7	71	Oregon	165.9	83
Arkansas	167.9	84	Massachusetts	183.8	84	Pennsylvania	171.2	76
California	180.2	68	Michigan	183.8	83	Rhode Island	192.9	78
Colorado	155.8	175	Minnesota	168.9	94	South Carolina	144.0	81
Connecticut	198.8	74	Mississippi	147.4	78	South Dakota	158.5	137
Delaware	196.3	74	Missouri	162.	79	Tennessee	157.3	85
Florida	183.2	70	Montana	162.6	119	Texas	163.5	60
Georgia	144.5	74	Nebraska	163.6	103	Utah	142.1	129
Hawaii	198.0	65	Nevada	170.7	102	Vermont	168.3	74
Idaho	152.6	115	New Hampshire	185.4	82	Virginia	170.8	76
Illinois	170.3	82	New Jersey	190.8	74	Washington	180.9	83
Indiana	165.0	76	New Mexico	147.1	139	West Virginia	152.1	83
Iowa	170.4	83	New York	184.0	82	Wisconsin	174.0	80
Kansas	163.7	83	North Carolina	150	89	Wyoming	143.3	175
Kentucky	161.5	79	North Dakota	165.5	92			

[a] Normalized cancer rate ($yr^{-1}/10^5$ population).
[b] Corrected external background radiation dose (mrem/yr).

Figure 1 presents this cancer and radiation data graphically. The data points present a picture similar to that previously shown by Frigerio (2), indicating a lack of correlation. Regression analysis of these data, assuming a linear relationship, indicates that the best-fit curve is described by the formula NCR = 190 - 0.27 CXBD, shown as a solid line in Fig. 1. The regression analysis reveals a somewhat negative correlation; however, the correlation coefficient, r = 0.39, indicates a very weak relationship.

Useful insight can nonetheless be derived by testing certain hypothetically assigned radiation-induced cancer rates against the

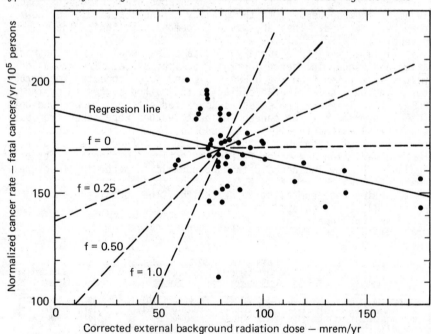

Fig. 1. Cancer incidence and background radiation.

data. This is done by determining mathematical relationships that (1) conform to the hypothesis and (2) best conserve the observed data. The hypotheses tested were:

1. The Fractional Hypothesis (a fraction, f, of all cancers are induced by exposure to external background radiation): NCR = 167(1 - f) + 2.1 f CXBD.
2. The Null Hypothesis (cancer and background radiation are unrelated): NCR = 167 (note: f = 0).
3. The Total Hypothesis (all cancers are induced by exposure to external background radiation and nothing else): NCR = 2.1 CXBD (note: f = 1.0).

These hypothetical relationships are shown as the dotted lines in Fig.1. In testing these hypotheses, the question is asked: Given that the hypothesis is correct, what is the probability that the actual data could have been observed?

The statistical method used for the calculation was the "t" test described by F.S. Acton (18), where:

$$t_{(n-2)} = \frac{(b - \beta) \sqrt{n - 2} \sqrt{S_{xx}}}{\sqrt{SSR}}$$

Knowing "t" (at 48 d.f.), it is possible to determine the probability of a fractional cancer incidence (f) equal to or greater than that hypothesized. This relationship is given in Fig. 2.

To relate "f" to the radiation-induced cancer rate, assume all cancer incidence is caused by external background radiation, in which case, at steady state in the U.S.A.,

$$\frac{167 \text{ cancers/yr}}{10^5 \text{ people} \times 0.079 \text{ rem/yr}} = 2.1 \times 10^{-2} \text{ cancers/man-rem.}$$

If, for example, the induction rate value were in fact 8×10^{-3} cancers/man-rem, then f = 0.38. In light of the observed data, the probability of an induction rate of 8×10^{-3} or higher is approximately 7×10^{-5}, or about one chance in 14,000. From

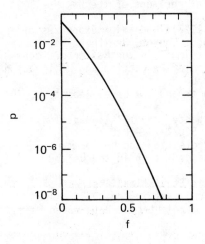

Fig. 2. Probabalistic asessment of radiation-induced cancer, where p is the probability that the indicated fraction or greater is valid, given observed data; and f is the hypothetical fraction of cancer incidence in U.S.A. attributable to background radiation.

this analysis it may be concluded that such high estimates of radiation-induced cancer rate, although not impossible, may certainly be considered highly improbable.

REFERENCES

1. Tamplin, A. and Cochran, T. (Dec. 1977): Uranium Impacts from Spent Fuel Reprocessing and Radioactive Waste, USNRC Hearing Docket No. Rm-50-3.
2. Frigerio, N.A., Echerman, K.E., and Stowe, R.S. (Sept. 1973): Carcinogenic and Genetic Hazard from Background Radiation, Environmental Statement Project, Argonne National Laboratory, Argonne, Ill., Rept. ANL/ES-26.
3. Jacobson, A.P., Plato, P.A., and Frigerio, N.A., (Jan. 1976): The Role of Natural Radiations in Human Leukemogenisis, Am. J. Pub. Health, 66.
4. International Commission on Radiological Protection (ICRP) (1966): The Evaluation Risks from Radiation, Publication No. 8.
5. Robinson, W.L. and Anspaugh, L.R. (1969): Assessment of Potential Biological Hazards from Project Rulison, Lawrence Livermore Laboratory, Livermore, Calif., Rept. UCRL-50791.
6. Gofman, J.W. and Tamplin, A.R. (Oct., 1969): Low Dose Radiation, Chromosomes, and Cancer, in Proc. of IEEE Nuclear Science Symp., San Francisco, Calif., GT-101-69.
7. Otway, H.O., Lohrding, R.K., and Battat, M.E. (Oct., 1971): A Risk Estimate for an Urban-Sited Reactor," Nuclear Technology.
8. Hull, A.P. (May 1971): Radiation in Perspective, Some Comparisons of the Environmental Risks of Nuclear and Fossil Fueled Power Plants, Nuclear Safety, 12(3), p. 185.
9. Report of the Advisory Committee on the Biological Effects of Ionizing Radiation (BIER) (Nov., 1972): The Effects on Populations of Exposure to Low Levels of Ionizing Radiation, Nat. Acad. of Sci., Nat. Res. Council.
10. Nuclear Regulatory Commission (1975): Reactor Safety Study, WASH-1400, Vol VI, App. G.
11. Nuclear Regulatory Commission Environmental Survey of the Reprocessing and Waste Management Portions of the LWR Fuel Cycle NUREG-0216 (March, 1977).
12. Mancuso, J., et al. (February, 1977): Radiation Exposures of Hanford Workers Dying from Various Causes, Health Physics.
13. Gofman, J. (December, 1977): Testimony in the Matter of Amendment of 10CFR, Part 51, Licensing of Production and Utilization Facilities, USNRC Docket No. RM 50-3.
14. Morgan, K.Z. (Sept., 1978): Cancer and Low-Level Radiation, The Bulletin.
15. U.S. Environmental Protection Agency (May 1976): Radiological Quality of the Environment, EPA-520/1-76 - 010.
16. National Council on Radiation Protection and Measurements (April 1975): Natural Background Radiation in the United States, NCRP Report No. 45.
17. U.S. Bureau of the Census (1975): Statistical Abstract of the United States: 1975 (96th edition), Washington, D.C.
18. Acton, F.S. (1959): Analysis of Straight-Line Data. John Wiley & Sons, Inc. - New York.

TWO DECADES OF RESEARCH IN THE BRAZILIAN AREAS OF HIGH NATURAL RADIOACTIVITY

T. L. Cullen, A. S. Paschoa, E. P. Franca, C. C. Ribeiro, M. Barcinski and M. Eisenbud

The geology of Brazil presents us with several areas of high natural radioactivity. Our university groups have identified and studied three natural theaters in a collaborative project over twenty years. The results were only recently presented in a symposium (1). The three areas are Guarapari, Araxa-Tapira, and Morro do Ferro.

GUARAPARI

With the slow weathering of the mountains that parallel the Atlantic coast, beaches have been formed with mottled patches of monazite sand. Monazite is a complex of rare earth phosphates with strong impurites of thorium and weaker uranium. Guarapari is a town of 12,000 in the monazite region.

The radiation levels were mapped. Patches on the beach showed levels up to 2.0 mR/hr; the streets averaged 0.09 mR/hr. A TLD survey of the population revealed that the dose rate ranged up to 2,000 mrem/year with a mean of 640 mrem/yr.

All the water and most of the food comes from normal areas. Any internal contamination might come from fine inhaled dust or from thoron and radon in the air. Whole body counting gave negative results, and analysis of placentae showed low values of internal contamination, as can be seen in Table 1.

The lack of reliable medical practise and records renders an epidemiological study impractical. With the relatively low dose rate and the small population, the somatic chromosomal aberrations in peripheral lymphocytes were selected as a biological parameter.

From the start a higher rate of chromosomal aberrations, especially the two-break type, were seen in Guarapari. A resumé of the data is given in Table 2. The number of aneuploid cells ($2n \neq 46$) and the chromatid type aberration are not considered as radiation induced, but are given as culture technique controls.

The double break phenomenon suggests an internal contaminant, an alpha emitter. Since tests for long-lived body burdens are essentially negative, it is thought that chronic exposure to higher values of thoron and radon are responsible.

TABLE 1. Guarapari: ^{228}Th and ^{226}Ra in Human Placentae.

Local	N	^{228}Th (pCi/g Ca)	^{226}Ra (pCi/g Ca)
Controls	17	0.09 - 1.36	0.05 - 0.76
Guarapari	10	0.10 - 5.96	0.11 - 1.38
Araxa	5	0.19 - 2.63	0.07 - 0.44
Tapira	8	1.00 -26.70	0.3 - 14.50

TABLE 2. Guarapari: Individual Means and Standard Deviations for Cytogenetic Data.

	Guarapari	Control
No. Cells	66.55 ± 22.75	61.23 ± 10.72
No. Aneuploid Cells	3.45 ± 3.81	3.46 ± 2.67
Chromatid Aberrations	2.00 ± 2.54	2.23 ± 2.92
Deletions	0.65 ± 1.01	0.52 ± 0.90
Dicentrics	0.07 ± 0.28	0.04 ± 0.19
Rings	0.02 ± 0.15	0.00
Total no. Breaks	0.85 ± 1.20	0.57 ± 0.93

ARAXA-TAPIRA

IN this region in the interior State of Minas Gerais, the soil is naturally fertilized by uraniferous apatite, a phosphate mineral that serves as a source of radium-226. Here the concern is naturally with the food chain as a contaminant. The highest concentrations are found in the staple foodstuffs, manioc and its flour, as well as in potatoes and citrus fruit. Radium-228 content reaches 2,720 pCi/kg and radium-226 81 pCi/kg.

A house to house inquiry gave clear information on the dietary habits of 28 families. Certain poor families live almost entirely on the contaminated produce. As one moves from the hot spots the radium content becomes more diluted. The small group has daily intakes of the range 20-40 pCi of radium-226 and 120-240 pCi of radium-228. This would result in body burdens of the order of 280-560 pCi of radium-226 and 1680-3360 pCi of radium-228. This is below the sensitivity of any whole body counter in Brazil.

It was found that only a small number of people are

affected, and only 196 individuals of the 1670 were selected for further investigation.

MORRO DO FERRO

In the region south of the city of Poços de Caldas an alkaline plug thrust its way 400-600 meters above the surrounding land. The inner portion lowered, and became highly mineralized. In the center Morro do Ferro (Iron Mountain) rises 140 m. above its base. Across it two dikes of magnetite have penetrated, with mixtures of titanium and manganese. A great variety of rare earth oxidation compounds are found here with strong percentages of thorium oxide and traces of uranium.

The harsh face of the mountain offers poor soil for vegetation, and only low grade grazing grass grows. The mountain was mapped with ionization chamber and portable scintillometer. Some 42,000 m^2 show levels above 1.0 mR/hr and small patches are above 3.0 mR/hr.

While it was considered an ecological laboratory many measurements were made of the uptake by plants, and of the concentration of thoron and radon in rat holes and termite mounds. Some of the data are given in Tables 3 and 4. In his study of the exposure to local rats Drew found that the greatest dose was to the trachea-bronchi area with an average dose of 200 rads/year and a maximum ten times that. Takahashi conducted a cytogenetic study of the scorpion found on the mountain.

TABLE 3. Morro do Ferro: Range of Radionuclide Concentration in Plants.

(pCi/kg)

Ra-228	169 - 10,303.
Th-228	68 - 2,200.
Ra-226	7 - 1,105.

TABLE 4. Morro do Ferro: Concentration of Thoron and Radon in Ground Holes

(pCi/l.)

	Average	Range
Thoron	16,790.	285 - 55,400.
Radon	3,300.	5 - 27,400.

The research work on Morro do Ferro has recently taken a different direction. This mountain is conservatively estimated to hold 12,000 tons of thorium. In view of the almost identical chemistries of thorium and plutonium, the deposit is now regarded as an analogue for modeling the transport of plutonium over geological time, once a depository has been breached. The mineral is near the surface of the mountain, and is being washed by tropical rains í70 cm per year. The equivalent amount of plutonium is greater than that which will be produced by the reactors in the United States up to the year 2050. It is thought that the deposit is from 60 to 80 million years old.

The drainage pattern is straightforward. The rain penetrates the surface through the mineral, reaches the water table some 70 meters below, and runs off through springs at the foot of the mountain. The sediments carried off are being analyzed. The water and the solids in suspension are also being measured. Early measurements show about 0.3 pCi/l in the water.

In the future the plume will be studied in detail. A portable X-ray fluorescent spectroscope will follow the pattern of the distribution of thorium, rare earths and uranium.

(1) Cullen, T.L. and Penna Franca, E. (ed.), "International Symposium on Areas of High Natural Radioactivity", Academia Brasileira de Ciências, Rio de Janeiro, 1977.

EVALUATION OF CANCER INCIDENCE FOR ANGLOS IN THE PERIOD 1969-1971 IN AREAS OF CENSUS TRACTS WITH MEASURED CONCENTRATIONS OF PLUTONIUM SOIL CONTAMINATION DOWNWIND FROM THE ROCKY FLATS PLANT IN THE DENVER STANDARD METROPOLITAN STATISTICAL AREA

C. J. Johnson

A plutonium-processing plant in Jefferson County, Colorado has released transuranic nuclides in exhaust plumes since 1953 (1,2). In addition to leaks, small particles can migrate through banks of high efficiency particulate air filters by diffusion, a quality shared with other alpha radiation emitters (3,4). This results in a "dissemination of the finest radionuclide particles throughout the area over a radius of several miles from the plant site", and "once airborne, these smallest particles are not noticeably reduced in number by gravitational settling up to three miles from the apparent point of origin and presumably reached much father afield" (5). Releases of Pu (Pu oxide) in exhaust from the plant (13 million m^3 daily from the main stack) ranged from an annual average concentration of 0.03 picocuries or 0.06 disintegrations per minute per cubic meter (pCi or dpm/m^3) in 1953 to 1.05 pCi or 2.33 dpm/m^3 in 1962 (6). Plutonium concentrations in air are consistently the highest in the U.S. Department of Energy 51 station worldwide monitoring network (7). Average daily concentrations of Pu in exhaust plumes were as high as 948 pCi/m^3, for the eighth day (12 mCi, or about 200 mg) after a fire in 1957 which burned out the filter system (8). There are no records of emissions for the seven day period during or after the fire, but those releases may have been 4 to 5 orders of magnitude greater. The releases of Pu and other transuranics in the fire represent the most important exposure to to the population near the plant during the period 1953-1971. The major route of exposure is the inhalation of airborne particles of Pu and other transuranium nuclides by people in the path of exhaust plumes from the plant, and, for those living near the plant, the inhalation of Pu in resuspended surface dust. Work in progress confirms the presence of Pu from the facility, identified by isotope ratio, in persons in the area (9).

A preliminary study of leukemia and lung cancer deaths compared eight census tracts around the facility with 19 census tracts with a similar population in the relatively uncontaminated part of the county. A higher age-corrected leukemia death rate was noted in the contaminated area (p=0.01) and the age-specific (45-64 years) death rate from lung cancer was about twice as great as for the control area (p<0.05) (10,11). A preliminary study of congenital malformations coded at birth found a rate of 14.5 per 1,000 births for a large suburban city near the plant compared with a rate of 10.4 for the remainder of the county, and 10.1 for the state of Colorado, a difference of interest (12).

In order to determine if exposure of a large population to small concentrations of Pu and other transuranics had produced a measureable effect on cancer incidence, the following investigation was conducted.

The incidence of cancer for the period 1969-1971 for each site was calculated for Pu isopleth areas of census tracts with decreasing concentration of Rocky Flats Pu (identified by isotope ratio) in soil, determined by an area-wide survey (core samples to a depth of 10 cm) carried out by the Atomic Energy Commission in the Denver area in 1972 (figure 1)(13,16). The position of the isopleths are approximate but useable in comparing the incidence of health effects between areas with decreasing environmental contamination around a point source of emission. The Pu content of soil was used as a surrogate measure of exposure through pathways other than those that originate from the soil (i.e. an indication of direction of exhaust plumes from the Rocky Flats plant since 1953). Census tracts divided by isopleths were counted in the area containing the major part. Age-specific cancer rates for whites, excluding persons with Spanish surname, (typical of the area near the plant) were calculated for the Denver SMSA, and sub-areas with similar population size and decreasing soil concentrations of Rocky Flats Pu were compared to the rates for the SMSA as a whole, yielding age-corrected cancer incidence rates per 100,000 per year, observed/expected case numbers, risk ratios and chi square values for males and females. Area I within the Pu concentration range 40-0.8 millicuries/km^2 (mCi/km^2), lies between 3 and 21 km from the center of the Rocky Flats plant site along the principal wind vector. Area II (0.8 to 0.2 mCi/km^2) extends from 21 to 29 km and Area III (0.2 to 0.1 mCi/km^2) from 29 to 35 km. Area IV, the unexposed population, had an age-adjusted cancer incidence virtually identical to that for the state for males, 269 and 270 per 100,000 respectively, and for females, 226 and 227 per 100,000 respectively (13). The risk ratio for Area IV was assumed to be 1.0 and the exposed populations (Areas I-III) were compared to Area IV.

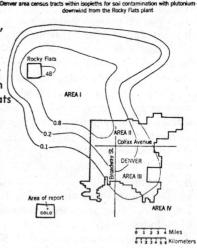

Fig. 1
Denver area census tracts within isopleths for soil contamination with plutonium - downwind from the Rocky Flats plant

Isopleths in millicuries of plutonium per square kilometer

The total incidence of cancer for the period 1969-1971 is summarized in Table 1 by isopleth area of Pu concentration. Compared to males in the unexposed area, there was an incidence of cancer 8%* higher (i.e. a proportionate morbidity ratio of 1.08) for males in Area III, most distant from the plant, 15%** higher in Area II, nearer the plant, and 24%** higher in Area I, nearest the plant. The corresponding values for females (all higher than expected) were 4%, 5% and 10%*, and for both, 6%, 10%** and 16%**.

The total population exposed to low concentrations of Pu (Areas I-III) can be compared to the unexposed population (Area IV) in Table 2. The incidence of cancer of lung and bronchus is higher for males* and for both sexes together**. The leukemia incidence is higher for both sexes*, and the incidence of lymphoma and myeloma for males*. The incidence of cancer of the tongue, pharynx, and esophagus was higher for males**, females**, and for both sexes**, as was cancer of the colon and rectum (males*, females** and both sexes**). The incidence of cancer of the liver and "biliary" was higher for males**. Both sexes had a higher

------

* critical X^2 value at 95% confidence level. **critical X^2 value at 99% conf. l.

incidence of cancer of the gonads (testis, 40/18** and ovary, 159/127**) with the greatest proportionate morbidity ratio for testis (2.22). This finding is consistent with the propensity of Pu to concentrate in gonads (17,18). The incidence for sites of cancer not listed ("other" in this table) was small in number for most sites but in toto was higher than the incidence in the unexposed population (males** and both sexes combined**).

Within the exposed area, the incidence of lung and bronchial cancer for males was about 33% higher** in Area I (a suburban area with more in-migration) than for males in uncontaminated Area IV (also predominantly suburban). This higher incidence persisted in Areas II and III (46%** and 13%, respectively). Females

Table 1
Census tract areas selected by decreasing soil concentrations of Rocky Flats plutonium, Anglo population size, median income and education, and total incidence of cancer, age-adjusted, for 46 sites, by sex, for the period 1969-1971 (a).

	Plutonium mCi/km^2	Anglo population male	Anglo population female	median education years	median income	Incidence of cancer compared to unexposed population					
						Male cases obs/exp(b)	% (c)	Female cases obs/exp	%	Total cases obs/exp	%
Area I	48 - 0.8	75,250	78,920	12.04	$8,891	644/519**	+24%	636/581*	+10%	1280/1100	+16%**
Area II	0.8 - 0.2	90,300	103,900	11.85	6,367	1086/947**	+15%	1154/1100	+5%	2240/2047	+10%**
Area III	0.2 - 0.1	117,370	129,530	12.69	12,094	1078/1000*	+8%	1149/1109	+4%	2227/2109	+6%*
Areas I-III	48 - 0.1	282,920	312,350	12.22	8,668	2808/2466**	+11%	2939/2790**	+5%	5747/5256	+9%**
Area IV	< 0.1	210,670	213,190	12.97	8,055	1114	0	1260	0	2374	0

Table 2
Summary: Anglo cancer incidence by site and sex in the Denver metropolitan area over a period of three years (1969-1971) by areas of census tracts with and without plutonium soil contamination by the Rocky Flats plant (a).

	Areas I-III 50 - 0.1 millicuries/kilometer^2						Area IV (unexposed)		
	Male		Female		Total		Male	Female	Total
Site	cases obs/exp(b)	pmr(c)	cases obs/exp	pmr	cases obs/exp	pmr	obs.	obs.	obs.
Lung and Bronchus	497/383*	1.30	128/120	1.07	625/503**	1.24	174	51	225
Other Respiratory	67/62	1.08	12/12	1.0	79/74	1.07	32	5	37
Leukemia	92/84	1.10	100/83	1.20	192/167*	1.15	45	38	83
Lymphoma, Myeloma	134/110*	1.22	109/123	0.89	243/233	1.04	59	56	115
Tongue, Pharynx, Esophagus	89/50**	1.78	41/17**	2.41	130/67**	1.94	24	7	31
Stomach	79/78	1.01	59/53	1.11	138/131	1.05	34	27	61
Colon, Rectum	379/333*	1.14	433/378**	1.15	812/711**	1.14	144	146	290
Liver and Biliary	52/31**	1.68	49/54	0.91	101/85	1.19	5	3	8
Pancreas	96/106	0.91	88/77	1.14	184/183	1.01	46	30	76
Testis	40/18**	2.22	---		---	2.22	13	-	13
Ovary	---		159/127**	1.25	---	1.25	-	63	63
Thyroid	22/28	0.79	80/71	1.13	102/99	1.03	18	42	60
Brain	40/48	0.83	39/34	1.15	79/82	0.96	27	20	47
Other	1221/1129**	1.08	1642/1570	1.05	2863/2699**	1.06	493	772	1265
All Cancer	2808/2466**	1.14	2939/2790**	1.05	5747/5256**	1.09	1114	1260	2374

(a) From the National Cancer Institute's Third National Cancer Survey: Incidence Data (21) (Calculated from age-adjusted rates, obtained by pooling the age-specific rates using the age distribution of the total 1950 population of the United States as a set of weights.) "Anglo" includes all whites except those with Spanish surname.

(b) $X^2 = \frac{(|obs. - exp.| - 0.5)^2}{npq}$ where n=population size, p=incidence of cancer, and q=1-p. The X^2 used with the variance = npq is a more conservative test than the Mantel-Haenszel X^2 (19)

* Critical X^2 value at a 95% confidence level. ** Critical X^2 value at a 99% confidence level.

(c) pmr - proportionate morbidity ratio observed/expected -1) x 100, compared to Area IV, the unexposed population.

did not have a higher incidence of lung and bronchial cancer near the plant, but did have a higher incidence in Areas II and III. A higher incidence (about 50% greater) of neoplasms of the nasopharynx and larynx ("other respiratory") was observed for males in Area I. The incidence of leukemia (all types) was higher for males in Area I near the plant (42%), but not for females, and higher for both males and females in Area III (9% and 58%**, respectively). The incidence of lymphoma and myeloma was 40% higher for males, and 12% higher for females in Area I. The higher incidence persisted in Areas II and III for males (29% and 12% respectively) but not for females. Cancer of the pancreas occurred with a greater than expected incidence for females in all three study areas (49%, 9% and 7% respectively) but not for males. There was a higher incidence of cancer of the thyroid for females in Areas I and II (33/26). The incidence of cancer of the stomach was higher for males in Area I, but not for females. The incidence of cancer of the colon and rectum was much higher for both males and females in Area I and II (42% higher** in Area I and 11% higher in Area II for both sexes). The incidence of cancer of tongue, pharynx, and esophogus was high for both sexes in all three study areas, especially in Area II (68/25* for both sexes).

Areas II and III include the Denver urban core, with the low income housing, lower educational and income level, and greatest air pollution (factors associated with a higher incidence of cancer) but have a lower incidence of cancer than Area I, a suburban area near the Rocky Flats plant similar to Area IV, the unexposed area with the same cancer incidence as the state of Colorado. The consistency of the increase in incidence of all cancer and of certain categories of cancer with increasing concentrations of Pu in soil supports the hypothesis that exposure of the general public to low concentrations of Pu in the environment may have an effect on cancer incidence.

References

1. Anon.(1977): Draft environmental impact statement, Rocky Flats plant site, Golden, CO. U.S.E.R.D.A. 1545-D.
2. Johnson, C.J., Tidball, R.R. and Severson, R.C. (1976): Science, 193, 488.
3. Hayden, J.A. et al (1972): Unpublished reports 317-72-165 (two reports) and ES-376-84-118 (1974). Dow Chemical Company, Rocky Flats Plant, Golden, CO 80401
4. McDowell, W.J., Seeley, F.G. and Ryan, M.T. (1977): Health Physics, 32, 445.
5. Nichols, H. (1977): Transactions of Meeting on Rocky Flats Buffer Zone, Rockwell International, Rocky Flats plant, Golden
6. Anon. (1963): Review of the exhaust air filtering and air sampling, Bldg. 71. Unpublished report, Rocky Flats Environmental Master File. The Rocky Flats plant, Golden, CO 80401.
7. Toonkel, L.E., Feely, H.E. and Larsen, R.J. (1979): Radionuclides and trace metals in surface air. Environmental Quart. the Environmental Measurements Laboratory. U.S. Dept. of Energy, New York, N.Y. 10014, pp C/1-160.
8. Anon. (1957): Report of investigation of serious incident in Bldg. 71, on September 11, 1957. Dow Chemical Company, Rocky Flats plant, Golden, CO 80401.
9. Cobb, J. (1978): Personal communication. University of Colorado School of Medicine, Denver, CO.
10. Johnson, C.J. (1977): Proceedings of the 105th Annual Meeting of the American Public Health Association, Wash., D.C.
11. Johnson, C.J. (1979): Proceedings of the 145th National Meeting of the American Association for the Advancement of Science, Houston, Tx
12. Johnson, C.J. and Van Deusen, K.V. (1978): Abstract. Proceedings of the 106th Annual Meeting of the American Public Health Association, San Francisco, Ca
13. Anon. (1975): Third National Cancer Survey: Incidence Data. National Cancer Institute Monograph 41, DHEW Pub. No. 75-787 U.S. DHEW Public Health Service, National Institute of Health, National Cancer Institute, Bethesda, MD 20014.
14. Berg, J. (1979): Unpublished work. The Colorado Regional Cancer Center, Denver, CO, N.I.H. grant #CA 25729-01.
15. Monson, R.R. (1974): Computers and Biomed Res. 7, 325-332.
16. Krey, P.W. (1976): Health Physics, 30, 209.
17. Smith, D.D. and Black, S.E. (1975): NERC-LV-539-36, National Environmental Research Center, EPA, Las Vegas, NV
18. Green, D., Howells, G.R., Humphreys, E.R. and Vennart, J. (1975): Nature 225, 17.
19. Snedecor, G.W. and Cochran, W.G. (1976): Statistical Methods, Sixth Ed., Iowa State Univ. Press, Ames, IO, p 212.
20. Acknowledgement: Assistance of Colorado Regional Cancer Center staff (Dr. John Berg and Dr. Jack Finch, N.C.I., N.I.H. grant CA 17060); the Division of Vital Statistics of the Colorado Department of Health; Kathryn Van Deusen, who assisted the author with the analysis of the data; Mr. James T. Martin, who provided statistical support; James J. Doyle of the U.S. Geological Survey, who assisted with the figure, and those who reviewed the manuscript. Supported by grant CA 25729 from the National Cancer Institute, National Institutes of Health.

HAZARDS OF RADON DAUGHTERS TO THE GENERAL PUBLIC

D. K. Myers, J. R. Johnson and A. M. Marko

Recent data from various countries in northern latitudes suggest that the most important source of public exposures to ionizing radiation may well be the inhalation of radon daughters which accumulate inside buildings; these exposures result in appreciable radiation doses to the bronchial epithelium and smaller doses to the pulmonary region of the lung, with little increment in radiation dose to other organs of the body (1).

The incidence of lung cancer resulting from inhalation of radon daughters by the general public is not known. Considerations based on vital statistics suggest that appropriate risk estimates for the general public probably lie in the range 0 to 200 fatal lung cancer per million working level months (WLM) (2,3). Risk estimates can be derived from analyses of data on excess lung cancer incidence observed in uranium miners who were in the past exposed to high concentrations of radon daughters. These data can be fitted by linear dose-response models, by "quasi-threshold" models involving, for example, a probit response on log dose (3) or by curvilinear relationships which exhibit a decreasing response per unit dose with increasing dose (4). The available data are not precise enough to distinguish between linear and "quasi-threshold" models as representing the most probable fit, but do indicate that the curvilinear relationship (4) is statistically much less probable than either of the first two models (3).

Assuming a linear dose-response relationship, and assuming that most of the fatal lung cancers induced in uranium miners occur between 10 and 25 years after the initial exposure to radon daughters (5), the available data from the U.S.A. and Czechoslovakia indicate about 100 fatal lung cancers per million·WLM in uranium miners, with a range from 50 to 200 in these estimates (3). The more limited data available from other countries including Canada (6) are not incompatible with this estimate.

If this risk estimate is directly applicable to the general public, and assuming that average public exposures due to accumulation of radon daughters in buildings are in the region of 0.08-0.16 WLM per year (1,7), then inhalation of radon daughters by the general public would be responsible for 8-16 lung cancers per million persons per year. This is equivalent to 3-6% of all lung cancers or about 0.8% of all fatal cancers in Canada in recent years. Other published risk estimates (1,4,7) would yield percentages which are considerably higher. However, even the lower numbers suggested above indicate that the public health hazards due to radon daughters are greater than those due to 0.1 rem whole body radiation per year either from natural background radiation (approximately 0.3-0.5% of all fatal cancers, based on the absolute risk model) or from medical diagnostic procedures. This conclusion remains the same whether absolute or relative risk models should prove to be most appropriate. All of the percentages are increased 2-3 fold when the relative risk model is used (7,8), even though the loss of life expectancy is essentially the same with the absolute or relative risk models (7,9).

The concentrations of radon daughters in open air, inside buildings and in air entrapped in soil are usually in the region of 0.0006, 0.003-0.005 and 0.3-10 WL respectively (1,7,10). The primary sources of radon and thus of radon daughters inside buildings are the building materials themselves and radon which enters the building from the soil through various openings in the building foundations; dissolved radon in water from wells may also form an appreciable source in some cases. The concentration of radon daughters inside buildings is strongly influenced by the ventilation rate (2). The possible health hazards of decreased ventilation in buildings and the cost of heating the air required to provide extra ventilation during winter months in a cold climate have been calculated on the basis of the following assumptions: (a) An average of 100 cubic metres of enclosed building space per person. (b) Approximately 5,000°C-days of heating required, primarily over six months of the year, as is true for the Ottawa region in Canada. (c) Approximately $120 (U.S.) per person required at current 1979 prices to heat 100 cubic metre of air at a ventilation rate of one change of air per hour over 5,000°C-days. (d) An average exposure of 0.14 WLM per year at one change of air per hour (2), assuming that 80% of a person's time is spent inside the building. (e) An increase in radon daughter concentrations at decreased ventilation rates in proportion to the values calculated by Cliff (2). (f) A risk of 100 fatal lung cancers per million WLM. The results of this calculation are shown in Table 1.

Current ventilation rates in Canadian homes during winter months are estimated to be about 0.3 to 0.5 air changes per hour (11,12). The cost of heating air to provide extra ventilation during the winter months is in the region of one to ten million dollars (U.S.) per fatal lung cancer avoided (Table 1) based on the assumptions listed above. Factors that would affect this result are:

a) If a heat exchange unit were installed between the outgoing warm air and incoming cold air in a forced ventilation system, costs of heating this air could be reduced by 30 to 40 percent. Offsetting this saving would be the cost of the heat exchange unit.

b) The actual radon daughter concentration in a house with a "forced air" central heating system, common to many Canadian homes, could be much less than taken from Cliff (2) for the same radon output. The rapid mixing of air in such a forced air system would result in increased plate-out of radon daughters onto the walls of the heating ducts and the coarse filters in the system. Wrenn et al. (13) have shown that this effect can reduce the concentrations in air by up to a factor of 10 under appropriate conditions.

c) The assumption that a linear exposure response relationship will extend from the high exposures accumulated by miners down to the exposure and exposure rate applicable in most buildings may overestimate the actual risks. There is some evidence to suggest that very low concentrations of alpha-emitters may not produce cancers within the normal life-span (14).

TABLE 1. Estimated benefits and costs of increased ventilation in buildings

(i) Increase in ventilation rate (changes of air per hour).	(ii) Decrease in exposure to radon daughters (WLM) for the six month heating period	(iii) Fatal lung cancers avoided (per million persons).	(iv) Cost of heating air to provide increased ventilation (millions of dollars per million persons).	(iv/iii) Cost/benefit (millions of dollars per lung cancer avoided).
from 0.1 to 0.2	from 0.94 to 0.46	48	12	0.25
from 0.2 to 0.5	from 0.46 to 0.16	30	36	1.2
from 0.5 to 1.0	from 0.16 to 0.07	9	60	6.7
from 1.0 to 2.0	from 0.07 to 0.032	4	120	30.

d) Risk estimates derived from studies of uranium miners may not be valid for the general population due to differences in smoking habits, age distribution, and exposure to dust and other factors.

This paper has attempted to quantify the cost-benefit relationship for decreasing radon daughter exposures in Canadian homes by increasing ventilation rates. It has identified the main components in this relationship and pointed out the uncertainties associated with some of them. The uncertainties in the calculated values appear to be related primarily to the most appropriate risk estimates for inhalation of radon daughters and to actual radon daughter concentrations inside buildings at various ventilation rates.

REFERENCES
1. United Nations Scientific Committee on the Effects of Atomic Radiation (1977): Sources and Effects of Ionizing Radiation. United Nations, New York.
2. Cliff, K.D. (1978): Phys. Med. Biol., 23, 696.
3. Myers, D.K. and Stewart, C.G. (1979): Some Health Aspects of Canadian Uranium Mining. Atomic Energy of Canada Limited, Report AECL-5970.

4. Archer, V.E., Radford, E.P. and Axelson, O. (1979): In: Conference/Workshop on Lung Cancer Epidemiology and Industrial Applications of Sputum Cytology, p. 324. Colorado School of Mines Press, Golden, Co.
5. Sevc, J. and Placek, V. (1973): In: Proceedings of the Sixth Conference on Radiation Hygiene, Czechoslovakia, p. 305.
6. Report of the Royal Commission on the Health and Safety of Workers in Mines (1976). Government of Ontario, Canada.
7. Ellett, W.H. and Nelson, N.S. (1979): In: Conference/ Workshop on Lung Cancer Epidemiology and Industrial Applications of Sputum Cytology, p. 114. Colorado School of Mines, Golden, Co.
8. Report of the Advisory Committee on the Biological Effects of Ionizing Radiation (1972): U.S. National Academy of Sciences, Washington, D.C.
9. Cohen, B.L. and Lee, I.S. (1979): Health Physics, 36, 707.
10. Report on Investigation and Implementation of Remedial Measures for the Reduction of Radioactivity Found in Bancroft, Ontario and Its Environs (1979): James F. MacLaren Limited, Willowdale, Ontario, Canada.
11. Myers, D.K. and Newcombe, H.B. (1979): Health Effects of Energy Development. Atomic Energy of Canada Limited, Report AECL-6678.
12. Smith, D. (1979): In: Second Workshop on Radon and Radon Daughters in Urban Communities Associated with Uranium Mining and Processing. Atomic Energy Control Board, Ottawa, Ontario, Canada.
13. Wrenn, M.E., Eisenbud, M. and Costa-Ribeiro, C. (1969): Health Physics, 17, 405.
14. Bair, W.J. and Thompson, R.C. (1974): Science, 183, 715.

LUNG DOSES FROM RADON IN DWELLINGS AND INFLUENCING FACTORS

E. Stranden

In recent years there has been a growing interest in the naturally occuring radiation, and in the possible biological effects of small exposures. One of the most interesting problems in this field seems to be the radon concentrations in dwellings and the associated lung cancer hazards to the population, and in this paper we will discuss some factors that have influence upon the population doses from inhalation of radon and its decay products in dwellings.

RADON IN DWELLINGS AND INFLUENCING FACTORS

There are several sources of radon to the indoor air, i.e., radon from building materials and the ground, well water and natural gas used in the household. In Norway, well water and gas is used very infrequently, but in some other countries such as Finland, radon supplied from water is a large problem (1).
The activity concentration of radon inside a room under steady state conditions may be expressed as:

$$C = \frac{E \cdot F}{V \cdot \lambda_v} \qquad (1)$$

where: C = radon concentration (Bqm^{-3})

E = exhalation rate of radon ($Bqm^{-2}h^{-1}$)

F = area of the radon source (m^2)

V = volume of the room (m^3)

λ_v = ventilation rate (h^{-1})

Recently we measured the radon exhalation from some main building materials. In table 1, the radon exhalation rate pr unit radium activity concentration is shown for some typical walls.
Measurements of the radium concentration of building materials from the whole country indicated the following values: Concrete: 28 $Bqkg^{-1}$, Brick: 63 $Bqkg^{-1}$, LECA: 52 $Bqkg^{-1}$. For houses with walls of concrete, the ratio F/V may be about 1.8 m^{-1}. Using this value, we find that the mean radon concentration inside concrete houses in

Norway may be between about 50 and 90 Bqm^{-3} for ventilation rates between 0.3 and 0.5 h^{-1}. These ventilation rates may be commonly found in modern houses in the Nordic countries. Recent measurements in Norwegian houses indicated a mean value of 74 Bqm^{-3} (2) in concrete houses, which seems to be in good agreement with the values found by exhalation measurements. The measured values in brick and wooden buildings were 37 and 48 Bqm^{-3} respectively.

Table 1. Radon exhalation rate pr unit radium activity concentration for some walls

Material	Typical wall thickness (cm)	Exhalation rate ($Bqm^{-2}h^{-1}/Bqkg^{-1}$)
Concrete	20	0.50
Brick	20	0.20
LECA	20	0.26

The radon concentration inside rooms is highly dependent on the ventilation rate. The ventilation rate is dependent on factors such as wind speed and direction and the temperature, while the exhalation rate is dependent on the atmospheric pressure. The radon concentration in houses will therefore show large variations (2,3). An example of this is shown in figure 1.

DOSES

The doses to the respiratory system is dependent on the radon daughter concentration. There will not be equilibrium between the radon and its daughters because of the ventilation and deposition of daughter products on walls, furniture etc. Measurements (2) have indicated that an equilibrium factor of about 0.5 may be representative for Norwegian houses. The mean radon concentration in Norwegian houses is about 50 Bqm^{-3} (2), and this indicates that the radon daughter concentration is about 0.007 WL (Working Levels). For a person spending 80% of his time within doors, the annual exposure to radon daughters is about 0.3 WLM (Working Level Months).

There are large discrepancies in the dose estimates for the respiratory tract. The risk estimates for miners are all given pr WLM, so this concept seems to be more useful in evaluations of the biological effects of inhaled radon daughters. In a recent study (4), we discussed the possible lung cancer incidence in Norway from inhalation of radon daughters in dwellings. From the lung cancer statistics, and from modified risk factors we concluded that about 30 cases pr year and 10^6 persons may be due to radon daughters in Norwegian dwellings. If we now use the ICRP concept of effective dose equivalent, this indicates that an exposure of 1 WLM in houses may be equivalent to an effective whole body dose equivalent of 10 mSv. If we adopt this, we may study the influence of reduced ventilation upon the population doses from radon daughters in dwellings. This is shown for a few ventilation rates in table 2. The values are calculated for concrete houses, using the exhalation rate given in table 1. The values are calculated for a radium concentration of 100 Bqkg^{-1}, which is proposed as an excempt limit by the OECD (5).

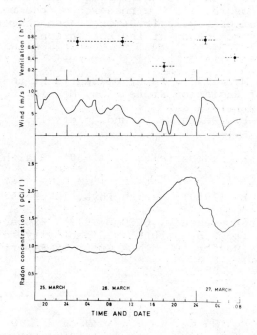

Figure 1. Measurements of wind speed, ventilation rate and radon concentration in a testhouse (ref. 2).

Table 2. The effective dose equivalent from inhalation of radon daughters in concrete houses for different ventilation rates. Radium concentration: 100 Bqkg^{-1}

Ventilation rate (h^{-1})	Daughter exposure (WLM/year)	Effective dose equivalent (mSv/year)
0.1	7	70
0.3	1.8	18
0.6	0.7	7
1.0	0.3	3

The population effective dose equivalent from inhalation of radon daughters in Norwegian houses today is about 3 mSv/year. This is the largest contributor to the population exposure, and the research on the radon problem should be given high priority among those concerned with the radiation protection of the population.

REFERENCES

1. Annamäki, M. (1978): Paper presented at the Nordic Society of Radiation Protection, Meeting 5, Visby.

2. Stranden, E., Berteig, L. and Ugletveit, F., (1979): Health Phys., 36, 413.

3. Stranden, E. and Berteig, L. (1980): Health Phys. in press.

4. Stranden, E. (1980): Health Phys. in press.

5. Nuclear Energy Agency, OECD, (1979): Exposure to radiation from the natural radioactivity in building materials, Report by an NEA Group of Experts, Paris May 1979.

ASSESSMENT OF BIOLOGICAL EFFECTS RESULTING FROM LARGE SCALE APPLICATIONS OF COAL POWER PLANT WASTES IN BUILDING TECHNOLOGY IN POLAND

J. Pensko and J. Geisler

INTRODUCTION AND RISK ESTIMATES

The aim of this work is to evaluate radiation-induced remote biological effects to the population of Poland /34.2 millions in 1975/ due to the hitherto progressing and the expected development of the building technology. From among the harmful biological effects only severe genetic damage, leukemia and malignant tumors are considered. Serious genetic damage is considered in two groups. One of them comprises the effects which would be expressed over several generations following the irradiation of the parents. The other group consists of effects which would be manifested in the first generation after the exposure, thus being virtually caused by dominant mutations. The value of 10.5 severe genetic defects in the first generation and 300 over all generations of the progeny of irradiated parents per 1 rad per 10 births are accepted as genetic risk factors of chronic irradiation at low dose rates. The risk of induction of harmful somatic effects is evaluated in terms of absolute risk, i.e. as the difference between the risk of the irradiated and non-irradiated population. On the basis of the linear model of the dose-response curve, the absolute risk is expressed as a number of excess cancer inductions or deaths observed in a given population per year and rad or rem. For calculating the absolute risk factors the radiation risk estimates inserted in BEIR-72 and UNSCEAR-72,-77 reports are used. Risk coefficients are determined as average values weighted for products of numbers of irradiated persons times years of observations.

The risk factors for leukemia are estimated in two age groups: in children 0-9 years old /0.96 excess deaths per 10^6 man-rems per year/ and in older people /2.22 excess deaths per 10^6 man-rems per year/. In accordance to the BEIR-72 report the latent period is taken for

2 years and plateau region for 25 years.

Absolute risk of death from cancer /excluding leukemia/ also varies with age at irradiation. Two age groups are distinguished, viz. people younger than 20 and adults, with the respective risk coefficients of 0.73 and 4.20. Latent periods of 15 years and a plateau regions of 30 years are accepted for both groups.

As the concentrations of Rn-222 and its daughters in the indoor air are usually higher than the natural levels it is reasonable to make an evaluation of the lung cancer risk as a specific kind of risk in a population exposed to these excess concentrations. In this case the risk factor of 0.79, a 15-year latent period and a 30-year plateau region are accepted.

The risk of death from cancer, attributable to the prenatal irradiation of an embryo or fetus, is evaluated for the first 10 years of life. A risk factor of 23 deaths per 10^6 pregnant mothers at risk per year rem is used. No latent period and a plateau region of 10 years are assumed.

METHOD OF CALCULATION

The general structure of types of buildings in Poland in each decade since 1950 up to the end of 2010 is determined with the use of statistical data concerning building development and consumption of building materials. Distribution of the population between the particular types of buildings is assumed to be consistent with the presupposed building structure. For four main types of buildings the average additional dose equivalent rates of gamma radiation in air, soft tissue, bone marrow and gonads as well as dose equivalent rates of alpha radiation in bronchial epithelium are estimated and considered as representative for the whole population. The residence time coefficient 0.8, the tissue screening factors for gamma radiation and the equilibrium factor $F = 0.5$ for radon daughters are taken into account. The estimated mean excess dose equivalents in various tissues, weighted for the population distribution among the various types of buildings, are within the range of 12 to 32 mrem/year for gamma radiation and of 276 to 1108 mrem/year for alpha radiation.

The average countrywide dose rate inside buildings varies from one decade to another because of variations in the building structure and the widespreading use of technologies exploiting industrial wastes for the production of building materials. The actual dose cumulated by the population is the sum of individual absorbed doses and its effects depend upon the age distribution of the population and the durations of the latent and the elevated incidence periods.

Somatic effects are calculated according to Johnson's method /1/ with the use of the above mentioned risk factors. As these values vary with the age of the irradiated persons, fractional and summarized risk factors are introduced for each 10-year periods. The appearence of the somatic effects of the irradiation extents for long periods which do not coincide with the calculation intervals. Therefore for each decade effective values of average annual dose equivalents to bone merrow, soft tissue and bronchial epithelium were calculated with allowance made for irradiation period and the plateau region. Basing on the above mentioned considerations and data the number of somatic effects occuring in a decade is calculated.

RESULTS AND DISCUSSION

Our assessments of risks indicate that the considerably growing use of power plant wastes for the production of building materials, creates an increased risk of death from neoplasms and genetic diseases. In the considered period /1951 - 2010/ the number of leukemias due to that reason are expected to increase more than twice /246 cases in the last decade/, the number of malignant neoplasms nearly three times /1049 cases in the last decade/ and the number of lung cancers nearly five times /8459 cases during 2001-2010/. By the same period only a relatively small increase of about 13 per cent should be observed in the serious genetic defects from the same cause. The malignant neoplasms in children caused by the excess irradiation of pregnant mothers during their stay inside buildings can be neglected as the irradiated population is small and the time of the irradiation reletively short.

According to the present evaluation, the serious somatic effects of the excess indoor irradiation, expressed in absolute numbers, amount in Poland through 1951 to 2010 to more than 31,000 cases. That figure comprised nearly 1000 leukemias and nearly 4000 other malignant neoplasms due to the elevated indoor irradiation, as well as more than 26,300 lung cancers caused by the excess concentrations of Rn-222 and its daughters in the air inside buildings. Deaths caused by the deleterious genetic effects during the same period are reletively small and should not exceed 260 cases in the first generation and 7,500 cases in the whole progeny.

Confrontation of the calculated number of these somatic effects with the death rate of neoplasms from all causes reveals the reletively high contribution of the indoor irradiation in the overall incidence of neoplasms. That contribution in the decade 1971-1980 should equal nearly 3 per cent of the cancers of the respiratory system and 0.7 per cent of all neoplasms in Poland.

TABLE 1. The expected numbers of leukemias, malignant tumors and severe genetic demages induced by the excess gamma-ray doses and numbers of lung cancers induced by the excess alpha radiation doses to the inhabitants of various types of buildings in Poland.

Biological effect	Decade						Total
	1951 1960	1961 1970	1971 1980	1981 1990	1991 2000	2001 2010	
Leukemias	108	121	138	174	209	246	996
Malignant tumors /excluding leucemias/	383	454	560	703	768	1049	3,917
Lung cancers	1826	2258	2999	4416	6336	8459	26,294
Malignant neoplasms after irradiation of embryo or fetus	2	2	2	3	3	3	15
Genetic effects in the first generation	46	33	41	44	45	51	260
Genetic effects in the whole progeny	1300	951	1159	1254	1289	1469	7,422

REFERENCES

1. Johnson, R.H., Bernhardt, D.E., Nelson, N.S., Calley, H.W. /1973/ : EPA-520/1-73-004, Washington.

Justification and Optimization

JUSTIFICATION AND OPTIMIZATION IN RADIATION PROTECTION

D. Beninson

1. INTRODUCTION

1. To meet the objectives of radiation protection the ICRP (1) has recommended the use of a system of dose limitation composed of the following requirements: 1) Justification of practices involving radiation exposures; 2) Optimization of the level of protection for such practices; 3) Individual dose limitation. The third requirement is individual-related, and is the continuation of previous recommendations limiting the risk to individuals from exposure to radiation. The first two requirements, on the other hand, are source-related. They apply even if all individuals are so well protected that their risk is negligible, requiring that the radiation detriment from a given source be reduced by increasing protection to the optimum level, and that the practice (with its remaining radiation detriment) be justified by benefits.

2. The ICRP has recommended the use of cost-benefit analysis in the assessment of justification of practices involving radiation exposures and in the optimization of radiation protection (1). The concept of "net benefit" from the introduction of a practice involving radiation exposures was defined in that publication in symbolic form as:

$$B = V - (P + X + Y)$$

where B is net benefit from the introduction of the practice, V is the gross benefit, P is all production costs excluding protection costs, X is the cost of achieving a selected level of protection and Y is the cost of detriment associated with that level of protection.

2. QUANTIFICATION OF THE RADIATION DETRIMENT

3. The application of cost-benefit analysis requires the assignment of quantitative values to X and Y in all cases, and for some applications to V and P. While P and X costs are readily expressed in monetary terms, V may contain components difficult to quantify. The quantification of Y, the cost of the radiation detriment, is regarded as the most problematic and the most controversial of the quantifications. Nevertheless, it is essential for the application of cost-benefit analysis to radiation protection.

4. Optimization of protection takes place in a region of low individual doses, always smaller than a fraction of the dose limits. Therefore, only the induction of somatic and genetic stochastic effects of radiation will contribute to the deleterious health consequences, as non-stochastic effects would be totally prevented. In order to deal with the risk of stochastic effects, the ICRP uses the quantity:

$$H_E = \sum_T w_T H_T$$

where H_E is a sum of weighted organ dose equivalents, called the "effective dose equivalent", w_T is a factor representing the fraction of risk resulting from tissue T when the whole body is irradiated uniformly, and H_T is the dose equivalent in tissue T. The recommended values of w_T are given in ICRP publication 26; and additional value for skin exposures has also been provided by the ICRP (2).

5. The "detriment" in an irradiated population group is defined as the expectation of the harm incurred, taking into account not only the probabilities of each type of deleterious effect but also the severity of the effects. If P_i is the risk of suffering the effect i, the severity of which is measured by a factor g_i, then the detriment G in a group of N persons is $G = N \sum_i P_i g_i$.

6. For stochastic effects it is assumed that increments of risk are proportional to increments of dose. Then p_T, the probability of suffering a stochastic effect in tissue T can be taken to be proportional to the average dose received in that tissue

$$p_T = r_T H_T$$

r_T being a risk factor per unit dose equivalent. When this is substituted into the equation for detriment, the detriment of one person is given by

$$G_1 = \sum_T r_T H_T g_T$$

Several approaches are possible to quantify the severity factors g_T (3). For radiation protection purposes it could be assumed as a first approximation that the detriment is dominated by the induction of fatal malignancies and of servere genetic effects in the first two generations, assigning a severity factor of one to all these effects. In this case, the effective dose equivalent would be proportional to the individual health detriment because

$$G_1 = \sum_T r_T H_T = R H_E,$$

where R is the total risk for whole body irradiation, and $w_T = \frac{r_T}{R}$. The value of R is taken to be 1.65×10^{-2} Sv^{-1} (3).

7. This first approximation, however, neglects the contribution to the detriment of subsequent generations after the second, and of non-fatal malignancies, which are not taken into account in the definition of effective dose equivalent. The contribution of subsequent generations to the detriment could be roughly taken into account by adding a term $w_{gon} H_{gon}$ to the effective dose equivalent. Non-fatal malignancies could probably be neglected when compared to fatal malignancies. Several attempts to quantify their contribution, which are very controversial, support the idea of neglecting such contribution to the detriment.

8. The detriment is an extensive quantity. The detriment from a given source is therefore the summation of the detriments of all individuals irradiated by the source, either at present or in

the future. It follows (4) that the detriment from the source k, G_k, is given by

$$G_k = R \sum_i N_i \overline{H}_{E,i} = R \cdot S^C_{E,k}$$

where R is the risk factor for whole body irradiation, N_i is the number of individuals receiving an average effective dose equivalent $\overline{H}_{E,i}$ from the source, and $S^C_{E,k}$ is the collective effective dose equivalent commitment from the source.

9. As the detriment is an expectation of death (and of serious genetic harm), the assignment of a cost to the detriment involves some valuation of human life. In fact, countless policy decisions affect the incidence of death and none tries to minimize this incidence regardless of cost. Implicit in any of such decisions, therefore, is some valuation of human life.

10. A key feature of the modern approach for taking account of life in cost-benefit analysis is that it does not value life as such, but only changes in the probability of death. Being the detriment a mathematical expectation of death, the assignment of a cost to the detriment would fit well with the quoted approach. As the detriment is proportional to the collective effective dose equivalent commitment, the problem reduces to the assignment of a monetary value to the unit of collective effective dose equivalent. Obviously, this assignment is a value judgement rather than a scientific determination. It has been attempted by assigning values to the increased probability of death, or by observation of the values society actually is willing to pay to reduce exposures in given practices.

11. With the first approach, values ranging from 20 to 200 dollars per man rem can be deduced from assessments of "cost of a statistical life" and a risk of 1 to 2 10^{-4} per rem. The second approach gives somewhat higher values for a man rem, up to a few hundred dollars. A value of about 100 to 200 dollars per man rem seems to be adequately representative, and could be used for planning purposes in those cases where the competent authority has not yet established the value to be used.

12. For the purpose of cost-benefit analysis in radiation protection, therefore, the cost of the detriment can be expressed as:

$$Y = \alpha \, S^C_E$$

where Y is the cost of detriment, α is the monetary cost assigned to the unit of collective effective dose equivalent and S^C_E is the collective effective dose equivalent commitment associated with the level of protection under consideration.

13. Problems associated with costs and detriments occurring over different time periods are frequent, especially when a practice leads to environmental contamination by long lived radionuclides and therefore to subsequent exposure in future populations. The concept of collective effective dose equivalent commitment allows the calculation of detriment in these cases giving the same weight to present and future detriments, which is not the usual practice in

other types of human judgements, which involve the traditional economical technique of discounting.

14. However, on ethical grounds it has been argued that discounting perhaps be properly applied within the time period of one generation, but that it should not be applied when a substantial part of the detriment will occur in future generations. Some have also expressed the opinion that it is not valid to discount the cost of the detriment (even if manifested in the future) committed from one year of practice, because only the present decision was relevant and the future harm was unavoidable. However, it would be legitimate to discount the cost of the detriment committed successively year after year of the practice.

3. OPTIMIZATION

15. A basic requirement of radiation protection is that all doses should be kept "as low as it is reasonably achievable", taking into account social and economical considerations. This requirement is usually called "optimization" of radiation protection and consists in reducing the collective dose (and thus the detriment) to a value such that further reductions are less significant than the additional efforts required to achieve such reductions.

16. Optimization, therefore, consists in an interplay of the cost of protection and the cost of the remaining detriment, in such a way that

$$X(w) + Y(w) = \text{minimum}$$

where X is the cost of protection, and Y is the cost of the radiation detriment, both at a level of protection represented by w (e.g., shielding thickness, ventilation rate, alternative options of protective equipment, etc.). It should be noted that w, and X(w) and Y(w), can in some cases be continuous, while in other cases they take only discrete values. It is obvious that the selection of the optimum pair of values for X and Y, would maximize the "net benefit" from the introduction of the practice, as defined in paragraph 2.

17. Some of the technical difficulties of optimization are related to the boundary condition introduced by the dose limits. As the limits apply to the combined exposure from all sources (except those specifically excluded), it is necessary to use a fraction of the limit as a boundary condition for the optimization of a given source. It is not the purpose of this paper to review the criteria to set such source upper bound, L, but to show its use as a boundary condition.

18. In the ideal optimization case, there is only one exposed group of individuals and one protection parameter or a simple set of protection options. Additionally, a basic requirement of this ideal case is the existence of a quantitative relationship between the collective effective dose equivalent commitment S, and the maximum annual effective dose equivalent, H^*, such as $H^* = f(S)$. Taking the detriment to be proportional to the collective dose (paragraph 12) and using the symbols defined previously, optimization in the ideal case can be expressed as the set of conditions (4)

Justification and Optimization

(1) $X(w) + \alpha S(w) = $ minimum
(2) $f(S) \leq L$.

19. The minimum for the first expression, usually called the objective function, can be obtained by differentiation and making the result equal to zero:

$$\frac{dX}{dw} = -\alpha \frac{dS}{dw}, \text{ or as usually presented,}$$

$$\left(\frac{dX}{dS}\right)_{S_o} = -\alpha$$

The optimized value S_o correspond to a given optimum protection parameter w_o and a given protection cost, S_o, because X can be expressed as a function of S, the function being called the constraining function.

20. The optimized value S_o must, however, comply also with the second condition of paragraph 18, namely the limit equation $f(S_o) \leq L$. Therefore, optimization is achieved at a value of collective effective dose equivalent commitment, S_o, such that

$$\left(\frac{dX}{dS}\right)_{S_o} = -\alpha$$

provided that $f(S_o) \leq L$, and at a value $S_o = f^{-1}(L)$ in all other cases.

21. Examples of application of this procedure of optimization have been published for radiation shielding and for ventilation design in installations handling radioactive materials, in uranium mines and in buildings (in relation to radon) (4)(5)(6). In many other practical cases of optimization assessments, the changes in protection levels are achieved in finite increments, both X and S being discrete instead of continuous variables. The decision of going from a level of control A to a more expensive level of control B would be taken if

$$-\frac{X_B - X_A}{S_B - S_A} \leq \alpha$$

Examples of application of this step by step procedure have been published relating to the control of release of radioactive effluents (4)(5)(6).

22. When exposures from a given source or practice can be regarded as composed of contributions of subsystems, each requiring appropriate protection measures, optimization implies that

$$\sum_j (X_j + \alpha S_j) = \text{minimum}$$

where X_j is the cost of protection of sub-system j, S_j is the collective effective dose equivalent commitment resulting from sub-system j when its cost of protection is X_j, and α is the monetary value per unit collective effective dose equivalent.

23. Optimization procedures in this situation can be complicated. In one case, however, the constraining functions in the optimization procedures can be readily established, namely when the sub-systems j are independent, in the sense that the control in one of them does not influence the collective effective dose equivalent commitments from the others (4). In this case, differentiating the objective function with respect to each S_j and making each result equal to zero, the following set of equations are obtained for j = 1, 2, ... n

$$\frac{dX_j}{dS_j} + \alpha = 0$$

because for all X_i and S_i where $i \neq j$, the derivatives are equal to zero ($\frac{dX_i}{dS_j} = 0$ and $\frac{dS_i}{dS_j} = 0$), due to the independence of the sub-systems.

24. As individual annual doses should not exceed the operational limit, a further set of equations (limit equations) are obtained (4)

$$f_j (S_j) \leq L$$

It follows from both sets j of equations that the optimization of control can be obtained by optimizing each independent sub-system taken separately. Similarly, the optimization for the combined exposures from several installations at a given site can be obtained by optimizing separately the protection at each installation, provided the condition of independence applies.

25. In cases where the sub-systems are not independent, optimization procedures can be difficult. The protection to be optimized can conceptually be divided into sub-systems while the exposed group can be conceived as composed by sub-groups. After establishing the Objective functions, Constraining functions and Limit equations, if the number of sub-systems and sub-groups is small, the solution can be obtained analytically (7). However, in most cases, the number of variables will not be too large and programming or direct search methods will have to be used (8).

26. A word of caution is necessary presenting the quantitative techniques of optimization. It should be recognized that optimization of radiation protection, as optimization in engineering in general, is basically an intuitive process (8). The quantitative techniques discussed above are a substantial aid to the process of optimization, but are not the complete process itself.

4. JUSTIFICATION

27. The justification of a proposed practice or operation involving exposure to radiation could be determined by consideration of the

advantages and disadvantages to ensure that there will be an overall net advantage from the introduction of the practice. Justification assessments would be required to decide the introduction of a given practice or to select one among many options. The first type of decision is really a particular case of the second, one of the options being not to change the present situation.

28. The decision among several options, the first being not to introduce any new practice, could conceptually be based on a cost-benefit analysis, as indicated in paragraph 2. The basic notion in the application of cost-benefit analysis to such decisions is very simple: a course of action is taken if the resulting net benefit exceeds those of the next best alternative, and not otherwise. Calling the options $i = 1, 2, \ldots n$ and noting with $i = 0$ the decision to introduce no change, then the options would be increasingly justifiable at increasing positive values of the net benefit B_i

$$B_i = (V_i - V_o) - (P_i - P_o) - (X_i - X_o) - (Y_i - Y_o)$$

where the symbols have the same meaning than in paragraph 2.

29. The justified option, B_j, would then be such that

$$B_j = \max (B_i)$$

In practice, the existence of intangible costs and benefits in many cases makes the analysis subjective. However, relative assessments comparing the justification of alternative procedures are simpler, because the same gross benefit is involved. It is apparent from the equations, that in the very simple case of only two options, differing only in the level of protection, the justification assessment becomes identical to optimization.

30. Acceptance of a practice or the choice between practices will depend on many factors, only some of which being associated with radiation. The role of radiation protection in justification procedures is to ensure that the radiation detriment is taken into consideration, and that the comparisons between practices are made after having applied the procedure of optimization to each of them.

5. REFERENCES

1. International Commission on Radiological Protection. Recommendations of the International Commission on Radiological Protection. ICRP Publication 26. Pergamon Press, Oxford (1977).
2. International Commission on Radiological Protection. Stockholm Statement, 1978.
3. International Commission on Radiological Protection. Problems involved in developing an index of harm. ICRP Publication 27, Annals of the ICRP, Vol. 1, N°4, Pergamon Press, Oxford (1977).
4. Beninson, D. Optimization of radiation protection in IAEA Topical Seminar on the practical applications of the ICRP recommendations and the revised IAEA Basic Safety Standards for radiation protection, IAEA-SR-36/53, 1979.
5. ICRP Committee 4 Task Group on Optimization. Principles and

methods of application of the optimization requirement as a part of the system of dose limitation. (In preparation).
6. OECD - NEA. Report of the expert group on radioactive effluents, 1979. (In publication).
7. Klein, B. Direct use of extremal principles in solving certain problems involving inequalities . Oper. Res. 3: 168 (1955).
8. Siddall, J.N. Analytical Decision-Making in Engineering Design. Prentice-Hall, Inc. Englewood Cliffs, New Jersey (1972).

THE COST OF OCCUPATIONAL DOSE

A. B. Fleishman and M. J. Clark

INTRODUCTION

The current system of dose limitation recommended by ICRP (1) is such that compliance with dose equivalent limits is a necessary but not sufficient criterion for radiological protection; the emphasis is instead on the concepts of justification and optimisation. The result is that the 'as low as reasonably achievable' (ALARA) principle has become a primary objective for radiation protection when dealing with justified sources of exposure.

For the vast majority of day to day problems concerning occupational exposure, the ALARA principle can be satisfied by using the intuitive judgement of operational health physicists. A formal analysis employing the cost benefit techniques suggested by ICRP will not be warranted by the scale of the problem. However in circumstances where there may be potentially large exposures and various possible ways to reduce them, the use of such an analysis can be a useful input into the required decision making. The cost benefit technique can, in theory, identify optimum exposures ie, the level of exposure below which further reductions would not be justified. Nevertheless, in order to perform the analysis in practice, a monetary valuation of radiation exposure is required so that the cost of detriment, Y, can be made directly commensurable with the cost of protection, X. The problems of assigning a cost to health detriment for public exposure have been examined by the authors elsewhere (2). In this paper some aspects of the corresponding costing for occupational exposure will be discussed.

THE VALUATION OF DETRIMENT

The ICRP have defined the health detriment from radiation exposure as a mathematical expected value ie, a summation of the product of the frequency of (stochastic) health effects and weighting factors for their severity. Assuming a linear dose response relationship for the stochastic health effects and the homogeneity of risk and severity factors in populations it is possible to show that the health detriment is proportional to the collective dose equivalent, S (3). There is therefore a simple proportional relationship between collective dose and the number of predicted health effects. However, to establish a relationship between the cost of the health detriment, Y, and collective dose, a separate judgement is required. It has generally been assumed that Y is also simply proportional to collective dose, ie. $Y = \alpha S$ where α is "the cost of the man Sv" in £ man-Sv^{-1}. The use of this relationship implies a single monetary valuation of stochastic health effects independent of the level of individual risks involved. Without questioning the assumption of proportionality between dose and health effects, it is important to note that this does not automatically lead to a proportional relationship between Y and S; other relationships are possible and the appropriate choice is a matter of judgement.

One alternative is to use a cost benefit approach to valuing risk changes (4) which explicitly considers the size of population at risk and the significance of the risk increment to individuals. This leads to a variable value for α for public exposure which increases with increasing per caput dose (2). Such an approach to the functional form of α will tend to concentrate (limited) protection resources in areas of high individual risk; it can therefore be shown to be consistent with equity considerations and has a strong intuitive appeal. The use of a variable value in optimising occupational protection could be justified on the same criteria, although there may be other criteria that need to be considered, both in general and on a case by case basis.

IMPLICATIONS FOR OPTIMISATION

As previously stated the choice between a fixed or a variable α to convert health detriment into monetary terms is a matter of judgement. Nevertheless this judgement can be shown to have important implications for the optimisation of occupational exposure which arises from a common set of operational conditions; namely where the principal mechanism for controlling individual exposure is to vary the number of workers, N, employed on a specific task. Typically one may assume that increasing the number of workers will reduce average individual doses, H, for example by reducing the average time necessary for each worker to spend in radiation areas. However it would appear that this increase in the number of workers will be accompanied by a general increase in doses resulting from non-productive work (5). In the previous example this might arise during the entry and exit from radiation areas. Increasing the number of workers will, in general, tend to increase both the total time required to complete any given task and the total non-productive dose. Assuming that there is a fixed dose associated with the task itself, this will typically lead to an increase in collective dose, S. In order to fulfil the ALARA principle under these conditions, it is necessary to assess what is the optimum exposure to be associated with the task.

In accordance with the formal optimisation procedure recommended by ICRP, the solution to this problem is that at which the sum of the protection costs and the detriment costs (X + Y) for each feasible level of manpower, is minimised. If a fixed value for α is employed in the analysis then Y must be at a minimum for the option which results in the lowest collective dose. On the basis of the general assumptions outlined above, this will occur where the minimum number of workers are assigned to the task and the average individual dose is at its highest (within the constraint of the dose limits). Moreover, as the costs of protection will generally increase if there are more workers requiring, for example, specialist training or protective equipment for the task, this option is likely to also minimise X, and will therefore appear to be optimum. Thus whenever these general assumptions concerning N,H,S and X apply to actual operational conditions, the optimisation procedure will consistently advocate options characterised by the smallest collective dose and the smallest feasible number of workers and will involve the highest resultant average individual exposure. Indeed if the required data validate these relationships between N,H,S and X for a given situation, then the analysis itself is unaffected by the precise value assigned to α.

The results of an optimisation using a variable value for α provides a significant contrast. Even where the same postulated relationships between N,H,S and X hold, the increasing valuation of α with increases in average individual dose precludes any automatic relationship between reductions in detriment costs and reductions in collective dose. Thus while the minimum S and highest H option will still minimise X, it may no longer minimise Y. The optimisation of any given task will therefore be crucially dependent on the specific relationship arising between N,H,S and X and the numerical relationship between α and H.

SUMMARY AND CONCLUSIONS

The implementation of the ALARA principle within the workplace is intended to oppose the attitude that a worker's dose limit represents a constraint on a 'resource' which until reached, can be fully utilised in an arbitrary manner. Thus while the method for formal optimisation is based on the parameter of collective dose, the ICRP suggest that individual exposure at or near the limit is only acceptable if justified by "a careful cost benefit analysis" (1). Such an analysis will generally consider the relationships between the number of workers involved, the costs of their protection and the resultant individual and collective doses. Moreover, it seems likely that the optimum solutions will be significantly influenced by the manner in which collective doses are converted into monetary terms. A fixed conversion factor will ignore individual dose levels, and in attempting to reduce both protection and detriment costs will tend to increase average individual doses. A variable conversion factor which increases with increasing average individual doses will, in contrast, explicitly account for the distribution of individual doses in the assessment of detriment costs, will discriminate against high individual doses, and will tend to select options on a more case by case basis. Such an approach is fully in the spirit of the ALARA principle as applied to all exposures, both individual and collective.

REFERENCES

1. Recommendations of the International Commission on Radiological Protection (1977) : ICRP publication 26, Pergamon Press, Oxford.
2. Clark, M.J. and Fleishman, A.B. (1979) : Topical Seminar on the practical implications of the ICRP recommendations (1977) and the revised IAEA basic standards for radiation protection. Vienna, IAEA-SR-36/3.
3. IAEA Safety Series No. 45 (1978) : Principles for establishing limits for the release of radioactive materials into the environment. IAEA Vienna.
4. Jones-Lee M.V. (1976) : The Value of Life. An Economic Analysis. Martin Robertson, London.

5. Warman, E.A., Wainio, K.M. and Eastman E.A.B. (1978) :
   Estimated additional number of workers and additional cumulative
   exposure due to reducing annual dose limits per individual.
   Stone and Webster Engineering Corp. Boston, U.S. RP-29.

OPTIMISATION ET CONTROLE DES REJETS RADIOACTIFS DES CENTRALES A EAU PRESSURISEE DU PROGRAMME ELECTRONUCLEAIRE FRANCAIS

J. Lochard, C. Maccia and P. Pagès

Les dernières recommandations de la CIPR[1] proposent d'introduire dans le processus de décision concernant la fixation des niveaux de protection dans les installations électronucléaires, la notion d'optimisation.

Dans l'état avancé de développement de certains programmes électronucléaires nationaux, et c'est le cas en particulier pour la France, les choix de radioprotection ont été arrêtés antérieurement et indépendamment de règles ou de directives issues plus ou moins directement de l'esprit de la CIPR. L'analyse de ces choix selon la démarche méthodologique préconisée par la CIPR n'est cependant pas dénuée de tout intérêt.

L'analyse des divers systèmes de traitement des effluents liquides et gazeux équipant actuellement les installations doit en effet permettre de situer les niveaux atteints en matière de protection par rapport à ce qu'il convient d'appeler les niveaux ALARA.

L'étude présentée ici concerne une telle analyse pour les systèmes de traitement des effluents radioactifs des centrales 1 300 MWe à eau pressurisée caractéristiques du programme éléctronucléaire français.

METHODE ET CADRE D'ANALYSE.

Afin de pouvoir comparer et classer les différents systèmes de traitement, il est nécessaire d'évaluer l'impact économique et sanitaire associé à chaque système. Ces évaluations ont été effectuées à l'aide d'une série de modèles reposant à la fois sur des données réelles et sur des hypothèses[2].

Les quantités rejetées de radionucléides dans l'environnement sont évaluées à partir des hypothèses concernant le fonctionnement du réacteur en marche normale et les efficacités théoriques des systèmes de traitement des effluents liquides et gazeux, telles que les admettent les exploitants dans les études préalables à la mise en route des réacteurs. (Rapport de Sûreté - Demande d'Autorisation de Rejets).

Le tableau 1 précise les principales caractéristiques de fonctionnement des réacteurs 1 300 MWe et les activités rejetées après traitement. Le tableau 2 donne pour les effluents traités le type de traitement associé, chaque traitement étant caractérisé par un facteur de décontamination (FD) et/ou une durée de stockage (D).

Pour l'évaluation des effets sanitaires associés aux rejets, un site particulier dans la vallée du Rhône, jugé représentatif de l'implantation actuelle des centrales françaises, a été retenu. Les concentrations puis les doses ont été calculées pour les principaux radionucléides (à l'exception du tritium pour lequel aucun traitement n'est prévu) et les voies d'exposition généralement retenues dans ce genre d'analyse. En ce qui concerne le site il s'agit de données réelles concernant les conditions météorologiques et la répartition géographique des populations et des productions agricoles.

TABLEAU 1. Caractéristiques principales du réacteur et du site.

REACTEUR	. Puissance électrique nominale	1 300 MWe
	. Durée de fonctionnement	8 000 h
	. Taux de rupture de gaine	0,25 %
	. Fuite primaire totale	14 kg/h
	. Fuite primaire-secondaire	4 kg/h
REJETS	. Total gazeux	15 000 Ci
	dont iode	0.9 Ci
	. Total liquide hors tritium	7 Ci
	dont iode	2 Ci
SITE	. Population (R $\leq$ 100 km)	2 10^6 hab.
	. Débit du fleuve	1 600 m^3/s

TABLEAU 2. Systèmes de traitement et voies d'effluents associés.

VOIE D'EFFLUENT	Symbole	TRAITEMENT
Drains + Effluents chimiques	DR	Evaporation thermique + stockage (FD = 100 D = 5 j)
Laverie	LV	Stockage (D = 30 j)
Dégazage eau primaire	TEG	Stockage sous pression (D = 60 j)
Fuites bâtiment réacteur	BR	Piégeage des iodes en balayage continu (FD = 10)

L'indicateur de dose retenu est l'engagement d'équivalent de dose effectif collectif dans un rayon de 100 km autour de l'installation. Le passage aux effets est effectué en utilisant le coefficient proposé par la CIPR de 1.65 10^{-2} effet (cancers mortels aux organes autres que la peau) par Homme-Sievert.

L'indicateur économique qui caractérise chaque système de traitement est le coût total actualisé, c'est à dire la somme actualisée sur 20 ans du coût annuel d'exploitation d'un système et de l'annuité d'amortissement de l'année correspondante. (Le taux d'actualisation retenu est de 9 %, les coûts exprimés en francs 1978). Le coût d'exploitation annuel ne tient pas compte du coût de traitement des déchets. Quant aux coûts d'investissement ils prennent en compte les coûts directs (matériel, montage, étude et transport) et indirects (intérêts intercalaires et frais d'ingenierie). Il est à noter que le système de traitement des eaux primaires et le système de traitement des purges des générateurs de vapeur sont considérés comme des systèmes de production indispensables au bon fonctionnement du réacteur et qu'en conséquence leur coût ne figure pas dans les coûts de protection.

RESULTATS

Le tableau 3 synthétise les résultats. Les systèmes sont classés dans l'ordre des rapports coût-efficacité croissants : l'efficacité étant définie par le nombre d'effets évités par le recours à un système.

TABLEAU 3. Coûts et efficacité des systèmes de traitement des effluents.

SYSTEME	COUTS (10^3 FF)			Effets évités(*)	ΔC/ΔE
	Inv.	Expl.	Tot. actual.		
TEG.	1 290	52	1 440	3.1	4.6 (2)
DR.	4 480	350	6 750	2.3 (-1) **	2.9 (4)
NR.	100	36	440	4.8 (-4)	9.1 (5)
LV.	300	88	1 160	1 (-4)	2.9 (6)

* 1,3 GWe x 20 ans   ** 2.3 (-1) = 0,23

Pour l'analyse coût-bénéfice, conformément à la présentation préconisée dans la publication 22 de la CIPR[3], il convient d'exprimer l'efficacité en effets résiduels. En l'absence de traitement l'ensemble des rejets conduisent à 3.5 effets pour 20 ans de fonctionnement du réacteur. Les valeurs de l'Homme-Sievert retenues sont respectivement de 400 FF et 4 000 FF. L'analyse de la Figure 1 montre que pour ces valeurs qui peuvent être considérées comme extrêmes, le niveau de radioprotection atteint se situe au-delà de la valeur qui minimise le coût social de radioprotection.

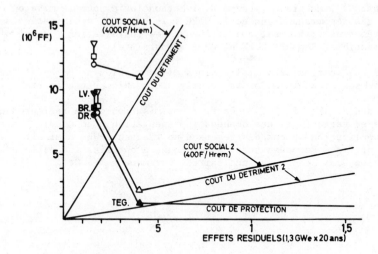

FIGURE 1. Analyse coût-bénéfice.

CONCLUSION

L'analyse a posteriori des choix de radioprotection permet comme on l'a montré, de situer voire de justifier ces choix. Elle permet d'autre part de donner les éléments nécessaires (le coût marginal de l'effet évité en particulier) pour une comparaison avec d'autres technologies présentant des risques en marche normale (énergie conventionnelle, industrie chimique, etc...).

C'est cependant dans une perspective décisionnelle que la méthodologie présentée s'avère la plus conforme aux recommandations de la CIPR : les résultats présentés démontrant que dans le domaine de la gestion des effluents des installations nucléaires, cette méthodologie est tout à fait praticable.

Il convient néanmoins de manier ce type d'analyse avec précautions. En particulier la seule dimension du risque prise en compte est ici le risque collectif, il est donc nécessaire de s'assurer par ailleurs que les limites de doses individuelles sont bien respectées. De ce point de vue des méthodes décisionnelles multidimensionnelles pourraient apporter un éclairage plus synthétique sur le problème de la gestion du risque radiologique.

D'autre part les données sur lesquelles reposent ces études sont caractérisées par une grande variabilité qu'il s'agit d'évaluer par le recours à une analyse de sensibilité approfondie, d'où la nécessité d'élaborer des outils adéquats[4].

REFERENCES

1. ICRP Publication 26 : Recommandations of the International Commission on Radiological Protection. Pergamon Press 1977.

2. Lochard J., Maccia C., Pagès P. : Méthodologie pour la mise en oeuvre des recommandations de la CIPR : le cas de l'optimisation des systèmes de traitement des effluents des installations du cycle du combustible. Rapport CEPN n° 29 - (A paraître).

3. ICRP Publication 22 : Implications of Commission recommendations that doses be kept as low as readily achievable. Pergamon Press 73.

4. Lochard J., Maccia C., Pagès P. : Optimisation de la radioprotection et variabilité des données : présentation d'un modèle général et application au cas des rejets d'un réacteur PWR. Séminaire Scientifique Européen sur "Les méthodes d'optimisation de la protection dans le domaine nucléaire". Luxembourg 3-5 Octobre 1979 - (A paraître).

OPTIMISATION DE LA PROTECTION ET EVALUATION DU RISQUE LE CAS DES ACCIDENTS DE TRANSPORT DE MATIERES DANGEREUSES

T. Meslin

Cette étude se propose de mettre en oeuvre les principes d'optimisation recommandés par la CIPR dans les publications n° 22 et 26, et traite du cas des risques encourus par le public du fait des accidents de transport de matières dangereuses.

L'objectif principal de l'étude est la rationalité des mesures de protection. Plusieurs sous-objectifs doivent être atteints préalablement :

- Analyse du processus accidentel
- Evaluation du risque sanitaire
- Evaluation économique des mesures de protection supplémentaires.

METHODE

Modèle physique d'évaluation du risque -

Ce modèle vise à calculer pour un programme de transport donné, le nombre de colis de matières dangereuses ouverts et le nombre d'effets sanitaires associés pendant une période de temps donnée.

L'évaluation du risque s'effectue en quatre étapes faisant appel à différentes méthodes mathématiques, statistiques ou physiques :

- Définition de l'ensemble des scénarios d'accidents probabilisés.
- Identification et détermination des indicateurs de gravité.
- Evaluation du nombre de colis ouverts étant donné un programme de transport.
- Nombre d'effets sanitaires dûs à un relâchement de matières dangereuses.

Modèle d'aide à la décision

La mise en oeuvre du modèle d'aide à la décision s'effectue suivant l'organigramme représenté sur la figure 1. Le choix d'un certain nombre de mesures visant à améliorer le niveau de protection du public est fait en tenant compte des contraintes de faisabilité, quelles soient de nature institutionnelle ou socio-économique.

L'analyse de cet ensemble de mesures sur la base de leur coût et de leur efficacité vis-à-vis d'un indicateur pertinent (risque résiduel) peut alors être réalisée. Par ailleurs, on peut comparer les mesures qui auront été jugées efficaces avec d'autres mesures affectant le risque dans divers secteurs industriels.

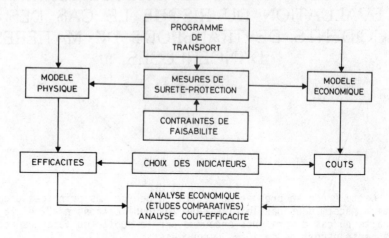

Figure 1. Organigramme fonctionnel

APPLICATION

Le programme de transport.

Le risque associé au transport de gros conteneurs est calculé à titre d'exemple pour un trafic de 100 000 véhicules x km dans le cas où le mode de transport est le chemin de fer. En faisant l'hypothèse d'un système de transport conforme aux pratiques en vigueur actuellement, notamment en ce qui concerne les mesures de protection, on a les résultats suivants quant au niveau de risque. Les effets sanitaires sont estimés à partir d'un relachement d'HF.

TABLEAU 1. Risque associé au programme de transport.

	Probabilité d'accident	Nombre d'ouvertures attendues	Risque résiduel (Nombre d'effets sanitaires)
Chemin de Fer	$.8 \cdot 10^{-2}$	10^{-4}	$2.5 \cdot 10^{-5}$

Optimisation de la Protection

Les options candidates

Cinq mesures de base sont prises en compte, susceptibles d'affecter le risque de différentes façons.
- M - Morcellement des expéditions (1 citerne par wagon au lieu de 2 ou 3).
- Q - Adoption de citernes renforcées pour l'ensemble des expéditions.
- D - Utilisation de train direct ne passant pas en triage.
- I - Contournement des grandes agglomérations.
- V - Limitation de la vitesse à 40 km/h.

Une option de protection est une combinaison compatible de ces cinq mesures, c'est ainsi qu'une quinzaine d'options sont candidates y compris la mesure "OPO" consistant à ne rien faire.

RESULTATS

Le classement coût-efficacité

Les options candidates sont symbolisées par les lettres identifiant les mesures de base qui les constituent. Le classement coût-efficacité, qui est indépendant du risque associé à l'option "OPO", puisqu'il ne tient compte que du risque résiduel après application de l'option et du surcoût associé est présenté figure 2 et explicité dans le tableau n°2.

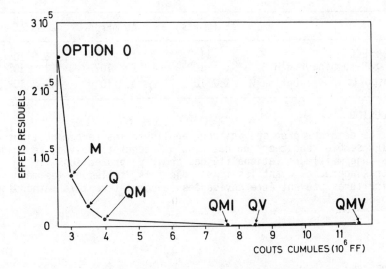

Figure 2. Courbe coût-efficacité

TABLEAU 2. Classement coût-efficacité

Classement	Option	Δ Coût (FF)	Δ Efficacité	ΔC/ΔE
1	M	$4.8\ 10^4$	$1.7\ 10^{-5}$	$2.7\ 10^9$
2	Q	$4.4\ 10^4$	$4.5\ 10^{-6}$	$9.8\ 10^9$
3	QM	$4.8\ 10^4$	$2.1\ 10^{-6}$	$2.3\ 10^{10}$
4	QMI	$3.7\ 10^5$	$7.4\ 10^{-7}$	$5.0\ 10^{11}$
5	QV	$8.3\ 10^4$	$1.1\ 10^{-7}$	$7.9\ 10^{11}$

Il apparaît ainsi que la mesure la plus coût-efficace est le morcellement des expéditions.

Comparaison avec d'autres mesures de protection.

On peut comparer les résultats obtenus ici avec ceux obtenus lors d'une étude coût-efficacité de certains systèmes de sûreté (ECCS = Emerging Core Cooling System,...) sur les réacteurs PWR /1/. Le tableau suivant n° 3, montre une relative cohérence entre les rapports coût-efficacité.

TABLEAU 3. Comparaison des rapports coût-efficacité

	ΔE (morts)	ΔC (FF)	ΔC/ΔE
PWR : ECCS	11	$7.5\ 10^6$	$6.8\ 10^5$
: Recombineur d'H_2	$2\ 10^{-5}$	$2\ 10^5$	$1\ 10^{10}$
Transports : Option M	$1.7\ 10^{-5}$	$4.8\ 10^4$	$2.7\ 10^9$

CONCLUSION

La démarche proposée ici pour appliquer les recommandations de la CIPR semble déboucher sur des résulats concrets pouvant se traduire par une meilleure rationalité des choix de protection.
D'autres approches ayant les mêmes objectifs, telles que les méthodes multicritères peuvent être envisagées et compléteront la méthodologie proposée.

REFERENCES

/1/ - E.P. O'Donnell, "What Price Safety ? A probabilistic cost-benefit evaluation of existing engineered safety features" in Nuclear Power Reactor Safety - ENS/ANS topical-meeting - Brussels - Octobre 1978.

APPLICATION OF THE PRINCIPLES OF JUSTIFICATION AND OPTIMISATION TO PRODUCTS CAUSING PUBLIC EXPOSURE

A. B. Fleishman and A. D. Wrixon

INTRODUCTION

Radioactive substances have been used in some consumer products for many years; other uses may be developed in the future. The possibility that the public might be unjustifiably exposed to radiation from these products is a reason for the development of systems of control whereby such products are required to be approved by national authorities before they can be supplied to the public. The basis of these controls should be the application of the ICRP principles of justification and optimisation of protection together with consideration of individual doses in comparison with dose limits. In this paper we explore the difficulties and judgements which are involved in applying these principles to radioactive consumer products. These problems should be explicitly recognised in the process of establishing radiological protection criteria for use by national authorities, or indeed, before attempting to harmonise international practices and thereby facilitate international trade.

JUSTIFICATION

ICRP has recommended cost benefit analysis as the 'ideal' mechanism for determining the acceptability of a proposal involving exposure of people to radiation, and have expressed the principles of justification and optimisation in mathematical notation (1). The aim of cost benefit analysis is to identify all the positive and negative aspects of a proposed practice, to quantify them in a common unit, usually money, and thereby determine whether the proposal provides a net benefit to society as a whole. In view of the quantitative nature of the technique, it might be regarded as a sufficient basis for objective decision making. This however is not the case. The major advantage of its application, in theory or practice, lies not in the removal of subjective judgements but rather in requiring their explicit recognition. It is therefore valuable in assisting the development of consistent policy criteria, but of itself does not necessarily provide a simple and unequivocal solution.

The analysis initially draws attention to the distribution of doses and benefits between both those involved in supplying the products and the public. However, frequently the only concern of the national authority is consumer protection as separate regulations cover occupational exposure. In defining the scope of the analysis the national authority is therefore likely to restrict its attention to estimating the net benefit to the public and ignore the occupational aspects.

The measure of net benefit to be attributed to the consumption of a product can, in the theory of cost benefit analysis (2), be derived from what is known as the "consumer surplus". This represents the sum of the maximum <u>additional</u> amounts of money that consumers would be prepared to pay for the product (ie above that actually paid). This is a measure of the net value to the consumers from their own viewpoints. However in practice, this perceived net benefit from a product is likely to be distorted both by advertising and misinterpretation (or ignorance) of the associated hazards. Indeed, it is partially because of this inability of consumers to correctly assess these radiological hazards that national authorities need to exercise control over such hazards. Thus even where the product buyers are the sole recipients of the benefits and the radiological detriment from a product, this measure of net benefit cannot automatically be assumed to reflect public preferences.

In most cases radioactive consumer products will expose non-users as well as users to radiological hazards, and the protection of these individuals provides a further reason for the control of such goods. However, if money is the common unit to be used in cost benefit analysis then this detriment will also need to be expressed in monetary terms. The difficulties and judgements of applying cost benefit analysis to value radiological detriment in this manner are comparatively well known and are described elsewhere (3). Once this conversion has been made, then for any product a measure of the net benefit to the public as a whole could be obtained, e.g. by subtracting the detriment cost from the consumer surplus. Even so the demonstration of an overall positive net benefit is not a sufficient criterion for the purposes of justification without some reference to the scale of the individual doses, particularly to non-users. Moreover, compliance with the ICRP dose equivalent limits may not be considered sufficient, and some lower maximum acceptable dose level may need to be specified using subjective judgements.

Quite apart from the subjective judgements involved in integrating the results from cost benefit analysis into a decision making framework, the technique requires a substantial amount of data. It requires realistic (as opposed to cautious) assessments of the dose distributions from normal use, misuse and accidents, and uncontrolled disposal, as well as extensive market research to obtain data on the likely demand for the product. The generation of these data, even with products already on the market, would involve considerable costs in terms of time and effort. These costs will normally be out of all proportion to the scale of the radiological problems associated with such products.

OPTIMISATION

Many of the comments made above are, of course, equally relevant for the application of the principle of optimisation to radioactive consumer products. This may be illustrated by reference to a worked example recently carried out to develop a standard for the quality control of gaseous tritium light sources (GTLSs) intended for use in liquid crystal digital (LCD) watches (4). Three manufacturers were involved in the study and each was already undertaking some form of quality control to detect inadequately sealed GTLSs. The changes in costs and benefits resulting from changes to their procedures were analysed. The following is a summary of the findings :

1. The analysis was only 'objective' in the numerical optimisation of parameters and the results were in general different for each manufacturer, depending on the relative cost effectiveness of their quality control procedures. In deriving a standard to be applicable to the whole industry subjective judgements were therefore required. One such judgement was that those parameters which had only a minimal effect on radiation exposure would be ignored.

2. While the primary concern was consumer protection the analysis focused on changes in manufacturing costs as a parameter reflecting changes in price and net benefit to consumers. In one case implementation of the derived optimum quality control parameter would have resulted in reduced costs to the manufacturer but increased detriment to the public. As it is not possible to ensure that the cost savings would be passed onto the public, it was considered undesirable to adopt this optimum parameter. However, in situations where implementation of the derived optimum quality control parameter would have resulted in reduced detriment to the public albeit at higher costs to the manufacturer, it was considered reasonable to adopt the optimum parameter; the manufacturer could then recover the higher costs by increasing the price for his GTLSs.

3. The results were very sensitive to both the monetary value assigned to the unit of collective dose equivalent and the assessed dose per unit of tritium evolved.

4. The effort involved in what initially appeared to be a relatively straightforward analysis was substantial. It was however, considered to be worthwhile in this instance as it highlighted some of the practical problems involved in optimisation and as there was the possibility of very wide scale distribution of LCD watches with GTLSs.

CONCLUSIONS

It may be concluded that the role of quantitative cost benefit analyses in determining the acceptability of consumer products is likely to be limited. A more pragmatic scheme is likely to be sufficient. For justification purposes, one such scheme could involve the restriction of doses to individual users and non-users perhaps in relation to a qualitative description of the benefits. For optimisation purposes, a more practical approach could again be adopted involving, for example, the selection of the least hazardous radionuclide in its intrinsically safest physical and chemical form (5,6). Such a basis for the control of radioactive consumer products would substantially reduce the effort required by the national authority. However if this control is to be applied in a consistent manner, it requires the development of specific criteria which will inevitably incorporate subjective judgements on, for example, acceptable levels of exposure. The explicit recognition of the judgements involved will clarify the development of such criteria both nationally and internationally.

REFERENCES

1. Recommendations of the International Commission on Radiological Protection (1977) : ICRP Publication 26, Pergamon Press, Oxford.

2. Mishan, E.J. (1975) : Cost Benefit Analysis. George Allen and Unwin, London.
3. Clark, M.J. and Fleishman, A.B. (1979) : Topical Seminar on the practical implications of the ICRP recommendations (1977) and the revised IAEA basic standards for radiation protection. Vienna, IAEA-SR-36/3.
4. Fleishman, A.B. and Wrixon, A.D. (1979) : Practical Application of Differential Cost Benefit Analysis to Products Causing Public Radiation Exposure. NRPB-M51. National Radiological Protection Board, Harwell, Didcot, Oxon.
5. Wrixon, A.D. and Freke, A.M. (1978) : Radioactivity in Consumer Products in the UK. Radioactivity in Consumer Products. NUREG/CP-0001. U.S. Nuclear Regulatory Commission, Washington D.C.
6. Hill, M.D. and Wrixon, A.D. (1977) : The Radiological Testing of Products which Irradiate the Public. IVth International Congress. IRPA Paris.

Radiation Protection Standards

SETTING STANDARDS FOR TRIVIAL CONCENTRATIONS OF RADIOACTIVITY

D. van As

1. INTRODUCTION

   Radioactivity is an integral part of the environment. The quantities produced and released to nature by man-made processes is but a small fraction of that allready present. Life has evolved more or less acceptably in an environment of natural radiation the levels and variation of which is known with considerable accuracy and the effects of which has been better studied than any other natural phenomenon. In spite of this, society is reluctant to consider man-made radioactivity in the same manner as natural radioactivity and to accept that the effects from these sources are no different.
   No material is totally devoid of radioactivity and it is clearly not the intention of any waste management practice that every material should be treated as a potential radioactive pollutant. In the legal profession there is a saying De Minimus non curat lex, meaning the law does not care for or take notice of very small or trivial matters. A similar philosophy should hold true for the radiological protection profession and the saying can be construed to mean that a quantity of radiation or an amount of radioactivity that is so low that it can not be distinguished from natural radiation should be regarded as trivial. Radiological protection should not be concerned with trifles.
   The principles of radiation protection as is incorporated in the recommendations of the ICRP, requires, among others, the balancing of costs and detriments of any practice involving radioactivity. In waste disposal practices the detriment due to radiation exposure is closely related to the total quantity of radioactivity while the cost is a function of the total mass of the waste to be disposed. If follows therefore that there can well be an optimum concentration value below which control is not cost effective.
   Radioactive wastes are often classified into three categories i.e. those requiring complete isolation from the environment over extended periods (thousands of years) through methods such as deep geological burial; secondly, isolation over shorter periods through methods such as shallow land burial - the period being that over which institutional control can be exercised and during which time the radioactivity will decay to trivial amounts; and finally wastes containing trivial amounts of radioactivity that would be exempt from special regulation and disposed of as normal waste into sewage systems, onto garbage dumps or into the ocean.
   A trivial level of radioactivity is that below which considerations other than those of radioactivity becomes of overriding importance and where the management should be concerned with these properties rather than with the radioactive content of the waste.
   A need exists to come to an agreement on a level or concentration below which the radioactive content can be considered trivial. Various approaches to the setting of radiation standards have been discussed, i.e. natural variation in radiation doses (1), relating a trivial dose to an insignificant risk (2).

2. EXISTING STANDARDS FOR TRIVIAL AMOUNTS OF RADIOACTIVITY

An NEA/OECD survey among its member countries of threshold values below which solid materials or effluents will be regarded non-radioactive showed that the value of 0,002 µCi/g are the most widely recognised limit. This value has its origin in the IAEA Basic Safety Standards of 1967 (3), where it is specified that only radioactive substances at concentrations exceeding 0,002 µCi/g or solid natural radioactive substances at concentrations exceeding 0,01 µCi/g shall be notified or registered by the competent authority. In the same publication maximum permissible activity values for exemption from notification for individual nuclides are given; these vary between 0,1 µCi and 1 000 µCi. The basis for these values is not explained.

Other values used are 10^{-5} µCi/g by the Radiochemical Inspectorate in the UK for man-made radioisotopes and is also in the USA by the Environmental Protection Agency (EPA) for transuranic nuclides. In the Netherlands, sources of radiation representing total activities between 0,1 and 100 µCi, depending on the radioactivity of the nuclide, can be disposed of as solid waste, probably in accordance with ref. 4 above. Others use a fraction of the ICRP maximum permissible concentration values in air or water as threshold values, below which the material is regarded as non-radioactive.

In order to be consistent, a basis should be found which can be used for the calculation of threshold values for all radionuclides and for all different disposal practices of radioactive material. Dose is the fundamental criterion for protection of man and therefore the correct basis to use when determining effects which can be considered trivial. An obvious approach is to define a certain trivial dose limit the detriment of which can be considered to be negligible and its contribution indistinguishable from the natural dose variation.

3. BASIS FOR THE CHOICE OF A TRIVIAL DOSE

3.1 Variation in natural radiation dose

The main contributors to the natural radiation dose are cosmic radiation, terrestial sources of which there are two important components, i.e. external gamma radiation and internal radiation from inhaled ^{222}Rn and its daughter products and ^{40}K.

Cosmic radiation dose varies with latitude and altitude. The dose is about 32 mrem/a at sea-level and mid-latitudes. The dose is highest at the poles and decreases by about 10 % at 40° latitude and by 25 % at the equator. The decrease is equivalent to 1 mrem/a for every 10° decrease in the latitude. The increase with altitude is around 2 mrem/a for every 100 m.

The radiation dose contribution from the three most important terrestial sources are 5,5 mrem/a from 1 ppm U_3O_8 in the soil, 1,9 mrem/a from 1 ppm ThO_2 and 13 mrem/a from 1 % of potassium in the soil. The large variations in soil composition give rise to whole-body radiation dose range for 95 % of the population living in "normal" areas of between 26 and 61 mrem/a (4).

Potassium is a major element present in the human body. The radiation dose from the natural radioactive isotope ^{40}K to various body organs varies with age, sex and natural concentration of potassium in the body; typically, gonads 9-21, lung 10-24 and red bone-marrow 16-38 mrem/a.

The variation in the dose to the lungs due to natural ^{222}Rn is even larger and can vary between 4 and 400 mrad/a (4).

These natural variations in the radiation dose to man of tens of mrem per annum is a fact which society accepts and do not consider when choosing a domicile.

Adler et al (1) makes a strong plea for using a standard deviation in natural radiation levels as a basis for setting standards from radiation resulting from human endeavours. This approach is based on the fact the human race has a long history of being exposed to these levels and that no correlation between these variations and any detriment can be shown. The value of the standard deviation in natural background is of the order 20 mrem/a in the USA.

3.2 A comparison of risks

All human activity entails a risk of some kind or another. The annual risk of death from natural causes in the prime of life is about 10^{-3} and never less then 10^{-4} (2). As it is assumed that the risk to health corresponds linearly with the dose received, a trivial dose would be that which corresponds to an insignificant level of risk to the recipient.

Through an analysis of voluntary and involuntary risks Webb and McLean (2) arrived at a value of 10^{-6} as an annual level of risk which is not taken into account by individuals in arriving at decisions as to their actions and which is therefore negligible. This corresponds to the value considered by USEPA as not being of regulatory concern. By applying the linear dose-risk relationship a value of 10 mrad/a is proposed by Webb and McLean as an insignificant or trivial dose to the individual. The risk from this dose has been compared to smoking 1½ cigarettes, drinking ½ liter of wine, travelling 16 km on a bicycle or 80 km in a car or 300 km by air.

4. THE APPLICATION OF A TRIVIAL DOSE

Dose is not a quantity of direct practical value when considering the disposal of radioactivity. It is necessary to convert dose into a quantity which can be measured such as a specific concentration or a rate of disposal.

4.1 Assessment of release rates

Depending on the particular practice i.e. atmospheric, aquatic or terrestrial disposal, a dosimetrie model, realistic in terms of the possible pathways to man, should be used to derive trivial quantities of radioactivity. Two important aspects i.e. multiple sources of exposure and the dose commitment due to the continuous nature of the practice must be considered to ensure that the continuation of this practice over many years and from many sources will not result in the tri=

vial dose limit being exceeded in any individual. The allocation of a percentage of the trivial dose to a particular practice can be used to resolve the first concern. Webb and McLean (2) propose that an annual dose commitment of 0,1 mrad be used and considers it very un= likely that any individual will receive a dose in excess of 10 mrad from the various practices that have been exempted in this manner.

An assumption as to the time span over which the practice will continue is needed so as to ensure that the dose commitment from fu= ture practices will not add up and exceed the trivial dose as defined. For short-lived nuclides the replacement value which is equal to the decay in one year can be used. For the long-lived nuclides this ap= proach becomes very restrictive and it would be more realistic to assume a finite time during which the practice will continue, for instance the time period for which the production of nuclear energy is expected to continue.

Assessments of this nature depend on the characteristics of the particular environment and requires realistic models, site specific data on transfer parameters through the environment and knowledge of the critical pathways to exposed groups. The result of the as= sessment specifies a release rate which is a quantity that can not always be controlled and is difficult to administer. In practice it may be necessary to convert the release rate to a concentration on the basis of the expected total mass of waste to be disposed.

4.2 Specific activity

A method which avoids these problems and allows for the cal= culation of trivial concentrations of radionuclides for general application under all circumstances and at all sites is one based on controlling the specific activity i.e. the ratio of the radio= nuclide to its stable isotope expressed as pCi/g stable element in the waste material.

The activity of a particular radionuclide that will result in a specified dose, i.e. the trivial dose to a particular organ or tissue, can easily be inferred from ICRP recommendations (5). By expressing this activity as a ratio to the total mass of the cor= responding stable element normally present in the organ or tissue, the trivial specific activity for a particular nuclide in the cri= tical organ is found.

The principle underlying this approach is that biological spe= cies do not distinguished between the radioactive and stable form of the same element. If therefore the radioactive and stable nuclides are in the same physical-chemical form this ratio will not be in= creased in its passage from the waste itself through the biological food chain to man. The specific activity will remain constant in the unlikely event of the waste being the only source of the parti= cular element to man or, more likely, decrease as mixing occurs with the vast quantities of the stable element which may be present in similar physical-chemical form. There is thus virtual assurance that the dose to man will be within the trivial dose value.

Despite the apparent attractiveness, there are reservations to which the method is subject. Firstly, the radionuclide and its stable element must be either in the same physical/chemical form, displaying the same biological availability from the outset or must equilibrate before the waste comes into contact with more of the radionuclide and its element. Secondly, the method is limited in its application to situations where internal pathways are critical and to those nuclides which have a stable element analogue. Finally, it cannot be used where the critical organ is the gastro-intestinal tract. With the introduction of the weighted whole body dose concept of the ICRP this latter problem might be overcome; however, the ICRP has now also refrained from specifing body or organ burdens of radioactivity on which, together with the stable element content of body organs, calculation of values of specific activity depends.

5. CONCLUSION

The ICRP system of dose limitation implies that a level of radioactivity exists below which the material should be considered for properties other than its inherent radioactivity. In waste management practice there is a need for a basis on which such a decision can be made. A trivial dose level from which trivial concentrations can be assessed appears to be the logical approach. A value of 0,1 mrad/a for a trivial dose commitment can be substantiated. The application of the specific activity approach may prove valuable for setting trivial concentration levels to wastes resulting from neutron activation processes such as, for example, structural components from reactor cores.

6. REFERENCES

1. Adler, H.I., Federow, H., Weinberg A.M. (1979): In: Proceedings of a Topical Seminar on the Implications of the ICRP Recommendations, IAEA, Vienna.
2. Webb, G.A.M., McLean, A.S. (197): In: Proceedings of the 4th International Congress of the International Radiation Protection Association, Paris.
3. International Atomic Energy Agency (1967): Safety Series no. 9 Basic Safety Standards for Radiation Protection, IAEA, Vienna.
4. United Nations Scientific Committee on the Effect of Atomic Radiation (1977). Sources and Effects of Ionizing Radiation Annex B. United Nations, New York.
5. International Commission for Radiological Protection (1959): Report of Committee II on Permissible Dose for Internal Radiation. Pergamon Press, Oxford.

INTERNATIONAL ELECTROTECHNICAL COMMISSION (IEC) AND RADIATION SAFETY REQUIREMENTS FOR MEDICAL X-EQUIPMENT

E. Koivisto and H. Bertheau

INTRODUCTION

The International Electrotechnical Commission (IEC) came into being in 1904. Its object is to facilitate the coordination and unification of national electrotechnical standards.

In 1966, the Technical Committee No. 62 of the International Electrotechnical Commission was established. Its first meeting was held in May 1968. The scope of the Technical Committee No. 62 of the IEC was defined as follows: "To prepare international recommendations concerning the manufacture, installation and application of electrical equipment used in medical practice. This also concerns surgery, dentistry and other specialities of the healing art".

The major task of IEC TC 62 is to reach an international concensus on the requirements for manufacture, installation, use and maintenance of electrical equipment used in medicine. One on the most important aspects is safety of the patients and users of the equipment. This means protection against electric hazards, radiation hazards, mechanical hazards and hazards that have to do with the particular kind of effects an electrical equipment is intended to produce.

The work of IEC TC 62 started with X-ray equipment because an urgent need for international harmonization in this field was apparent.

IEC TC 62 Sub-committee B deals with X-ray equipment operating up to 400 kV and accessories, and Sub-committee C with high energy equipment and equipment for nuclear medicine.

LIASONS WITH INTERNATIONAL ORGANIZATIONS

The Central Office of the IEC in Geneva, Switzerland, maintains liason with the following international organizations in the field of electrical equipment in medical practice:

International Electrotechnical Commission (IEC)

ICRP	International Commission on Radiological Protection (particularly with Committee 3 on External Exposure)
ICRU	International Commission on Radiation Units and Measurements
IAEA	International Atomic Energy Agency
IFMBE	International Federation for Medical and Biological Engineering
IFIP	International Federation for Information Processing (particularly with Committee 4, Information Processing in Medicine)
WHO	World Health Organization
SIC	Société Internationale de Cardiologie
FDI	Fédération Dentaire Internationale
OIML	Organisation Internationale de Métrologie Légale
ISO	International Standards Organization, particularly with Committees:
/TC 42	Photography
/TC 85	Nuclear Energy
/TC106	Dentistry
/TC121	Anaesthetic equipment and medical breathing machines
/TC150	Implants for Surgery

IEC PUBLICATIONS ON RADIATION PROTECTION

In 1973 IEC Publication 407: Radiation Protection in Medical X-ray equipment 10 kV to 400 kV was published. In 1975 this was supplemented by publication 407A: Radiation protection in dental X-ray equipment.

At the present time a new publication 601, supplement A to part 1 is in preparation to be published. It deals with the radiation safety of medical X-ray equipment and is a part of publication 601 that deals with the general philosophy of the safety of electrical medical equipment.

The purpose of this paper is to present some main features of this new standard.

SCOPE OF THE NEW STANDARD

This document concerns the radiation protection of diagnostic X-ray equipment including computed tomography (CT) and dental equipment. High voltage generators and X-ray therapy equipment are covered by separate standards.

CATEGORIES OF EQUIPMENT

All diagnostic equipments are devided into categories according to the imaging arrangements as follows:

Radioscopy and radiography with spot film device

Radiography
- image receptive areas, FID or both variable
- image receptive area and FID fixed

Radioscopy
- direct
- indirect

Special purpose X-ray equipment

Special purposes X-ray equipment are listed separately:
- mass chest survey
- transportable, radiography
- transportable, for indirect radioscope
- mammography
- therapy simulator
- reconstructive tomography
- diagnostic, above 200 kV
- dental, with intraoral receptor
- dental, panoramic tomography
- dental, panoramic radiography with intraoral X-ray tube
- cephalometric radiography

OBJECT OF THE STANDARD

The object of this standard is to protect both patient and staff from unwanted radiation. The philosophy of ICRP has been followed as closely as possible.

TECHNICAL ASPECTS

In the following some technical aspects of this new standard are discussed.

Filtration

Recommendation of ICRP are respected. However, the user may remove or add special filters.

Alignment of radiation beam

Limitation of the radiation beam to a minimum is an important factor affecting both the image quality and the radiation dose. Therefore, exact requirements concerning the alignment of the X-ray field are given.

International Electrotechnical Commission (IEC)

Leakage radiation

KERMA in air from leakage radiation: max 0.87 mGy (=100 mR) in one hour is allowed at 1 m distance from the focal spot.

Focal spot to skin distance

This is limited to minimum 0.38 m.

Radioscopy

During radioscopy maximum KERMA rate allowed is 50 mGy (=5.7R) per minute. The apparatus shall be operable only on this condition.

Material between the patient and image receptor

Maximum absorption values are specified as follows:

Cassette holder	
Film changer	max 1.0 mm Al equivalent
Spot film device	
Patient support	max 1.5 mm Al equivalent
Cradles	max 2.0 mm Al equivalent

Normal operation location

In this document models are given for calculating the radiation dose received by the operator of the equipment assuming a normal operation location for each type of X-ray equipment.

CONCLUSIONS

Recommendations of the ICRP have been used as a basis for setting up the requirements of the IEC standards according to which compliance of an equipment or device can be tested and verified.

The new standard is of basic importance for the manufacture and installation of X-ray diagnostic equipment. Its aim is to ensure adequate protection of the public as well as personnel using ionizing radiation.

REFERENCES

1. Radiation protection in medical X-ray equipment 10 kV to 400 kV. Publication 407. International Electrotechnical Commission. Geneva 1973.
2. Radiation protection in dental X-ray equipment. Publication 407A. International Electrotechnical Commission. Geneva 1975.

THE LABORATORY APPRAISAL OF IONISATION CHAMBER SMOKE DETECTORS

B. T. Wilkins and D. W. Dixon

A paper at the Paris conference in 1977 described the approach which the National Radiological Protection Board (NRPB) was to adopt in the United Kingdom with regard to the radiological testing of products that irradiate the public (1). The paper summarised early results of tests on ionisation chamber smoke detectors (ICSDs). Since that time, there has been a steady supply of detectors for evaluation. At the time of writing 47 detectors have been received, most of which utilised americium-241 foil sources. One detector received utilised krypton-85 and some utilised radium-226 foil sources. The test programme which was originally proposed in 1976 (2) has been modified in the light of experience, and the results have formed part of the input to the Nuclear Energy Agency (NEA) in the drafting of recommendations on these devices (3). These were published in 1977 and the NRPB evaluation procedure follows that described in the document. The present paper describes some of the Board's experimental methods and experiences and its interpretation of some of the NEA criteria. The evaluation consists of four sections; visual inspection, dose rate measurements, surface contamination measurements and destructive testing. These will be discussed in turn.

The visual inspection is intended to detect any shortcomings in design, with particular regard to access to the source(s). The recommendations require that under normal conditions of use, direct access to the sources shall be impossible and that the construction of the ionisation chamber for single-station ICSDs shall be sufficiently tamper-proof. While none of the detectors examined so far permit direct access to the sources, several have been considered to be insufficiently tamper-proof. The interpretation of this term within the UK is that the source(s) should only be accessible by means of a special tool or by damaging the detector. For example, the securing of the ionisation chamber by the use of plain screws would be considered unacceptable although the use of special screws would be acceptable. The use of rivets, solder, glue or certain types of plastic clip would be considered acceptable if their removal constituted significant damage to the device. Construction is of particular importance in the domestic situation since there can be no statutory control over use.

The requirement of the recommendations with regard to dose rate is that irrespective of the radionuclide the dose equivalent rate at any accessible point may not exceed 1μ Sv h^{-1} (0.1 mrem h^{-1}) at 0.1m from the surface of the device. Dose equivalent rates from those detectors containing radium-226 have been measured directly using a calibrated GM counter, and that from the detector utilising krypton-85 was measured using thermoluminescent dosimetry. However, for detectors which utilise americium-241 there is a significant contribution to the dose equivalent from low energy x-rays. In consequence the total dose equivalent rate is dependent upon the materials which shield the source, the methods of construction and the internal dimensions of the device. It is therefore necessary to measure the photon spectrum using a Si(Li) detector, in order to calculate the total dose equivalent rate. The assessed dose equivalent rates for different types of detector which have the same source activity vary by a factor of three. Most of the detectors examined have dose equivalent rates of less than 10% of the recommended maximum, the two exceptions being the ones with the most active sources. The detectors which contained activities of radium-226 in quantities comparable with those used for americium-241 sources exceeded the limit as did detectors utilising krypton-85. The dose equivalent rate at 10cm from the surface of most americium-containing detectors is below the detection limit. In consequence measurements are routinely made at 0.05m and the results extrapolated to 0.1m. Some of these results are compared in Table 1.

TABLE 1. Dose rates at 0.05m from outer casing

Nuclide	Activity kBq	Dose rate μSv h^{-1}
Am-241	33.3	6.0×10^{-3}
	37.0	8.5×10^{-3}
	33.3	6.1×10^{-3}
	37.0	1.0×10^{-2}
	37.0	1.3×10^{-2}
	33.3	1.8×10^{-2}
	29.6	2.4×10^{-2}
Ra-226	37.0	1.5+
	1.85	1×10^{-1}
Kr-85*	18500	2×10^{-1}

The standard deviation in the results for Am-241 is 1.2×10^{-3} μSv h^{-1}
* Dose rate measured at 0.1m from the surface of the device
+ β dose rate 2.0 μSv h^{-1} at 0.05m

The preferred method of surface contamination assessment is the wipe test, which permits several components of each detector to be independently checked. The wipes are performed using an alcohol-moistened cotton wool swab. A comprehensive wipe-testing programme results in a large number of samples, and for this reason the activity transferred to each swab is measured by liquid scintillation counting, which allows an automatic throughput of a large number of samples with a detection limit of 0.2 Bq (5 pCi). The results obtained on the inactive areas have almost invariably been below the limits of detection.

Wipe tests also play a major part in the integrity assessment during the final part of the evaluation, the destructive testing programme. Where possible, the source and holder are wiped separately both before and after the test, but where dismantling might invalidate the subsequent test, only the post-test wipes are carried out. Most of the tests (impact, puncture, drop, pressure, temperature, vibration) have produced few problems with regard to wipe testing, although they were designed to simulate conditions of normal use and credible abuse.

The current test programme differs from the earlier proposals (2) in several aspects. It was found that the sulphur dioxide corrosion test was unrealistically severe. The same conclusion was reached by NEA and the test was not included in the recommendations. However, it was apparent that in tests such as corrosion and fire, there was a likelihood of inactive deposits on the source. A policy of carrying out two consecutive wipe tests was therefore adopted for these tests. In addition, in the case of corrosion tests, activity was frequently found in the liquid beneath the sample, indicating that wipe tests alone were an insufficient criteria of leakage.

The earlier programme contained two fire tests - one at 600°C to simulate a domestic fire and one at 1200°C to simulate a hot industrial fire.

With the 600°C experiment it was possible to define a pass-fail criterion based on leakage and wipe tests and to identify material incompatibility problems. The 600°C test continues to provide the most interesting results. The use of a closed flow system and representative samples continues to give reproducible results. Standard counting methods are used to determine total leakage, and further qualitative information about source behaviour is obtained by alpha spectrometry and autoradiography. Alpha spectra in particular correlate well with the results of wipe testing. The results of 600°C fire tests are summarised in Table 2. The data serves to illustrate the effects of different holder materials, methods of fixing the source and different types of plastic. The range of results quoted refer to the first wipe taken.

TABLE 2. Summary of the results of the 600°C fire test

Holder material	Range of activity transferred to wipes, Bq	Comments
Stainless steel	0.3 - 370	No dispersion
	3589	Plastic housing contains fire retardant
Aluminium	0.6 - 163	No dispersion
Tin plated	148 - 3145	Extensive dispersion in tube and some in filters
Brass	555 - 12950	Extensive dispersion throughout apparatus
Source soldered on brass holder	81	Activity found in the vapour trap

The higher temperature in the 1200°C test was sufficient to melt the source and cause some dispersion of activity within the combustion tube. Several experiments were performed at the higher temperature, whereupon it was decided to curtail the test since no adequate pass-fail criteria could be defined. The NEA included a test at this temperature, defining a criterion in terms of activity which escapes from the combustion tube. This test is being incorporated into the NRPB programme and those samples which were subjected to it proved satisfactory in terms of the NEA criterion. Activity was detected remote from the combustion tube in only one instance, when 28 Bq was found in the vapour trap.

At the time of writing, the Board continues to act in an advisory role with regard to consumer products, although it is likely that this will change in the future. Nevertheless, manufacturers and distributors have willingly submitted samples for evaluation, normally prior to distribution within the UK, were equally willing to make modifications as a result of the Board's findings, and were agreeable to the results being published in a recent Board report (4).

ACKNOWLEDGEMENTS

The authors have pleasure in acknowledging the contributions made to this work by Mr. D.L. Bader, Miss D.E. Smith and Miss E.J. Bradley.

REFERENCES

1. M.D. Hill and A.D. Wrixon, Paper presented at the 4th International Congress of IRPA, Paris, 1977. The Radiological Testing of Products which irradiate the public.
2. M.D. Hill, A.D. Wrixon and B.T. Wilkins, NRPB-R42 (1976). Radiological Protection tests for Products which can lead to exposure of the public to ionising radiation.
3. Nuclear Energy Agency, Paris (1977). Recommendations for Ionisation Chamber Smoke Detectors in implementation of radiation protection standards.
4. B.T. Wilkins and D.W. Dixon, NRPB R-85 (February 1979). The Radiological testing of consumer products : 1976-78.

OCCUPATIONAL DOSE EQUIVALENT LIMITS

E. P. Goldfinch

The International Commission on Radiological Protection, in its publication No. 9, adopted by the Commission in 1965, recommended a maximum permissible dose to the critical organs, gonads and red bone-marrow, of 50 mSv in a year. It was acceptable for up to half of this, 30 mSv, to be received in one calendar quarter. The Commission accepted that to provide flexibility it may be necessary to repeat the quarterly exposure each quarter of the year on some occasions but recommended a cumulative limit of 50(N-18) mSv to the age N years. The Commission believed that these cases would occur infrequently. It was also recommended that because any exposure involves a degree of risk, all unnecessary exposures should be eliminated and all doses kept as low as is <u>readily</u> achievable, economic and social considerations being taken into account.

During the last decade new information became available which necessitated a review of the basic recommendations. This review, published as ICRP 26 was adopted by the Commission in January 1977. Certain clarifications were made at the 1978 meeting of the Commission in Stockholm and published within ICRP 28. The system of dose limitation now recommended by ICRP is essentially, that no practices shall be adopted unless their introduction produces a positive net benefit, all exposures shall be kept as low as <u>reasonably</u> achievable, economic and social factors being taken into account, and the dose equivalent to individuals shall not exceed the appropriate limits recommended by the Commission.

In its review of the dose equivalent limit the Commission has found that with the previous annual limit of 50 mSv, the distribution of annual dose equivalents in large occupationally exposed groups has very commonly fitted a log-normal function with an arithmetic mean of about 5 mSv, with very few values approaching the limit. The Commission has also found that the resulting average risk within such occupational groups is comparable with other safe industries and therefore considers that a dose equivalent limit of 50 mSv should be retained. The Commission assumes that workers who are exposed near the dose equivalent limits are unlikely to be so exposed every year, but recognises that some individuals may have a higher than average risk. The Commission recommends too that continued exposure of a considerable proportion of the workers at or near the dose equivalent limits will only be acceptable if a careful cost benefit analysis shows that the resultant risk is justifiable.

The Commission indicated in ICRP 9 that the dose limits were intended for planning design and operation in foreseeable conditions of exposure. In its review in May 1978 the Commission specifically eliminated reference to the use of dose equivalent limits for the purpose of planning.

Legislators and designers must take account of the advice offered by ICRP but nevertheless use it in a practical way. ICRP

offers no advice on the size of the occupational group over which the
achievement of an average dose equivalent equal to one tenth of the
dose equivalent limit would be acceptable. The development of public
and industrial awareness in many countries suggests that perhaps some
quantification of acceptable lifetime individual risks is required.
In the development of an index of harm (1) ICRP specifically rejects
the practicability of controlling the cumulative dose equivalent of
workers according to age and sex, regarding the achievement of an
adequate standard of protection to be more reliably ensured if a
single dose equivalent limit can be used for all workers. The
success of this philosophy is entirely dependent on the distribution
of dose equivalent within groups of workers remaining unchanged.
Furthermore, this philosophy allows individuals a radiation risk
comparable with occupational risks in high risk industries such as
construction and mining. This radiation risk is additional to conventional risks in the individuals occupation and does not necessarily result in a corresponding benefit to that individual.

This paper considers methods of limiting individual radiation
risks by recognising the variation of risk with age at exposure,
taking into account both somatic and genetic risks and proposes a
simple formula for controlling individual cumulative exposure and
hence risk.

VARIATION OF RISK WITH AGE AT EXPOSURE

Two components of radiation risk can be identified, namely
genetic and somatic. In both cases dose equivalent received early in
life carries a greater risk than that received later. In the case of
the genetic component the risk becomes zero after completion of child
production. For somatic effects the latent period has the effect of
reducing the risk from dose equivalents received later in life. In
quantitative terms data on the variation of risk with age is sparse.
However, such data as is available has been used by ICRP (1) to produce a curve showing variation of radiation risk with age at exposure
including both genetic and somatic components. The curve, Fig. 4 in
(1) may be approximated to a straight line represented as a function
of N:

$$R(N) = (2.82 - 0.0425N)10^{-5} \text{ mSv}^{-1}$$ ... where N is age at exposure

A cumulative dose equivalent limit, $D(N)$, expressed as a
function of N may be proposed, where $D(N)$ may be a continuous
function (e.g. $D(N) = (N-16)^2$ mSv), or have a discontinuity
(e.g. $D(N) = (N-16)^2$ mSv with a superimposed over-riding annual limit
such as 50 mSv).

If the resulting dose equivalents in years N and N-1 are D_N
and D_{N-1} respectively, then the cumulative risk to age N is:

$$10^{-5} \int_n^N (D_N - D_{N-1})(2.82 - 0.0425N) \, dN$$

where n is the age at commencement of occupational exposure.

Figure 1 shows cumulative risks with age for a range of dose
limit functions and Figure 2 shows the corresponding cumulative
limits of dose equivalent. Table 1 shows the resulting lifetime
risks for the same range of cumulative dose equivalent limit functions,
expressed as percentages, assuming exposure to age 65.

The risk and cumulative dose equivalent curves are identified as in Table 1. In the case of the cumulative dose equivalent functions 3, 4 and 6 the over-riding limits of 50 mSv, 30 mSv and 20 mSv per year are dominant from ages 44, 34 and 26 respectively. The latter two functions give very little benefit in terms of risk and also lead to very restrictive exposure regimes.

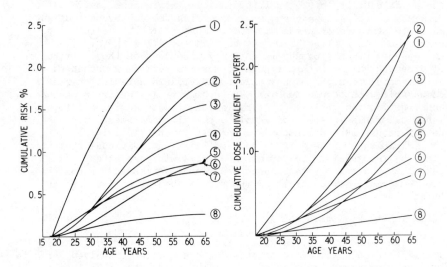

Fig 1 Cumulative risks with age. Fig 2 Cumulative dose limits with age.

TABLE 1 Lifetime risks for a range of dose equivalent limit functions.

Dose equivalent limit function	Identity	Lifetime risk (%)
50 mSv per year	①	2.4
$(N-16)^2$ mSv to age N	②	1.7
$(N-16)^2$ mSv with 50 mSv limit	③	1.5
$(N-16)^2$ mSv with 30 mSv limit	④	1.15
$0.5(N-16)^2$ mSv	⑤	0.85
$(N-16)^2$ mSv with 20 mSv limit	⑥	0.8
15 mSv per year	⑦	0.75
5 mSv per year	⑧	0.25

DISCUSSION

It will be seen from Table 1 that the lifetime risk to persons annually exposed to 5 mSv is 0.25%. This represents the average risk within a group of persons if their average exposure is one tenth of the current dose limit of 50 mSv. It is suggested here that it is undesirable for individuals to be exposed to a lifetime risk much greater than five times the average risk. Inspection of Table 1 shows that this can be achieved by limiting individual cumulative dose equivalent to $(N-16)^2$ mSv with a 50 mSv superimposed annual overriding limit. If on the other hand the individual risk is limited to three times the average then this is achieved by a cumulative limit of $0.5 \, (N-16)^2$ mSv with no individual yearly limit.

The advantages of either scheme are:
  i) Retention of the facility to receive dose equivalent of 50 mSv in individual years.
  ii) Reduction of individual radiation risks.
  iii) Limited dose equivalent from the age of 16 for training purposes.
  iv) Higher dose equivalents in the second half of the working life when work skills are fully developed.

It is suggested also that planned special exposures should be limited to the difference between exposures already received and the ceiling given by the cumulative limit without any further need for restriction within that planned exposure.

PRACTICAL APPLICATION

Either scheme may be applied by means of a complete table of permitted cumulative exposure and age. It may in some circumstances be more convenient to limit the average exposure over groups of years. In the latter case the formula $(N-16)^2$ mSv with a 50 mSv annual limit gives almost exactly the same risk as allowing 10 mSv per year from age 16 to 25, 30 mSv per year from age 25 to age 40 and 50 mSv per year above age 40.

CONCLUSIONS

On the basis that the current dose equivalent limit of 50 mSv has been shown to give an average dose equivalent of 5 mSv within a large group of workers, the individual risk within such a group can be limited to no more than three times the average if a cumulative individual limit of $0.5 \, (N-16)^2$ mSv is applied or to five times the average if a limit of $(N-16)^2$ mSv with an over-riding annual limit of 50 mSv is applied, where N is the age of the individual.

ACKNOWLEDGEMENTS

The permission from the CEGB to publish this paper is gratefully acknowledged. The views expressed in it are the authors and are not necessarily those of the CEGB. Inspiration and assistance from colleagues is acknowledged, especially Mr. P.F. Heaton and Mr. H.C. Orchard.

REFERENCES

(1) ICRP Publication 27, Problems Involved in Developing an Index of Harm - Annals of the ICRP Vol. 1 No. 4 1977.

THE DEVELOPMENT OF THE AMERICAN NATIONAL STANDARD "CONTROL OF RADIOACTIVE SURFACE CONTAMINATION ON MATERIALS, EQUIPMENT AND FACILITIES TO BE RELEASED FOR UNCONTROLLED USE"

J. Shapiro

In January, 1976, the Health Physics Society Standards Committee (HPSSC) submitted to the secretariat of the American National Standards Institute (ANSI) Committee N13 (Radiation Protection) the completed standard, "Control of Radioactive Surface Contamination on Materials, Equipment, and Facilities to be Released for Uncontrolled Use," with the recommendation that it be forwarded to the Board of Standards Review (BSR) of ANSI for final processing and publication.

CHRONOLOGY

This standard had gone through the many steps required in the development of a standard under ANSI. The first was the appointment of a chairman in September, 1971, with the charge to organize a subcommittee on surface contamination under HPSSC to develop the standard. The subcommittee held its first meeting in July, 1972 at the annual meeting of the Health Physics Society to discuss the approach to the standard. Four months later, the second meeting of the subcommittee was held at offices of the Atomic Energy Commission in Washington, D.C. The subcommittee first met with regulatory personnel of the AEC for an exchange of views and a review of their experience and policies on contamination control. This was followed by a session to prepare an outline for submission to ANSI as the first step in the development of an ANSI standard. The outline attempted to cover all the basic considerations in hazard evaluation and control of surface contamination and as a result was quite ambitious. Drafts prepared by the subcommittee members were reviewed at the third session during the annual meeting of the Health Physics Society in June, 1973, and led to a first draft of a standard one month later. This draft emphasized basic properties and data on surface contamination rather than specific conditions for release of contaminated equipment. It appeared to be of limited practical use and was rewritten as a performance standard. The approach was based in part on a position paper prepared by the Standards Committee of the Southern California Chapter of the Health Physics Society. The draft was submitted to ANSI N13 and N42 (Nuclear Instruments) for comment in September, 1973. The subcommittee continued to work on the standard during the same period, in particular on limits for surface contamination. It met at the Atomic Industrial Forum office in New York City in December, 1973 to consider the comments. The Atomic Industrial Forum at that time served as the secretariat for the development of standards under N13. Comments were reviewed and voted upon, and those accepted were incorporated into the standard. All persons who submitted comments were informed of the subcommittee action and reasons. The revised standard was approved unanimously by the subcommittee, and sent to HPSSC in July, 1974 for transmittal to ANSI. That same month, HPSSC submitted the standard to ANSI N13 for letter ballot action. The standard was sent out by ANSI for letter ballot action in September, 1974. The returns included three negative ballots. Two of these were resolved by changes in the standard. The third, which was cast because the limits were felt to be too low, could not be resolved. All action on the balloting was completed by the end of 1975 and in January, 1976, the standard was sent by HPSSC to the ANSI N13 secretariat (which by that time had been transferred from the Atomic Industrial Forum to the Health Physics Society) with the request to send it to BSR for final processing. The voting action reported was 28 affirmative and 1 negative. Three committee members did not respond despite follow up.

The standard provided criteria for the release for uncontrolled use of materials, equipment and facilities contaminated or potentially contaminated with radioactivity. Permissible contamination limits were specified as well as methods for assessing the levels of contamination. While more precise phrasing was given in the standard, the limits applied essentially to the following categories of radionuclides.

GROUP	TOTAL $dpm/100\ cm^2$	REMOVABLE $dpm/100\ cm^2$
1. Long lived alpha emitters except natural uranium and thorium	100	20
2. More hazardous beta-gamma emitters	1000	200
3. Less hazardous beta-gamma emitters	5000	1000
4. Natural uranium and thorium	5000	1000

The condition for placing a radionuclide in Group 1 was that the nonoccupational maximum permissible concentration in air (MPC_{air}) applicable to continuous exposure of members of the public be 2×10^{-13} Ci/m^3 or less and the nonoccupational MPC_{water} be 2×10^{-7} Ci/m^3 or less. (There had been considerable discussion on whether to use 4×10^{-13} or 1×10^{-13} for the MPC_{air} and 2×10^{-13} was adopted as a compromise. The value chosen determined whether all, or only a portion of the more hazardous alpha emitters would be in Group 1.) The upper limits for Group 2 were 1×10^{-12} Ci/m^3 air and 1×10^{-6} Ci/m^3 water. Acceptable sources for MPC values were those published by ICRP, NCRP, or NRC. The standard specified that the levels could be averaged over 1 m^2 provided that the maximum activity in any area of 100 cm^2 was less than 3 times the limit value. The criteria for the standard put some beta emitters in Group 1 and ^{210}Po in Group 2, but otherwise, the breakdown was as shown above.

As the standard was being processed through the final stages prior to promulgation as an official standard, the chairman of N13 changed his ballot from affirmative to negative after concluding that the alpha limits were too low and not readily measurable with state of the art detectors. It was sent out for reballot to ANSI N13 in September, 1976. During this period, the limits were adopted in

Regulatory Guide 1.86 of the Nuclear Regulatory Commission and the
standard was sent out to various laboratories of the Energy Research
and Development Administration (ERDA) for implementation on a "trial
and use" basis. In April, 1977 ANSI N13 transmitted the standard to
BSR with a report of 20 affirmative votes, 3 negative votes, and 4
unreturned ballots. This was followed by a period of public review
and comment which ended in August, 1977. Because of the unresolved
negative votes, HPSSC requested that it be issued as a Draft Standard
for a one year trial and use period. This period began in January,
1979. The limit of 100 dpm/100 cm^2 for the most hazardous group of
alpha emitters was changed in the Draft Standard to "non detectable,"
with an accompanying footnote stating that the instrument utilized
for the measurement was to be calibrated to measure 100 pCi of any
Group 1 contaminants uniformly spread over 100 cm^2. The total
activity limit for Group 2 beta or gamma emitters was also changed
to read "non detectable" with an accompanying footnote that the
instrument used for the measurement was to be calibrated to measure
1 nCi of any group of beta or gamma contaminant uniformly spread over
an area equal to the sensitive area of the detector. The limits for
removeable contamination were unchanged. The change in the wording
addressed the concerns of those who did not want to have to account
for a specific limit which they felt could not be measured accurately.

COMMITTEE DELIBERATIONS

At its first meeting in 1972, the subcommittee made plans to
produce a standard that would serve as a source document rather than
simply as a control document. As such, it would provide basic information on the nature of contamination, transport through the environment, and resultant doses. The standard would also deal with the
determination of the contamination potential from both the history of
use and the interpretation of monitoring results; criteria for release, test instrumentation and procedures; and decontamination
methods. Other proposed sections were cost-benefit analyses and an
annotated bibliography. The result would be an authoritative treatment of exposure risks associated with given types of surface contamination that would provide regulatory agencies with the guidance
needed to set numerical limits for contamination levels in specific
cases. Additional views on the features of a standard were obtained
from representatives of regulatory agencies. Numbers proposed for
contamination limits should be related to dose; the relationship
could be based on experience factors. Healy's "decision level"
approach was recomended for consideration. The need to keep levels
as low as reasonably achievable(ALARA) should be incorporated. Did
ALARA require some decontamination in all cases where contamination
was found, or only at levels which were above the limits? It was
possible that the requirements of industry, such as the photographic
industry, could be limiting rather than the hazard to people. Should
the levels depend on the number of people at risk? The standard
should be readily incorporated into normal practice by industry. It
should develop as a consensus rather than as a decree. Survey
techniques presented should be adequate and clear. But how specific
should the standard be? Should the number of measurements, or wipes
be specified? Was it really desireable to present numerical limits?

The Development of the American National Standard

Shouldn't one rather rely on best decontamination practice as shown by experience? The committee was advised to take a fresh look at setting limits without being influenced by the old numbers and to also determine what was as low as practicable.

It was obvious that the initial goals required a much greater effort than the committee could undertake. Also, the highly technical nature of the proposed approach to contamination evaluation would turn every case into a research project. In the end, the committee came to the conclusion that a workable standard had to be a performance standard. It was not possible to present a truly representative contamination level-dose curve. The standard would have to present specific limits and test procedures that could allow for uniform survey techniques. Technical analyses could be presented as a backup but not as a substitute for limits.

The following considerations were the basis for the preparation of the standard:

(1) The relative hazards of surface contamination produced by different radionuclides were given by the MPC's in air and water.

(2) For practical purposes, radionuclides were assigned to a small number of groups with given contamination limits, but this did not preclude the use of a graded scale based on the MPC's in air and water.

(3) A contamination reference level of 1000 dpm/100 cm^2 for ^{90}Sr was set as the basis of assigning limits to radionuclides presenting an ingestion hazard and other radionuclides were grouped based on the values of their MPC_{water} relative to ^{90}Sr.

(4) The contamination limit for ^{239}Pu was chosen as the basis for assigning limits based on MPC_{air} to radionuclides presenting an inhalation hazard. Values of 100 and 200 dpm/100 cm^2 were considered and 100 was adopted in the standard.

(5) An upper limit to surface contamination (i.e. the limit for the least hazardous group) was set on the basis of practicability of achievement and control rather than on MPC. Values of 2000-10000 dpm/100 cm^2 were considered, and a value of 5000 selected for the standard.

(6) Values of maximum scanning speeds for survey instruments were specified to provide the needed detection sensitivity. However, when contamination was detected, survey instruments had to be held stationary when recording readings.

Suggested values for removeable contamination varied between 10 and 20 percent of the total level. A value of 20 percent was adopted in the standard.

The current trial use of the standard should result in useful comments and documentation on the practicability of cleaning equipment to the limits in the standard, and on detecting those limits. It should be noted that the promulgation of the standard does not preclude release at levels above the limits. However, appropriate controls and restrictions would have to be observed until and if the limits in the standard could be satisfied.

Measurement Techniques

A SIMPLE METHOD FOR COLLECTION OF HTO FROM AIR

A. N. Auf der Maur and T. Lauffenburger

About 50 plants process luminous paint in Switzerland. The contamination of the air by tritium constitutes a major problem in these plants. 250 workers in these plants are professionally exposed to tritium and have an average concentration of about 7 µCi/lt tritium in the urinary excretion corresponding to annual doses of 1,2 rem. Furthermore watch factories and stores might have locals where the stock of watches or dials with luminous paint might also lead to an excessive content of tritium in the air. This contamination of the air is due to the fact that luminous paint loses annually 5 to 10 % of the tritium to the environment. The tritium in air is in the form of HTO.

It was therefore desirable, to adoperate a rapid and practical method to check the HTO content in air in order to estimate the tritium intake by workers. Ideally this method should cover the range of three interesting limiting values:

10 µCi m^{-3}, 300 nCi m^{-3}, 30 nCi m^{-3}.

These three values correspond to the upper limit of concentration for working places of professionally exposed persons, for places accessible by single persons of the population and for places with general access, respectively.

So far 3 methods are described in the literature (1), which allow to measure in the range stated above:

- Freezing or condensing the atmospheric moisture
- Bubbling the air through water
- Absorbing the atmospheric moisture

We describe now an alternative method which also covers the interesting measuring range. It is less sensitive than the methods mentioned above, but it is much faster and simpler and therefore it was adopted as our standard method.

MEASURING METHOD

If a liquid scintillation counter is available the rest is very simple. One goes to the place where the measurement of the content of HTO in the air has to be made, opens a soft plastic bottle and squeezes it several times in order to get the original air out and the environmental air into the bottle. Afterwards one pours quickly some water into the bottle and the bottle is immediatedly closed. Of course the water was prepared beforehand and brought along in a small flask in order to avoid contamination. The HTO and other soluble gases containing tritium will go from the gaseous to the liquid phase in the bottle. At home this liquid will be poured into the vial and mixed with the cocktail for liquid scintillation counting. the determination of the efficiency of the liquid scintillation counter is the only step which requires calibration.

CHOICE OF BOTTLE SIZE AND AMOUNT OF ADDED WATER

A first choice which has to be made is the size of the bottle which serves for the collection of air. To increase sensitivity one would choose the size of the bottle as large as possible. However, there are some limitations:

- The amount of water in the vapour phase and the water remaining on the walls of the bottle should be small compared to the amount of added water.
- One should be able to squeeze out a considerable fraction of the bottle.
- As one bottle for each sample is needed, the size should not be unpractically large.

We decided to use half litre bottles, although for conditions where increased sensitivity is needed, larger bottles could be used. We found that the volume of the bottle is actually 2 % larger than the nominal value. When squeezing the bottle for air sampling, each time about 40 % of the volume is exchanged. After squeezing the bottle 10 times, less than 1 % of the original air remains in the bottle.

The subsequent addition of 10 ml of water serves to absorb the HTO into the water phase. This process can be accelerated by agitating. But there is no problem as the time constant for diffusion over a distance of 10 cm in air is of the order of 10 min. Usually the bottle is left 24 hours before pouring the water into the counting vial. When this is done, some water is lost on the wall of the bottle. For the half litre bottle this loss amounts to 90 µl with a standard deviation of 30 µl. As we were adding 10 ml of water, this loss can be neglected. Also the proportion of HTO remaining in the gaseous phase will be negligible. Even the amount of the added water needs not to be exactly known.

A Simple Method for Collection of HTO from Air

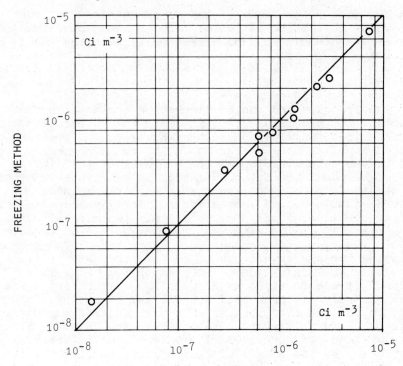

Figure 1
Agreement between bottle method and freezing method.

When mixing 10 ml water with 10 ml of Insta-Gel we find in our laboratory a counting efficiency of 25 %. this efficiency is calibrated and controlled by the measurement of the external standard ratio. The background count is 25 cpm. For 20 min. of counting time the limit of detection is about 3 cpm. This corresponds to

$$400 \text{ Bq m}^{-3} \quad \text{or} \quad 10 \text{ nCi m}^{-3}.$$

RESULTS

An alternative method to measure HTO in air was used to check the bottle method. We collected frost on a can filled with crushed dry ice, scraped the frost off and measured its tritium concentration in the liquid scintillation counter (2).

Simultaneously the humidity had to be measured in order to calculate the concentration of tritium with respect to the air. Fig. 1 shows good agreement between the two methods.

DISCUSSION

The freezing and the bottle method are both absolute methods, the precision being given by the calibration of the liquid scintillation counter. However, the freezing method involves more measurements, of which the most critical is the determination of the humidity of the air. On the other hand, the freezing method is nearly three orders of magnitude more sensitive than the bottle method.

The bottle method proved its reliability and simplicity during one year of use. One needs only two minutes to take an air sample, while with the freezing method it takes 20 min. However, one precaution must be taken: The person sampling the air should not be contaminated with tritium, otherwise the measurements can be much too high. This happens if one first looks around in a laboratory where luminous paint is processed and thereafter measures the tritium outside the plant. The freezing method is less vulnerable in this respect as the sampling takes much more time and one can go away during this time. The freezing method may be the method of choice for measuring tritium in the environment. The bottle method is adequate for measurements in laboratories or for a quick check in the environment in order to decide whether the concentration is above or below the legal limit.

REFERENCES

1. NCRP (1976): NCRP Report no. 47, Washington, D.C.
2. Koranda, J.J., Phelps, P.L., Anspaugh, L.R. and Holladay, G. (1971): In: Rapid Methods for Measuring Radioactivity in the Environment, p. 587, IAEA, Vienna.

A PROCEDURE FOR ROUTINE RADIATION PROTECTION CHECKING OF MAMMOGRAPHY EQUIPMENT

L. G. Bengtsson and I. Lundéhn

In Sweden, screening for mammography is only permitted in demonstration projects in a few parts of the country including Gävle. After evaluation around 1983 it is possible that screening might be widely used at the 60 installations where mammography equipment is available. Should this occur, standardized checking of the equipment will be an essential radiation protection task.

As radiation protection physicists can be consulted fairly simply at all equipment sites, we wanted to design a checking system that could easily be handled by them on-the-spot. The system should include adequate instruction about criteria for remedial action.

Techniques in Sweden are fairly standardized and the main difference between screening and clinically indicated examinations is in the number of projections, where screening normally employs only the oblique one. Almost all mammography units have molybdenum anodes, automatic exposure control and 0.03 mm Mo filter. Only low-dose film-screen systems are used. Potential differences range from 24 kV to 37 kV with a tendency towards the lower end, and typical tube charges range from 5 mAs to 200 mAs.

RADIATION PHYSICS PROPERTIES OF BREASTS

The distribution of breast thickness in screening situations measured in three locations with the following result:

	Number of measurements	Brest thickness, mm		
		min	max	mean
Location 1	111	15	75	41
Location 2	413	15	80	48
Location 3	486	15	95	50

In all locations the relative standard deviation of the breast thickness was about 27%. The differences in mean thickness are thus significant at the 99% confidence level.

For the estimation of energy imparted to the total breast, the breast area was derived from the exposed x-ray films for 200 films at one location. To a good approximation, the area was proportional to the breast thickness, being 150 cm^2 at 50 mm breast thickness. The relative standard deviation of the area for a given thickness was about 30%.

We have also looked at the attenuation properties of breasts compared with those of water, polymethylmetacrylate and 85% solution of ethyl alcohol in water. The following results were obtained for the average ratio of the tube charges required to trip the automatic exposure control for breasts of a given thickness and for polymethylmetacrylate of the same thickness:

Breast thickness, mm	30	50	70
Tube charge ratio breast/polym.			
Location 1	1.35	0.9	0.4
Location 2	1.0	0.8	0.6
Location 3	1.3	1.1	0.9
Mean	1.17 ± 0.15	0.91 ± 0.15	0.60 ± 0.3

The uncertainly interval assigned to the mean covers the full range of variation between the three locations. This variation may have been due to different screen-film systems having different energy dependence, and to the employed tube potential differences. An error of 10-20% may have been caused since the cassette used at the phantom measurements did not represent exactly the mean of all cassettes used in the breast exposures.

Smaller breasts approached water in linear attenuation properlies, larger breasts approached alcohol/water solution. Polymethylmetacrylate seemed the best approximation when the entire range of breast thicknesses was considered, but obviously it can overestimate the breast dose by more than a factor of 2 at large breast thicknesses.

RECOMMENDATIONS

The recommended checking method includes the use of the Kodak mammography image quality phantom (3) placed above 40 mm of polymethylmetacrylate. Reference pictures will be included in the set of recommendations, and it is recommended that remedial action be taken if the actual image quality subjectively appears to be significantly poorer than that of the reference film. This qualitative recommendation has been thought to be sufficient since the high professional competence in diagnostic radiology in Sweden makes poor image quality rather unlikely.

Radiation dose is to be based on phantom measurements using 50 mm polymethylmetacrylate. Detailed instructions are given for the measurements using either thermoluminescent LiF dosimeters (4 dosimeters per measurement), ionization chambers or a specially designed plastic scintillator (1) with diameter 25 mm and height 50 mm. The exposure at the phantom surface should be multiplied by the following conversion factors to give mean absorbed dose in the breast:

Potential difference, kV	25	28	31	34
Conversion factor, mGy/R	1.19	1.28	1.34	1.39

According to a preliminary recommendation, a mean breast dose of 2 mGy indicates a need for remedial action. Suggestions will be given for countermeasures.

DISCUSSION AND CONCLUSIONS

Since the range of potential differences is rather small, the TLD, ionization chamber and plastic scintillator methods are about equally accurate and expected to yield the same alleged mean absorbed dose to the breast within about $\pm$ 10%. A larger range of potential differences might be handled most accurately using the plastic scintillator. The three month reproducibility (double standard error with a given exposure) will according to long term tests be about $\pm$ 3% with the ionization chamber and plastic scintillator (mean of two measurements) and considerably higher with TLD (mean of 4 dosimeters).

The day-to-day variations of the x-ray equipment for a given exposure setting may give doses outside a range of $\pm$ 10% from the mean. In practice it is difficult to make the dose with the cassette used at phantom measurements represent the average dose for all cassettes within less than $\pm$ 10%.

The country-wide consistency may thus be rather poor for the reasons just discussed. In national dose intercomparisons a standards laboratory could provide an assessment of "true" dose using the methods recommended. Several centers would then be asked to irradiate the phantom as if it were a 50 mm breast and state the mean breast dose as assessed by them using a recommended method. It is believed that this dose might be up to $\pm$ 30% off from the "true" dose. This variability would mainly be due to errors associated with the automatic exposure, and similar variability would be inherent in the normal breast exposures.

In addition, there are the problems of representativity. Results obtained at any installation using the phantom measurements are expected to represent the mean for a screened population within -30% and +50% for the mean breast dose, and within -40% and +60% for the mean energy imparted to the breast. These errors are mainly due to the previously discussed variations in radiation physics properties of breasts. They have little relevance for the levels selected for remedial action but should be kept in mind when it comes to assessment of population risk. The thickness 50 mm was selected as the best compromize with representativity criteria for mean energy imparted and mean breast dose. A thickness of 40 mm would have been a much poorer compromize and 60 mm a somewhat poorer. If representativity is sought only for the mean energy imparted to the breast as compared with the mean breast dose, a thicker phantom results.

Measurements by one team employing the recommendations at about half of the 60 Swedish mammography installations (2) gave a mean breast dose of 1.6 mGy, a minimum of 0.5 mGy and a maximum of 4 mGy. About one fifth of all installations had doses above 2 mGy, mostly because of different film-screen systems, sub-optimal developing procedures or unusually high optical density of the films.

REFERENCES

(1) Studsvik Energiteknik AB, (1979): private communication with Erik Dissing.

(2) Leitz, W and Eklund, S (1979): private communication

HEALTH PHYSICS ASPECTS OF THE IN VIVO ANALYSIS OF HUMAN DENTAL ENAMEL BY PROTON ACTIVATION

F. Bodart and L. Ghoos

INTRODUCTION.

The technique of fluorine analysis by low energy nuclear reaction has often been described (1, 2) and the application to fluorine concentration determination is now routinely used in various laboratories. Usually the sample is placed in a chamber and maintained under vacuum. This technique was adapted to non-vacuum analysis (3) allowing the handling of the samples at atmospheric pressure, and thus permitting in-vivo analysis of human teeth.

As described in a previous paper (4), the protons, having an initial energy of 3 MeV, emerge in air through a tantalum window in which they lose part of their energy; the distance in air between the foil and the impact point on the surface of the tooth being 0.8 cm. The residual energy of the proton beam is 2.725 MeV at the surface of the enamel and the beam diameter is 2 mm.

When the beam crosses the tantalum foil, X-rays and γ-rays from Ta(p,p'γ) reaction are produced. Other nuclear reactions take place when the protons interact with fluorine, phosphorous, sodium etc... contained in the tooth enamel. A typical spectrum is shown in Fig.1.

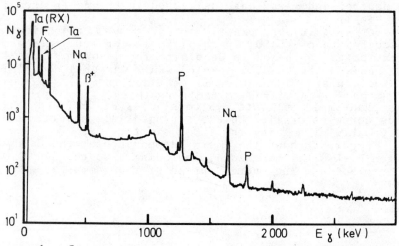

Figure 1 : Gamma-rays emitted during proton bombardment of human teeth and recorded with a Ge(Li) detector.

These γ-rays are used for the determination of the absolute concentrations of these elements in the surface of the tooth. Each measurement takes 1 min, with a beam intensity of 20 nA, which corresponds to a power dissipation of 0.055 W. No heat can be felt by the patient during the bombardment.

The technique of fluorine measurement by prompt activation allows repeated determinations on the same area of tooth enamel both before and after topical application of fluoridated compounds.

EXPERIMENTAL PROCEDURE.

In tooth enamel, ionization, produced by charged particles losing their energy, is very intense over a short range, 30-40 μm. This process can produce local destruction of the enamel and emission of soft X-rays that are completely screened by the tooth and the surrounding frame; hence they are not observable. Neutrons are not observed because the proton energy is below the neutron threshold for the majority of nuclei present in the teeth. Any dose rate is thus due mainly to hard X-rays and γ-rays induced in atomic and nuclear reactions. Due to the short irradiation time, as mentioned above, the dose delivered to the patient is very small : a few μrad. Preliminary measurements have been taken during a 60 min.run made on an extracted tooth to obtain the order of magnitude of the dose around the irradiated tooth. Neutron dose rates were measured with a neutron REM counter (Nuclear Enterprise type NM1), and the dose rate was less than the minimum observable, i.e. 0.1 mrem/hr. To estimate the absorbed dose of gamma-rays, LiF chips were placed around the mouthpiece surrounding the tooth, and an ionization chamber of 500 cm^3 with a tissue equivalent wall of 300 $mg.cm^{-3}$ (Babyline type Nardeux) was used 5 cm behind the beam spot. The dose rate observed with the ionization chamber was less than 1 mrad/hr, but too small to be measured with the LiF chips, indicating that the neutron and γ-ray dose delivered in the mouth area is very small.

To estimate more exactly this very small radiation rate we decided to use $CaSO_4$:Dy chips in longer irradiations on an extracted tooth and to determine the doses to the skin and inside the head, a phantom skull was obtained. This phantom contains a human skull embedded in a tissue-equivalent medium contoured to human features.
This phantom is cut into 2.5 cm slices (Fig.3), and each slice contains drilled holes, arranged in a matrix, which are capable of holding the TLD dosimeters.

Figure 2 shows the energy dependance of the $CaSO_4$:Dy teflon disks (0.4 mm thick), including a correction to take into account the energy spread of the gamma-ray spectrum.

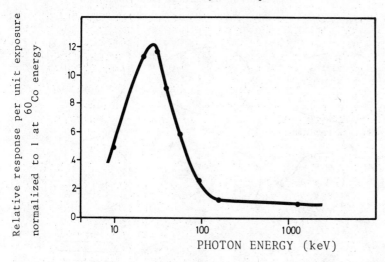

Figure 2 : Energy dependance of the $CaSO_4$:Dy teflon disks

Table 1 summarizes the percentage of the total photon yield for various energy intervals.

TABLE 1 : Percentage of photons emitted during the fluorine determination versus energy.

Energy (keV)	Mains origins of the photons	Percentage normalized to 100 for the entire spectrum.
10-100	Ta X-rays and Compton γ-rays	97.2 %
100-200	γ-rays from Na and F	1.6 %
200-1000	γ-rays from Na and β^+ annihilation	0.5 %
1000-2000	γ-rays from Na and P	0.4 %
2000-8000	high energy γ-rays from F	0.3 %

The calibrated TLD dosimeters were placed in the matrix at appropriate levels of the head which was then placed in the same position as the patient for an 8 hr irradiation under identical experimental conditions.
For each slice, a map of the exposure data was obtained. This data is presented in Figure 3, where the doses (in μrad) correspond to those which would be received after a 1 min. irradiation time.

SUMMARY.

The health physics aspects of the in-vivo analysis of human dental enamel by fast proton activation have been studied. Doses to the head including surface skin and in-

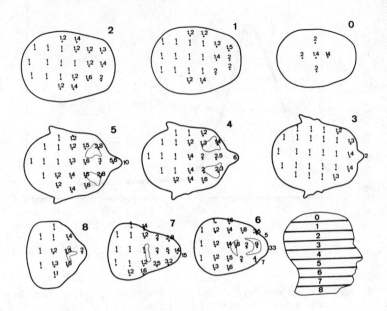

Figure 3 : Doses, in micro-rad, received by the patient during the in-vivo analysis (corresponding to 1 min. irradiation time).

terior regions of the brain and mouth were determined using TLD materials in a phantom head. The maximum dose to the head occurs on the right part of the lips and is of the order of 30µrad. In the mouth, behind the tooth analysed, the average dose is 10µrad, and the average over the whole head is of the order of 1µrad; this is twice the dose received by personnel who spend the same period of time in the counting room a considerable distance from the beam line.

As shown in a previous paper (4) no residual destruction has been found in the first 30µm of the enamel, which is the total range of the 2.7 MeV protons penetrating hydroxyapatite. Indeed, even more than 2 years after in-vivo analysis, there is no apparent destruction of tooth enamel.

REFERENCES.
1. Mandler, E., Moler, R.B., Raisen, E. and Rajan, K.S., (1973) : Thin Solid Films, 19, 165.

2. Rytomaa, I., Keinomen, J. and Antilla, R., (1974) : Archs. Oral. Biol., 19, 553.

3. Deconninck, G., (1976) : In. : Proceedings, IV Conf. Scient. and Ind. Appl. of Small Acc. p 533.

4. Baijot-Stroobants, J., Bodart, F. and Deconninck, G., (1979) : Health Physics, 36, 423.

A RELATIVELY FAST ASSAY OF Sr-90 BY MEASURING THE CHERENKOV EFFECT FROM THE INGROWING Y-90

B. Carmon and Y. Eliah

In the year 1978 alone, over 30 publications were registered dealing with the biological effects and radiation hazards due to the intake and adsorption of Sr-90 (1). Some of these publications include experimental results obtained from radioanalyses of Sr-90, which by themselves are mainly based on measurements of the short-lived, separated daughter nuclide Y-90. A delay of 10 to 15 days is often necessary prior to the radioassay, in order to insure a large enough ingrowth of Y-90 ($T_{\frac{1}{2}}$ = 64.1 h). This time interval can be considerably reduced if the assay is carried out by Cherenkov counting of the ingrowing yttrium. Cherenkov radiation is produced in aqueous solutions by high energy β- particles and has been successfully used for the determination of P-32 (2), Na-24 (3), K-40 (4) and Rb-86 (5). This work is an improvement of a procedure on the assay of Sr-89 and Sr-90 in presence of each other, in which the equilibrated Y-90 is separated from the parent before counting; this part of the procedure is superfluous if only Sr-90 is present (6).

METHOD

The Cherenkov effect induced by the 2.27 MeV β-emitter Y-90 can be measured with good efficiency using conventional liquid scintillation counters (6). The weak contribution of the soft β-s from the parent Sr-90 is eliminated by suitable adjustments of the gain setting and the base-line window opening in the counting channel. Any detectable activity in a freshly separated Sr-90 sample is thus due solely to the ingrowing Y-90. The activity of the strontium is calculated by use of the following expression:

$$\mu Ci(t_o) = \frac{cpm\ Y\text{-}90\ (t_b) \times \exp \lambda_{Sr}(t_a - t_o)}{y_{Sr} \times \varepsilon_Y \times 2.22 \times 10^6 \{1 - \exp[-\lambda_Y(t_b - t_a)]\}} \quad [1]$$

where : t_o = the reference time of sampling;
t_a = the final Sr/Y separation time;
t_b = the Y-90 counting time;
λ_{Sr} and λ_Y = the decay constants of the two radionuclides (0.0243 y^{-1} and 0.0108 h^{-1} respectively);
y_{Sr} = the radiochemical yield of strontium;
ε_Y = the counting efficiency of Y-90.

THEORETICAL ASPECTS (summarized from 7,8)

Cherenkov radiation is produced when charged particles (in our case β-s) pass through a transparent medium at a velocity greater than that of light in the same medium. This radiation is emitted at an angle θ with respect to the direction of the inducing particle according to (7) cos θ = 1/βn, where β is the velocity ratio v/c and n the refractive index of the medium. For relativistic electrons, β is related to the energy E (keV) of the particle, by (7)

$$\beta = \left(1 - \left[\frac{1}{(E/511) + 1}\right]^2\right)^{1/2}$$

The threshold condition for Cherenkov radiation is βn = 1, so that if n for water equals 1.332 then β must be > 0.751. This lower energy threshold is 263 keV, below which no Cherenkov effect occurs. However only β-emitters above 1 MeV give sufficiently high yields to be of any practical significance (8).

Cherenkov radiation is one-directional, weak in intensity and has a continuous spectrum mainly in the ultraviolet, with only a small portion extending into the visible region (7).

RESULTS AND DISCUSSION

All the measurements were conducted with 20 ml aqueous solutions in standard polyethylene counting vials. Two optional liquid scintillation spectrometers were used, the Packard Tricarb Model 3390 and the older Model 3214; the working conditions for both instruments are summarized in table 1.

TABLE 1. Working conditions for the Cherenkov counting of Y-90

	Tricarb 3390	Tricarb 3214
Gain %	40	30
Window opening	100 - 1000	50 - 1000
Counting efficiency	ε = 0.43	ε = 0.39
Background	9 - 11	11 - 13

The data in table 2. are a measure of the feasability of assaying a sample before the Sr/Y equilibrium is reestablished. It gives the magnitude of the experimental error that can be anticipated if the Y-90 counting is carried out at various time intervals after the final separation and purification of strontium. Activities as low as 10^{-3} µCi per vial can be measured with reasonable accuracy after only 2 days following the separation. Lower activities require longer periods of time for the Y-90 ingrowth, but even for 10^{-5} µCi 5 days seem to be sufficient.

The radiochemical yield was determined with the help of the γ-emitting Sr-85. Additions of up to 2000 dpm were not detectable at our working conditions. With activities higher than .01 µCi larger aliquots of the marker Sr-85 are desirable (6).

The lower limit of detection is about 3×10^{-6} µCi at conditions of secular equilibrium (≈ 18 d), which in practice amounts to 2 or 3 cpm above background. An assay of such very low activities requires long counting times (about 500 counts per sample), if the measurements are to be statistically significant. In addition, too early counting of such samples would introduce an error due to an increase in the activity of the ingrowing Y-90, during the measurement itself. Very low activities should therefore be assayed only after the Sr/Y equilibrium has been reestablished.

TABLE 2. Estimated overall errors for Sr-90 activities, measured at various time intervals after the separation from Y-90 (counting times per sample : 100 minutes or less)

Y-90 ingrowth time $(t_b - t_a)$ (days)	± 95% confidence limit, in % of activity :			
	.01 - .15 (μCi)	.003 - .009 (μCi)	8×10^{-5} - .001 (μCi)	10^{-5} (μCi)
< 1	30.	-	-	-
1 - 2	4.4	8.1	-	-
2 - 3	4.5	5.9	34.	-
3 - 5	5.6	5.7	23.	-
5 - 7	6.0	7.7	9.0	15.

RECOMMENDED PROCEDURE (short summary)

1. About 5 to 10 mg of inactive carrier and a known activity of Sr-85 tracer ($10^{-3} - 10^{-4}$ dpm) are added to the original sample. The strontium is separated and purified by a standard radiochemical procedure.

2. The aqueous sample is transfered into a polyethylene counting vial and made up to the desired volume (20 ml). It should be ready for measurement 24 hours after the final Sr/Y separation step. The gain settings and window openings on the liquid scintillation spectrometer should be arranged in such a way as to eliminate the contribution of the parent nuclide Sr-90; only the ingrowing Y-90 is counted.

3. If necessary, the measurements are repeated on consecutive days until the desired precision is achieved. The radiochemical yield is determined by comparison with a blank Sr-85 sample.

REFERENCES
1. INIS - Atomindex, (1978) : 9 (pt.3), subject index, pp 855, 1022.
2. Frič, F. and Finoccharo, V. (1976) : Radiochem. Radioanalyt. Letters, 25, 187.
3. Braunsberg, H. and Guyver, A. (1965) : Analyt. Biochem., 10, 86.
4. Mertl, F. (1973) : Jad. energ., 19, 43.
5. Läuchli, A. (1969) : Int. J. appl. Radiat. Isotopes, 20, 265.
6. Carmon, B. (1979) : Int. J. appl. Radiat. Isotopes, 30, 97.
7. Ross, H.H. (1969) : Analyt. Chem. , 41, 1260.
8. Elrick, R.H. and Parker, R.P. (1968) : Int. J. appl. Radiat. Isotopes, 19, 263.

DECONTAMINATION OF TRITIATED WATER SAMPLES PRIOR TO TRITIUM ASSAY

B. Carmon, S. Levinson and Y. Eliah

The liquid scintillation assay of tritiated water samples in radiation protection control is often inaccurate and unreliable due to the presence of other radionuclides. A simple and efficient separation procedure based on distillation and recovery of the tritiated water was published not long ago (1). In the work presented here various aspects of the above method were investigated, with the view of improving the purification and decontamination of the sample from interfering nuclides. The technique is especially useful in the tritium assay of liquid waste effluents from hospitals, biological and veterinary institutes, as well as in the production facilities for labelled compounds.

METHOD

Measured aliquots of tritiated water samples, containing other additional radionuclides, are introduced into one of the distillation apparatuses shown in the illustrations. The larger apparatus (fig.1) consists of a test tube for 25 ml samples and is provided with a small aluminum condenser. The sample is heated to boiling in a glycerol bath and the distillate recovered into a 20 ml plastic vial. About 300 mg of a hold-back carrier are added prior to the distillation; it consists of a dry mixture of $AgNO_3$ (40%), NaI (20%), CuS (10%), anhydrous Na_2CO_3 (10%) and anhydrous $Sr(NO_3)_2$. CuS cannot be exchanged with sodium thiosulfate, which forms a soluble complex with AgI, but $NaHSO_3$ can be used. Salts containing crystallization-water must be absent.

A device for smaller samples is useful when only limited amounts of tritiated water are available, or if volatile compounds of radioiodine are present (fig. 2). The glass-tripod with a 4 ml cup on top is placed into a 100 ml beaker and covered with a watch-glass (2); the water sample at the bottom of the beaker is slowly evaporated on a sand bath at 70°C and in presence of the hold-back carrier, until the cup is filled with a sufficient amount if distillate.

In both apparatuses most of the interfering nuclides are completely removed by the hold-back carrier, except for some traces of radiocesium and radioruthenium, which sometimes appear in the distillate. They are removed by shaking the sample with (< 100 mg of) $K_2Co[Fe(CN)_6]$ before or after the distillation, and discarding the solid residue (3).

The sample should be distilled to dryness whenever time and the presence of foreign nuclides permit it. The separation lasts up to one hour with the larger apparatus and about 4 hours with the smaller one. Twelve samples can be easily separated in one run. The purified sample is now mixed with the liquid scintillator and assayed.

RESULTS AND DISCUSSION

The method was tried out on aqueous solutions of sixteen of the most common nuclides (in ordinary water, no H-3 !), in order to determine their "decontamination factors". The following tracers (mostly as chlorides or nitrates) were used: Am-241, C-14 (as carbonate or uridine), Cl-36, Co-60, Cs-134, Fe-55, gold-198, Hg-203, I-125, Mn-54, Na-22, Ru-103, Sr-85, Tc-99, Th-230 and Zn-65; their activities ranged from 10^{-3} to 1.1 µCi. They were subjected to the distillation procedure either individually or in prepared mixtures. As expected, most of the radionuclides were absent from the distillate except for some traces of radioiodine, Hg-203 and Ru-103; decontamination factors of these nuclides are presented in table 1.

TABLE 1. Decontamination factors for I-125, Hg-203 and Ru-103; R = cpm(distillate)/cpm(sample)

Nuclide	sample (ml)	distillate (ml)	hold-back carrier	R
I-125				
0.3 µCi	10	5	Hamiltons (1)	4×10^{-5}
"	10	4	" without CuS, but with $Na_2S_2O_3$	10^{-3}
"	10	4	Hamiltons, no CuS, but with $NaHSO_3$	7×10^{-5}
I-125				
0.05 µCi	25	first 8	Hamiltons	3×10^{-4}
"	25	second 8	"	4×10^{-4}
"	25	last 9	"	0.034
Hg-203				
0.02 µCi	18	first 15	"	2×10^{-5}
"	18	last 3	"	10^{-4}
Ru-103				
0.02 µCi	20	first 7	"	0.027
"	20	second 7	"	7×10^{-3}
"	20	last 6	"	0.026

All the activities were measured by liquid scintillation counting using the Instagel scintillator. A Packard Tricarb Model 3390 spectrometer was used, at optimum gain settings and window openings.

OPTIMAL FIGURES OF MERIT FOR TRITIUM COUNTING

The "figure of merit" (FM) in tritium assay is obtained by multiplying the sample volume with the counting efficiency ε. The highest FM with polyethylene vials (20 ml) were with 7 ml of sample and 9 ml of Instagel ($\varepsilon = .18$); with the small (7 ml) vials FM was highest for a 1:1 ratio in a 5 ml total volume ($\varepsilon = .16$).

Figure 1.

Figure 2.

REFERENCES

1. Hamilton, R.A. (1974) : Determination of Tritium in Waste Processing Effluents by Distillation and Liquid Scintillation Emulsion Counting. - USAEC report no. ARH-SA-188.
2. Frenkler, K.L. (1977) : KFA Jülich GmbH, private communication.
3. Boni, I.L. (1966) : Analyt. Chem., 38, 89.

EVALUATION OF THE SPECTRAL DISTRIBUTION OF X-RAY BEAMS FROM MEASUREMENTS ON THE SCATTERED RADIATION

E. Casnati* and C. Baraldi**

INTRODUCTION

The knowledge of photon or energy flux density distribution with respect to quantum energy is of certain importance when X-ray beams are produced by tubes supplied with voltages below 100 kV. The reason for this is threefold: 1) the interaction parameters show a high energy gradient below 100 keV, 2) secondary electrons produced by photons fall in an energy range which seems very effective in radiobiological damage (1), 3) beams of such kind are by far the largest source of irradiation for man (radiodiagnostic uses). In these cases, however, spectral distribution measurements present difficulties arising from the very high flux densities attained during the current pulse. Therefore, a number of contrivances have been proposed in order to make the nuclear spectrometers usable. The most important of these are: a) a very narrow beam collimation, b) a large distance between the focal spot and the detector, c) a very low current through X-ray tube.

PRINCIPLES

None of the above mentioned methods is free of criticism. For this reason a different approach was regarded as worthy of more detailed study. A very partial experiment on this spectrometry method is described by Greening (2). The principle of the method is based on the advantage which can be gained from the physical attenuation which the beam undergoes in its partial scattering over a thin sheet in order to solve the difficulty raised by high density fluxes. Therefore, a calculation procedure is required, suitable for the evaluation of the radiation incident on the scatterer from the instrumentally measured distribution. On the grounds of both theoretical and experimental considerations it may be inferred that as far as such kind of beam spectral analysis is concerned the most appropriate direction for measuring the scattered radiation involves crossing the beam axis at about $\pi/2$.

The evaluation of the true spectrum incident on the scatterer needs a procedure complicated on the one hand for the coexistence

----
* I.N.F.N.,Sezione Sanità, Rome (Italy).
** I.N.F.N.,Laboratori Nazionali, Legnaro (Italy).

of coherent and incoherent scattering and on the other hand for the presence of distortions caused by collateral physical effects. The exact solution with respect to the first problem can be obtained by measuring spectra of the same beam as scattered by two thin metallic sheets made of pure elements of different atomic number. An alternative procedure which appears equally satisfactory and much simpler both in theory and in experience, requires the measurement of the spectrum scattered by only one scatterer provided it is made of a pure element of a very low atomic number for which the interference between coherent and incoherent scattering is pratically avoided. The collateral effects are due partly to the response function of the detector and partly to distortions in the beam energy distribution caused by a number of unavoidable attenuators. There are no less than ten corrections to be considered.

Even if all of these effects are taken into account in the calculations, which is possible, the above mentioned spectrometry technique has a cost which is low and acceptable only for quantum energy below about 100 keV. In fact, apart from the resolution loss connected to the finite aperture of the beam which can be rendered negligible by a suitable choice of geometrical arrangement, that resolution loss coming from the broadening of the energy interval of the reconstructed discrete spectrum in comparison with the direct one should be pointed out. Indeed it is possible to show that the reconstruction necessarily involves the incoherent scattering which changes the energy interval breadth.

RESULTS

Two distributions are used in order to check the theory: a) the first one is the highly filtered spectrum produced by a Be window tube supplied with a high voltage of 60 kVp, an average current of 4 mA and filtered by an additional 0.5 mm Cu, b) the second one is the spectrum emitted by the same tube supplied with a high voltage of 30 kVp, an average current of 3 mA and filtered by the Be window 0.5 mm thick. While the first irradiation technique allows the comparison of the reconstructed spectrum with the direct one, the second technique requires a theoretical evaluation of the spectral distribution for comparison.

Being practically indistinguishable the results obtained from the measurement done with only one Be scatterer or from the two in which a Be and an Al scatterer are used respectively, no graphs comparing such results are given. In figure 1 the spectral distributions of the 60 kVp beam are drawn. The continuous line represents the spectrum measured on the direct beam at a suitably large focus to detector distance. The dotted line is the reconstructed spectrum. Besides the above mentioned corrections, in this example special attention must be paid to the difference in the dead time resulting from the two measurement conditions. The two spectra are normalized to the same area. Some differences are present and their amount is significant. This preliminary result, however, is still under investigation and a better understanding of such small discrepancies should be achieved very soon. In order to complete the evaluation of this X-ray spectrometry method a check is required at the lowest limit of the

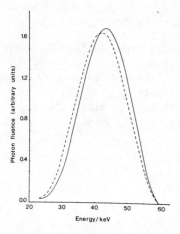
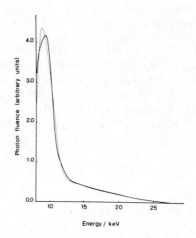

Fig.1: Distributions, twice smoothed, obtained at 60 kVp nominal high voltage: —— primary spectrum, ----- reconstructed spectrum.

Fig.2: Distributions once smoothed, obtained at 30 kVp nominal high voltage: —— calculated spectrum, ----- reconstructed spectrum.

pratical energy range. The second irradiation technique satisfies this requirement with its large, low energy interval. The very high flux density and the presence of very low quantum energies do not allow the measurement on the direct beam. The comparison is then possible only by calculation of the spectral distribution. This is achieved by means of the procedure described by Birch et al. (3) for the continuous spectrum and by means of the one worked out by Casnati et al. (4) for the characteristic lines. The experimental reconstructed distribution (dotted lene) is compared with the theoretical spectrum (continuous line) in figure 2. The last one clearly includes the convolution with the response function of the spectrometer distorted for the scattering dependent energy interval change. Normalization criterion is like the one used for 60 kVp spectra but applied to the continuous component alone. Comments on this result are quite similar to those relevant to figure 1.

CONCLUSION

On the ground of the results so far gained the scattered radiation spectrometry seems a promising technique for a better description of low energy X-ray beams as their interaction properties and their large diffusion would need.

REFERENCES

1. Paretzke, H.G. (1976): In: Proceed. 5th Symp. on Microdosimetry, p.41. Eds.: J. Booz, H.G. Ebert and B.G.R. Smith. EUR 5452 d-e-f.
2. Greening, J.R. (1972): In: Topics in Radiation Dosimetry, p. 262. Ed.: F.H. Attix. Academic Press, New York.
3. Birch, R. and Marshall, M. (1979): Phys. Med. Biol., 24, 505.
4. Casnati, E., Baraldi, C. and Galvani, F.: In preparation.

EFFICACITE DE COMPTAGE DE GELS SCINTILLANTS

M. Chauvet-Deroudilhe, M. Dell'Amico, M. Bordeaux and C. Briand

Nos travaux antérieurs (1) sur des gels scintillants contenant du dioxanne, PPO, Diméthyl POPOP, et du naphtalène et dont la viscosité était augmentée par addition de quantités croissantes de Cab-o-sil, avaient mis en évidence une augmentation parallèle du rendement de scintillation. Les résultats obtenus en présence d'inhibiteurs chlorés (CCl_4, $CHCl_3$) ont permis de conclure que la silice ne modifie pas l'efficacité du transfert d'énergie. Par contre l'augmentation du rendement de scintillation serait due à l'augmentation du rendement quantique de fluorescence de PPO quand la viscosité du milieu devient plus élevée. Pour confirmer cette hypothèse nous avons étudié une autre série de gels scintillants de même composition hormis la silice et de viscosité croissante, obtenus par addition au liquide de base de différentes quantités de HP_{55} (hydroxypropyl-méthyl-phtalate de cellulose), seul produit parmi les nombreux testés, qui conduisait à un milieu macroscopiquement homogène. D'autre part, nous avons élargi notre travail à d'autres gels de silice scintillants, différents par la nature du soluté primaire utilisé.

TECHNIQUES EXPERIMENTALES

HP_{55} de poids moléculaire moyen égal à 20 000 daltons est fourni par la Shin Etsu chemical company. Toutes les autres conditions expérimentales ont été rapportées par ailleurs (1, 2).

RESULTATS ET DISCUSSION

- Etude de gels scintillants à base de HP_{55}.

Les premiers essais effectués à concentration de $HP_{55} < 10^{-4}$ M ne produisent pas de variation détectable de la viscosité, ni de variation du rendement de scintillation lors d'irradiation γ. Pour des concentrations plus élevées ($0,1\ 10^{-3}$ M $< HP_{55} < 1,5\ 10^{-3}$ M), qui conduisent à une phase d'aspect gélifié, nous observons (figure n° 1) une diminution du rendement de scintillation, alors qu'à viscosité identique, la silice provoquait une augmentation de ce paramètre. Les densités des deux milieux étant très voisines, cette différence ne peut être imputée à une différence d'absorption du rayonnement ionisant.

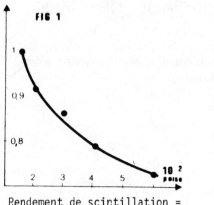

Rendement de scintillation = f (viscosité)

Pour tester l'activité de HP_{55} sur les mécanismes de transfert d'énergie dans le gel, nous avons mesuré les variations du rendement de scintillation en présence de $CHCl_3$ ou de CCl_4. Pour des concentrations fixes de ces inhibiteurs chlorés comprises entre 0,5 et 3 10^{-2} M, les courbes de rendement de scintillation en fonction de la concentration molaire en agent gélifiant sont parallèles entre elles, leurs pentes négatives sont identiques pour une même concentration de HP_{55}. (exemple figure n°2). De même les courbes des rendements de scintillation en fonction de la concentration en inhibiteur chloré, pour différentes concentrations de HP_{55} (non présentées) sont parallèles. Il n'y a donc aucune compétition entre HP_{55} et les dérivés chlorés en ce qui concerne la capture d'énergie par voie ionique ou excitonique, résultat identique à celui obtenu avec la silice.

Nous avons alors étudié la fluorescence du naphtalène, de PPO et de diméthyl POPOP en solution dans le dioxanne en présence de HP_{55}. L'absorption lumineuse de ces composés à leur maximum d'excitation n'est pas modifiée, de même que l'allure de leur spectre d'émission. Par contre, l'intensité de fluorescence au maximum d'émission normalisée par rapport à celle obtenue en l'absence de HP_{55} est nettement diminuée

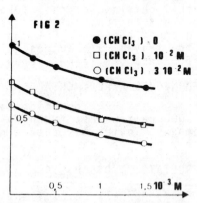

Rendement de scintillation = f $[HP_{55}]$

dans le cas du naphtalène (figure 3), de façon moins importante pour PPO (figure 4) alors quelle ne varie pas pour diméthyl-POPOP.

Nous avons conclu précédemment (1) que la modification de viscosité du milieu était responsable des

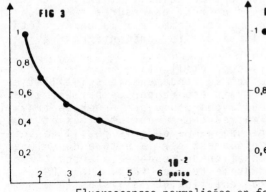
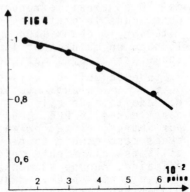

Fluorescences normalisées en fonction de la viscosité
naphtalène (λ_{ex}=322nm, λ_{em}=370nm)   PPO (λ_{ex}=322nm, λ_{em}=370nm)

modifications du rendement de fluorescence de PPO. Il est alors logique de penser que cet effet persiste en présence de HP_{55} pour une viscosité identique. Il faut donc en conclure que l'inhibition observée dans ce cas est en fait le résultat de deux mécanismes qui s'opposent : une exaltation de la fluorescence de PPO due à la viscosité et une inhibition propre à HP_{55}. Cette inhibition se manifeste aussi bien sur le naphtalène que sur PPO mais elle est apparemment plus importante sur le premier car il n'y a pas alors d'effet d'exaltation de fluorescence par viscosité. Les variations des rapports $\phi_0 \cdot \phi^{-1}$ de fluorescence en présence respectivement de silice et de HP_{55} sont rapportées sur la courbe 5, en fonction de la concentration en HP_{55}. Les résultats sont du même ordre de grandeur pour les deux composés. Les droites obtenues par regression linéaire présentent une pente trop élevée pour pouvoir être considérées comme des droites de Stern-Volmer. HP_{55} inhiberait la fluorescence du naphtalène ou de PPO par un mécanisme de quenching statique. Ainsi l'étude de gels scintillants contenant

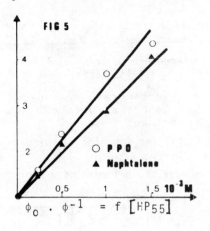

$\phi_0 \cdot \phi^{-1} = f [HP_{55}]$

HP$_{55}$, qui certes sont moins intéressants du point de vue pratique que ceux à base de silice du fait de leur moindre rendement de scintillation, nous ont permis de confirmer l'importance de la viscosité. Les valeurs plus élevées de ce paramètre augmentent le rendement quantique de fluorescence de PPO et ainsi le rendement de scintillation. Il convient donc, dans la pratique, d'opérer les mesures de radioactivité sur l'échantillon à doser comme sur la gamme d'étalonnage en maintenant la viscosité constante pour éviter toute erreur systématique.

- Etude de gels de silice scintillants.

L'augmentation du rendement de scintillation par addition de silice avait été rapportée par Germai(3,4) dans le cas de PPO. Nous avons voulu prolonger ce travail aux autres solutés primaires couramment utilisés et nous avons donc étudié toute une gamme de gels de silice scintillants, différents uniquement par la nature du soluté primaire. Aucune augmentation du rendement de scintillation n'a été observé dans le cas de αNPD, αNPO, PPO, CPO et PFD lors de l'augmentation de la viscosité du milieu par addition de silice et parallèlement ces solutés ne présentent pas de variation de rendement de fluorescence dans ces conditions. Par contre PBD et butyl PBD se comportent comme PPO (figures 6,7) et ces 3 solutés primaires sont donc les plus intéressants pour la

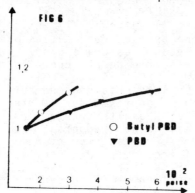
Fluorescence normalisée= f(viscosité)

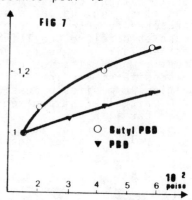
Rendement de scintillation= f(viscosité)

fabrication des gels scintillants. On comprend dès lors leur emploi très fréquent à cet usage qui apparemment ne résultait que de constatations empiriques.

Remerciements : nous remercions H.BOUTEILLE pour son aide technique.

Bibliographie
1. M. Chauvet, M. Dell'Amico, M. Bourdeaux, C. Briand :Int.J.Nucl. Med. Biol. (sous presse).
2. M. Chauvet: thèse Sci. Univ. Marseille (1978)
3. Germai G.: Bull. Soc. Royal Sciences Liège (1963),11-12, 863.
4. Germai G.: Radiochem. Radioanal. Letters (1970),3-4, 281.

STUDY AND MEASUREMENT OF THE ATMOSPHERIC POLLUTION BY ^{85}Kr

G. Eggermont, J. Buysse, A. Janssens and F. Raes

The continuation of the total release of ^{85}Kr in the reprocessing plants, for the expected growth of nuclear power up to the year 2000, would cause the largest specific nuclear background increase, for the next decades. This irreversibel pollution would yield a significant β-radiation skin dose to the world population, which justifies retention techniques on cost-benefit grounds (1). This conclusion could be invalidated by application of ICRP 26. However, the actual insufficient knowledge on skin cancer risc at low dose, necessitates a conservative policy. Further research is necessary, and should take synergism between ionizing and ultra-violet radiation (2) into account. Secondly, the micro- and macro-climatological consequences of aerosol distrurbancies by ^{85}Kr should be investigated. Our laboratory has started a research project on this subject.

Since different reprocessing centres are sited in Western Europe (Fig.1), the follow-up of ^{85}Kr activity is very important in this area. In our laboratory, a method, similar to the E.P.A. method (3) for measurement of low ^{85}Kr activities in air samples, has been worked out succesfully. The measuring cycle has been calibrated. The results of a large number of measurements performed in Ghent during 1979, are presented and discussed.

Fig.1

SAMPLING AND COUNTING METHOD

The set-up for sampling the air, chromatographic separation of Kr, and condensation in a scintillation vial (a), has been reproduced in Fig.2. About 1 m^3 of air is sucked to and absorbed on an activated charcoal trap (IV), passing cooling trap I, for water removal, molecular sieve MS II, for H_2O and CO_2, and a cooling trap III for final CO_2 removal. Immersion in liquid nitrogen is applied for any cooling. Then the charcoal trap IV is heated to 100°C and purged with helium, such that all gases are transferred to MS V. This 5Å molecular sieve is 1.5 m long, with 6.5 mm inner diameter. After removal of the cooler, trap V is eluted in a helium carrier flow, to a thermal conductivity detector. The sequence of gases, Kr being identified with a mass spectrometer, is the same as in (4). The chromatographic resolution between Kr and methane is

---

This work is sponsored by the Interuniversity Inst. of Nucl. Sc., and by the C.E.C., Contract 275 - 79 - 1 Bio B.

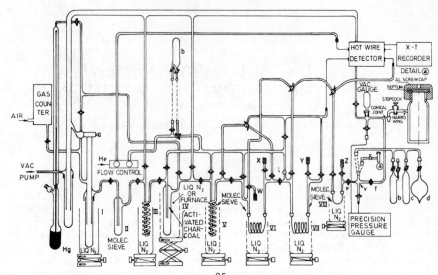

Fig.2 : Set-up for measurement of ^{85}Kr activity in air

about 1.5. Switching of the taps allows to direct the Kr fraction to trap VI, the residual gases being vented (W). This procedure is repeated for transfer to traps VII and VIII, thus augmenting the purity. Traps VI and VII, 3 m long, 2 mm I.D., contain MS 5Å, 60-80 Mesh. Helium is pumped off from the cooled trap VIII, but there remains a residual He pressure, which has to be corrected for (see Calibration Method). The gases on trap VIII are expanded at room temperature to a cross of glas tubes (V), with precisely known volume. The pressure is measured with a precision pressure gauge (0 to 19 psi, Texas Instr.). The Kr gas is then condensed in a scintillation vial and the narrowing is fused. The Instafluor scintillator is injected through a Teflon coated septum. Counting the ^{85}Kr activity (with efficiency 72%), and calculating the Kr mass, from the known temperature, pressure and volume, allows to determine the ^{85}Kr concentration relative to the known amount of Kr in air (1.14 cm^3/m^3). The Kr recovery varies between 55 and 70%.

CALIBRATION AND ERROR DISCUSSION

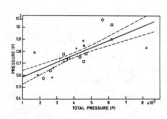

Fig.3 : He pressure versus total (He + Kr) pressure

The calibration is performed with a ^{85}Kr source with known activity (8% accuracy, all uncertainties given for 99% confidence level). A fraction of this source is introduced in the scintillation vial for the determination of the counting efficiency. The total dilution uncertainty is 2.8%. The He residual pressure (Fig.3) on trap VIII has been found to be dependent on the Kr content, probably through covering of the pores in the molecular sieve by Kr. The net residual

He pressure, together with the Kr recovery, have been measured by mixing a known mass and activity of ^{85}Kr from bottle b (Fig.1) in the He stream, and directing this stream to trap V and consequently through the same path as for a atmospheric measurement. The ^{85}Kr partial pressure is derived from the measured activity. The sets of measured He pressures, obtained by substraction, have been reproduced on Fig.3, as a function of the total pressure. A straight line fitting allows to correct the total pressure in an atmospheric measurement with an uncertainty of about 2.2%. The computed error on the measured amount of Kr + He, due to errors on temperature, pressure and volume, is 1.7%. The scintillation counting error is 9.7% for normal counting rates, about twice the scintillation background. The total uncertainty is 13% (99% conf. level).

The installed equipment thus allows measurement of the atmospheric ^{85}Kr concentration, within quite narrow error limits. The accuracy of the scintillation measurement has been verified by counting the activity of Kr from a bottle, bought in september 1971. The gas is expanded in glas tubes V and in the scintillation vial, at the same pressure existing in the vial in case of atmospheric measurements. The measured activity, corrected for decay, is 0.46 Bqm^{-3}, which is in good agreement with the values measured by Stevenson (3) in the same year, ranging from 0.43 to 0.50 Bqm^{-3}.

MEASUREMENTS AND DISCUSSION

In the course of the year 1979, a large number of atmospheric measurements have been performed, which are reported on Fig.4. The base line of the activity is rather constant, and equals about 0.70 Bqm^{-3}. There is thus a strong increase of the ^{85}Kr background, from 0.46 Bqm^{-3} in 1971 to 0.67 Bqm^{-3} in september 1977 (measured with a gas bottle of that year) and 0.70 Bqm^{-3} in 1979. This increase must be essentially related to the world reprocessing of nuclear fuel in this period. There is only a small increase from 1977 to 1979, and no significant change of activity during 1979. Our measurements will be continued in the next years, to allow monitoring of the evolution.

On Fig.4 a number of peak measurements can be observed, which can be related to the transport of released activities in reprocessing

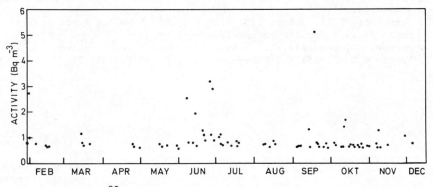

Fig.4 : Measured ^{85}Kr activities in 1979

centres, under favourable meteorological conditions. In regard of the predominant S-W wind direction, during measurement of the increased activities, their origin should be situated in La Hague. Pilot calculations of the meteorological transport, and realistic estimations of the releases, indicate that this correlation is justifiable. Thus, if the detailed source data would be available, which is only the case for Karlsruhe, these measurements of the ^{85}Kr activity would become important for the study of meteorological transport on mesoscale, ^{85}Kr serving as a tracer.

CLIMATOLOGICAL IMPACT OF ^{85}Kr

It is well known (5) that a number of radio-chemical reactions, in particular with SO_2, lead to formation of aerosols from the gas phase. Reaction vessels (20 l), filled with pure air and known concentrations of SO_2, NO_2 and H_2O, are being irradiated with U.V. light and with a γ-source. The total particulate and SO_4^{2-} concentrations are measured in function of the different parameters. In the near future, experiments will be performed in large volumes (m^3) filled with ^{85}Kr at realistic concentrations. Aerosol distribution measurements will be used for the assesment of transformation rates of the precursors. Finally the effect of homogeneous nucleation induced by ^{85}Kr on ambient air will be measured, and its relevance for smog formation and acidification of the environment will be evaluated.

ACKNOWLEDGEMENTS

The authors acknowledge H. Bultynck (S.C.K. Mol) for helpful information on atmospheric transport and also their director Prof.Dr. A.J. Deruytter, and student D. Van Peteghem, for their assistance. We thank L. Schepens for typing the manuscript, and R. Verspille for the drawings.

REFERENCES

1. Eggermont, G., Jacobs, R., Janssens, A. (1976) : Ann. Belg. Ver. Stralingsbescherming 1, n°3, 275
2. Chadwick, K.H., Leenhouts, H.P. (1978) : VIth Symposium on Microdosimetry, II, 1123
3. Stevenson, D.L., Johns, F.B. (1971) : Int. Symp. on rapid measurements of radioact. in the environment, 68
4. Aubeau, R., Champeix, L., Reiss, J. (1961) : J. Chromatog., 6, 209
5. Vohra, K.G. (1975) : IAEA Symposium, IAEA SM 197/35, 209

NEUTRON SPECTRA AND DOSE EQUIVALENT INSIDE REACTOR CONTAINMENT*

G. W. R. Endres, L. G. Faust, L. W. Brackenbush and R. V. Griffith

INTRODUCTION

The purpose of this study is to measure, characterize and evaluate neutron radiation dose equivalent rates and neutron energy spectra at selected commercial nuclear facilities where operating plant workers may be exposed to neutron radiation fields. Improved understanding and control of occupational neutron exposure should result from this study.

Neutron exposures to operating plant workers have not been observed on dosimeters in the past due to the lack of sensitivity of the personnel dosimeters used and the energy ranges suspected to be present in commercial nuclear facilities, particularly PWR plants. Most of the facilities presently use nuclear track emulsions to detect fast neutrons. The emulsions and most of the newer dosimeters based on track-etch techniques are not sensitive to the neutrons in the energy ranges of the leakage spectra which may be present in the commercial power reactor plants. Average energies are expected to be below 500 keV.

Recently, albedo type neutron dosimeters have become available for use at these facilities and relatively large neutron dose equivalents are being observed, especially when workers enter containment during full power operation.

At the present time, the albedo type of personnel dosimeter is the only available dosimeter which seems to have adequate sensitivity for neutrons in this energy range.

This study was designed to provide measurement data from a minimum of six nuclear power sites, which were selected to include reactors manufactured by each of the four U.S. NSSS vendors and nuclear plants with at least four different architect-engineers. The measurement data include: 1) a determination of neutron dose and dose equivalent rates inside and outside of containment; 2) neutron spectral and flux distributions at selected locations both inside and outside of containment; 3) special monitoring with currently available neutron dosimeters, i.e., albedo, film and fission fragment, at each site; and 4) correlate instruments and dosimeter data to flux and spectral distributions to determine their proper response and interpretation. Lawrence Livermore Laboratory personnel have assisted with measurements at two of the sites and analysis of multisphere data taken at all the sites.

To accomplish this study, tissue equivalent proportional counters (TEPCs) were used to measure the neutron dose and dose equivalent rates.

* Work performed on this project was sponsored by the U.S. Nuclear Regulatory Commission.

**R. V. Griffith employed by University of California, Lawrence Livermore Laboratory, Livermore, CA, U.S.A.

At the same time, currently used neutron survey instruments, such as the Snoopy, were used for dose equivalent rate measurements. The multisphere technique and other spectrometers were used to estimate the neutron spectrum at selected locations inside the outside containment and were supplemented by TEPC data. Currently available dosimeters were then placed on phantoms to provide data which can be used to establish their capability for adequate personnel dosimeter measurement requirements as stated in Regulatory Guide 8.14.

The personnel dosimeters used were obtained from vendors who normally supply this service to licensees and include film, TLD and track etch detectors. We used an improved version of the Hanford Multipurpose Dosimeter (MPD). A combined track-etch-albedo neutron dosimeter developed by Hankins and Griffith of Lawrence Livermore Laboratory (1) was also used.

Neutron calibration exposures were conducted with a bare Cf-252 neutron source, and with spheres of Al, H_2O and D_2O surrounding the Cf-252 source. Neutron monitoring instruments, which use moderated BF-3 detectors, tend to read high in the moderated Cf-252 spectra when they are calibrated with the bare Cf-252. The tissue equivalent proportional counter correctly measures all absorbed dose and dose equivalent for event sizes larger than about 5 keV/μm. Albedo TLD dosimeters were exposed on phantoms to the various Cf-252 source configurations. When calibrated with bare Cf-252, the albedo dosimeters read a factor of 2 to 3 high for the moderated Cf-252 sources.

EXPERIMENTAL RESULTS

The multisphere spectrometer system consists of five polyethylene spheres of various sizes plus a bare and cadmium covered neutron detector. For a measurement point, the thermal neutron count rate was measured using a ^{6}LiI(Eu) detector in seven different configurations. A spectrum unfolding computer program, known as LOUHI (1), is used to determine the neutron spectrum from each set of data. Response functions for each of the detector geometries are included in the computer program. The response functions for each of the sphere sizes, 7.6, 12.7, 20.3, 25.4 and 30.5 cm diameter, were calculated by Sanna (2).

Typical results for spectra measured inside containment at a PWR nuclear power plant are shown in Figure 1.

Because of the "low" resolution of the multisphere system, the unfolded spectra show no sharp peaks or edges. The computer program provides 26 neutron energy groups from thermal to about 20 MeV.

The differential flux, integral flux, integral dose equivalent, energy band width, and flux density are provided for each energy. Kerma rate, dose rate and average energy for each spectrum is also given. The average energy for the spectrum is 62 keV at Location 3. The range of average energies for all locations at all the PWR sites was from about 10 keV to 90 keV. Readings taken with standard neutron survey instruments at the same locations as the multisphere system show a response about a factor of 2 higher than the multisphere.

Tissue equivalent proportional counters (TEPCs) were used at several locations inside containment to measure absorbed dose directly for comparison with "Snoopy" and Rascal monitoring instruments. At most locations a 12.7 cm diameter spherical counter filled to 1 μm equivalent size with tissue equivalent gas was used.

FIGURE 1. Typical Spectra Inside Containment at a PWR

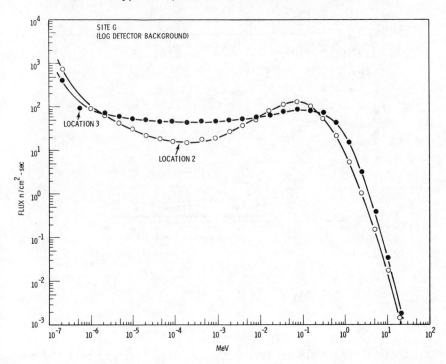

A typical event size spectrum from the TEPC is shown in Figure 2. Data are shown as they appear on a multichannel analyzer for Location 4 at Site I. The sharp drop in the number of events per channel at about channel 200 corresponds to 28 keV/μm, the maximum energy loss a proton recoil can have in the TEPC.

This drop point is often called the proton drop point and is used to calibrate the measurement system. Absorbed dose is obtained by multiplying the number of events at each energy by the energy and integrating over the spectrum.

The absorbed dose rate for the spectrum in Figure 2 is 0.51 Gray/hr. The quality factor determined by Rossi (3) analysis is 10.8. In this case the analysis of the data for quality factor includes all events greater than 5 keV/μm. At this location the dose equivalent is about 560 mrem/hr. Measurements with monitoring instruments gave readings of 1300 to 1600 mrem/hr.

FIGURE 2. Typical Event Size Spectrum from TEPC

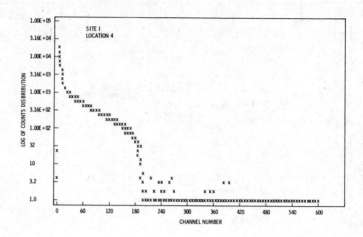

Both the multisphere system and the TEPC indicate that the monitoring instruments are a factor of 2 or so high in the well-scattered neutron fields inside containment.

REFERENCES

1. Routti, J. T. (April 1969): High Energy Neutron Spectroscopy with Activation Detectors Incorporating New Methods for the Analysis of Ge(Li) Gamma Ray Spectra and the Solution of Fredholm Integral Equations. Lawrence Berkeley Laboratory Report UCRL-18514, Berkeley, CA, U.S.A.
2. Sanna, R.S. (1973): Thirty-One Group Response Matrices for the Multisphere Neutron Spectrometer Over the Range Thermal to 400 MeV. USAEC Health and Safety Laboratory Report HASL-267.
3. Attix, F. H., Roesch, W. C. and Tochilin, E. (editors) (1968): Radiation Dosimetry - Volume 1, pp 43-90. Academic Press, New York.

DOSIMETRY OF CRITICALITY ACCIDENTS USING ACTIVATIONS OF THE BLOOD AND HAIR*

D. E. Hankins

INTRODUCTION

The evaluation of the dose that a person received in a criticality accident can be difficult. Most accidents have occurred when the person was not wearing nuclear accident dosimetry and since the NRC no longer requires these dosimeters, future dose evaluations may have to be based on body activations and gamma-to-neutron dose ratios. To aid in a dose evaluation we have compiled in a table (available from the author) the results from numerous criticality accident studies using 10 different critical assemblies, each with different neutron leakage spectra. There are several problems involved in applying these results accurately, the most significant problem being the determination of the configuration of the fissile material at the time of the accident. Other problems include a lack of information concerning the location, orientation, and possible shielding between the person and the accident assembly.

DOSIMETRY STUDIES

The literature contains a number of criticality accident dosimeter studies,[1] including a mock-up study of the accident which occurred in the Y-12 facility at Oak Ridge, Tennessee and four studies sponsored by the International Atomic Energy Agency. Also studied were the Health Physics Research Reactor (HPRR) at the Oak Ridge National Laboratory and five different types of critical assemblies at the Los Alamos Scientific Laboratory.

These dosimetry studies were made with reactors and critical assemblies having extremely diverse configurations. Schematics of these assemblies, drawn to scale in Fig. 1, show major features that are important in modifying the neutron leakage spectrum. Leakage spectra vary greatly, and as a result the blood or hair activations for the same neutron dose will be very different from each assembly.

SODIUM ACTIVATION OF BLOOD AND ^{32}P ACTIVATION OF THE HAIR

The most useful activations of the body for dose estimation following a criticality accident are those of the blood and the hair. Activation of ^{23}Na in the body is determined by counting the ^{24}Na produced in about 20-cm^3 of blood with a NaI or GeLi detector. The 1.369-MeV gamma ray is used to determine the ^{24}Na activity in μCi of ^{24}Na per mg of ^{23}Na. The probability of neutron capture by sodium in the human body is fairly constant for neutron energies from thermal to 5 MeV,[2] but the kerma dose delivered by

---

*Work performed under the auspices of the U. S. Department of Energy by the Lawrence Livermore Laboratory under contract No. W-7405-ENG-48.

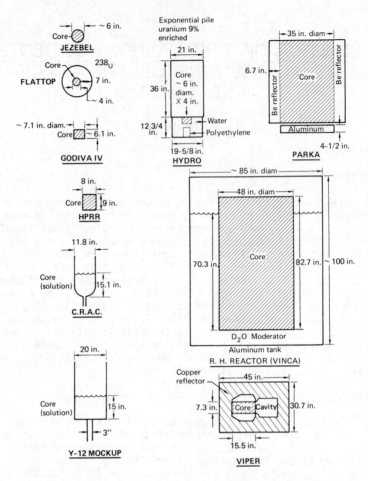

Figure 1. Schematic of the critical assemblies and reactors used in dosimetry studies (drawn to scale), showing the core configuration and components that affect the leakage neutron spectrum.

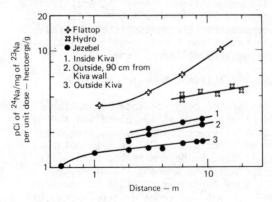

Figure 2. Blood sodium activation per hectoerg/g of neutron dose as a function of distance from the assembly. Jezebel results with assembly located inside, outside, and near the outside wall of a building are given.

neutrons is predominantly from fast neutrons. Consequently, ^{24}Na activation in the blood is not proportional to the neutron dose.

Figure 2 shows blood sodium activation from the Jezebel assembly as a function of the distance from the assembly, when the assembly is (1) in a building (kiva), (2) outside, and (3) near a concrete wall. Increases in activations are caused by scattered, low energy neutrons, which activate the blood but contribute little to the neutron dose. Similar curves for the Hydro and Flattop assemblies are shown.

Sodium activation in the body can be detected by placing a G-M instrument in the abdomen. The G-M reading is proportional to the Na activation and the resulting curves (available from the author) are very similar in shape to the curves given in Fig. 2.

Activation of the sulfur in the hair by fast neutrons (>2.9 MeV) produces the beta emitter ^{32}P. If the individual has not been contaminated, the technique preferred, because of accuracy and simplicity, is the direct counting of the hair with a beta counter following the procedure described in detail by Hankins.[3] The counter is calibrated using a ^{90}Sr-^{90}Y source, and the count rate of the hair in counts/min/g is determined and divided by 1.77 to obtain the fast neutron dose in rads (above the 2.9 MeV threshold for sulfur). Unfortunately, the percent of the neutron kerma dose that is from neutrons having energies of 2.9 MeV varies from 2.1 to 48%.

RELATIVE GAMMA AND NEUTRON DOSES

The ratio of gamma-to-neutron doses can be used to determine the total dose if either the gamma or the neutron dose is known. Unfortunately, the estimation of dose based on gamma-to-neutron ratios may not be accurate since this ratio for the various critical assemblies varied from around 0.09 to 2.9. The effect of shielding material is also significant. For example, the HPRR ratios vary from 0.19 for a bare assembly to 1.9 for a polyethylene-shielded assembly.

EVALUATION OF THE DOSE

Following an accident the configuration of the fissile material at the time of the excursion is established, if possible. This configuration can then be compared with those shown in Fig. 1, and if it is similar to one of them, the experimental results (available from the author) obtained with the assembly most closely resembling the excursion can be applied to evaluate the doses.

The neutron leakage spectrum and subsequent activations of hair and blood are also affected by other factors which include: shielding the exposed person; whether the exposed person was indoors or outdoors; where he was with respect to walls or floors; his orientation; the angle of exposure; and his distance from the assembly. Several of the assemblies shown in Fig. 1 have been used to evaluate the effect of many of these factors.

DOSE EVALUATION USING A COMBINATION OF THE BLOOD AND HAIR ACTIVATIONS

The most serious problem following an accident is a lack of information on the configuration of the fissile material which makes it impossible to find in Fig. 1 an appropriate critical assembly. Furthermore, there is often a lack of other information necessary to accurately assess the dose. The dose can still be evaluated by using a combination of blood and hair activations.

The activation of the blood and hair is determined as described previously and the ratio of the sulfur fluence to the blood sodium activation is calculated. In Fig. 3, we have plotted the ratio of sulfur fluence to blood sodium activation as a function of blood sodium activation and have drawn two curves.

To evaluate the dose using this procedure we first read the blood activation (in pCi ^{24}Na/mg^{23}Na per rad of fast neutrons) from the curve at the point corresponding to the measured sulfur-fluence/blood-sodium-activation ratio. Then, we divide the blood sodium activation by that quantity to obtain the neutron dose that the individual received.

This procedure is independent of the neutron spectrum, hydrogenous shielding, victim orientation and distance, and room scatter. Where there were thick metal shields (>10 cm) either associated with the assembly or between the person and the assembly, the curve on the left must be used. Fortunately, thick metal shields are not commonly used. A dose estimate accurate to within ±20-30% should be obtainable using this procedure.

REFERENCES

1. Hankins, D. E. (1979): "Dosimetry of Criticality Accidents Using Activations of the Blood and Hair," submitted to Health Physics Journal.
2. Delafield, H. J. (1974): "The Neutron Capture Probability for Sodium Activation in Man Phantoms," AERE Harwell, UK, AERER-7728.
3. Hankins, D. E. (1969): "Direct Counting of Hair Samples for ^{32}P Activation," Health Phys. 27, 740.

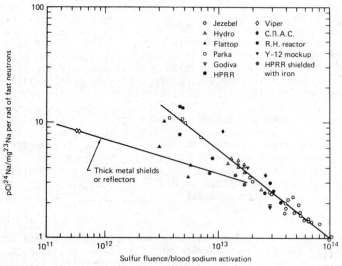

Fig. 3. Curve used to determine the neutron dose using a combination of blood and hair-activation data.

A PASSIVE MONITOR FOR RADON USING ELECTROCHEMICAL TRACK ETCH DETECTOR

G. E. Massera, G. M. Hassib and E. Piesch

INTRODUCTION

Radon gas and its airbone daughters are a health hazards not only for the workers in uranium mining but also for the population in dwellings. Short-term fluctuations of the radon level outdoor and in houses have been found to be in the order of one magnitude dependent on the emanation rate from the ground, the walls or the floor, the ventilation rate in the room and the meteorological conditions.

In working level meters the decay products of radon are collected by an aerosol filter or an electrostatic field precipitation. For longterm monitoring mainly cellulose nitrate track etch detectors [1] or thermoluminescent detectors [2,3] are applied to registrate α-particles emitted from the filter. In recent studies Makrofol polycarbonate was found to be a promising electrochemical track etch detector (ECED) for the detection of radon daughters [4,5].

For the long-term estimation of the inhalation dose from radon daughters inside buildings a single inexpensive passive radon dosimeter was developed which consists of a Makrofol track etch detector inside a diffusion chamber similar in principle to one described before [6]. It provids a time-integrated indication of the mean dose to the lung with a sensitivity of 130 mrem.

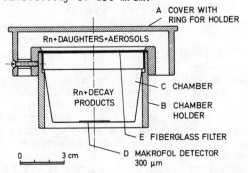

Fig.1 Cross section of the radon diffusion chamber

DOSEMETER DEVICE

As it is shown in Fig.1 the diffusion chamber (C) consists of a plastic cup which is closed at the top by a fiberglass filter (D) through which radon may pass by diffusion while radon daughters and aerosols are retained at the surface of the filter. On the bottom of the cup a Makrofol foil (E) of 300 μm thickness is placed to registrate α-particles reaching the detector with energies between zero and the maximum value. A special cover (A) placed in front of the filter is an excellent fit for the filter (D) between the chamber holder (B) and the cover (A). The cover avoids any damage on the filter and the deposition of heavy dust particles on its surface.

ETCHING TECHNIQUE

A pre-etching technique immediatly before the ECE is applied which removes a surface layer of the Makrofol foil in order to reduce background tracks and to reveal etchable tracks from high energy α-particles [7]. A solution 4:1 of ethyl alcohol and 6N KOH shows a layer removal rate of 2.34 μm in Makrofol after 1 hour of pre-etching at room temperature. The ECE was performed in a 6N KOH solution containing 20% by volume of alcohol by applying a high voltage at 800 V_{eff} and 2 kHz for 3 hours at room temperature.
After different periods of pre-etching the optimum condition corresponds to a layer removal of 1.17 μm and a background of 37 ± 15 tracks/cm (Fig. 2).

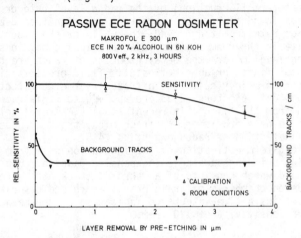

Fig. 2 Rel. sensitivity and background tracks vs. layer removal

ESTIMATION OF RADON EXPOSURE

The exposure of the population in houses is caused by the inhalation of radon daughters as free ions or attached to aerosols in the atmosphere. Inhaled radon daughters result in an inhomogenious irradiation of the various parts of the human respiratory tract primary in the bronchial region which depends on the way of breathing the rate and depth of respiration and the translocation and clearance of the deposited activity.

For the estimation of the radon exposure, the working level WL is defined as the potential alpha energy associated with 100 pCi/l of ^{222}Rn in radioactive equilibrium with its short-lived decay products and can be calculated from the ^{222}Rn concentration C_{Rn} and an equilibrium factor F between Rn and its short-lived daughters. In adequately ventilated rooms a value of F = 0.5 corresponding to a relative activity concentration ^{222}Rn:RaA:RaB:RaC of 10:9:5:3.5 [8] is a good estimate and agree with short-term experimental indoor results [9]. This value was adopted for our calibration. F varies with the ventilation rate in the room resulting in values of 0.7 for 1 cycle/hour and 0.4 for 2 cycles/hour. For the usual ventilation conditions in uranium mines F values between 0.2 and 0.3 have been found [10].

The passive radon dosimeter has been exposed for 20 hours in a Rn concentration of 4.04 nCi/l measured by means of a surface barrier α-spectrometer [11] resulting in 808 WLh and a corresponding reading of $2.2 \cdot 10^3$ tracks/cm^2 (Fig.3). Additional calibrations have been performed in a closed room with a high radon concentration. The track diameter was found to be in the order of 150 μm.

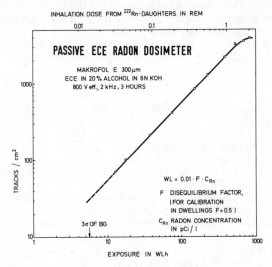

Fig. 3   Track etching detector reading vs. exposure

With regard to the mean dose to the lung UNSCEAR [8] recommend a dose factor 0.2 rad (WLM)$^{-1}$. The working level month (WLM) is defined as he accumulated exposure of 1 WL concentration during a period of 170 hours. Taking into account a quality factor of 20 for α-particles, we applied a conversion factor 4 rem·(WLM)$^{-1}$. This conservativ value agree with new epidemiological data on radiogenic lung cancer resulting in a conversion factor of 6 rem·(WLM)$^{-1}$ with 1.4 rad·(WLM)$^{-1}$ and a rem per rad of about 4 [12].

APPLICATION

The lower detection limit of the radon dosimeter given by the 3σ value of the background tracks was found to be equivalent to 5.6 WLh or a corresponding mean dose in the lung of 130 mrem (Fig. 3). After an exposure period of 3 month (2160 hours) a mean radon concentration of 0.3 pCi/l can be detected.

During storage periods up to 1 month at 50°C no significant fading effect was found for the detection of α-particles [14] which confirmes earlier fading results for neutron induced recoils.

The passive radon dosimeter discriminates the detection of thoron and its daughters because of the short half-life of thoron compared with the diffusion time necessary to pass the fiberglass filter.

The passive radon dosimeter was applied in a uranium mine as well as for a 3 month monitoring period in houses. Comparisons of the radon dosimeter results with the short-term results of an instant working level meter and a time-dependent decay of radon daughters collected for a period of 3 min on the surface of a filter [13] agree within 20%. In contrast to the radon measurement by means of the passive dosimeter these techniques are based on the measurement of the potential α-energy. Differences in the results are expected from additional thoron contents in the atmosphere and the approximation of the equilibrium factor F. The radon dosimeter described here is applied for a long-term radon exposure study in buildings to estimate the real inhalation dose of the population in Germany.

We wish to thank Dr. Wicke, Justus-Liebig-University Gießen and Mr. Urban for the calibration of our radon dosimeter.

REFERENCES

[1] A.M.Chapuis, P.Dupont, P.Zettwog, NEA Proc.Specialist Meeting Paris, 1978,87
[2] A.J.Breslin, A.C.George, see 1, 1978, 133
[3] J.Huber, B.Haider, W.Jacobi, see 1, 1978, 139
[4] G.M.Hassib, E.Piesch, see 1, 1978, 35
[5] G.M.Hassib, E.Piesch, G.E.Massera, Proc. 10th Int.Conf.Track Detectors, Lyon 1979
[6] A.L.Frank, E.V.Benton, Nucl.Track Detection 3, 1977, 149
[7] G.M.Hassib, Nuclear Tracks, 3, 1979, 45
[8] United Nations Scientific Commitee on the Effects of Atomic Radiation, 1977 Report
[9] E.Stauden et al, Health Physics 36, 1979, 413
[10] T.Domanski, Health Physics 36, 1979, 448
[11] A.Wicke, Dissertation, University Gießen 1979
[12] P.J.Walsh, Health Physics 36, 1979, 601
[13] M.Urban, to be published
[14] G.E.Massera, E.Piesch, to be published

BETA DOSIMETRY
WITH SURFACE BARRIER DETECTORS

M. F. M. Heinzelmann, H. Schuren and K. Dreesen

INTRODUCTION

In the vicinity of small, unshielded radioactive substances, the dose rate due to beta radiation may be substantially higher than the dose rate produced by gamma radiation. However, for radiation protection monitoring, a small dosimeter for β-radiation is still needed, by means of which it will be possible to determine the dose rate on the surface of radioactive substances in an energy-independent manner.

A very small dose rate meter for β-radiation has been proposed by G. Nentwig (1) (2). He uses scintillators of 10 - 25 mg/cm² thickness and counts the pulses exceeding a suitable discriminator threshold. A low-energy beta particle, passing through the scintillator in the same way as a β-particle of high energy while giving off only part of its energy, generates a larger pulse, since the linear stopping power dE/dx increases with dropping β-energy. In connection with a suitable discriminator threshold, a pulse generated by such a particle has a higher probability of surpassing the threshold than a pulse produced by a β-particle of higher energy. By variation of the scintillator thickness and discriminator threshold, Nentwig obtains a small β-dosimeter featuring an energy-independent indication. To our knowledge, this interesting proposal made by Nentwig has not been further pursued.

Nentwig's idea has been transferred by us from scintillators to semiconductor detectors (3). In the case of surface barrier detectors, the thickness of the sensitive layer is changed with the aid of the detector voltage. At low voltages, very small detector thicknesses can be obtained.

EXPERIMENTAL DESCRIPTION

By means of a surface barrier detector, the count rate was determined for various detector voltages and discriminator thresholds. As described by Nentwig (1) (2), the pulse rate surpassing a set discriminator threshold was determined integrally. In addition, a single-channel analyzer was used, and the channel width was adjusted optimally as a further parameter.

---

*This study was sponsored by the Commission of the European Communities under Contract No. 107-77-1 PST D

Measurements were conducted using a ND-7-S[1] detector with an active surface of 7 mm². The detector was covered additionally with a light-tight plastic foil of 1 mg/cm². Measurements were taken at several distances from different β-emitters. For the individual measuring points, the dose rate had been determined by means of an extrapolation chamber featuring a front electrode thickness of 7 mg/cm².

The β-emitters used were one nuclide of low peak energy (Pm-147), one of medium peak energy (Tl-204), and one of high peak energy (Sr-90/Y-90). In the case of Sr-90/Y-90, the Sr-90 radiation could be absorbed by plexiglass. Additional measurements were carried out on the secondary normal for β-radiation developed by PTB (Physikalisch Technische Bundesanatalt, Braunschweig) (4).

EXPERIMENTAL RESULTS

The sensitivity[2] was determined for various detector voltages, discriminator thresholds and for 2 time constants of the amplifier (0.25 μs and 0.5 μs). In connection with measurements using the single-channel analyzer, the channel width is varied additionally.

When using the discriminator, counting all of the pulses which surpass a set threshold as was done by Nentwig (1) (2), an energy-independent determination of the β-dose rate was only possible with the time constant of 0.25 μs (Fig. 1). For further measurements, the discriminator threshold of 200 mV was generally selected. The sensitivity to β-radiation is then equal to 6.7 cpm/mrad/h within ± 20 %. In the course of earlier investigations (3), we had not succeeded in determining the β-dose rate in an energy-independent manner without the single-channel analyzer.

The use of a single-channel analyzer permits an energy-independent determination of the β-dose rate with both time constants.

Fig. 2 shows the energy dependence of the sensitivity to γ-radiation. For Cs-137-γ-radiation, the sensitivity is equal to that for β-radiation.

The sensitivity as a function of the dose rate is shown in Fig. 3. With Sr-90-β-radiation it was only possible to measure up to 600 rad/h. When measuring with the time constant of 0.25 μs, the sensitivity to γ-radiation only decreased by 20 % at 1400 rad/h as

---

[1] Canberra Elektronik GmbH

[2] Sensitivity denotes the ratio of count rate : dose rate below 7 mg/cm² for measurements using the discriminator, and the ratio of count rate in the channel : dose rate below 7 mg/cm² for measurements using the single-channel analyzer.

compared to values at low dose rates. A corresponding decrease in sensitivity already occurs at 300 rad/h in connection with measurements with 0.5 µs and the single-channel analyzer.

Measurements of the dose dependence are shown in Fig. 4 on a second detector. The sensitivity to radiation from 2 nuclides for 2 settings of the electronic system was determined in each case after irradiation with a predefined dose. Up to doses of 50 000 rad, no sensitivity change could be observed.

One disadvantage of the detectors lies in the fact that they become easily defective. For this reason, further measurements are planned using ion-implanted silicon detectors.

REFERENCES

(1) NENTWIG, G., β-Dosimetrie mit Szintillationszählanordnungen
    Dissertation, Dresden, 1967

(2) WEBER, K.H., NENTWIG, G., Auslegeschrift 1287 705
    (1969) (German Patent)

(3) HEINZELMANN, M.F.M., SCHÜREN, H.
    Betadosimetry with Surface Barrier Detectors
    Advances in Radiation Monitoring, IAEA-Symposium
    Stockholm 26.-30.06.1978

(4) BÖHM, J., Sekundärnormal für die Energiedosis durch Betastrahlung in Gewebe, in: E. Piesch, D.F. Regulla
    FS-79-19-AKD (1979) p. 87

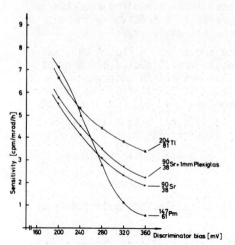

Fig.1: Sensitivity as a function of the discriminator bias for β-radiation of different nuclides

Detector: ND-7-S; number 1998
Detector voltage: -20 V
Gain: 100 x
Shaping time constant: 0,25 μs

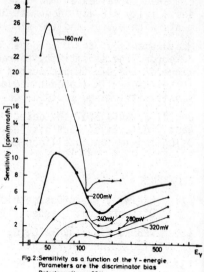

Fig.2: Sensitivity as a function of the γ-energie Parameters are the discriminator bias
Detector voltage: -20 V
Shaping time constant: 0,25 μs

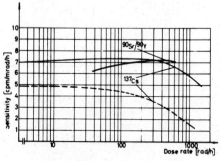

Fig.3: Sensitivity as a function of the dose rate for $^{90}Sr/^{90}Y$ β-rays and ^{137}Cs γ-rays

Detector: ND-7-S
Detector voltage: -20 V

— Discriminator, bias 200 mV, shaping time constant 0,25 μs
-- Single channel analyser, bias 240 mV, channel width 130 mV
   Shaping time constant 0,5 μs

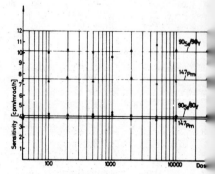

Fig.4: Sensitivity as a function of the dose for $^{90}Sr/^{90}Y$ - and ^{147}Pm-β-radiation
Detector: ND-7-S; number 1999
Shaping time constant 0,25 μs
· Discriminator, bias 200 mV
▲ Single channel analyser, bias 230 mV, channel width 100 mV

^{230}TH ASSAY BY EPITHERMAL NEUTRON ACTIVATION ANALYSIS

R. L. Kathren, A. E. Desrosiers and L. B. Church

^{230}Th (Ionium) is an important member of the ^{238}U decay chain that has recently been shown to be the major contributor to the environmental dose from actinides. (1) Earlier studies (2-4) have shown that following inhalation of ^{238}U + daughters, the uranium is cleared relatively rapidly from the lung via ciliary action and the gastrointestinal tract, while the ^{230}Th daughter is removed much more slowly. This produces a "biological enrichment" of ^{230}Th in later fecal samples and further suggests that ^{230}Th may be of greater import from the standpoint of dose than heretofore considered.

The lack of a simple, inexpensive, rapid, and sensitive assay method for ^{230}Th has limited studies of the metabolism and fate of this nuclide within biological systems. The most widely used available methods involve time-consuming, complex, and often tedious chemical separations followed by alpha counting; accuracy and sensitivity of these methods may be wanting, and interferences from uranium, other actinides or isotopes of thorium may further complicate the assay.

The nuclear properties of ^{230}Th, however, suggest that neutron activation analysis (NAA) may provide a simple, inexpensive, rapid and highly sensitive method of assay. This nuclide has a 1010 barn epicadmium resonance absorption cross-section for activation to 25.52 hour ^{231}Th; ^{231}Th has a complex gamma ray spectrum with the most prominent energies being an 84.4 keV complex with a yield of 6.55% (5-6). Although ^{231}Th emits numerous other photons, yields are significantly (i.e. one or more orders of magnitude) lower, with the exception of a photon at 89.95 keV, which has a yield of 0.95%, or about 1/7 that of the complex line at 84.4 keV.

The suitability of the NAA method in the presence of the ^{238}U parent and other members of the natural uranium decay chain was examined by activating a quantity of various uranium ores with both a reactor (TRIGA MKI) thermal neutron spectrum having a cadmium ratio of approximately 10. The ore was exposed bare and wrapped in 0.5 mm cadmium sheet to eliminate the subcadmium (i.e. thermal) neutrons. The resultant spectra were counted on a 14% coaxial GeLi detector and showed prominent peaks from uranium activation and fission products, with a "window" where no (or insignificant) peaks were observed between about 75 and 96 keV. This is in part as expected, for natural activity from ^{231}Th (daughter of ^{235}U) is negligible, and would not interfere, and the experiment showed no other potentially interfering peaks. Thus the NAA method appeared feasible as a means of determining ^{230}Th in the presence of natural uranium.

Pure ^{230}Th was obtained and irradiated in the reactor shielded by 0.5 mm of cadmium. The resultant spectrum (Figure 1) showed a pronounced broad peak at 84.6 keV, and secondary peaks at approximately 73.5 and 90 keV. Both the 84 and 90 keV peaks are attributable to ^{231}Th; the other peaks were not identified, but may be attributable to impurities or K x-rays. The presence of ^{231}Th was verified by observing the decay of the peaks; with the exception of the peaks at 68, 73.5 and 145 keV, all peaks decayed with a half-life of approximately 25.5 hours. A combined sample of ^{230}Th and uranium ore was also counted together after irradiation. The results clearly showed ^{231}Th (Figure 2).

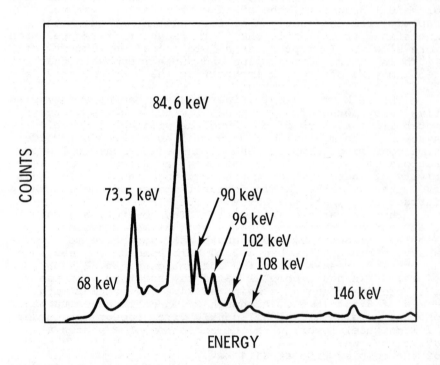

Figure 1. Observed spectrum energy over the range 60 to 150 keV from ^{231}Th obtained by epicadmium neutron activation of ^{230}Th.

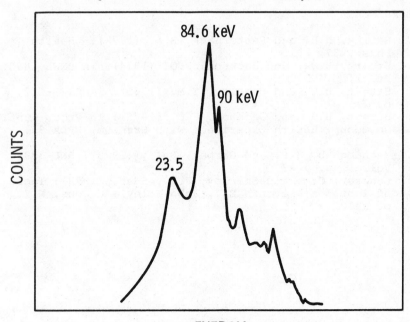

Figure 2. Spectrum in the region of 60-120 keV from mixture of natural uranium and ^{230}Th following epicadmium neutron irradiation. The peaks identified at 84.6 and 90 keV are from ^{231}Th; the 73.5 keV peak is from ^{239}U.

The above preliminary work demonstrates the feasibility of epicadmium NAA as a means of assay of ^{230}Th, and the data suggest a sensitivity of at least 10 ng for a 10 g sample, 100 minute counting time, and epicadmium fluence of 3×10^{12} n/cm^2 (5×10^{11} n/cm^2-sec for one minute). While interferences from other substances will be minimized through the use of epicadmium neutrons and a brief post-irradiation delay to permit shortlived activation products to decay, sodium, present in large quantities in feces and other biological materials, may produce significant counts in the low energy channels from Compton scattering of the high energy photons associated with the decay of its activation product ^{24}Na. The 15 hour half-life of this nuclide is sufficiently close to that of ^{231}Th to preclude holding the sample for decay. Hence, pre-irradiation sodium removal may be desirable. Further studies along these lines, as well as to improve the method by use of thin intrinsic germanium detectors are now in progress.

REFERENCES

1. Harley, N.H. and Pasternack, B.S. (1979): Health Phys., 37, 291.
2. Stuart, B.O., and Jackson, P.O. (1974): In BNWL-1850, Pt. 1, p. 97.
3. Stuart, B.O. and Beasley, T.M. (1965): In BNWL-122, p. 21.
4. Stuart, B.O. and Jackson, P.O. (1975): In Proc. Conf. on Occup. Health Experience with Uranium, ERDA 93, p. 130.
5. Mughabghab, S.F. and Garber, D.I., (1973): BNL-329, Vol. 1, p. 90-3.
6. Lederer, C.M. and Shirley, V.S., Eds., (1978): Table of Isotopes, Seventh Ed., John Wiley and Sons, N.Y., p. 1418.

CALIBRATION OF RADIATION PROTECTION INSTRUMENTS AT SSDL LEVEL IN ISRAEL

M. E. Kuszpet, A. Donagi and T. Schlesinger

INTRODUCTION

The application of ionising radiation in medicine, industry and research in Israel has reached the same level relative to its population as in other industrialized countries. In order to achieve the maximum benefit from the use of ionising radiation, it is essential that dosimeters should be routinely calibrated with respect to the correct dose or dose rate indication. For this purpose, a Secondary Standard Dosimetry Laboratory (SSDL) was established in Israel in 1976. The main tasks of the laboratory have been in the field of therapy level dosimetry, and at present all radiotherapy facilities in Israel have at least one dosemeter calibrated at the laboratory.

In the last years there has been an increased demand for the calibration of protection level instruments. According to unofficial figures, there are in Israel several thousand instruments of this kind. In spite of the fact that the accuracy requirements for these instruments have been relaxed by one order of magnitude compared with therapy level instruments, the need for dose rates ranging from very low to very high values, and other technical problems make calibrations a difficult task.

The present work describes the actions taken at the Israeli SSDL in order to extend its activities down to the low dose region. Furthermore, preliminary results of the calibration of several typical instruments are presented.

FACILITIES

The set-up used for the calibration of protection level dosimeters is esentially the same as that used for therapy level instruments, except for the fact that a distance of 2 meters between the X-ray focus and the ionisation chamber is adopted. In order to allow continuous variation of the high tension and current of the X-ray generator, special modifications have been carried out. After the modifications, the X-ray generator can be operated at currents down to 0.5 mA, the stability of the high tension being better than 0.5%. In this way, constancy of the output up to ±1% is achieved.

Since the laboratory should be able to calibrate instruments over a wide range of intensities, different series of radiation qualities were adopted (Table 1).

TABLE 1. FILTERED X-RAY RADIATIONS USED AT THE ISRAELI SSDL

Series	Generating Voltage (kVp)	Additional filtration (mm)				HVL (mm)
		Pb	Sn	Cu	Al	
Low Exposure Rate (about 50 mR/h)	25			0.2		1.3 Al
	50			1.0		5.0 Al
	75		0.8	0.25	1.0	0.6 Cu
	100		2.5	0.4	1.0	1.4 Cu
	125		4.5	0.4		2.1 Cu
	150	0.9	3.0	2.0	0.8	3.0 Cu
	200	3.0	2.0	0.4	0.8	4.5 Cu
Narrow Spectrum (about 500 mR/h)	25	Inherent (about 2 mm Al)				1.0 Al
	40			0.2		2.2 Al
	60			0.5		5.0 Al
	80			2.0	2.0	0.6 Cu
	100			5.0	1.6	1.2 Cu
	120		1.0	4.8	0.8	1.8 Cu
	150		2.5	0.4	1.0	2.5 Cu
	200	0.9	3.0	2.0	0.8	4.1 Cu
	250	3.0	2.0	0.4	0.8	5.4 Cu
Wide Spectrum (about 5000 mR/h)	60			0.2		3.4 Al
	80			0.5		0.3 Cu
	110			2.0	2.0	1.0 Cu
	145		0.8	0.25	1.0	1.8 Cu
	200		2.5	0.4	1.0	3.3 Cu
	250		4.5	0.4	1.0	4.3 Cu

The qualities adopted are similar to those recommended by ISO (1), except in the 10-50 keV region where filtered radiations are used instead of fluorescent radiations. A detailed discussion about the uncertainties introduced when fluorescent radiations are not used can be found elsewhere (2).

The reference instrument used at the laboratory is a 2550 NPL Protection Level Secondary Standard Dosimeter.

The instrument was originally calibrated at the National Physical Laboratory (UK), but owing to a leak in the chamber it was necessary to replace the rubber balloon, thus invalidating the calibration. The instrument with its new chamber was recalibrated at our laboratory, by comparing it against a 2560 Therapy Level Dosemeter. A high quality 30 c.c. chamber was used as the transfer instrument. It has been estimated that the uncertainties introduced during the transfer process are unlikely to exceed 4% at the 99% confidence level.

Since the use of radioactive sources for checking linearity of the instruments is involved with a considerable cost, it was decided to check this parameter by changing the current of the X-ray machine. For this purpose, a special gold coated transmission monitor chamber was designed by the German factory PTW. This chamber allows the monitoring of beam intensities down to 1 mR/h.

PRELIMINARY RESULTS

The following parameters have been measured during the setting up of the laboratory:
a) HVL
b) Beam homogeneity.
c) Reproducibility of the positioning of the instruments.
d) Reproducibility of the monitor system.
e) Energy dependence of the monitor chambers.
f) Long term reproducibility of the reference instrument.

Parameters of the instrument under test that will be evaluated are as follows:
a) Energy response
b) Linearity
c) Saturation characteristics
d) Directional dependence
e) Temperature dependence
f) Overload characteristics
g) β response
h) Neutron response
i) Transient characteristics

As an example, Figure 1 shows the energy response of some typical instruments used in Israel, while Figure 2 shows the saturation characteristics of a Nuclear Enterprises 30 c.c. chamber. It is evident that owners of radiation protection dosimeters should be aware of these characteristics before choosing the most suitable instrument for a particular application. For instance, the use of an Elscint GSM-1 Geiger Counter might introduce an error of up to 300%, while the use of a Nuclear Enterprises 30 c.c. for the measurement of pulses of radiation (for instance the output of an X-ray machine) will also introduce a significant error.

FUTURE ACTIVITIES OF THE ISSDL

The Soreq branch of the ISSDL is installing a Manganese bath for the absolute calibration of neutron sources. In addition its facilities are scheduled to perform calibrations of radiation protection monitors for neutron and beta radiation. A set of sources produced by Buchler (FRG) and calibrated

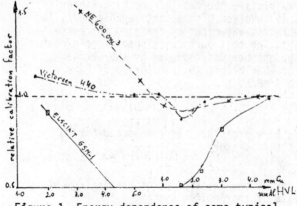

Figure 1. Energy dependence of some typical instruments.

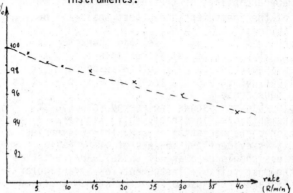

Figure 2. Saturation characteristics of Nuclear Enterprises 30 c.c. chamber (Polarising voltage: 300 V)

at the PTB will serve as the secondary standard for beta radiation.

Absolute calibrations of beta-gamma sources will be performed with a 4π proportional counter and a sodium iodide spectrometer system using beta-gamma coincidence techniques.

REFERENCES

1. Draft International Standard ISO/DIS 4037, X and γ reference radiations for calibrating and determining the energy dependence of dosimeters and doserate meters (1976).
2. Thomson I.M.G., International Standard Reference Radiations and Their Application to the Type Testing of Dosimetric Apparatus, In: National and International Standardization of Radiation Dosimetry, IAEA, 1978.

ATMOSPHERIC DISPERSION STUDY WITH ^{85}Kr AND SF$_6$ GAS

O. Narita, Y. Kishimoto, Y. Kitahara and S. Fukuda

^{85}Kr concentration in air around the reprocessing plant was measured when it was released from the plant stack during the hot-test operation treating PWR spent fuels. Charcoal absorption method was devised for this experiment as a rapid, simple and inexpensive technique for the measurement of ^{85}Kr concentration at about 10^{-9} µCi/cm^3. Special air samplers were located at few kilometers distant from the stack and the air was collected for an hour at a flow-rate of 1.5 liters per minute. About thirty air samples were taken and analyzed for ^{85}Kr concentration successfully. The performance of sampling and analytical methods employed was good, although some improvements were found to be necessary. An example is that an air-collection bag should be made of "saran" so that the leakage be minimized. Usually in the atmospheric diffusion study, the sulfer hexafluoride gas (SF$_6$) is used as an air tracer, and analytical procedure for the gas concentration down to 10^{-3} ppb is well established. In this experiment SF$_6$ gas was injected into the stack during the discharge of ^{85}Kr and the concentration in sampled air was also determined. The results show a good correlation between ^{85}Kr and SF$_6$ concentration in sampled air.

METHODS AND RESULTS

PNC reprocessing plant located near the sea-shore of the Pacific-Osean where is about 0.5 km distance from the exhaust point of air tracer. Air tracer gas of SF$_6$ was released continuously during air sampling duration. Fig. 1 shows air tracer exhaust system. SF$_6$ concentration was detected by the E. C. D. gaschromatgraph. ^{85}Kr gas was concentrated with cold-charcoal-trap and detected by GM tube detector showing in Fig. 2 and 3. Atmospheric diffusion profile and concentration of air tracer observed on neutral stability condition were good correlation with one calculated by Gaussian plume model equation, but in case of the unstable condition they were not like, as shown in Fig. 4, and relative cross-wind integrated concentration (χcic.U/Q) of air tracer is shown in Fig. 7. As shown in Fig. 5 atmospheric diffusion parameter (σ_y, σ_z) based on air tracer concentration are somewhat like on neutral condition pattern in spite of stability class. The relation between dilution factor of ^{85}Kr and SF$_6$ were good correlation as shown in Fig. 6.

Annual mean concentration with annual weather data was not far different between the value estimated with dilution factor based on this study and calculated with Gaussian plume model equation.

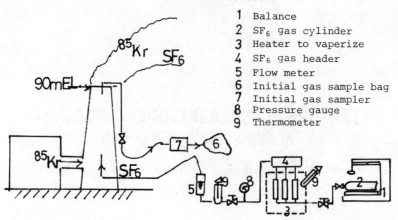

Fig. 1   Air tracer gas (^{85}Kr and SF$_6$) exhaust system schematic

1 Balance
2 SF$_6$ gas cylinder
3 Heater to vaporize
4 SF$_6$ gas header
5 Flow meter
6 Initial gas sample bag
7 Initial gas sampler
8 Pressure gauge
9 Thermometer

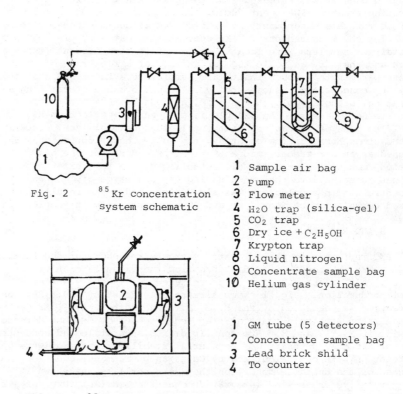

Fig. 2 ^{85}Kr concentration system schematic

1 Sample air bag
2 Pump
3 Flow meter
4 H$_2$O trap (silica-gel)
5 CO$_2$ trap
6 Dry ice + C$_2$H$_5$OH
7 Krypton trap
8 Liquid nitrogen
9 Concentrate sample bag
10 Helium gas cylinder

1 GM tube (5 detectors)
2 Concentrate sample bag
3 Lead brick shild
4 To counter

Fig. 3 ^{85}Kr detection system schematic

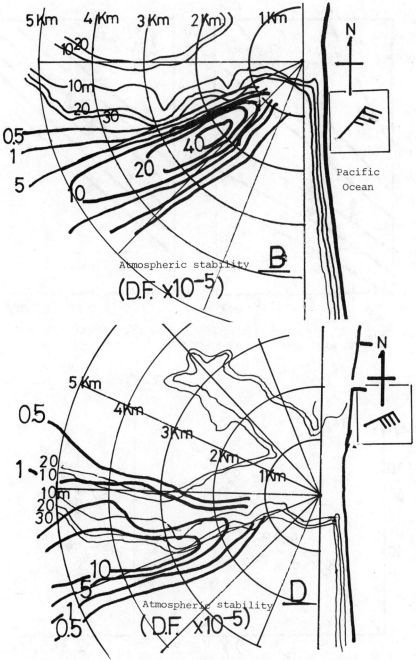

Fig. 4   Horizontal dilution factor (D.F.) contour of the air tracer on the ground surface (at 1m EL.)

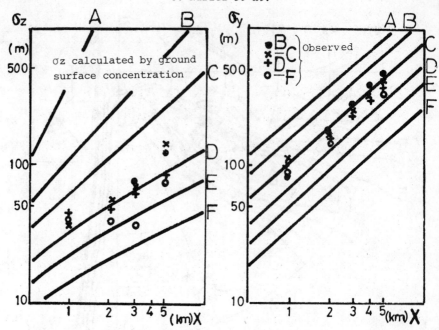

Fig. 5 Atmospheric diffusion parameter (σy, σz) based on observed air-tracer concentration profile.

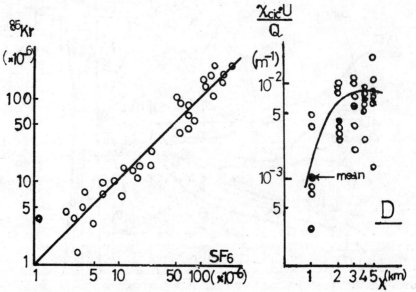

Fig. 6 Relation of dilution factor of ^{85}Kr and SF_6

Fig. 7 Crosswind integrated concentration for neutral condition

METHODS OF I-129 ANALYSIS FOR ENVIRONMENTAL MONITORING

T. Nomura, H. Katagiri, Y. Kitahara and S. Fukuda

Among the radioiodine isotopes discharged from nuclear facilities I-129 has the longest half-life of 1.7×10^7 years and is accumurated in the environment for a long time period, therefore, it is one of the most important nuclides in the environmental monitoring arround a nuclear fuel reprocessing plant.

Low level contamination of enviornmental samples with I-129 may cause considerably high thyroid dose to the population. For instance, only a tenth pico-curies of I-129 per liter of fresh milk may give one millirem of thyroid dose. Accordingly, in considering the I-129 discharged to the atmosphere or the marine environment, it is important to establish a methodology for evaluating the environmental impacts caused by the long-term accumulation and to develop the measuring techniques of the environmental samples having very low radioactivity.

This paper presents the methods of the analysis of low-level I-129 in the environmental samples such as milk, vegetations, sea weeds and soils.

PROCEDURES

Leafy vegetables, sea weeds and soils are dried with a low-temperature oven and are ground to powders. Milk is pulverized by freeze drying method.

The iodine is separated from the dried or pulverized samples by ignition at high temperature (about 1000°C) in a quartz combution apparatus with a stream of oxygen. Figure 1 shows the apparatus for ignition used in the experiments. Furnace 1 moves from the edge of the combustion tube to a sample slowly and the sample is heated gradually, finally ignited at 1000°C. The off-gas from the sample is burned completely while passing through Furnace 2. The iodine carried with the off-gases is trapped on a small bed of activated-charcoal. The iodine is then recovered from the charcoal to a dryice cooled quartz tube by heating in vacuum. The cooled end of the quartz tube is sealed off to make a irradiation ampule.

The quartz ampule is irradiated in a reactor for several ten minutes at thermal neutron flux of $10^{13} n/cm^2 \cdot sec$.

After the irradiation the iodine is purified through the solvent extraction method using carbon tetrachloride. Iodine is finally precipitated as AgI and counted with a Ge(Li) detector. Each activities of I-126, I-128 and I-130 are calculated from the gamma-ray spectra. The chemical yield of this method are calculated by counting I-125 which has been added to the sample prior to the ignition as an yield tracer.

RESULTS

Several samples were analized on I-129 by the method mentioned above and no I-129 concentration higher than detection limits were found. The results of analysis for typical food samples collected near the fuel reprocessing plant of Tokai Works are given in Table 1. The detection limit of I-129 by this method is about 10^{-2} pCi for a 10 g dry sample. Stable iodine I-127 is simultaneously determined and atom ratio of $^{129}I/^{127}I$ are calculated in order to evaluate thyroid dose by the specific activity method and long term environmental impacts by I-129 discharged from nuclear facilities.

REFERENCES

1) J. J. Gabay and C. J. Paperiello et al ; Heal. Phys., 26, 89-96, 333-342 (1974)

2) A. D. Mathews and J. P. Riley; Anal. Chim. Acta. 51, 295-301 (1970).

3) J. C. Daly and C. J. Paperiello et al ; Heal. Phys., 29, 753-760 (1975)

4) R. R. Edwards; Science 137, 851 (1962).

5) F. P. Brauer, J. K. Soldat, et al.; IAEA-SM-180/34 "Environmental Surveillance Around Nuclear Installations" Vol. II, P. 43-66 (1974), IAEA.

6) F. P. Brauer and H. Tenny; BNWL-SA-5287 (1975).

Table 1 Results of Analysis

Sample	I-127 * (μg/g)	I-129 (10^{-3} pCi/g)	$^{129}I/^{127}I$ Atom Ratio (10^{-7})
Seaweed	210 ± 11	< 5.1	< 1.5
Cabbage	4.2 ± 1.3	< 0.9	< 15
Rice	34 ± 1.2	< 2.1	< 3.6

* Determined by counting on I-126

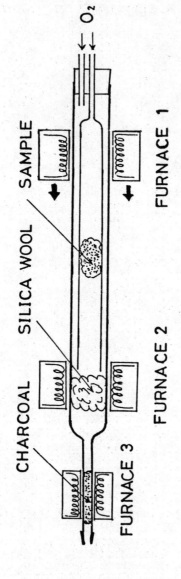

Figure 1 Diagram of Sample Combustion Apparatus

A NEW TECHNIQUE FOR NEUTRON MONITORING IN STRAY RADIATION FIELDS

E. Piesch and B. Burgkhardt

INTRODUCTION

At reactors, accelerators and therapy facilities including linear accelerators there is the need to monitor and interpret low level stray radiation fields. The techniques applied in neutron monitoring today is based mainly on the measurement of the dose equivalent by means of rem counters. The response function of the different rem counter types, however, has been recently found to overestimate intermediate neutrons by a factor 6 (Leak type) or factor 8 (Anderson-Braun type) and to underestimate thermal neutrons up to a factor 0.34 (polyethylene sphere 30 cm diam) [4]. Both the Bonner multi spheres and spectrometers, on the other hand, are sophisticated techniques not applicable for routine work.

The new approach in neutron monitoring described here is based on a single sphere technique and passive thermoluminescence detectors which allows to measure
- the total dose equivalent of neutrons and gamma rays,
- the dose equivalent components of thermal neutrons < 0.4 eV, epithermal neutrons between 0,4 eV and 10 keV and fast neutrons above 10 keV,
- the effective neutron energy E_{eff} of the fast neutron component in unidirectional or isotropic stray radiation fields.

MEASUREMENT TECHNIQUE

The single sphere technique applied for the measurement of the dose equivalent and for the interpretation of the neutron spectrum makes use of a passive rem counter (polyethylene sphere of 30 cm diam) and a TLD600/TLD700 detector in the center (see Fig. 1). The rem counter sphere serves also as a phantom for two albedo dosimeters.

The Karlsruhe albedo dosimeter [1] designed as an analyser detector type contains three TLD600/TLD700 detector pairs inside a boron-loaded plastic capsule allowing a separate indication of albedo neutrons (detector i), incident thermal neutrons from the field (detector a) and epithermal neutrons (detector m).

By means of TLD600/TLD700 detector pairs, thermal neutrons are measured via the reaction $^6Li(n,\alpha)^3H$ and the neutron dose reading is given by the difference of TLD600 and TLD700. The TL detectors are calibrated in a ^{137}Cs gamma field which results in a gamma equivalent neutron dose reading presented here in the unit R.

In stray radiation fields the neutron detection is directionally independent for the rem counter sphere and in a first approximation also for the albedo dosimeter system if the corresponding readings in the opposite position at the phantom surface are summed up.

The response R of the albedo dosimeter i found by calculation [2] and calibration with monoenergetic neutrons [3] is presented in Fig. 1 as a function of neutron energy. The response function of the passive

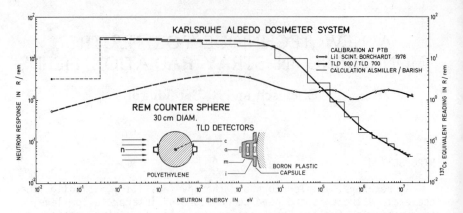

Fig. 1: Response of the Karlsruhe Single Sphere Albedo Dosimeter

TLD rem counter is expected to be equal to that of the rem counter with the LiI scintillation detector [4] and was related to a response of 1.6 R/rem found with Am-Be neutrons. The detector c in the sphere represents the true neutron dose equivalent H_n in a good approximation and the dose reading ratio $\alpha(i)/H_n$ the response $R(i)$ of the albedo dosimeter. The high change in response in the energy range above 10 keV compared to the flat response of the rem counter is the basis for an estimation of an effective neutron energy in stray radiation fields.

INTERPRETATION OF THE STRAY RADIATION FIELD

In practice the effective response $R_{eff}(i)$ of the albedo dosimeter may vary by one order of magnitude around one facility mainly caused by local changes of the thermal neutron fluence and/or the moderation of the fast neutrons.

For the interpretation of the neutron spectrum a computer program is used taking into account the response function of the rem counter (detector c) and of the detectors i, a, m in the Karlsruhe albedo dosimeter. In a neutron stray radiation field the neutron dose reading of the detectors in the albedo dosimeter can be interpreted on the basis of three energy components by the following response matrix

$$\alpha(a) = R_{th}(a) \cdot H_{th} + R_e(a) \cdot H_e + R_f(a) \cdot H_s$$
$$\alpha(m) = R_{th}(m) \cdot H_{th} + R_e(m) \cdot H_e + R_f(m) \cdot H_s$$
$$\alpha(i) = R_{th}(i) \cdot H_{th} + R_e(i) \cdot H_e + R_f(i) \cdot H_s$$

$\alpha(a), \alpha(m), \alpha(i)$ neutron dose reading in the ^{137}Cs equivalent unit R for the detectors a, m, i measured in the stray radiation field

H_{th}, H_e, H_f neutron dose equivalent fraction in the unit rem due to thermal, epithermal and fast neutrons with the total neutron dose equivalent $H_n = H_{th} + H_e + H_f$

$R_{th}(k), R_e(k), R_f(k)$ neutron response in R/rem for thermal, epithermal and fast neutrons for the detector k = i, a or m.

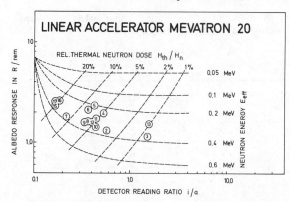

Fig. 2: Albedo dosimeter calibration at the Linac, Karlsruhe Vincentius Hospital

Taking into account the calibration data for $R_{th}(k)$, $R_e(k)$ and the relation $R_f(a) = R_f(m) = \varepsilon \cdot R_f(i)$ the response matrix can be solved resulting in analytical indications of the dose equivalent data H_{th}, H_e, H_f, H_n. For the estimation of the energy dependent response $R_f(i)$ the term H_f is calculated by $H_f = H_n' - A \cdot H_{th} - H_e$ taking into account the underestimation (factor A) of the rem counter reading H_n' for thermal neutrons.

The estimation of E_{eff} is based on the correlation between neutron energy and the albedo response $R_f(i)$ (see Fig. 1). In contrast to the calibration results for monoenergetic neutrons presented in Fig. 1 the applied program is fitted for neutron stray radiation fields assuming a homogeneous energy distribution for fast neutrons (Gaussian normal distribution).

In addition the statistical errors of all computed results are calculated based on the statistical error of each dose reading given by previous data of a reader test [5] (standard deviation vs. exposure) and caused also by the discrimination of the gamma dose reading in the detectors i, a, m and c [6].

APPLICATION IN STRAY RADIATION FIELDS

For personnel monitoring by means of albedo dosimeters an extended field calibration with the single sphere technique is applied at various locations in the stray radiation field of each neutron facility [7-9]. The application in personnel monitoring is based on individual correction factors for energy dependence based on the correlation between the dose reading ratio $\alpha(i)/\alpha(a)$ and the experimental response $R_{eff}(i)$ found by field calibrations before.

In the neutron/gamma stray radiation field of a 20 MV electron linear accelerator, for instance, the albedo response varies between 1 and 2.8 R/rem (Fig. 2). The value for H_{th}/H_n increases from 1% in 1 m source distance up to 20% at the shielding's entrance. The neutron field interpretation results in E_{eff} values between 0.2 and 0.4 MeV in agreement to results found at a similar accelerator by means of activation detectors and a computer unfolding calculation [10].

The single sphere albedo method is applied as a standard technique to interpret low level stray radiation fields [11]. Some experimental results are presented in Table 1. First applications at power reactor

Table 1: Interpretation of neutron stray radiation fields by means of the single sphere albedo technique

Facility		$\dot{H}_n$ mrem/h	H_n/H_γ	Rel. Neutron Dose in % H_{th}/H_n	H_e/H_n	H_f/H_n	E_{eff} keV	$R_{eff}(i)$ R/rem
^{252}Cf in air	2m	336	15.7	0.1	0.0	99.1	1900	0.31
20 MV Lin. Electr.	1m	13910	0.99	1.1	1.2	97.7	295	2.1
Accelerator	5m	1496	0.70	8.6	2.3	89.1	260	2.7
Mevatron 20 entrance		48	3.21	30.4	4.7	64.9	260	3.6
Linac SL 75-20	1m	64410	3.40	1.3	0.6	98.1	345	1.7
	entrance	613	4.21	13.5	3.7	82.8	240	3.2
	behind shield	13	1.55	37.3	5.8	56.8	120	5.1
14 MeV 'KARIN'	1m	4360	11.8	1.9	0.5	97.7	1140	0.67
Theraphy Fac.	9m	430	9.8	4.5	0.8	94.7	831	1.0
Compact Cyclotron	1m	500	20.2	2.2	0.4	97.4	1220	0.63
d(d,n), DKFZ	5m	146	13.6	5.0	1.0	94.0	720	1.2
	10m	138	5.1	14.6	2.2	83.3	400	2.2
	15m	16	1.7	43.9	4.9	51.2	410	3.5
Oak Ridge	Bare	6575	17.3	0.3	0.0	99.7	880	0.63
HPRR	Lucite	4715	12.2	4.3	0.4	95.3	850	0.88
	Concrete	4281	10.0	3.1	0.7	96.2	590	1.2
Jülich FRJ-1	Beam	62	2.33	3.0	1.2	95.8	235	2.5
Reactor	3m	8.7	4.08	20.6	5.1	74.3	140	4.6
GKN Power	in reactor cavity	5423	4.2	14.0	5.9	80.1	205	4.1
Reactor	23m	11	0.67	19.2	5.0	75.9	200	4.0
	64m	2.4	0.38	20.7	5.5	73.8	215	4.0
	entrance to	0.06	0.15	70.7	2.8	26.5	200	3.8
	sump	2.6	0.02	52.8	4.3	42.9	240	3.9
Kahl Exper.	containment	0.5	3.26	26.4	8.1	65.4	174	3.1
Reactor	steam heat exchanger	2.4	46.2	13.6	5.3	81.0	216	3.9
	valve in steam pipe	14.6	11.9	14.4	5.0	80.6	173	4.2

sites show that the local neutron spectrum varies only to a small extent for fast and epithermal neutrons but H_{th}/H_n may change from 5% to 70%.

The specific properties of the Karlsruhe albedo dosimeter was found in an effective response which is equal for thermal neutrons and neutrons with E_{eff} between 100 and 200 keV, therefore, at reactor sites only a small change of the dosimeter response $R_{eff}(i)$ is expected.

REFERENCES

[1] E. Piesch, B. Burgkhardt, Proc.IAEA.Symp. Stockholm 1978, to be published
[2] R.G. Alsmiller, J. Barish, Health Physics 33, 1974, p. 13
[3] E. Piesch, B. Burgkhardt, I. Hofmann, Report KfK 1979
[4] W.G. Alberts et al., PTB-ND-17, 1979
[5] E. Piesch, B. Burgkhardt, FS-78-17 AKD, 1979
[6] B. Burgkhardt et al., Nucl.Instr.Meth. 160, 1979, p. 533
[7] E. Piesch, B. Burgkhardt, in report
[8] B. Burgkhardt, E. Piesch, S. Schmid, Proc.Med.Physik 1979, p. 67
[9] B. Burgkhardt, D. Krauss, E. Piesch, Proc.Med.Physik 1979, p. 61
[10] McCall Associates Reports from 24.9.1977, 20.11.1977, private communication
[11] E. Piesch, B. Burgkhardt, to be published

MINIDOSIMETRY OF ALPHA-RADIATION FROM 239-Pu IN THE SKELETON

E. Polig

Plutonium which enters the bloodstream (via one of the main intake routes) deposits primarily in the skeleton. The distribution pattern is largely nonuniform owing to the fact that this radionuclide belongs to the class of the so-called surface seekers which concentrate on all periosteal, endosteal and trabecular bone surfaces (6). Radiation doses derived from assays of whole bones assume a uniform distribution and therefore tend to underestimate local doses to the cell populations at risk. For osteosarcoma induction the endangered cell populations are believed to be the osteogenic cells lying close to bone surfaces (1). To obtain a realistic correlation between the radiation doses delivered at a cellular level and the observed pathological effects the determination of local dose rates on a microscopic scale is essential. Further, this information can be linked to existing quantitative histological data to provide estimates of hit frequencies for cellular targets and of the probability of malignant transformation.

EXPERIMENTAL PROCEDURE

One year old female rats were injected intravenously with 18 kBqkg^{-1} 239-Pu-citrate and killed at various times after injection. Lumbar vertebral bodies from all animals were embedded in Methylacrylate and cut, on a sawing-microtome (Leitz, Germany) with a center-hole diamond blade, into sections of 150 μm thickness. The calcified tissue in these bone sections was stained selectively with Alizarin-red (30 min in 1 % solution).

After staining the sections were mounted in contact with cellulose nitrate foils (LR115, Kodak Pathé, France) which served as radiation detectors. To ensure alignment between detector surface and bone structure, landmarks (100 μm dia.) were drilled into the detector-section sandwich. After about 9 months exposure at -38°C the detectors were removed from the bone samples and etched in 2.5 N NaOH for 1 hr. Then both the detector and the bone sample were scanned with a microscope photometer using 20x20 μm scan fields (MPV II, Leitz, Germany) controlled by a PDP 8/E minicomputer system (DEC, USA). Thereby quantitation of the track density distribution and the bone structure was achieved in the form of light absorption patterns.

Using a series of FORTRAN programs the two files containing the raw scanning data were processed to generate a digital image of the bone structure, to calculate background levels on the autoradiographs, erase artifacts and, finally, to combine the two files into one. These processed files then constituted the input for an evaluation program that calculated mean dose rates in specific anatomical regions, dose rate distributions, the variation of these parameters, the mean burial depth etc.. Fig. 1 shows a computer generated image of a piece of trabecular bone with its associated light absorption pattern caused by alpha-particle tracks. Images of this kind are derived from the processed files.

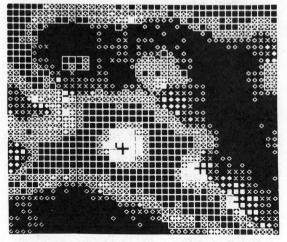

Figure 1. Computer generated display of bone structure (grid) with associated the track distribution expressed as the percentage of light absorbed by the tracks.
blank: 0-5 %, x: 5-10 %, ◊: 10-15 %, ◆: 15-20 %, ■: > 20 %.

RESULTS

Radiation dose rates were calculated according to a formula derived by Mays (4) under the assumption that surface deposits are infinitely thin and planar. At sites where no 'hot line' could be detected near a bone surface, the distribution of radioactivity was assumed to be locally uniform (5). Fig. 2 displays the frequency distribution of calculated dose rates in three adjacent bands of 10 µm width each at 28 days after injection. These distribution curves represent the composite variation of activity concentrations and burial depths of 239-Pu deposits and the geometrical locations within the bands.

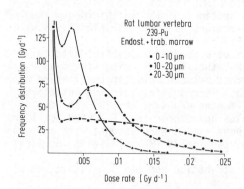

Figure 2. Dose rate distributions in three 10 µm wide bands 28 days after injection of 18 kBqkg^{-1} 239-Pu. The position of the bands relative to the bone surfaces is indicated on the graph.

In the 0-10 µm band 17 % of all sites have a dose rate below 1.25×10^{-3} Gy d^{-1}. The corresponding figures for the 10-20 µm and 20-30 µm bands are 19 % and 30 %, respectively. In a considerable part of the 0-10 μm band dose rates above .025 Gy d^{-1} can be found, whereas in the other two bands such dose rates are very infrequent. Comparison with results from later times after injection reveals that the distributions in Fig. 2 remain fairly constant in shape during the time of observation, with a slight increase from a section average of .0062 Gy d^{-1} (28 days) to .007 Gy d^{-1} (168 days). This indicates a low remodelling activity in the lumbar vertebra, which is expected in adult females. The increase could be attributed to recirculating 239-Pu released from other parts of the skeleton. A spatial dose rate spectrum at 14 different locations is shown in Fig. 3 for early and late times after injection. Initially the mean periosteal dose rates are about 4 times lower than the endosteal and trabecular ones. This relationship however changes in time as the dose rates in periosteal marrow are increasing much more rapidly than those on the endosteal and trabecular surfaces. The enhancement ratio shown in Fig. 3 also indicates that the dose rates in deep bone and deep marrow rise nearly 3-fold and 2-fold, respectively.

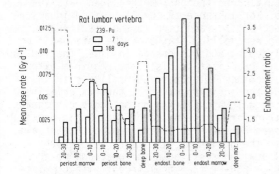

Figure 3. Mean dose rates in three 10 μm wide bands on both sides of periosteal and endosteal + trabecular surfaces and in deep marrow and deep bone. The dashed line represents the enhancement ratio of the values at 168 days after injection relative to 7 days.

The meaning of the dose rate distributions becomes more obvious if they are considered in terms of event frequencies at a cellular level. Specifically, if one considers the cell nucleus of proliferating osteogenic cells close to bone surfaces as the sensitive target and that a certain number of hits by alpha-particles are required to produce harmful effects like cell killing or non-lethal transformation to a malignant state, it can be shown (5) that a conversion from dose rate distribution to hit probabilities can be readily carried out provided the distribution functions are nearly constant in time. If a cell nucleus is approximated by a sphere of 5 μm diameter (2) one has for the fraction of cells f_n receiving n hits up to time t after injection

$$f_n = \frac{(ct)^n}{n!} \int \dot{D}^n \exp(-c\dot{D}t)\delta(\dot{D})d\dot{D}$$

where $\delta(\dot{D})$ is the distribution function of dose rates $\dot{D}$ (5) (c= .882 Gy^{-1}). Curves calculated according to the above expression are shown in Fig. 4 for the 0-10 μm band at periosteal and endosteal + trabecular surfaces. A considerable fraction of the targets receive one and two hits at the interior surfaces, less in the periosteal region.

A large fraction of those nuclei receiving one or more hits are likely to be killed.

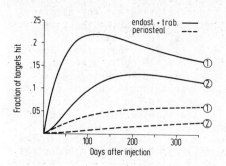

Figure 4. Fraction of spherical targets (5 µm dia.) uniformly distributed throughout a 10 µm band in marrow adjacent to bone surfaces that are hit until time t. Numbers specify the number of hits.

DISCUSSION

The extraction of results from the photometric scan data involves some approximations. For instance the problem of oblique surfaces was not taken into account. In addition to the results shown above information about the remodelling status can be obtained from the distribution of burial depths and its change in time. The linking of minidosimetric results to quantitative histology requires extension of present knowledge about cellular parameters like size and shape of sensitive targets, the position of these targets relative to bone surfaces (3) and the relevant time t in the above expression. With regard to malignant transformation the calculated 2-hit curves may be interpreted as indicating the size of the pool of potentially transformed cells. Moreover the notion of hit frequency perhaps provides a better means of scaling risks down to lower and more realistic skeletal burdens of 239-Pu than the concept of accumulated radiation doses.

REFERENCES

1. ICRP Publication 26 (1977), p. 10. Editor: F.D. Sowby, Pergamon Press, Oxford.
2. James, A.C. and Kember, N.F. (1970): Phys. Med. Biol. 15, 39.
3. Kimmel, D.B. and Jee, W.S.S. (1977): Report COO-119-252, 152.
4. Mays, C.W. (1958): Report COO-217, p. 161.
5. Polig, E.: In preparation.
6. Vaughan, J., Bleaney, B. and Taylor, D.M. (1973): In: Uranium, Plutonium, Transplutonic Elements, p. 349. Editors: H.C. Hodge et al., Springer, Berlin.

DECONTAMINATION AND MODIFICATION OF LIQUID SCINTILLATORS

S. R. Sachan and S. D. Soman

INTRODUCTION

Liquid scintillators (LS) are effectively employed for evaluation of soft beta (H-3, C-14, P-32, I-125) and alpha emitting nuclides. (1) The monitoring of environmental samples at the source of release of tritium (Atomic Power Reactors and nuclear installations) is done very frequently. This involves consumption of large amount of LS and hence generation of active waste. The disposal of liquid waste poses a problem. It would be doubly advantageous to decontaminate the used LS and recycle it.

This paper discusses the new techniques of decontaminating and recycling the used active LS. A modification of hydrophobic scintillator for use with aqueous samples is also described. Both aliphatic and aromatic LS are effectively decontaminated. 1, 4 Dioxane based LS is decontaminated by extraction with NaOH. Single extraction gives a decontamination factor (DF) of about 90% and thus 3-4 extractions decontaminate the LS to background level (2). Aromatic LS Tritol which is a cocktail of 1:2 Triton x-100 and toluene scintillator (3) has also been similarly decontaminated.

EXPERIMENTAL

Aromatic LS (Tritol)

It is spiked with tritiated water and counting efficiency is determined by liquid scintillation spectrometer (LSS 3255). The counted LS is transferred to separating funnel and shaken with solid NaOH for about 20-30 minutes and kept for settling for about two hours. The two phases separate out. The concentration of NaOH should be about 15% in the aqueous phase. The bottom aqueous layer contamining tritium is collected. The top organic phase which is intact LS is transferred to vails. The orginal amount of distilled water is added (to maintain constant water activity. This process is repeated 3-4 times depending upon initial level of activity.

Hydrophobic LS

The modification of toluene scintillator which does not hold water is attempted for solubilizing aqueous samples. Toluene alcohol solution of varying concentrations from 75% to 30% are prepared prior to addition of Naphthalene 100 gm/1, PPO 7.0 gm/1 and POPOP 0.3 gm/1. The counting efficiency and water holding capacity (WHC-maximum percent of water added to LS without phase separation) are evaluated for each ratio of toluene alcohol using LSS 3255. The counted LS is transferred to separating funnel and water to LS ratio is increased to 1:2 by addition of water. The mixture is shaken vigorously and kept over-night for settling. The bottom aqueous layer containing tritium is collected. The top organic layer is taken out and made up to orginal valume with ethyl alcohol as major portion of the alcohol goes along with water. The decontaminated LS is again examined for remaining activity after adding initial amount of water. The counting efficiencies and flourescence characteristics of recovered and fresh LS are compared.

RESULTS

Atomatic LS

It is observed that about 90% of the spiked activity is transferred with orginic phase after first extraction. Figure 1 shows the degree of decontamination of Tritol with number of extractions. Three to four such extractions make the LS compeltely decontaminated.

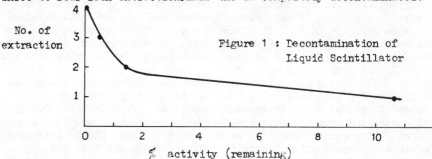

Figure 1 : Decontamination of Liquid Scintillator

The recovered LS is subjected to investigations for reuse. The counting efficiencies of fresh and recovered LS are compared by spiking these with equal amounts of tritiated water. Table 1 shows efficiencies of fresh and recovered LS.

TABLE I Comparison of Scintillators

Fresh Scintillator			Recovered Scintillator		
Bkg. cpm	Activity cpm	Efficiency percent	Bkg. cpm	Activity cpm	Efficiency percent
13.6	2811.0	25.2	21.0	2626.0	23.4
13.9	2847.0	25.5	20.3	2513.0	22.4
16.2	2623.0	23.5	22.2	2201.0	19.6

The wholesomeness of recovered LS and its suitability is examined by fluorescence characteristics also. It is observed that fluorescence characteristics of both fresh and recovered LS are identical except some loss of fluorescence yield in the case of recovered one. This explains the loss in counting efficiency.

Hydrophobic LS

The water holding capacity of modified toluene/alcohol LS decreases with increasing concentration of toluene in LS. Figure 2 shows linear relationship between water holding capacity and toluene concentration. WHC increases from 3% to 13% with decrease in toluene concentration from 75% to 30%.

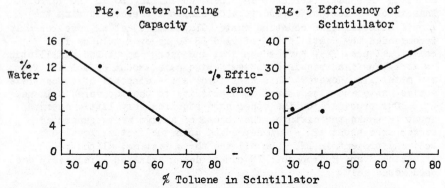

Fig. 2 Water Holding Capacity

Fig. 3 Efficiency of Scintillator

% Toluene in Scintillator

The counting efficiencies for various toluene/alcohol ratios are determined and shown in figure 3. The efficiency increases with increasing concentration of toluene in scintillator at their maximum WHC. The efficiency increases from 16% to 35% with increase in toluene concentration from 30% to 75%.

The counting efficiency **increases with % toluene in LS while WHC decreases**.. The optimum working range is selected with highest figure of merit (4) which takes into account background, efficiency and volume of the sample. FM is found to be maximum around 50% toluene concentration.

The decontamination of used modified LS is acheived by single washing with excess amount of water. After proper dilution the background is checked prior to spiking. It is observed that the recovered LS has almost similar background as that of fresh one. Table II shows the relative counting efficiencies of 50% toluene/alcohol LS before and after decontamination. The counting efficiency of decontaminated LS is about 92% of the fresh LS. The quality of recovered modified LS is also compared on the basis of fluorescence characteristics. The spectral characteristics of both the LS are identical except slight loss in fluorescence yield in case of recovered one.

TABLE II Decontamination and Recycling

	Fresh Scintillator			Recovered Scintillator	
Bkg. cpm	Activity cpm	Efficiency percent	Bkg. cpm	Activity cpm	Efficiency percent
51.6	1283.0	23.8	59.6	1191.0	21.9
54.7	1265.0	23.4	50.2	1191.0	21.8
50.8	1298.0	24.1	62.0	1209.0	22.2
56.0	1320.0	24.4	47.0	1242.0	23.1

DISCUSSION

It is, therefore, possible to recycle the same LS after decontamination. It will not only save expenditure on LS but solve the disposal problem also, as the activity is contained in aqueous phase with reduced volume. Gaylord (5) has suggested the disposal of used LS vials into sea after crushing them and filling LS in the varrels. This involves further expenditure on disposal besides loosing the liquid scintillation counting waste. Glaycamp (6) et al have recently recommended the distillation of used LS to recover toluene for commercial use. They have shown that the energy input for distillation is much less than required to symthesise same amount of toluene from raw materials. However, this will involve the disposal of concentrate active waste and also impart some activity to the recovered toluene.

Our procedure on the other hand requires very little energy input to shake the mixture. The volume of active waste generated ranges from about 12% - 47% depending upon the initial level of activity present in LS. Thus all the three systems, aliphatic LS, Aromatic LS and Hydrophobic LS can be decontaminated effectively and used repeatedly.

REFERENCES

1. Mcklveen J.W. and W.R. Johnson (1975)
   Health Phys. **28**, 5
2. Sachan S.R. and Soman S.D. (1979)
   Health Phys. **36**, 67
3. Patterson M.S. and Green R.C. (1965)
   Analy. Chem. **37**, 854
4. Loevinger R. and Berman M. (1951)
   Nucleonics **9**, 26
5. Gaylord M.T. (1976)
   Health Phys. **30**, 499
6. Claycamp H.G., Cember H. and Prot E.A. (1978)
   Health Phys. **3**, 716

TRITIUM - IS IT UNDERESTIMATED?

G. D. Whitlock

INTRODUCTION

The Whitlock Tritium Meter was first introduced to the IRPA as a simple and satisfactory method for the direct measurement of Tritium surface contamination, at the 4th International Congress, Paris 1977. It was shown then that though instrumentally the measurements can be made simple, the fundamental characteristics of Tritium air absorption and self-absorption must be clearly understood, if the results of measurements are to be meaningful.

Practical experience of the author in the use of the Whitlock Tritium Meter in various laboratories and industrial establishments throughout the world since its first development has shown that:
a. Measurements by smear/wipe tests can often be in error by three orders of magnitude or more.
b. Sub-visual scratches (8μ deep) are radiologically important.
c. Volatile forms of Tritium exist in 20% to 30% of establishments visited.

The author questions the widespread use of smear/wipe techniques for the assessment of 3H surface contamination based on the assumption that 10% of removable activity is collected by the smear/wipe.

Tritium surface contamination assessed as "fixed" can contain volatile fractions with a hazard potential which may be considerably greater than the hazard from removable activity at present covered by the Maximum Permissible Level (MPL) recommendations.

The Whitlock Tritium Meter has now been used for radiation protection purposes for six years in all manner of working environments, from Atomic Power Station, to Pharmaceutical laboratories. In addition, the author has demonstrated its use in several hundred establishments for direct measurement of suitable surfaces [1] and the rapid assay of smears.

During this time it has been of concern to observe a discrepancy in measurements which indicate that smear sampling techniques can be in error by a factor of 1000 or more. Of particular significance were surveys of several working surfaces [2] in separate establishments which had been shown clean or well below the MPL (Maximum Permissible Level) - notes[1] and [2] by smear surveys carried out by independent authorities a short time previously.

Direct measurement of these surfaces, using the Whitlock Tritium Meter, revealed a residual activity estimated to be in excess of 1000 times the Maximum Permissible level. A reduction of this activity was observed when surfaces were de-contaminated with detergent and water. Successive cleanings reduced the total activity in a quasi exponential way.

The potential error in the assay of activity on the smears by Liquid Scintillation counting would be very unlikely to be greater than a factor of 2 or 3.

With regard to the smear, at the time when direct measurements were made, the surfaces appeared clean. The activity detected was not in the form of dust particles, and at that moment of time could be considered bound i.e. not easily removable activity. Thus we could conclude that the activity was fairly firmly fixed to the surface and as such the contamination indicated by smear was within the recommended levels.

However, consideration must be given to the following points:
1. With such a high activity present should only 0.1% become unbound by the day or week following the survey, the Maximum Permissible Level even in the terms of removable surface contamination would be exceeded.
2. The assumption that the activity was "bound" is not necessarily correct. It was observed that large areas of the working surfaces surveyed by smears were covered with almost sub-visual scratches and activity in these may not be readily collected by a smear.
3. The isotope we are dealing with is Tritium, and the exposure pathways include absorption through the skin (3) as well as inhalation of dust particles.
4. In addition we have potential inhalation of volatile fractions (described later).

The main point of concern relative to these observations is that the present reliance on the smear technique to give reliable information as to the actual contamination hazard is injustified, even if the technique is only applied to smooth working surfaces as defined in the Codes of Practice. They would also suggest that radiological protection consideration of Tritium contamination should be based upon a direct measurement of surfaces contamination, if necessary by the provision of a suitable surface in locations where surfaces suitable for direct measurements (1) would not normally be available. Assumptions for the surrounding work surfaces can then be made relative to factual information.

If our findings are representative of working surfaces everywhere, they illustrate a potentially hazardous situation on a world wide scale, for they apply not only to laboratories where high levels of Tritium are used, but even low level laboratories, because the problem is not due to a "one-time" large spill, but a gradual history

Tritium - Is It Underestimated?

of build-up due to poor house-keeping, as a result of highly inaccurate measurements giving a false sense of security.

The high inaccuracies in measurements also extend to certain types of Tritium sensitive survey instruments, capable of making direct measurements of very small areas of the working surface which, because of the very small sensitive areas (1.6 to 10 cm^2) of the detectors, encourage the survey of a surface by a sweeping or scanning technique commonly adopted for isotopes emitting more energetic betas than Tritium, enabling the use of geiger counters. Here the problem of the integration time constant of rate meter instruments is often forgotten, and areas are covered more quickly than the probe area really permits. This is compounded by the very severe absorption by air of Tritium betas (range typically 0.5mm for 5.6 keV betas [1]. Lack of appreciation of the severe limitation imposed by this fundamental characteristic leads to the conduct of surveys where the detector is held outside the range of a significant number of emitted betas which, in turn, produces a potential error of infinity. As is the case with smear measurements, direct measurement by hand held devices can result in the conclusion that it is safe to continue working in the areas surveyed with the erroneous conviction that the contamination is below the Maximum Permissible Level.

To this situation we must add that in 20 - 30% of the establishments in which we have made direct surface measurements, using the Whitlock Tritium Meter, we find the presence of volatile Tritium. At first sight this would appear to be of academic interest, but on further consideration the radiological hazard could be considerable, and most likely specific to one individual among the personnel in the working environment. The Whitlock Tritium Meter detects the presence of volatile Tritium by virtue of the fact that a vacuum of approximately half of an atmosphere is established in the measuring chamber for each measurement. If volatile fractions do exist, they are "driven off" and can be identified by successive measurements (10 seonds each) of the same area (100 cm^2). It is clear that we do not live in an environment where the atmospheric pressure suddenly changes 50%, so the condition is artificially emphasised, nevertheless, it is quite within the bounds of possibility that working surfaces will be subjected to temperature changes; for instance, sunlight through a window sweeping across the surface, or more likely, in consequence of a hot object, such as a human hand, or a recently heated beaker, being placed on the surface. In such circumstances, the volatile fraction given off will form an active "cloud" local to the person carrying out the work.

As we have seen from the previous discussion the source from which it comes can quite easily be several thousand times greater than the Maximum Permissible Level.

The active cloud would not readily be detected by installed Tritium in air monitoring equipment, due to the dilution volume of the room, or by urine tests, because of subsequest dilution (usually) in the total body water. The hazard seems likely to be the same as inhalation of dust particles from surface contamination.

CONCLUSION

Evidence suggests that the question of Tritium surface contamination should be re-considered in its entirety as the actual hazard may widely exceed acceptable levels. In particular the question of calibration of the smear technique, the effect of surface condition and the hazard of volatile fractions need investigation.

REFERENCES

1. G D Whitlock - "Tritium Contamination Measurement A Simple and Satisfactory Method " - Proceedings 4th Congress IRPA Vol 3 Pages 809 -812
2. Manufacturers trade marks - Formica, Wearite
3. J A Gibson and A D Wrixon - "Methods for the Calculation of D.W.L's - for Surface Contamination by Low-toxicity Radionuclides" - Health Physics Vol 36 (March) Page 318

Note 1: The M.P.L. (Maximum Permissible Level) referred to in the text = $10^{-4} \mu Ci/cm^2/100cm^2$

Note 2: There is no derived working limit (DWL) for Tritium

Internal Dosimetry

INHALATION DU RADON ET DE SES PRODUITS DE FILIATION DOSES ET EFFETS PATHOLOGIQUES

D. Mechali

Les études sur la toxicologie du radon et de ses produits de filiation présentent un intérêt particulier tant en hygiène professionnelle qu'en hygiène publique car ces produits constituent dans les deux cas une nuisance qui pose souvent de difficiles problèmes de protection.

I. LE RADON ET SES PRODUITS DE FILIATION A VIE COURTE

Le radon est un gaz rare qui descend du radium 226. Il donne naissance à un certain nombre de produits de filiation à vie courte qui s'enchaînent successivement pour aboutir au plomb 210 dont la période est de 21 ans et auquel on ne s'intéressera pas dans cette étude.

1 - FORME PHYSIQUE

Les produits de filiation du radon peuvent se trouver dans l'atmosphère sous forme d'ions ou d'atomes libres ou être fixés sur les particules en suspension dans l'air. D'assez nombreuses mesures de la fraction libre ont été effectuées dans les mines (1 - 4) et on dispose de quelques données sur l'importance de cette fraction à l'air libre ou à l'intérieur des habitations (5, 6). Une excellente étude théorique de Jacobi (7) met en évidence l'importance respective des différents facteurs qui déterminent dans les locaux ou les mines la répartition des produits de filiation du radon entre forme libre et forme fixée.

D'une façon générale, dans les exploitations minières, la fraction libre, bien que variant avec la ventilation et les opérations effectuées, reste toujours faible. Le modèle théorique de Jacobi donne des résultats qui concordent très bien avec les résultats expérimentaux : dans des zones de travail minier, avec une ventilation modérée (3 à 5 h^{-1}), la fraction libre est de l'ordre de 1 à 3 % pour le Ra A et de l'ordre de 0,2 à 0,5 % pour le Ra B.

Dans les habitations (5, 6), bien que le taux de renouvellement de l'air soit plus faible, la faible concentration de l'air en particules fait que la fraction libre est plus élevée que dans les mines. Les mesures effectuées donnent des résultats qui varient de quelques pour cent à près de 40 %.

2 - EQUILIBRE

Les produits de filiation du radon ne sont généralement pas en équilibre radioactif avec le radon lui-même. Dans l'étude déjà citée (7), Jacobi met en évidence les facteurs qui conditionnent les activités de chacun des produits de filiation par rapport à celle du radon : taux de ventilation, concentration de l'air en particules, dépôt sur les parois. Le déséquilibre est généralement plus marqué dans les mines que dans les habitations ou à l'air libre, en raison surtout des taux de ventilation plus élevés. Les valeurs relevées dans les mines varient avec les opérations effectuées et les lieux de prélèvement, mais le déséquilibre est toujours prononcé. Des rapports 1 : 0,5 : 0,2 : 0,1 représentent une situation moyenne (8).

Dans les habitations, le déséquilibre est nettement moins marqué; des valeurs de l'ordre de 0,7 à 0,9 pour le Ra A, 0,6 à 0,8 pour le Ra B, 0,4 à 0,6 pour le Ra C sont souvent rencontrées.

3 - EXPRESSION DE L'EXPOSITION

On sait depuis longtemps maintenant que l'irradiation de l'appareil respiratoire par le radon lui-même est négligeable devant celle qui résulte du dépôt dans les voies respiratoires de ses produits de filiation présents dans l'air. La concentration de l'atmosphère en radon ne traduit pas la nuisance réelle, puisque l'équilibre entre le radon et ses produits de filiation successifs peut varier dans de très larges proportions, et cette nuisance ne peut être évaluée qu'en fonction de la concentration dans l'air de chacun des produits de filiation.

L'irradiation par les rayonnements β émis par le Ra B et le Ra C est, compte tenu de leur efficacité relative, négligeable devant l'irradiation par les rayonnements α émis par le Ra A et le Ra C'. Un atome de Ra A donne naissance, au cours des désintégrations successives des produits de filiation à vie courte, à deux particules α alors qu'un atome de Ra B, de Ra C ou de Ra C' ne produit qu'une particule α : de là est née l'idée d'énergie potentielle α, l'énergie potentielle α d'un atome de Ra A étant près de deux fois plus élevée que celle d'un atome de Ra B, de Ra C ou de Ra C'.

La nuisance de l'ensemble des produits de filiation du radon dans l'air peut être exprimée par la somme de leurs énergies potentielles α par unité de volume. L'unité adoptée correspond à l'énergie potentielle α totale de 3,7 Bq/l (100 pCi/l) de radon en équilibre avec ses produits de filiation ou à n'importe quelle combinaison de produits de filiation ayant la même énergie potentielle, c'est-à-dire $1,3.10^5$ MeV α par litre. On a donné à cette unité, en raison de son utilisation dans les mines d'uranium, le nom de niveau de travail.

L'exposition intégrée s'exprime alors généralement en niveaux de travail-mois (WLM) correspondant à l'exposition pendant 170 heures de travail dans une atmosphère où la concentration des produits de filiation est égale à un niveau de travail. L'incorporation de produits de filiation correspond à une énergie potentielle α de $2,65.10^{10}$ MeV et équivant à l'inhalation de 0,75 MBq (20 μCi) de radon en équilibre avec tous ses produits de filiation à vie courte.

II. EVALUATION DE L'IRRADIATION PAR LES PRODUITS A VIE COURTE DU RADON

Un certain nombre de modèles ont été utilisés pour évaluer l'irradiation de l'appareil respiratoire en fonction de l'activité inhalée de la fraction des produits de filiation sous forme libre et de la granulométrie des particules présentes dans l'air (9 à 15).

La plupart de ces modèles (9 à 14) visent à évaluer la dose dans les différents étages de l'arbre bronchique. D'après le dernier d'entre eux, celui de Harley et Pasternack (14), les bronches segmentaires sont celles où la dose est la plus élevée ; pour une atmosphère où

les produits de filiation du radon seraient dans les rapports
1 : 0,61 : 0,29 : 0,21 et la fraction libre du Ra A de 4 %, la dose dans les bronches segmentaires serait de 3,6 mGy, soit d'environ 70 mSv, par niveau de travail-mois (WLM).

En 1972, Jacobi (15) se fonde sur les principes de la C.I.P.R. qui considère que la valeur significative de la dose est la valeur moyenne dans l'organe ou le tissu considéré. Il suit le modèle pulmonaire de la C.I.P.R. en considérant que les produits de filiation du radon sont très transférables (classe D) et calcule la dose moyenne dans les régions trachéo-bronchiques (TB) et parenchymateuse (P). Il admet que dans la région TB se déposent 50 % des atomes libres inhalés - le reste se déposant dans le naso-pharynx - et 8 % des atomes fixés sur des particules, les dépôts correspondants dans la région P étant de 0 % et 50 %. En raison de l'élimination d'une fraction du dépôt, essentiellement par passage dans le sang, l'énergie effectivement absorbée est, pour la région TB, très sensiblement différente de l'énergie potentielle déposée, la fraction cédée aux tissus étant de 0,17 pour Ra B et d'environ 0,4 à 0,5 pour Ra A et Ra C.

Pour une atmosphère présentant les caractéristiques habituellement rencontrées dans les mines (ventilation de l'ordre de 3 à 5 h^{-1} constante de fixation pour les particules de l'odre de 1 000 h^{-1}), l'énergie absorbée dans l'arbre bronchique est de l'ordre de 0,1 mJ par niveau de travail-mois, soit pour une masse de 40 g une dose d'environ 2,5 mGy (50 mSv). Les valeurs correspondantes pour la région parenchymateuse sont de 2 mJ et 2 mGy (40 mSv). Pour des atmosphères peu ventilées (1 h^{-1}) et peu empoussiérées, les valeurs sont respectivement de 0,4 mJ et 1,5 mJ.

Il est intéressant d'appliquer à l'inhalation de radon et de ses produits de filiation les méthodes recommandées dans les publications 26 et 30 (16) de la C.I.P.R.. Ceci revient à pousser un peu plus loin la méthode utilisée par Jacobi en 1972 en évaluant l'équivalent de dose effectif à partir des équivalents de dose dans les différents organes. En ce qui concerne le poumon, on considère que la région TB, la région P et les ganglions lymphatiques pulmonaires constituent un organe de masse égale à 1 000 g et on calcule l'équivalent de dose moyen dans cet organe composite.

En utilisant les valeurs données par Jacobi (15) en ce qui concerne la :
- la fraction déposée dans l'arbre bronchique et dans la région P, selon qu'il s'agit d'atomes libres ou de particules
- l'énergie absorbée par atome ou par Becquerel déposé dans la région TB ou la région P selon qu'il s'agit de Ra A, de Ra B ou de Ra C,
il est possible de calculer l'équivalent de dose engagé pour le poumon en fonction de l'énergie potentielle inhalée sous forme de Ra A, de Ra B ou de Ra C et de la fraction de l'activité f se trouvant sous forme libre.

Si l'on calcule l'équivalent de dose engagée au poumon pour une énergie potentielle inhalée de $2,65.10^{10}$ MeV, ce qui correspond à un niveau de travail-mois (WLM), on obtient les résultats suivants pour les différents produits de filiation du radon :

- Ra A = (43,4 − 25,0 f) mSv
- Ra B = (41,0 − 33,8 f) mSv
- Ra C = (44,7 − 25,3 f) mSv.

On peut remarquer :
- que la dose engagée varie en sens inverse de f, mais que ce facteur n'a qu'une importance limitée dans la gamme des valeurs habituellement rencontrées. Par exemple, l'équivalent de dose résultant de l'inhalation de Ra A ne diminue que d'environ 10 % quand f passe de 0,02 à 0,25
- que, pour une même valeur de f, la dose engagée par niveau de travail-mois varie peu d'un produit de filiation à l'autre.

A titre d'exemple, on a calculé l'équivalent de dose par niveau de travail-mois pour deux atmosphères très différentes et correspondant l'une à une situation fréquente dans les mines, l'autre à une situation que l'on peut rencontrer dans des habitations.

Dans le premier cas, on a admis que les produits de filiation étaient par rapport au radon dans les proportions 1 : 0,5 : 0,2 : 0,1, que la fraction de l'activité sous forme libre était de 0,02 pour le Ra A, de 0,003 pour le Ra B, et négligeable pour le Ra C. Dans ces conditions, l'équivalent de dose pour le poumon, par niveau de travail-mois est de 42 mSv.

Dans le deuxième cas, on a admis que les produits de filiation étaient dans les rapports 1 : 0,8 : 0,6 : 0,4 et que les fractions de l'activité sous forme libre étaient de 0,25 pour le Ra A et de 0,03 pour le Ra B. L'équivalent de dose pour le poumon est alors de 41 mSv par niveau de travail-mois.

La dose au poumon qui résulte de la présence de radon dans les voies respiratoires est négligeable devant ces valeurs. Pour un volume pulmonaire moyen de 3 litres, le séjour dans une atmosphère où la concentration moyenne du radon serait de 3,7 Bq/l, l'équivalent de dose en un mois serait d'environ 0,1 mSv.

La dose au poumon est donc d'environ 40 mSv par niveau de travail mois, ce qui correspond pour cette composante de l'équivalent de dose effectif à environ 5 mSv par niveau de travail-mois.

On ne peut complètement négliger l'irradiation des autres organes par les produits de filiation du radon transférés au sang dans l'appareil respiratoire. Le transfert se fait essentiellement dans l'arbre bronchique, la période d'épuration de la région parenchymateuse était longue devant les périodes radioactives. Pour une même énergie potentielle inhalée, le transfert est d'autant plus important que le dépôt dans l'arbre bronchique et donc la fraction sous forme d'atomes libres sont plus élevées. Pour une atmosphère à l'équilibre radioactif et des fractions libres égales à 0,40 pour le Ra A, 0,10 pour Ra B et 0,05 pour Ra C, et en suivant le modèle métabolique décrit dans la publication C.I.P.R. 30 (16) pour le polonium et ceux décrits par Adams et al. (17) pour le plomb et le bismuth, l'équivalent de dose effectif résultant de l'irradiation des organes autres que le poumon est d'environ 0,5 mSv par niveau de travail-mois.

L'équivalent de dose effectif, compte tenu de l'irradiation du poumon par les produits de filiation déposés et de celle des autres organes par les produits transférés au sang, est donc d'environ 5 à 5,5 mSv par niveau de travail-mois.

III. LES EFFETS PATHOLOGIQUES RESULTANT DE L'INHALATION DES PRODUITS DE FILIATION DU RADON

L'irradiation de l'appareil respiratoire par les produits de filiation du radon peut provoquer des effets non stochastiques et des effets stochastiques.

Les effets non stochastiques résultent de la sclérose et de l'emphysème pulmonaire liés à l'irradiation du parenchyme. Des lésions de ce type ont été observées chez l'animal soumis à des expositions élevées (18) et jouent certainement un rôle dans le raccourcissement de la durée de vie que l'on observe dans ce cas. Chez l'homme, on a observé (19) chez des mineurs ayant travaillé plus de 10 ans au fond, une augmentation de la mortalité par maladies respiratoires non cancéreuses, mais il n'est pas possible de mettre clairement en évidence le rôle respectif du radon, des poussières de silice, du tabac et des fumées de diesel dans ce processus.

Mais ces effets n'apparaissent qu'à des niveaux élevés et le risque essentiel résultant de l'inhalation des produits de filiation du radon est celui d'une augmentation de la fréquence des cancers pulmonaires. On n'examinera ci-dessous que les résultats des deux enquêtes épidémiologiques menées sur des mineurs d'uranium, l'une aux Etats-Unis (19 - 21), l'autre en Tchécoslovaquie (22 - 25).

Les deux enquêtes sont assez comparables : même ordre de grandeur des échantillons étudiés ; échantillon constitué entre 1950 et 1960 parmi des mineurs travaillant déjà depuis un certain nombre d'années au fond, dans le premier cas ; échantillon constitué de mineurs ayant commencé à travailler au fond entre 1948 et 1952, dans le deuxième cas; traitement des données arrêtées en 1968 (21) puis 1974 (19), dans le premier cas, les dates correspondantes étant de 1973 (23, 24) et de 1975 (25) dans le deuxième cas. Dans les deux cas, les incertitudes statistiques sur la mortalité sont fournies ou peuvent être calculées ; il n'en est pas de même pour les incertitudes peut-être importantes sur l'évaluation de l'exposition, ce qui soulève un problème difficile à résoudre sur les marges d'incertitudes qui affectent les estimations des relations dose-effet.

La méthodologie suivie dans les deux cas n'est pas identique, ce qui fait que les résultats ne sont pas directement comparables, mais il est possible de traduire avec une exactitude suffisante les résultats de l'enquête américaine dans les mêmes termes que pour l'enquête tchécoslovaque. D'autre part, les auteurs tchécoslovaques n'ont pas essayé d'individualiser le rôle du tabac, en admettant que la consommation était identique dans la population de mineurs et dans la population de comparaison. Pour comparer les résultats des deux enquêtes, on admettra qu'il en est de même aux U.S.A., mais cette hypothèse peut dans les deux cas biaiser l'interprétation.

Il aurait été intéressant, pour essayer de juger de l'évolution dans le temps de la fréquence surajoutée, de comparer dans l'enquête effectuée aux U.S.A. les résultats du traitement des données arrêtées en 1968 et de celles arrêtées en 1974, six ans séparant ces deux points. Mais cela est impossible car la population témoin diffère dans les deux cas (population mâle des Etats de l'Arizona, du Colorado, du Nouveau Mexique et de l'Utah dans le premier cas, population mâle de l'ensemble des U.S.A. dans le deuxième cas).

On se fondera donc essentiellement sur les données arrêtées en 1975 pour l'enquête tchécoslovaque et en 1974 pour l'enquête américaine.

Dans les deux cas, les données ne sont pas incompatibles avec une relation linéaire passant par l'origine. La pente de la droite correspondrait à une augmentation de fréquence des cancers de 4.10^{-5} par niveau de travail-mois pour l'enquête effectuée aux U.S.A. et de 2.10^{-4} par niveau de travail-mois pour l'enquête effectuée en Tchécoslovaquie.

Comme cela est habituellement le cas quand des enquêtes épidémiologiques destinées à évaluer l'action d'une nuisance sont menées sur des populations différentes, les conclusions quantitatives des deux enquêtes ne sont pas identiques mais elles convergent de façon suffisante pour évaluer l'ordre de grandeur de l'augmentation de fréquence des cancers pulmonaires. Celle-ci serait sur la base des données actuelles, d'environ 100 cas par million et par niveau de travail-mois. Exprimée en augmentation de fréquence par sievert sur la base des calculs de dose effective, elle serait actuellement d'environ $2,5.10^{-3}$ par sievert, c'est-à-dire voisine de la valeur indiquée dans la publication C.I.P.R. 26 (2.10^{-3} Sv^{-1}).

Il faut noter que ces évaluations souffrent d'erreurs systématiques de deux ordres qu'il est difficile d'apprécier et qui jouent en sens opposé :
- les populations étudiées sont suivies depuis environ 25 ans mais une partie importante de chacune d'elle est encore en vie. S'il est difficile de prévoir l'évolution de la différence entre le nombre de cas observés et le nombre de cas attendus, durant la période qui s'étend jusqu'à l'épuisement de chacune des deux cohortes, il est néanmoins certain que les évaluations actuelles de la fréquence surajoutée seront majorées ;
- dans les deux enquêtes, l'augmentation de fréquence des cancers pulmonaires est attribuée à la seule action du radon alors que les populations étudiées ont été également soumises à d'autres nuisances, radioactives (irradiation externe et inhalation de poussières de minerai) ou non radioactives (fumées de diesel en particulier). Les évaluations de l'action du radon devraient donc être minorées et ce relativement plus, peut-être, dans la zone des expositions faibles que dans celle des expositions élevées. Dans le même esprit, l'hypothèse d'une consommation identique de tabac dans les populations de mineurs et les populations de comparaison devrait être vérifiée.

Enfin il faut rappeler l'importance des incertitudes, connues en ce qui concerne les fréquences surajoutées, mais non en ce qui concerne l'évaluation des expositions.

De nombreuses études sur l'induction de cancers pulmonaires par les produits de filiation du radon ont été également menées chez l'animal, qui (26, 27 par exemple) ont apporté des résultats intéressants mais il n'est pas possible de les analyser dans cette trop brève revue.

CONCLUSION

La poursuite des enquêtes déjà entreprises, la mise en route de nouvelles enquêtes sur des populations de mineurs exposés à une époque plus récente et dont l'exposition serait donc connue de façon plus précise, la poursuite de l'expérimentation animale enfin permettront sans aucun doute, de connaître de façon plus précise dans l'avenir les relations entre l'exposition au radon et ses modalités d'une part et l'augmentation de fréquence des cancers du poumon d'autre part. Mais la convergence raisonnable des évaluations fondées soit sur l'approche dosimétrique, soit sur les données actuelles de l'approche épidémiologique laisse penser que les connaissances dont nous disposons permettent déjà de définir de façon raisonnablement satisfaisante le niveau de protection nécessaire.

REFERENCES

1. Billard F., Miribel J., Madelaine G. et Pradel J. (1963): Méthodes de mesure du radon et de dosage dans les mines d'Uranium : In : Radiological Health and Safety in mining and milling of nuclear materials, Vol. I, p. 411 AIEA, Vienne.
2. Fusamura N., Kurosawa R. and Maruyama M. (1967): Determination of f-value in uranium mine air : In : Assessment of airborne radioactivity in nuclear operations, p. 273 AIEA, Vienne.
3. Pradel J., Chapuis A., Lopez A., Cabrol C. et Billard F. (1970): Sur les caractéristiques des aérosols radioactifs présents dans les mines françaises d'uranium, Second IRPA International Congress, Brighton.
4. George A.C. and Hinchliffe L. (1972):Measurements of uncombined radon daughters in uranium mines, Health Phys. $\underline{23}$, 791.
5. George A.C. (1972):Indoor and outdoor measurements on natural radon and radon decay products in New-York City air : In : The natural radiation environment, vol. 2, p. 741, US.ERDA report CONF-720805.
6. Duggan M.J. and Howell D.M. (1969):The measurement of the unattached fraction of airborne Ra A, Health Phys. $\underline{17}$, 423.
7. Jacobi W. (1972):Activity and potential alpha-energy of ^{222}Radon and ^{220}Radon-daughters in different air atmospheres, Health Phys. $\underline{22}$, 441.
8. Lopez A., Chapuis A., Fontan J., Billard F. and Madelaine G. (1970):Mesure de l'état d'équilibre entre le radon et ses descendants dans les mines d'uranium, J. Aerosol Sci. $\underline{1}$, 255.
9. Chamberlain A.C. and Dyson E.D. (1956) The dose to trachea and bronchi from the decay products of Radon and Thoron, Brit. J. Radiol., $\underline{29}$, 317.
10. Bale W.F. and Shapiro J.V. (1955): Radiation dosage to lungs from Radon and its daughter products : In : Proc. U.N. Intern. Conf. on Peaceful uses of Atom Energy, Vol. 13, 233.
11. Altshuler B., Nelson N. and Kuschner M. (1964): Estimation of the lung tissue dose from inhalation of radon and daughters, Health Phys. $\underline{10}$, 1137.
12. Jacobi W. (1964):The dose to the human respiratory tract by inhalation of short-lived ^{222}Rn and ^{220}Rn-decay products, Health Phys. $\underline{10}$, 1163.
13. Haque A.K. and Collinson A.J. (1967):Radiation dose to the respiratory system due to radon and its daughter products, Health Phys. $\underline{13}$, 431.
14. Harley N.H. and Pasternack B.S. (1972):Alpha absorption measurements applied to lung dose from radon daughters, Health Phys. $\underline{23}$, 771.
15. Jacobi W. (1972):Relation between the inhaled potential alpha-energy of ^{222}Rn and ^{220}Rn daughters and the absorbed energy in the bronchial and pulmonary region, Health Phys. $\underline{23}$, 3.
16. ICRP Publication 30, Part I (1979): Limits for intakes of radionuclides by workers, Oxford, Pergamon Press.

17. Adams N., Hunt B.W. and Reissland (1978) : Annuals limits of intakes of radionuclides for workers, NRPB - R-82
18. Chameaud J., Perraud R., Lafuma J., Masse R., Chrétien J. (1976): Résultats biologiques expérimentaux et relation dose-effet du radon avec ses produits de filiation. In : Proc. NEA specialist meeting, Elliot-Lake, p. 49, OCDE Paris.
19. Archer J.E., Gillam J.D. and Wagoner J.K. (1976): Respiratory disease mortality among uranium miners. An. N.Y. Acad. Sci, 271, 280.
20. Lundin F.E., Wagoner J.K. and Archer V.E. (1971) : Radon daughter exposure and respiratory cancer : Quantitative and temporal aspects, NIOSH and NIEHS Joint Monograh. no. 1, NTIS, Springfield.
21. Archer V.E., Wagoner J.K. and Lundin F.E. (1973) : Lung cancer among uranium miners in the United States, Health Phys. 25, 351.
22. Sevc J., Placek V. (1973) : Radiation induced lung cancer : Relation between lung cancer and long term exposure to radon daughters. In : Proc. 6th Conf. on Rad. Hyg., CSSR, p. 305.
23. Sevc J., Kunz E. and Placek V. (1976) : Lung cancer in uranium miners and long term exposure to radon daughters products, Health Phys. 30, 433.
24. Kunz E., Sevc J., Placek V. (1978) : Lung cancer in uranium miners (methodological aspects). Health Phys. 35, 579.
25. Kunz E., Sevc J., Placek V. and Horacek J. (1979) : Lung cancer in man in relation to different tissue distribution of exposure. Health Phys. 39, 699.
26. Chameaud J., Perraud R., Masse R., Nénot J.C., Lafuma J. (1976) : Cancers du poumon provoqués chez le rat par le radon et ses descendants à diverses concentrations. In : Biological and environmental effects of low-level radiation, Vol. 2, p. 223 AIEA, Vienne.
27. Lafuma J. (1978) : Cancers pulmonaires induits par différents émetteurs α inhalés : évaluation de l'influence de divers paramètres et comparaison avec les données obtenues chez les mineurs d'uranium. In : Late biological effects of ionizing radiation, Vol. 2, p. 531 AIEA, Vienne.

RECENT PAST AND NEAR FUTURE ACTIVITIES OF ICRP COMMITTEE 2*

R. C. Thompson

The presence of this paper on the program is, I am sure, related to the publication last year of the long-awaited report of ICRP Committee 2 on "Limits for Intakes of Radionuclides by Workers" (1). I will spend most of my time discussing that report, but I think it also appropriate to say something about the committee itself.

RECENT PAST

It has been my privilege to serve as a member of Committee 2 since 1970, and for some years prior to that as a member of its Task Group on Plutonium and Related Elements. Even before 1970, the Committee was at work on the revision of its earlier report on "Permissible Dose for Internal Radiation", which was published as ICRP Publication 2, in 1960 (2). By 1973, the last year of Karl Morgan's tenure as chairman of the committee, the revision was in a draft form covering the major items which eventually appeared last year in Publication 30. The year 1974 marked the beginning of Jack Vennart's tenure as chairman, and was the year the Main Commission adopted its policy of summing risks by means of weighted organ dose, as a replacement for the "critical organ concept." This decision removed the last technical obstacle to completing the report, and we began to actually believe our annual prediction that the report would be published "next year." In fact, publication had to await the issuance of the Commission's basic "recommendations" as included in Publication 26 (3), and the completion of voluminous machine calculations.

I have mentioned the names of Karl Morgan and Jack Vennart, who served so capably as chairmen of Committee 2. I must also acknowledge the major contribution of our late colleague, Geoffrey Dolphin who served as Secretary of the Committee during compilation of the final drafts of Publication 30. The contributions of Norman Adams and Michael Thorne, who assisted the Secretariat in putting the data together, must also be acknowledged. But above all, Walter Snyder must be remembered for the unstinting and patient ministration of his vast knowledge of internal dosimetry to the Committee and to the Committee's publications. I must name one other person--Mary Rose Ford-- who assumed the calculational chore after the death of Dr. Snyder in 1977, and averted a catastrophe that everyone feared.

I would not like to leave the impression that Committee 2 was concerned, for the 20 years since 1960, only with the preparation of Publication 30. A supplement to Publication 2 was published in 1964 as a part of ICRP Publication 6 (4). Dosimetry models for the gastrointestinal tract (5,6) and for the respiratory tract (7) were published in 1966 by task groups of Committee 2. A joint task group of Committees 1 and 2 prepared, "A Review of the Radiosensitivity of the

---

*Work supported by U.S. Department of Energy Contract EY-76-C-06-1830.

Tissues in Bone," which was published in 1968 as ICRP Publication 11 (8). Publication 19, "The Metabolism of Compounds of Plutonium and Other Actinides," appeared in 1972 (9). Publication 20, "Alkaline Earth Metabolism in Adult Man," followed in 1973 (10). The monumental "Report of the Task Group on Reference Man," ICRP Publication 23 (11), appeared in 1975, another memorial to the chairman of that task group, Walter Snyder. Other unpublished task group reports have been important in the development of Publication 30.

The current, and recent past membership of Committee 2 is shown in Table 1.

Table 1. Membership of ICRP Committee 2

Current	Recent Past*
J. Vennart, Chmn.	G.C. Butler
R.C. Thompson, Vice Chmn.	B. Chr. Christensen
W.J. Bair	G.W. Dolphin
L.E. Feinendegen	M. Dousset
M.R. Ford**	M. Izawa
A. Kaul**	W. Jacobi
C.W. Mays	J. Lafuma
J.C. Nenot**	J. Liniecki
B. Nosslin**	L.D. Marinelli
P.V. Ramzaev	W.G. Marley
C. Richmond**	K.Z. Morgan
N. Veall**	P.E. Morrow
	J. Müller
	V. Shamov
	W.S. Snyder
	C.G. Stewart

*Since work on Publication 30 began in 1967
**Appointed subsequent to final drafting of Publication 30, Part 1.

NEAR FUTURE

Only Part 1 of Publication 30 has thus far appeared in print, and this includes limits and metabolic data for the radioisotopes of only 21 elements (Table 2). Limits for an additional group of 31 elements (Table 3) are in the final stage of compilation and will be soon published as Part 2 of Publication 30. Michael Thorne, for the Committee, has compiled the metabolic data for an additional 18 elements and is working on 25 more. These will be considered by the Committee at its meeting next week, in Brighton, England, and will eventually be included in a Part 3 of Publication 30. Perhaps a Part 4 will be required to complete the elements.

Table 2. Elements included in ICRP Publication 30, Part 1 (with atomic number)

1	Hydrogen	38	Strontium	53	Iodine	90	Thorium
15	Phosphorus	40	Zirconium	55	Cesium	92	Uranium
25	Manganese	41	Niobium	58	Cerium	94	Plutonium
27	Cobalt	42	Molybdenum	84	Polonium	95	Americium
36	Krypton	52	Tellurium	88	Radium	96	Curium
						98	Californium

Table 3. Elements to be included in ICRP Publication 30, Part 2 (with atomic number)

9	Fluorine	26	Iron	45	Rhodium	76	Osmium
11	Sodium	29	Copper	47	Silver	77	Iridium
16	Sulfur	30	Zinc	48	Cadmium	79	Gold
17	Chlorine	35	Bromine	49	Indium	80	Mercury
18	Argon	37	Rubidium	54	Xenon	82	Lead
19	Potassium	39	Yttrium	56	Barium	83	Bismuth
20	Calcium	43	Technetium	61	Promethium	93	Neptunium
24	Chromium	44	Ruthenium	75	Rhenium		

For each "Part" of Publication 30, there will be issued a "Supplement". These supplements will be reproduced directly from computer printouts, and will tabulate the data employed in arriving at the recommended values of ALI and DAC. These data are necessary for the derivation of limits for exposure to mixtures of radionuclides, and to particles varying from the one micrometer diameter assumed for the tabulated ALI's and DAC's. Finally, there will be published a separate report tabulating the radionuclide decay schemes employed in deriving these limits. Much remains to be done before we have finished with Publication 30, but another year or two should see its completion.

While Committee 2 has, in the past, dealt only with problems of internal exposure, since November of 1977 its official title has been "Committee 2 on Secondary Limits," and it is the intention of the Main Commission that Committee 2 should have the responsibility for derivation of secondary limits for external exposure as well as for internal exposure. Committee 3, which formerly dealt with external exposure, is now Committee 3 on Protection in Medicine. However, the Main Commission has acknowledged that, for the immediate future, Committee 2 will be fully concerned with the preparation of secondary limits for internal irradiation.

Aside from the completion of Publication 30, identified future activities of Committee 2 in the area of internal exposure include the following. A task group of Committee 2, with Dr. Nosslin as chairman, is preparing a report on Dose to Patients from Radiopharmaceuticals. This effort is not concerned with establishing limits, but only with estimating patient dose from unit intake; questions of philosophy and medical ethics lie in the domain of Committee 3. Committee 2 will also be concerned with the improvement of internal dosimetry models, in particular those concerned with bone and lung. The bone model of Publication 30 does not take account of the burial of surface deposited radionuclides, and the lung model is applicable only to inhaled particulates. Finally, the Committee will address itself to the establishment of radionuclide exposure limits for members of the public. The exact manner of formulating these limits has not been determined; however, it has been agreed that an exhaustive appraisal of the internal dosimetry of all radionuclides at all ages is impracticable, and that the approach will involve the application of a correcting factor to the occupational limit, this correcting factor probably being different for different radionuclides.

ICRP PUBLICATION 30, PART 1

Let me now return to a more detailed consideration of Part 1 of Publication 30 (1). I will concentrate on specific aspects of the Publication, which seem most important to its understanding and application. A most obvious change in Publication 30 is the absence of "Maximum Permissible Concentrations (MPC) or Maximum Permissible Body Burdens (MPBB), to which we had become accustomed in Publication 2. They are replaced by Annual Limits on Intake (ALI) and Derived Air Concentrations (DAC); the DAC being equivalent to the old MPC for air, but renamed to avoid the connotation that it should never be exceeded. The new limits are expressed in SI units, without even parenthetical microcuries.

Previous internal radiation exposure standards, as formulated in ICRP Publication 2, were based on limiting the dose equivalent received by the critical organ after a period of 50 years of continuous exposure, critical organ being defined as "that organ of the body whose damage by the radiation results in the greatest damage to the body" (2). The exposure standards derived in Publication 30, on the other hand, limit the annual effective dose equivalent commitment, thus differing in two respects from Publication 2: (1) the limit is on *annual commitment* rather than on ultimate realization, and (2) the limit is on *effective* dose equivalent, i.e., the sum of weighted organ or tissue dose equivalents. What are the implications of these changes?

The change from a dose rate achieved after 50 years of exposure to an annual dose commitment has no effect at all in a mathematical sense--one arrives at the same exposure limit by either approach. This is best illustrated graphically, as shown in Figure 1, where A_n, B_n, C_n, and D_n represent the dose in successive years resulting from the exposure in year n. For simplification, this illustration assumes that dose contributed beyond the fourth year is insignificant and that a steady-state total dose is therefore achieved after four years of constant exposure. It should be apparent from the graph that the annual dose commitment, $A_1 + B_1 + C_1 + D_1$, is numerically identical to the total dose in the 50th year, $A_{50} + B_{49} + C_{48} + D_{47}$. Concern has been expressed, however, by those who must apply these limits, that controlling to an expressed annual limit on intake may prove more restrictive than the old practice of controlling to a fraction of a permissible body burden. Thus, for a radionuclide tenaciously retained in the body, like plutonium, an accidental exposure to several Annual Limits on Intake might be considered an "overexposure", and as such might limit the work status of the exposed individual in future years, even though the actual radiation dose, received or projected during any year, is never more than a small fraction of that allowed on an annual basis. It must be emphasized that such

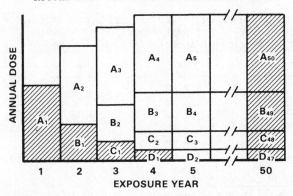

Figure 1. Illustration of dose commitment concept (see text)

an application of the dose commitment concept exceeds the intention of the ICRP, which employs it to calculate ALI's for the control of the environment in which people work, and not to determine the work status of exposed individuals.

The change from the "critical organ concept" to the concept of an effective dose equivalent, has more profound consequences. Instead of applying an assigned dose limit to a single organ or tissue considered critical, the total body limit is applied to the sum of doses received by all significantly exposed organs or tissues, each weighted in accord with its presumed contribution to the total risk of whole body exposure. This is certainly a more logical approach to limit setting and is, in fact, essential to a system based on the limitation of risk. Unfortunately, our knowledge of the biological parameters required to implement such a system is not entirely adequate, and the assumptions required because of that inadequacy result in uncertainties with the new system that are probably as large as the more obvious inaccuracies in the old system. And the calculational complexities introduced are formidable. The new system is, however, a more logically consistent one, and if we lack the information to implement it most effectively, it at least focusses attention on these shortcomings. I would only caution that our numbers are not as good as the refinement of the calculational procedures might suggest.

A vestige of the old "critical organ concept" still remains in Publication 30, in the guise of an overriding "non-stochastic" limit of 0.5 Sv (50 rem) per year, applicable to any organ or tissue except the lens of the eye, where the non-scholastic limit is 0.3 Sv (30 rem) per year. The weighted organ dose system of ICRP-26 would, in certain instances, allow doses to single organs in excess of 0.5 Sv (50 rem) per year, but this is prevented by the non-stochastic limit.

Let me now illustrate some of the previous discussion by considering, as an example, the derivation of the ALI for ingested ^{239}Pu. For this exercise I have gathered in Table 4 information which will appear in the supplement to Publication 30, Part 1. Table 4 lists values of committed dose equivalent per Bq ingested, for the tissues

Table 4. Committed dose equivalent (H_{50T}) and weighted committed dose equivalent ($W_T H_{50T}$) per unit ingestion of soluble ^{239}Pu in units of Sv/Bq, and derivation of annual limits on intake (ALI)

Tissue	H_{50T}	W_T	$W_T H_{50T}$
Ovaries	2.6×10^{-8}	0.25	0.6×10^{-9}
Red Bone Marrow	$17. \times 10^{-8}$	0.12	$20. \times 10^{-9}$
Bone Surfaces	$210. \times 10^{-8}$	0.03	$63. \times 10^{-9}$
Lower Large Intestine	5.4×10^{-8}	0.06	3.2×10^{-9}
Liver	$44. \times 10^{-8}$	0.06	$27. \times 10^{-9}$
			119.8×10^{-9}

Non-stochastic ALI (bone surfaces):

210×10^{-8} Sv/Bq ⇌ 0.5 Sv/<u>240</u> kBq

Stochastic ALI:

$\Sigma_T W_T H_{50T} = 119.8 \times 10^{-9}$ Sv/Bq ⇌ 0.05 Sv/<u>420</u> kBq

of significance. These values result from calculations based on metabolic models and dosimetric models which we will have more to say about later. The largest committed dose equivalent is calculated for the bone surfaces, which would have been considered the critical organ in the old system. We must still calculate a non-stochastic ALI for bone surfaces, which turns out to be 240 kBq; i.e., 240 kBq will deliver the non-stochastic committed dose equivalent limit of 0.5 Sv.

The committed dose equivalent values in Table 4 are multiplied by the weighting factors, W_T, as defined in ICRP-Publication 26, to give the weighted committed dose equivalents, which are summed to yield the effective committed dose equivalent. This sum is compared with the committed dose equivalent limit of 0.05 Sv for stochastic effects, leading to an ALI of 420 kBq. Since the stochastic limit is higher than the non-stochastic limit, the non-stochastic limit of 240 kBq is taken as the ALI for ingested soluble plutonium. For many radionuclides, non-stochastic limits are controlling and we have, in fact, a limit still based on a single critical organ. We have in the process, however, considered all organs and tissues to insure that a summation of organ risks would not have led to a more restrictive value.

I thought it best to dwell at some length upon the preceding basic changes in philosophy, but it leaves us time to consider only briefly a number of other important changes.

Dose equivalent (H) is now defined as the product of absorbed dose (D), a quality factor (Q), and the product of any other modifying factors (N). Publication 2 employed a conceptually similar "RBE dose", which was the product of absorbed dose, relative biological effectiveness (RBE) and a relative damage factor (n), which applied only to non-radium alpha and beta emitters in bone. Though somewhat differently defined, the new quality factor serves the same function as the old RBE, and the old "n" factor might be considered a specific modifying factor which could be part of N. However, we no longer calculate average dose to bone, so the "n" factor is no longer needed. A significant change has been made in the numerical value of the quality factor for alpha particles, which is now 20 rather than 10.

Dose equivalent commitment in a given "target" organ or tissue ($H_{50}T$) is now calculated by taking into account the "crossfire" from radiations originating in all other significant "source" organs or tissues, as well as the radiation originating in the target organ itself. This is a complex process, some of the intermediate stages of which will be detailed in the supplements to Publication 30.

The distribution and retention of radionuclides among and within the various source organs is determined by application of suitable metabolic and dosimetric models. These models, for most elements, have become considerably more complex during the interval between Publication 2 and Publication 30. General dosimetric models for the respiratory system, the gastrointestinal tract, for bone, and for submersion in a radioactive cloud, are described in Publication 30, Part 1. The model for the respiratory system is that developed by the Task Group on Lung Dynamics (7), as modified in ICRP Publication 19 (9). The model for the gastrointestinal tract is based on the model developed by Eve and Dolphin (5,6). The bone model estimates dose to red marrow and to bone surfaces, for cortical and trabecular bone; radionuclides being assumed to deposit either uniformly throughout bone or on bone surfaces. For most elements, detailed information on distribution within bone is not available. In the absence of more specific information it is assumed that: (1) alkaline earth radionuclides with radioactive half-lives greater than 15 days are uniformly distributed throughout the volume of bone, (2) radionuclides with radioactive half-lives of less than 15 days are uniformly distributed on bone surfaces, (3) radionuclides on bone surfaces are equally distributed between trabecular and cortical bone, and (4) radionuclides uniformly distributed throughout the volume of bone are present 20% in trabecular bone and 80% in cortical bone.

Specific metabolic models are employed for each radionuclide, and these are not restricted to any particular mathematical form, although most are systems of first order differential equations with constant coefficients. More than one value for the absorption coefficient from the gastrointestinal tract and/or lung may be employed, to represent different compound forms; the respiratory tract model also provides for three different classes of compounds, based upon their assumed clearance rate from the lung. The metabolic models and briefly summarized supporting data are presented separately, for each element in Publication 30, just preceding the tabulated values of ALI and DAC.

Finally, let us look at an example of the actual limit values, as tabulated for plutonium. In Table 5, values are shown for ^{239}Pu; Publication 30 lists values for 12 isotopes of plutonium--from ^{234}Pu to ^{245}Pu. Oral ALI's are listed for compounds exhibiting two absorption fractions, the value of 10^{-5} applying to oxides and hydroxides and the value of 10^{-4} applying to other commonly occurring compounds. It is indicated that the ALI's are based on a non-stochastic limit for irradiation of bone surfaces and that in the absence of such a limit the stochastic limit would have been the value shown in parentheses. Inhalation ALI's are listed for two compound classes, Class Y being applicable to plutonium oxide

Table 5. ALI (Bq) and DAC (Bq/m^3) (40 h/wk) values for ^{239}Pu as listed in ICRP Publication 30

	Oral		Inhalation	
			Class W	Class Y
	$f_1 = 1 \times 10^{-4}$	$f_1 = 1 \times 10^{-5}$	$f_1 = 1 \times 10^{-4}$	$f_1 = 1 \times 10^{-5}$
ALI	2×10^5	2×10^6	2×10^2	5×10^2
	(4×10^5)	(3×10^6)	(4×10^2)	(6×10^2)
	Bone surf.	Bone surf.	Bone surf.	Bone surf.
	$[15 \times 10^5]$*	$[9 \times 10^6]$*	$[2 \times 10^2]$*	$[27 \times 10^2]$*
DAC	---	---	8×10^{-2}	2×10^{-1}
			$[7 \times 10^{-2}]$*	$[15 \times 10^{-1}]$*

*Values derived from ICRP Publication 2.

and Class W to other commonly occuring compounds; the fraction absorbed from the gastrointestinal tract after clearance from the lung is different for the two classes. Again the ALI's are based on a non-stochastic limit. Values of the DAC are shown for the two compound classes.

As a matter of interest, I have listed in brackets, in Table 5, the values for ALI and DAC which one would obtain from the old limits of Publication 2. The DAC values are derived from the old (MPC)$_a$ values by a simple change of units. The ALI's are derived from the MPC values by multiplying by the assumed annual intake of water or air, and by changing units. It will be seen that, except for inhaled Class W compounds, where there is no significant change, all ^{239}Pu limits have become more restrictive by a factor of about six. Similar comparisons, involving a representative isotope of each element, suggest that about 1/3 of Publication 2 limits have become more restrictive, by as much as a factor of 100; about 1/3 have become less restrictive, by as much as a factor of 25; the remaining 1/3 have changed by less than a factor of 2.

REFERENCES

1. ICRP Publication 30, Part 1 (1979): Limits for Intake of Radionuclides by Workers. Pergamon Press, Oxford.
2. ICRP Publication 2 (1960): Report of Committee II on Permissible Dose for Internal Radiation. Pergamon Press, Oxford.
3. ICRP Publication 26 (1977): Recommendations of the International Commission on Radiological Protection. Pergamon Press, Oxford.
4. ICRP Publication 6 (1964): Recommendations of the International Commission on Radiological Protection. Pergamon Press, Oxford.
5. Eve, I.S. (1966): Health Phys., 12, 131.
6. Dolphin, G.W. and Eve, I.S. (1966): Health Phys.,12, 163.
7. Task Group on Lung Dynamics (1966): Health Phys.,12, 173.
8. ICRP Publication 11 (1968): A Review of the Radiosensitivity of the Tissues in Bone. Pergamon Press, Oxford.
9. ICRP Publication 19 (1972): The Metabolism of Compounds of Plutonium and Other Actinides. Pergamon Press, Oxford.
10. ICRP Publication 20 (1973): Alkaline Earth Metabolism in Adult Man. Pergamon Press, Oxford.
11. ICRP Publication 23 (1975): Report of the Task Group on Reference Man. Pergamon Press, Oxford.

LES EFFETS CANCERIGENES COMBINES DES RADIATIONS IONISANTES ET DES MOLECULES CHIMIQUES

J. Lafuma

Ce problème est difficile à aborder car les données humaines sont rares et les expériences qui ont été pratiquées sur les animaux l'ont été dans des buts différents et en utilisant des doses élevées et des produits chimiques très divers. Il est impossible de trouver dans la littérature des données permettant de se faire une idée même approximative de ce que peuvent être les risques combinés aux faibles niveaux de dose.

Peu de données humaines sont disponibles en ce qui concerne l'action combinée de deux agressions sur le taux de cancers. Dans le domaine des radiations ionisantes, la combinaison avec le tabac a été étudiée dans différents groupes de mineurs d'Uranium.
Les données U.S. montrent qu'il existe une synergie d'action qui se traduit par le raccourcissement du temps de latence chez les individus qui fument (1). Ce raccourcissement du temps de latence est accompagné d'une augmentation de la fréquence des cancers.
Dans un autre domaine, le même résultat a été trouvé chez les travailleurs de l'amiante pour lesquels il existe une indiscutable synergie entre les fibres minérales et la fumée de cigarettes (2).
Bien que rares, les données humaines montrent le double aspect de la synergie: action sur la fréquence des cancers et accélération du processus cancéreux. C'est sous ce double aspect qu'il faudra analyser les données expérimentales.

Les données expérimentales proviennent de sources très variées et donnent l'impression d'une très grande dispersion. En général, les expérimentateurs ont plus essayé de comprendre des mécanismes que d'étudier la synergie sous l'angle de la relation dose-effet. Ceci se comprend car, pour de telles études, on est amené, si l'on veut obtenir des résultats rigoureux à utiliser un nombre d'animaux infiniment plus grand que celui nécessaire pour l'étude d'un seul agent.

En effet, en plus de la combinaison des doses, il faut étudier l'influence de la chronologie car certaines synergies ne s'observent que si l'un des facteurs est donné après l'autre (effet de promoteur).

La dispersion des données expérimentales ne permet pas de faire une synthèse. Leur analyse ne peut se faire que par fractions séparées. Le classement peut se pratiquer de différentes façons, soit en fonction de l'objectif de l'expérience ou bien en fonction du type de cancer. C'est cette deuxième solution que nous avons choisie: étant entendu que toutes les expériences ne sont pas prises en compte.

LEUCEMIES et LYMPHOMES:

La combinaison la plus utilisée est celle des rayons X délivrés in toto et de l'urethane, chacune des deux agressions produisant des lymphomes.

Dans une expérience avec des souris, les doses de rayons X allaient jusqu'à 300 rad et étaient données avant les traitements par l'urethane. Le rôle de l'âge et des différentes modalités de combinaisons ont été testés.

L'incidence de lymphomes, faible chez les témoins et chez les animaux traités soit par les rayons X seuls, soit par l'urethane seule, n'a été que peu augmentée (3).

Dans une expérience, la sensibilité de trois souches de rat a été testée en combinant Urethane et Rayons X administrés simultanément 5 fois. La dose totale était de 835 rad. L'effet de la combinaison des agressions différaient suivant les souches allant d'une certaine potentiation à un antagonisme (4).

Une troisième expérience sur des souris, où l'effet de l'âge était étudié a montré que l'administration d'urethane chez le jeune suivie d'une irradiation chez l'adulte aboutissait à une synergie alors qu'en inversant l'ordre des agressions on n'obtenait aucun résultat (5).

Enfin, une autre méthode a montré que seule l'irradiation, précédant l'urethane permettrait d'observer un effet synergique (6).

Les Effets Cancerigenes

Dans toutes ces expériences, ni les doses d'irradiation, ni la façon dont elles sont délivrées, ni les modalités de l'administration de l'urethane ne sont comparables. La seule conclusion que l'on puisse en tirer est que dans certaines conditions expérimentales, il existe une synergie entre l'urethane et les rayons X administrés à des doses toujours supérieures à 100 rad et délivrées en quelques minutes.

FOIE.

Plusieurs expériences ont été pratiquées pour étudier la synergie des molécules chimiques et des radiations au niveau du foie, organe particulièrement radio-existant.

Dans une première expérience, l'action du N - N' - 2,7 Fluoroenylenebisacetamide a été combinée à celle des rayons X. Les doses locales étaient de 10000 R et la molécule chimique était donnée per os, chaque jour. Dans un groupe, le début du traitement per os précédait l'irradiation, et le suivait dans un autre groupe. C'est dans ce seul groupe qu'un effet synergique a été observé. (7).

Une autre expérience a combiné l'irradiation neutronique et le Tetrachlorure de Carbone et a comparé l'effet des neutrons et celui des Rayons X.

Les doses de neutrons étaient comprises entre 150 et 300 rad, celles de rayons X de 500 rad (une seule séance) et les animaux recevaient ensuite une seule injection de CCl_4.

La potentiation Neutron - CCl_4 est importante et surtout on note une accélération du processus cancéreux (8).

Enfin, dans une expérience, on a combiné DAB et Cerium-144. Les doses de Ce-144 (20μCi ou plus) sont très élevées. Des animaux mâles et femelles ont été utilisés et divers protocoles réalisés en faisant varier l'ordre d'administration et les doses. Les résultats sont complexes et aucune synthèse ne peut être tirée (9).

THYROIDE.

Trois expériences combinent l'Iode-131 et le Methylthiouracyl. Les doses d'Iode-131 sont extrêmement élevées (de 10 à 100 microcuries) et dans les trois expériences on observe une synergie qui se traduit par un raccourcissement du temps de latence et une augmentation de la fréquence (10, 11, 12). Dans l'une d'entre elles on voit qu'à la dose la plus élevée (100 microcuries), la fréquence des cancers chute brutalement.

POUMONS.

Différents expérimentateurs combinent aujourd'hui les inhalations d'aérosols radioactifs avec les molécules chimiques. Des expériences sont en cours pour étudier un éventuel effet de synergie dans l'atmosphère des mines d'Uranium.

Les premiers résultats montrent que la fumée de cigarettes agit en synergie avec le Radon. Cette synergie se traduit par un raccourcissement du temps de latence, une augmentation de la fréquence et un accroissement de la malignité des tumeurs (13).

Dans une autre expérience, l'Oxyde de Beryllium et le Plutonium ont été combinés. Plusieurs doses de Beryllium et le Plutonium ont été utilisés. Le Beryllium modifie l'épuration des particules de Plutonium mais n'a pas d'action sur l'incidence des tumeurs (14).

L'Oxyde de Plutonium et le Benzopyrène ont été également combinés. L'effet synergique est important. La fréquence des tumeurs est augmentée et surtout la taille de celles-ci est impressionnante (plus de 10 grammes). Le temps de latence est considérablement raccourci (15).

Si l'on combine Plutonium inhalé et Dimethylnitrosamine per os, on n'observe pas de synergie. Ceci n'est pas étonnant car la dimethylnitrosamine a une action surtout hépatique (15). L'effet combiné n'existe que si les deux facteurs agissent sur le même organe.

Dans une expérience, on a combiné le Radon inhalé et l'administration intrapleurale de chrysotile. Le Radon seul ne donne jamais de cancer de la plèvre. Administré avant le chrysotile, il n'en a pas moins accéléré le développement des cancers pleuraux induits par ce dernier (16).

Enfin, dans une autre expérience, on a combiné le Radon et une molécule non cancérigène, la 5-6 Benzoflavone, administrée par voie systémique. L'effet synergique est considérable. Le temps de latence passe de 15 à 3 mois et les tumeurs épidermoïdes sont très développées (17).

AUTRES ORGANES.

D'autres expériences ont étudié des combinaisons diverses portant sur de nombreux organes et tissus. Parmi ceux-ci, on note la

Les Effets Cancerigenes

peau, les glandes mammaires, la vessie, le peritoine et la machoire.

Divers effets ont été observés dépendant des protocoles choisis.

Si l'on veut résumer les expériences animales, on voit que les protocoles choisis sont très variés, les doses sont en général très élevées et que pratiquement tous les résultats ont été acquis sur des rongeurs.

L'antagonisme n'a que rarement été observé, l'additivité est souvent mentionnée, mais les nombres absolus sont souvent trop faibles pour que la conclusion soit indiscutable.

La potentiation existe, elle a été mise en évidence. Ce qui s'observe le plus souvent c'est une accélération du processus tumoral conduisant à un raccourcissement du temps de latence avec augmentation considérable de la taille et de la malignité des tumeurs.

Il ne semble pas exister aujourd'hui de méthodologie précise pour l'étude des effets combinés. Celle-ci ne peut être basée sur les données actuelles et devrait s'appuyer sur ce que l'on sait des mécanismes de la cancérogénèse.

Si l'on schématise le mécanisme d'action des cancérigènes, on considère qu'il existe deux étapes. L'une, l'initiation, irréversible, est assimilée à la transformation maligne cellulaire. Pour les cancérogènes chimiques et l'irradiation on admet qu'elle est due à une action directe sur la cellule.

L'autre, la promotion, est l'étape qui s'écoule entre l'irradiation cellulaire et l'apparition de la tumeur (temps de latence).

En principe, l'initiation s'effectue en un temps très court, indépendant de la dose. Celle-ci n'intervient que sur le nombre de cellules transformées. Par contre, la promotion a une durée variable, d'autant plus courte que la dose est plus faible. Pour un cancérogène, ce processus promotionnel est irréversible.

BERENBLUM a montré qu'il existait des molécules chimiques n'ayant qu'une activité de promotion. Ces molécules n'agissent que si elles sont administrées après le cancérogène. Leur action est réversible et pour agir elles doivent être administrées de façon répétée et étalée dans le temps. On ne connaît pas d'initiateur pur.

Cette théorie est schématique, mais peut servir de base pour des études d'effets combinés.

Si l'on veut étudier l'action combinée d'un cancérogène chimique et des radiations ionisantes possédant tous deux la double fonction d'initiation et de promotion, il est nécessaire pour interpréter correctement les résultats d'inverser l'ordre des administrations. De plus, il est préférable de sélectionner une modalité de cancérogénèse chimique qui aboutisse à l'apparition de cancers spécifiques soit par leur localisation soit par leur nature histologique. Par exemple, on peut étudier le fibrosarcome local induit par une injection intramusculaire de benzopyrène et par une irradiation totale qui n'a qu'une faible probabilité d'induire un tel cancer à cet endroit précis. On peut aussi combiner l'injection pleurale de fibres minérales et l'inhalation de Radon ou l'irradiation pulmonaire.

En faisant varier l'ordre des facteurs et les doses de ceux-ci on peut espérer étudier comment la combinaison des effets promotionnels évolue en fonction de deux doses.

Si l'on veut étudier l'action des radiations combinées avec celle d'un promoteur, il faut tout d'abord vérifier que l'inversion de l'ordre des facteurs modifie totalement la réponse. Ensuite, on peut étudier la relation entre les combinaisons de doses et les effets.

Ces deux approches qui cherchent à ne faire que comme s'il n'y avait qu'un seul cancérigène permettent une analyse plus simple des données. En effet, expérimenter avec deux éléments suscetibles à eux seuls d'induire le même cancer exige de travailler sur de grands nombres d'animaux pour vérifier si le résultat est ou non une simple additivité.

Nous avons actuellement en cours des expériences des deux types. L'accélération des processus est évidente dans tous les cas. L'augmentation de la fréquence est toujours une potentiation car l'un des deux facteurs ne produit pas à lui seul le cancer étudié. Cancérogènes vrais et promoteurs se dossocient très bien.

Quantitativement, l'accélération du processus dépend d'une combinaison des deux doses, c'est à dire que l'importance du phénomène décroît très rapidement quand les doses diminuent et ceci est important car il est possible qu'en dessous de certaines doses ou de certains débits de doses, la synergie n'existe plus ou ne soit plus observable.

En conclusion, nos connaissances sur les effets combinés sont aujourd'hui très fragmentaires. Aux fortes doses, l'irradiation par son absence de spécificité cellulaire et tissulaire peut avoir une action synergique avec de très nombreuses molécules chimiques. Aux faibles doses et lorsque la synergie porte sur l'effet promotionnel, l'existence d'un seuil est une hypothèse raisonnable car il est difficile de concevoir que cette action complexe puisse provenir de la simple transformation de deux cellules.

REFERENCES.

1. ARCHER V.E., GILLIAM J.D., WAGONER J.K.
   Respiratory disease mortality among uranium miners. Ann. N.Y. Acad. Sci. 271. 280-293 (1976)

2. FRANKS Arthur L., - Public Health Significance of smoking - Asbestos Interactions - Ann N.Y. Acad. Sci. 330 791-794 (1979)

3. GOLDFEDER A.
   Urethan and X-ray effects on Mice of a Tumor-resistant strain, X/Gf^1
   Cancer Research 32, 2771-2777, December 1972

4. MYERS D.K. - Effects of X-Radiation and Urethane on Survival and Tumor Induction in Three strains of Rats, Radiat. Res. 65, 292-303 1976

5. VESSELINOVITCH S.D., SIMMONS E.L., MIHAILOVICH N., LOMBARD L.S. and RAO K.V.N. - Additive Leukemogenicity of Urethan and X-irradiation in Infant and Young Adult Mice
   Cancer Research 32, 222-225, February 1972

6. BERENBLUM I., CHEN Louise, TRAINUM N. - A quantitative study of the Leukomogenic action of whole body X-irradiation and urethane. Israël - J. Med. Sci. Vol. 4, p 1159-1163, 1968

7. NAGAYAO T., ITO A., YAMADA S. - Accelerated induction of hepatoma in rats fed N, N'-2,7 Fluorenylenebisacetanide by X irradiation to the target area - Gann 61: 81-84 (1970)

8. COLE L.J., NOWELL P.C. - Accelerated induction of hepatomas in fast neutron-irradiated mice infected with carbon tetrachloride. Ann. New-York Acad. Sci. 114: 259-267 (1964)

9. MALHLUM D.D.. - Hepatic tumor development in rats exposed to ^{144}Ce and dimethylaminoazobenzene p. 159-167 in Radionuclide Carcinogenesis (C.L. SANDERS, R.H. BUSCH, J.E. BALLOU et al. eds.) U.S. Atmic Energy Commission, Office of Information Services (1973).

10. DONIACH I. - The effect of radioiodine alone and in combinaison with methylthiouracil and acetylaminoflourene upon tumor production in the rat's thyroid gland. Brit. J. Cancer 4: 223-234 (1950)

11. DONIACH I. - The effect of radioiodine alone and in combinaison with methylthiouracyl upon tumor production in the rat's thyroid gland. Brit. J. Cancer 7: 181-202 (1953)

12. CHRISTOV K., and RAICHEV R. - Thyroid carcinogenesis in hamsters after treatment with 131-iodine and methylthiouracil. Z. Krebsforschung 77: 171-179 (1972)

13. CHAMEAUD J., PERRAUD R.,CHRETIEN J., MASSE R., LAFUMA J. Experimental study of the combined effects of inhalation of radon daughter products and tobacco smoke.
19 th annual "PULMONARY TOXICOLOGY OF RESPIRABLE PARTICLES "
Hanford Life Science Symposium _ Octobre 22/24 1979

14. SANDERS C.L., CANNON W.C. and POUERS G.J. - Lung carcinogenesis induced by inhaled high fired oxydes of Beryllium and Plutonium Health Physics, Vol. 35 - p. 193-199

15. METIVIER H.. MASSE R.. L'HULLIER I., LAFUMA J.- Etude de l'action combinée de l'Oxyde de PLutonium inhalé et de deux cancérogènes chimiques de l'environnement. - Colloque International sur les EFFETS BIOLOGIQUES DES RADIONUCLEIDES REJETES PAR LES INDUSTRIES NUCLEAIRES - Vienne - 26/30 Mars 1979

16. MORIN M., QUEVAL P., LAFUMA J. - Etude expérimentale de l'action co-carcinogénique du Radon-22 et de la benzoflavone -
IAEA-SM-224/904 - Ds: AIEA Late Biological Effects of ionizing radiation - Vienne AIEA 1978 -
Vienna 13 /17 Mars 1978 - Vol. 2 - p. 423/427

17. LAFUMA J., MASSE R. - Mesothelia induced by intrapleural injection of different types of fibres in the rat. Synergestic effects of other carcinogens. C.I.R.C. - Symposium on the biological effects of mineral fibres - LYON - 25/27 Septembre 1979.

Regulatory Aspects

FLEXIBILITY IN RADIATION PROTECTION LEGISLATION - THE UK APPROACH

P. F. Beaver and J. R. Gill

It is usually assumed that flexibility in radiation protection legislation is a desirable objective and that there is no need to argue the case. However, the opposite view, on a number of occasions, has been expressed. For example, managements have suggested that if legislation is clear and unequivocal with no room for opinion then they can get on with their task of management secure in the thought that they cannot be criticised by workers, trade unions or enforcing authorities. In the same way, trade union officials have expressed the view that it is easier for them if the legislation is precisely framed and hence departures from its standards are obvious. There is certainly a view from some parts of the enforcing authority that legislation should be specific and absolute in its requirements. This will, it is claimed, lead to the most effective enforcement using minimal numbers of specialist inspectors.

Flexibility is taken to mean that the objectives of the legislation are clearly spelt out but the means whereby those objectives are to be achieved are left open, or qualified by terms such as "where reasonably practicable" or a variety of options are offered.

The pressure for flexibility comes, in the first instance, from the knowledgeable to whom a variety of options is appealing in its mind stretching potential, to whom a cost benefit analysis is a common technique and to whom the fact that a means to achieve the objective might not be specified is seen as a challenge to their ingenuity. The pressure from this group is echoed, for quite different reasons, by other groups. The enforcing authorities, desirous of simple legislation requiring minimal interpretation, are attracted by the concept of simple regulations specifying objectives and in a field as wide as radiation protection this can lead to a very effective coverage. We see trade unions and managements as less convinced but gradually appreciating the merits of a flexible system and becoming increasingly confident in their ability to operate within it and that flexibility can lead to improved safety standards at marginally extra cost.

PARTICIPATION AND CONSULTATION

In 1974 a substantial new piece of safety legislation was enacted in the UK; it was called the Health and Safety at Work Act (1) and apart from creating a new enforcing authority, the Health and Safety Executive (HSE), it laid the foundations for a new framework of specific safety legislation. The Act allows that <u>Regulations</u> may be approved by the UK Parliament and also allows the preparation of <u>Codes of Practice</u> which, having been approved by the Health and Safety Commission (HSC), becomes official guidance on the means which should be followed in order to meet the objectives specified in the Regulations. Regulations and approved Codes of

Practice may be supported by <u>Guidance Notes</u> which have no legal status but nevertheless give detailed advice on the interpretation and the objectives of codes and regulations.

In order for a Code of Practice to enjoy a special legal status it must first be approved by the HSC. The HSC is a representative body, representing workers' organisations, management organisations and local government authorities. In addition to its role of approving codes it also directs the general policy of the HSE. The Health and Safety at Work Act requires that persons who might be affected by proposed legislation are consulted. New radiological protection legislation is necessary because of the fragmented nature of the existing legislation, the need to conform with the Euratom Directive (3) binding on the UK as a member state of the European Community and because of the new recommendations contained in ICRP 26. Since there is a substantial body of informed opinion we sought to involve that opinion in discussion and debate even before the stage of formal consultation. In the participation phase, some 17 working parties were organised, each dealing with a single aspect of the proposed regulatory package. In this way some 200 experts in varied fields of radiological protection were brought together and participated with the HSE in establishing the general content of the legislative material which was subsequently published as a Consultative Document (2). The extensive participation of experts in the formulation of the Document did not totally inhibit all comment on its contents. However, the comment was quite limited, confined to a few areas where those commenting either had proposals for change or where they found the proposals to lack clarity.

REQUIREMENTS FOR NOTIFICATION OF USE

Like many of the other requirements of the proposed regulations those for notification to the enforcing authority of the "production, processing, handling, use, storage, transport and disposal of natural and artificial radioactive substances and of any other activity which involves a hazard arising from ionising radiations" stems from provisions in the Euratom Directive.

Many exemptions are possible. In particular there may be a small quantities exemption based upon the classification of radionuclides into toxicity groups. These groupings reflecting early IAEA work,(4) needed updating to take account of new data on Annual Limits of Intake, and of the use of SI Units. Other exemptions for particular types of apparatus will undoubtedly be necessary and a procedure for type testing such pieces of equipment will be devised for use in the UK.

It is very tempting to set up a notification of use system in which substantial amounts of information are called for and in which constant updating is mandatory. It needed no consultation for the enforcing authority to realise that in a country such as the UK where there was a widespread use of ionising radiations and substantial turnover of users, that the administrative burden of such a system would be considerable. Analysis of the reasons why an enforcing authority needs to know of use and discussions of what action might be taken by enforcing authorities when they do know about use has led us to the conclusion that a general notification containing only a superficial description of the work to be undertaken is all that can properly be expected. It has become clear that the inspections of the enforcing authority both in terms of depth and frequency should be based on conditions as found and that notification merely acts as a trigger to cause an inspector to visit the premises. Thus, notification can take a very simple form and can minimise the administrative burden on occupiers of premises.

CRITERIA FOR CONTROLLED AREAS

Controlled areas have always played an important role in UK legislation. The reason is the route by which a Category A Radiation worker is defined is to consider his access to such areas. The objective can be simply stated, that is to define an area which encompasses all those workers liable to exceed 3/10ths of the Annual Dose limit. It was the means whereby to reach that objective that caused debate and discussion. It was felt by many that a simple boundary condition such as 7.5 microsieverts per hour, as read on a dose rate meter, was highly enforceable, easily determined and demonstrably in accord with the objective. However, such a boundary condition can give rise to larger controlled areas than are necessary to meet the objective, particularly obvious if the radiation emission is intermittent. Introduction of a flexible approach clearly requires some expert consideration of the possible doses that might be received by persons working in the controlled areas. This is, of necessity, a time consuming exercise requiring a professionalism which not every employer is prepared to pay for. The solution for the proposed UK legislation is that a Code of Practice will give alternative guidance as to the ways by which the objective of defining the controlled area might be achieved. The first procedure will be based on demarcating areas where the dose rates exceed 7.5 microsieverts per hour averaged over any one minute. The alternative procedure involves the Radiation Protection Adviser (see below) considering the work situation and using his professional judgement to determine the boundaries of the area encompassing all workers likely to exceed 15 millisieverts (mSv) per year. Even then to afford some safeguards it will be necessary to give him some guidelines; the most important of which is that in determining areas he should not take into account occupancy times less than 40 hours per week and he should assume an optimum (but not maximum conceivable) operation of the facility. This accords with ICRP 26 paras 163-6. If all entrants to the area were then to be treated as radiation workers this would lead to a large number of such workers being designated who would, in fact, not exceed 15 mSv per year. Again, flexibility is to be introduced by the Radiation Protection Adviser exercising his professional judgement. It is proposed that he prepare house rules or schemes of work in which he clearly identifies individuals or groups of workers who, albeit that they may need to enter specified controlled areas from time to time, will not, because of the nature of their work and their occupancy time, receive more than 15 mSv per year.

THE RADIATION PROTECTION ADVISER (RPA)

The requirements for the appointment and selection of an RPA in the proposed legislation are that the employer should make an appointment of a suitable person if his operations are such that workers enter a controlled area - in this way users of essentially harmless sources are exempt. The requirement would go on to state that such appointments must be notified to the enforcing authority, who in turn must "recognise the capacity to act" of the individual (or corporate body) appointed as an adviser. It is not the intention of the enforcing authority to enter into an elaborate scrutiny and certification scheme for advisers for the reason that it is seen as proper that in the first instance employers must consider and exercise their own judgements as to who they should employ. To help them in making such decisions a number of professional bodies in the radiation protection field are setting up certification schemes whereby the professional competence of potential advisers is examined and certified by a committee of the professional society concerned. In this way it seems unlikely that the enforcing authority will ever feel it necessary to reject an application for recognition. The conceptual basis for the use of a RPA as an expert advising users comes from the Euratom Directive. However, the proposed UK Regulations exploit the concept in the interests of flexibility and self inspection. Thus, the RPA will have tasks as varied as definition of areas, advice on selection of workers as radiation workers, checking and calibration of instruments, advice on training and instruction, etc. Particularly he is envisaged as having substantial responsibilities to advise the employer on the effectiveness of the application of the principle of ALARA (as low as reasonably achievable). The application of the principle is seen as a continuous exercise before the work activity begins at any prospective level of individual dose or collective dose and after the event as a retrospective examination of whether or not the procedures did, in fact, result in doses being optimised. Serious consideration is being given at the time of writing to setting specific levels of individual exposures at which the RPA carries out formalised retrospective consideration of the effectiveness of the ALARA principle.

In the UK, workers' involvement in safety matters has found expression by the statutory appointment of workers' safety representatives and the creation of joint Worker/Management safety committees. It is seen as a vital part of the operations of the RPA that his advice to the employer is also transmitted to the safety representative and the safety committee so that they too may express a view on such advice and identify with it.

REFERENCES

1. Health and Safety at Work etc Act 1974.
2. Ionising Radiations: Provisions for Radiological Protection HMSO 1978.
3. European Communities Council Directive, June 1976, OJ Vol 19, No. L187.
4. A Basic Toxicity Classification of Radionuclides, IAEA 1963.

A REVIEW FROM THE REGULATORY POSITION OF THE CONTROL OF OCCUPATIONAL EXPOSURE ASSOCIATED WITH THE FIRST 20 YEARS OF THE UNITED KINGDOM COMMERCIAL NUCLEAR POWER PROGRAMME

B. W. Emmerson

LICENSED NUCLEAR INSTALLATIONS

The Nuclear Installations Inspectorate (NII) was established in 1959 to implement and administer the licensing and inspection of all nuclear installations in the UK, except those operated by the Crown or the United Kingdom Atomic Energy Authority (UKAEA). The UKAEA was already supplying power from the first of the gas cooled, Magnox type power stations at Calder Hall (1956) and Chapelcross (1958) and was also responsible for the manufacture and processing of nuclear fuel. In 1971, responsibility for the operation of these power stations and for the fuel manufacturing, enrichment and reprocessing plants located at Springfields, Capenhurst and Windscale, respectively, was transferred from the UKAEA to a newly formed Company, British Nuclear Fuels Ltd (BNFL) and consequently these plants became subject to licensing.

When the NII commenced work in 1960, the Central Electricity Generating Board (CEGB) and the South of Scotland Electricity Board (SSEB), together, had four gas cooled, Magnox type nuclear power stations under construction with five more in various stages of planning, all of which became subject to licensing. The first of these commenced operations in 1962, and currently the Boards have nine twin reactor Magnox type stations and (since 1976) two twin reactor Advanced Gas Cooled (AGR) type stations in operation. Three other AGR stations are approaching completion.

CONTROL OBJECTIVES

Licensed nuclear installations are regulated through a system of conditions attached to the site licence, which cover all aspects of nuclear safety, including radiological protection. The safety principles against which these installations are assessed have been outlined by Gronow and Lewis (ref 1), a fundamental criterion being that an installation should not cause any person to exceed the exposure limits recommended by the International Commission on Radiological Protection (ICRP); currently those incorporated into the Euratom Directive on Radiological Protection (ref 2). The licensee is required to make arrangements covering safety policy and practice and these are assessed, and in some cases formally approved, by the NII. Control extends over the design, construction, operation, maintenance and eventual decommissioning stages of any licensed nuclear installation, and the introduction of any new plant or process on an existing licensed site.

B. W. Emmerson

PRINCIPAL SOURCES OF OCCUPATIONAL EXPOSURE - Power Generation

Annual radiological exposure data associated with the operation of nuclear power stations in the UK are summarised in fig 1. The two main occupational dose components are chronic exposure from plant background, and chronic and acute task oriented exposures. The former depends on the particular plant design and operating power. Earlier steel pressure vessel designs with heat exchangers outside the main biological shield give backgrounds, which although reduced on later designs, are difficult to eliminate. In recent stations, including the AGR's, this background component is virtually eliminated by enclosing the reactor and boilers in a single pre-stressed concrete pressure vessel. The effect of this design development can be seen in fig 2 which shows the annual whole body dose equivalent distribution, expressed as a percentage, and averaged over the years 1971-78, for typical stations at each development stage.

The main task oriented exposures are usually associated with workers carrying out routine operations and maintenance of "on-load" refuelling plant and irradiated fuel storage facilities. A number of chronic and acute radiation and contamination exposure control problems have arisen which were not foreseen at the design stage (ref 3). In particular, the longer than anticipated pond storage time of irradiated Magnox fuel prior to despatch to the reprocessing plant continues to pose difficulties in controlling exposure for this group of workers. To-date, however, these operations have not resulted in any significant pattern of internal exposure. For those tasks not directly associated with the fuel handling route, including those from the internal inspection of gas ducts and boilers during off-load maintenance, the dose contribution remains relatively small.

In general, the radiological control procedures applied at nuclear power stations have proved effective in limiting individual occupational exposures. Annual (50mSv) and quarterly (30mSv) whole body dose equivalent limits are seldom exceeded, the majority of workers not exceeding an annual dose of 15mSv (5mSv at pre-stressed concrete pressure vessel stations). Fig 1 shows that although the annual collective whole body dose equivalent has remained sensibly constant since 1971 there has been a steady reduction in the annual average whole body dose equivalent, due primarily to a significant increase in the occupationally exposed workforce (5007 to 9132). A better indication of the efficacy of dose management is the annual collective whole body dose per unit of electrical energy supplied (Sv/Gwh). Fig 1 indicates a steady reduction in this value for CEGB and SSEB stations; much of this reduction is probably attributable to the **increase in the unit size of new stations.**

PRINCIPAL SOURCES OF OCCUPATIONAL EXPOSURE - Fuel Processing

Fuel processing is carried out by BNFL at its factories at Springfields (fuel element production), Capenhurst (uranium enrichment) and Windscale (irradiated fuel reprocessing). These processes and their associated radiological problems have been discussed in detail by Clarke et al (ref 4). When the Company was licensed in 1971 much of its plant had already operated for over a decade. Although complying with the recommendations of ICRP-9, it was not reasonably practicable to limit the annual whole body dose to 50mSv, particularly at the irradiated fuel reprocessing plant, and control was based on quarterly rather than annual limits, provided that the 5(N-18) lifetime dose limit was not exceeded. Progressive improvements of existing plants and in administrative procedures for controlling occupational exposures were made by the Company so that in 1977, the radiological protection conditions attached to BNFL site licences were made consistent with those for other licensed sites, and in particular, by limiting the annual whole body dose to 50mSv.

Annual average and collective whole body dose equivalents for the three processing sites are shown in fig 1. This indicates a significant decrease in the average dose, particularly at the irradiated fuel reprocessing plant. The annual collective dose is sensibly constant, with the exception of the reprocessing plant where the occupationally exposed workforce has doubled since 1972 to 6000, partly to deal with increased plant throughput but, also, to reduce the number of persons exceeding 50mSv per annum (fig 3). Some of the resultant increase in annual collective dose between 1971 and 1976 was incurred in the improvement of the plant radiological environment, which in turn should help achieve the longer term objective of reducing the collective dose per unit of plant throughput.

The main potential for internal exposure is at the fuel reprocessing plant, either from chronic or acute intakes of fission products or transuranics (mainly plutonium). Whole body monitoring is used to measure γ emitting fission products and lung retained plutonium, systemic plutonium being assessed by urinalysis. To avoid possible over-exposure of any person who has received $>50\%$ of the maximum permissible body burden for any of the higher toxicity transuranic radionuclides, it is the Licensee's policy to control any subsequent exposure to within the non-occupational dose limits.

Although particular groups of occupationally exposed workers can be identified as receiving above average external exposures, in contrast to power station operation, the annual whole body dose equivalent received by maintenance and health physics staff at the reprocessing plant tends to be significantly less than that received by the plant operators.

CONCLUSIONS

This review indicates an improving trend in the radiological exposure pattern, despite increasing power generation, fuel burn-up and processing plant throughput. The initial objective of ensuring that individuals do not exceed the statutory dose limits has largely been achieved. There is still, however, a need to ensure that the collective dose for particular working groups within the fuel cycle represents the practical minimum. To achieve this, optimisation procedures, taking into account all the associated detriments and benefits, may have to be employed. A bank of task oriented exposure data will be an essential prerequisite. The NII, therefore, may need to place greater emphasis on more detailed assessment and recording of such task oriented exposure data.

REFERENCES

1. Gronow, W.S. and Lewis, G. (1978) : Radiation Protection in Nuclear Power Plants and the Fuel Cycle, BNES, London.
2. Euratom Directive (76/579 Euratom) 1976.
3. Emmerson, B.W., Goldfinch, E.P. and Skelcher, B.W. (1971) : Health Physics, Vol. 21, P. 643, Pergamon Press.
4. Clark, L., Emmerson, B.W. and Wojcikiewicz, E.A. (1978) : Radiation Protection in Nuclear Power Plants and the Fuel Cycle, BNES, London.

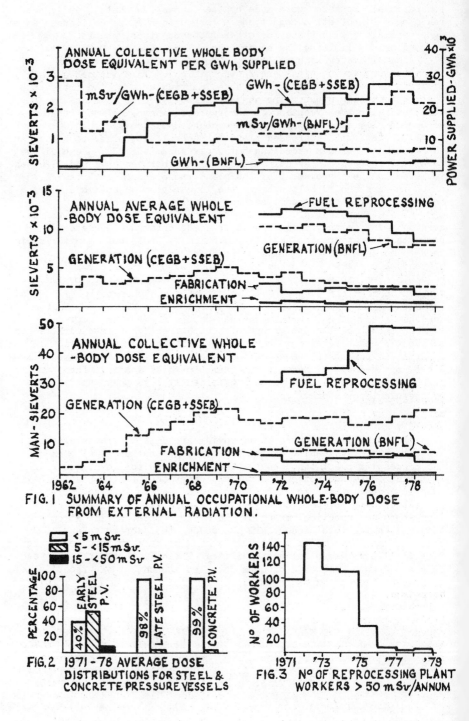

FIG. 1 SUMMARY OF ANNUAL OCCUPATIONAL WHOLE-BODY DOSE FROM EXTERNAL RADIATION.

FIG. 2 1971-78 AVERAGE DOSE DISTRIBUTIONS FOR STEEL & CONCRETE PRESSURE VESSELS

FIG. 3 N° OF REPROCESSING PLANT WORKERS > 50 mSv/ANNUM

OCCUPATIONAL RADIATION PROTECTION LEGISLATION IN ISRAEL

J. Tadmor, T. Schlesinger, A. Donagl and C. Lemesch

A committee of experts appointed by the Minister of Labour and Social Affairs has proposed a comprehensive draft regulation, concerning the legal aspects of occupational radiation protection in Israel. The main sections of the proposed regulation are:
1. Control - general
2. Personal monitoring
3. Medical supervision

CONTROL - GENERAL

The first section of the proposed regulation sets forth guidelines for control in facilities where workers handle radioactive materials or radiation equipment. The managers of such places should take the following steps:
a) Nominate a radiation protection officer
b) Advise the inspector of the Ministry of Labour and Social Affairs of all unusual occurrences
c) Procure the equipment necessary for shielding and monitoring of radiation.
d) Restrict access to hazardous areas
e) Ensure compliance with the regulations for the safe operation of the facility
f) Train the radiation workers
g) Advise the officials of radiation exposure in excess of the maximum recommended doses.

Table 1 shows the maximum radiation doses recommended for normal operation.

During special jobs, which cannot be performed under the limitations specified for normal operation, radiation workers may be exposed to twice the annual recommended radiation dose, or - once in a lifetime- to the annual dose multiplied by a factor of five. After such an exposure, further exposure of the worker should be avoided if the integrated dose exceeds (N-18) x the annual dose limit, where N is the age of the radiation worker. The doses given in the table are actual doses, and should not be considered as design base doses for normal operation. Design base doses should be kept as low as reasonably achievable. The annual limits do not include medical exposure. In case of a simultaneous exposure of several tissues, the calculated overall equivalent of the whole body dose should not exceed the limit indicated for the whole body. The doses should be reduced by a factor of 10 for 16 to 18 year-olds, whose exposure should be allowed only if it is connected to professional training, for which a special permit should be required.

TABLE 1. Maximum recommended doses for radiation workers during normal operation

Tissue	Dose to individual workers (Rem/y)*
Whole body	5
Gonads	20
Breast	25
Thyroid	50
Bone	50
Bone marrow	25
Lungs	25
Eyes	30
Other single organs	50

*1 Rem = 10 millisievert

PERSONAL MONITORING

The second section deals with the monitoring regulations for radiation workers who may be exposed to doses in excess of 500 mRem/y. The regulations stipulate that these workers should:
 a) wear radiation badges, to measure external exposure
 b) be checked for internal radioactive contamination
 c) report on all employment in which they may be exposed to additional radiation.

MEDICAL SUPERVISION

Medical check-ups are required for all applicants for work which involves radiation. Also, radiation workers are required to undergo routine periodical examinations, the type of the examination to be determined by the type of work performed. Radiation workers must also be examined following overexposure or accidents.

A. Routine examinations

Compulsory routine examinations should include the following:
 a) Complete clinical check-up
 b) General urine analysis
 c) Blood count: hemoglobin, white count, differential and thrombocyte count
 d) Complete anamnesis, including medical and occupational history.

B. Special examinations

Special examinations of radiation sensitive organs and tissues should be performed according to the type of work and circumstances. These examinations include:
  a) Blood analysis, including bleeding and coagulation time, in case of whole body exposure to radiation
  b) Skin examination, in case of external exposure
  c) Periodical eye examinations, usually once in five years, or once in three years for X-ray machine operators and in special cases, such as exposure to neutrons
  d) Chest X-ray, investigation of the performance of the lungs, liver and kidneys, and analysis of internal contamination, in case of internal exposure and contamination.

The results of the medical check-ups should be recorded in a health report, which should be kept for 30 years by the medical authorities. The main results are also recorded on a personal health card. The following information should be recorded on the card:
  a) Date and purpose of check-up
  b) Any positive finding of the medical check-up or laboratory tests
  c) Occupational disease or effects detected
  d) Decision on whether the worker is medically fit to work with radiation
  e) Date of next medical check-up
  f) Name and signature of physician.

ADDITIONAL RECOMMENDATIONS

In addition to the draft regulations, the committee proposed several codes of practice encompassing the principles of radiation protection, compliance, inspection and licensing.

A series of recommendations were also made by the committee, to the Minister of Labour and Social Affairs, indicating the need to:
  a) Nominate a national advisory committee for ionizing radiation, safety and hygiene
  b) Establish a central medical authority, to deal with emergency cases of high exposure to radiation
  c) Publish safety rules for work with ionizing radiation
  d) Spell out the training and experience required of persons who install and maintain ionizing radiation machines
  e) Establish curricula for the training and instruction of radiation workers, according to the type of work
  f) Spell out standards for ionizing radiation machines and radiation measurement instruments.

ACKNOWLEDGEMENTS

In addition to the authors, the following members of the committee contributed to the draft regulation: Dr. J. Caftori, Dr. M. Friedman, Prof. Z. Fuchs, Mr. G. Nativ, Mr. M. Naveh, Dr. M. Ronen, Mr. N. Rosenthal, Mr. K. Schniedor and Mr. J. Talmon. Their contribution is gratefully acknowledged.

LEGAL PROVISIONS CONCERNING THE HANDLING AND DISPOSAL OF RADIOACTIVE WASTE IN INTERNATIONAL AND NATIONAL LAW

W. Bischof

One of the current main problems of the peaceful use of nuclear energy and of radiation protection is the handling and disposal of radioactive waste. The solution of this problem is not only a technical and economic task but also a mission for the legislative bodies, international and national, to provide by legal instruments that damage to the general public and to radiation workers does not occur by the harmful effect of nuclear waste materials and that any danger caused by these substances should be compensated. This paper gives a short survey on the situation of international legislation (I) and of national legislation in countries where nuclear installations are in operation (II) concerning the radioactive waste handling and disposal (1).

I. INTERNATIONAL LAW

Until now there is no special international multilateral convention which governs exclusively the handling and disposal of radioactive waste. Nevertheless we find special rules on the disposal of nuclear waste in a number of conventions on the protection of the marine environment and of the high sea against pollutions (2):

- Convention on the High Sea (Geneva Convention) of April 29, 1958, esp. article 25 (3);
- Convention on the Prevention of Marine Pollution by Dumping of Wastes and Other Matter (London Convention) of December 29, 1972 (4);
- Convention on the Protection of the Marine Environment of the Baltic Sea Area (Helsinki Convention) of March 22, 1974 (5);
- Convention for the Prevention of Marine Pollution from Land-based Sources (Paris Convention) of June 4, 1974 (6);
- Convention for the Protection of the Mediterranean Sea Against Pollution (Barcellone Convention) with Protocol for the Prevention of Pollution of the Mediterranean Sea by Dumping from Ships and Aircraft of February 16, 1976 (7).

In addition and in implementation of the mentioned London Convention of 1972 the International Atomic Energy Agency (IAEA) has published in 1974 Provisional definition and recommendations concerning radioactive wastes and other radioactive matter referred to in Annexes I and II to that convention (8) which has been revised in 1978 (9).

Legal Provisions

Taking in consideration the international conventions on the protection of the marine environment, especially the London Convention of 1972, the Organization for Economic Cooperation and Development (OECD) has set up within its Nuclear Energy Agency (NEA) a multilateral consultation and surveillance mechanism for the sea-dumping of radioactive waste by Decision of the OECD-Council of July 22, 1977 (10). In addition to this decision NEA has published in April 1979 Recommended Operational Procedures for Sea-Dumping of Radioactive Waste and Guidelines for Sea-Dumping Packages of Radioactive Waste.

It should be mentioned that the International Atomic Energy Agency in its Safety Series has enacted recommendations "Radioactive Waste Disposal into the Sea" and "Methods of Surveying and Monitoring Marine Radioactivity" (11).

Concerning the Antarctic Region the disposal of radioactive waste materials is absolutely prohibited by article V of the Antarctic-Treaty of December 1, 1959 (12). In 1975 the parties of that Treaty have recommended again that their governments continue to exert appropriate efforts to the end that no one disposes of nuclear waste in that Antarctic Treaty Area.

Sometimes one may read of proposals to dispose radioactive waste into the outer space. Until now we have no special rules - neither international nor national - on radioactive waste disposal in such a way. The Treaty on Principles Governing the Activities of States in the Exploration and Use of Outer Space Including the Moon and Other Celestial Bodies of January 27, 1967 (13) does not mention that problem. By art. IX of that Treaty in the exploration and use of outer space the State Parties shall be guided by the principle of co-operation and mutual assistance; they are obliged to conduct all their activities in outer space with due regard to the corresponding interests of all other Parties. There is until now no absolute prohibition of the radioactive waste disposal in outer space, but the States are responsible for such activities (article VI of the Outer Space Treaty; and Convention on International Liability for Damage Caused by Space Objects of March 29, 1972).

"Radioactive products and waste" are also subject of the international conventions on third party liability in the field of nuclear energy. By the Paris Convention of 1960 (art. 3) (14) and the Vienna Convention of 1963 (art. II) (15) the operator of a nuclear installation shall be liable in accordance with the provisions of these conventions for nuclear damage upon prove that the damage has been caused by a nuclear incident involving nuclear fuel or radioactive products or waste in his installation or coming from it.

For the nine Member States of the European Atomic Energy Community (EURATOM) the Treaty establishing that Community (16) contains a special provision (art. 37) that each Member State shall provide the Commission with such general data relating to any plan for the disposal of radioactive waste in whatever form as will make it possible to determine whether the implementation of such plan is liable to result in the radioactive contamination of the water, soil or airspace of another Member State. By art. 3 of the EURATOM-Basic Safety Standards of June 1 1976 (17) each Member State shall make the reporting of the disposal of natural and artificial radioactive substances compulsory; each Member State may decide that such disposal activities shall be subject to prior authorization by the competent authority.

In addition to its recommendations concerning the waste disposal into the sea the IAEA has published some other guidelines for waste disposal (Safe handling of radionuclides, 1973 edition, disposal of radioactive wastes into fresh water; the management of radioactive wastes produced by radioisotope users and technical addendum; radioactive waste disposal into the ground; basic factors for the treatment and disposal of radioactive wastes; management of radioactive wastes at nuclear power plants; disposal of radioactive wastes into rivers, lakes and estuaries; management of wastes from the mining and milling of uranium and thorium ores) (18).

Occasionally waste disposal is subject of bilateral treaties, for instance the Technical Exchange and Co-operation Arrangement between the USAEC in the Federal Ministry for Research and Technology of F.R.G. of December 20, 1974 (19).

II. NATIONAL LEGISLATION

Provisions on the handling and disposal of radioactive waste have been enacted in many countries, particularly during the last years. It is not possible to give here a comprehensive survey, but the following legal provisions should be mentioned:
1. Austria: Radiation Protection Act, June 11, 1969; Radiation Protection Decree, Jan. 12, 1972;
2. Belgium: Radiation Protection Regulations, Febr. 28, 1963 (with amendments) (Sec. 33-37);
3. Denmark: Radiation Protection Regulations, Nov. 20, 1975, (Sec. 8);
4. France: Décret no. 66-450, June 20, 1966; décret no. 63-1228; Dec. 11, 1963, and décret no. 73-405, March 27, 1973; Arrêté, Nov. 7, 1979;
5. Germany, F.R.: Atomic Energy Act 1959/1976 (Sec. 9a); Radiation Protection Ordinance, Oct. 13, 1976 (Sec. 47);
6. Germany, D.R.: Radiation Protection Ordinance and First Executive Order, November 26, 1969; Guidelines on the centralized management of radioactive waste, March 28, 1974;
7. Israel: The Supervision of Supplies and Services (Construction and Operation of Nuclear Reactors) Order,

No. 5735, Sept. 27, 1974;
8. Italy: Act No. 1860 of the Peaceful Use of Nuclear Energy, Dec. 31, 1962; Radiation Protection Regulations No. 185, Febr. 13, 1964 (Sec. 104-107);
9. Luxembourg: Radiation Protection Act, March 25, 1963; Radiation Protection Regulations, Febr. 8, 1967;
10. Netherlands: Decree on nuclear installations, fuel and ores, Sept. 4, 1969; Decree on radioactive substances, Sept. 10, 1969;
11. Switzerland: Atomic Energy Act, Dec. 23, 1959; Revision of the Atomic Energy Act, Oct. 6, 1978; Radiation Protection Ordinance, June 30, 1976; Decree on the compilation and delivery of radioactive wastes, March 18, 1977;
12. Spain: Nuclear Energy Act, April 29, 1964;
13. United Kingdom: Radioactive Substances Act, June 2, 1960 (Sec. 6-10);
14. USA: USNRC Regulations, 10 CFR Part 20 (Sec. 20.301-20.305).

REFERENCES

1. Cf. Strohl, P., Legal, administrative and financial aspects of long term management of radioactive waste, Atomic Industrial Forum International Conference on Regulating Nuclear Energy, Bruxelles 1978.
2. Pelzer, N., Rechtsprobleme der Beseitigung radioaktiver Abfälle in das Meer, Göttingen 1970; Courteix, S., and Pontavice, E. de, in: AIDN/INLA, Nuclear Inter Jura '75, Aix-en-Provence 1975, p. 71 and 119.
3. United Nations Treaty Series (UNTS) 450, p. 11, 169.
4. IAEA/INFCIRC/205.
5. International Legal Materials (ILM) 1974, p. 546.
6. Official Journal (OJ) of the European Communities 1975, No. L/194/6; ILM 1974, p. 352.
7. OJ of the European Communities 1977, No. L 240/3.
8. IAEA/INFCIRC/205/Add. 1.
9. IAEA/INFCIRC/205/Add. 1/Rev. 1.
10. OECD-Doc. C (77) 115 Final; Nuclear Law Bulletin No.20 (1977), p. 37; cf. P. Strohl, in: AIDN/INLA, Nuclear Inter Jura '77, Florence 1977, p. 344.
11. IAEA Safety Series No. 5 (1961) and 11 (1965).
12. UNTS 402, p. 71.
13. UNTS 610, p. 205.
14. Bischof/Goldschmidt/Greulich, Internationale Atomhaftungskonventionen, Textsammlung, Göttingen 1964, p. 131.
15. Bischof/Goldschmidt/Greulich (Fn. 14), p. 13.
16. United Kingdom Treaty Series No. 17 (1979).
17. OJ of the European Communities 1976, No. 187/1.
18. IAEA-Safety Series Nos. 1, 10, 12, 15, 19, 24, 28, 36, 44.
19. Bundesgesetzblatt 1975 II 269.

Waste Disposal

MANAGEMENT OF RADIOACTIVE WASTES IN THE UNITED STATES OF AMERICA

F. L. Parker

I. INTRODUCTION

The problem of radioactive wastes is so minescule that in two (2) days a city the size of New York generates sufficient solid municipal waste to equal the total United States of America defense radioactive high level waste generated in the past thirty-five (35) years (1). In the United States, the radiation dose over the next twenty-five (25) years from commercial power is expected to be about ½ of 1% of the natural background we will be exposed to in that time (2). Yet, the waste is sufficiently radioactive so that, if the accumulated commercial high level wastes (2300 MTHM)* were uniformly diluted in the Great Lakes, it would take the volume of another 16 Great Lakes to dilute them sufficiently to reach permissible drinking standards (3).

It is not a question of "Can We Manage Radioactive Waste?" but the fact that we must manage radioactive wastes because we have large quantities present now (270,000 m^3) which will not go away except by radioactive decay. The existence of this large amount of highly toxic material raises health and safety questions about their somatic effect on the present generation and genetic effects in future generations. The first question we would like to ask is, are we safe?; and, second, are future generations safe? This means, of course, we must define "safe". "Safety" is a societal decision on how much risk we are willing to accept. Consequently, though scientists and engineers can try to quantify the risks (probabilities of occurrences times the consequences) only a societal decision can be made as to whether these risks are acceptable or not (safe or not).

However, the main thrust of this paper is to outline the risks involved in radioactive waste management and the means for reducing some of them. It is obvious from the definition of the problem that a societally satisfactory solution to the nuclear waste problem is essential to a viable nuclear energy program. In fact, a number of states, including California, and a number of nations, including Sweden and West Germany, have made the solution to the waste problem a precondition to the licensing of nuclear power plants.

II. WASTE CHARACTERISTICS

To analyze the radioactive waste problem, it is necessary to look at the waste arising from the complete fuel cycle; from mining and milling, processing, reactor operation, reprocessing, waste disposal, decontamination and decommissioning. The volume of wastes generated in manufacturing plutonium for weapons, wastes already generated in the commercial nuclear power fuel cycle, and wastes generated in a multitude of other uses in medical, industrial, and

*Metric Tons Heavy Metal

educational facilities are shown in Table 1 (4).

TABLE 1. Quantities of Existing Waste

	Commercial	Defense
High Level		
Thousand cubic meters	2.3	270
Million curies Sr-90 and Cs-137	600	600
Kg Pu	19000	630
Transuranic (kilograms)	120	1100
Low Level (million cubic meters)	0.5	1.4
Spent Fuel (tons heavy metal)	2300	-
Mill Tailings (million tons)	-	4	-

We may use the Sr-90 and Cs-137 curie content and the kg of Pu as surrogates for the radioactive content of these wastes and, the hazardousness of the wastes. They are also shown in Table 1 (5). Most of the radioactive material from commercial activities is still retained within the spent fuel elements. The President of the United States has indefinitely postponed the reprocessing of spent commercial fuel because of his belief that this increased the likelihood of diversion of fissile material for illicit bomb manufacturing. Consequently, we shall consider both spent fuel and the solidified first raffinate from the reprocessing step as high level waste. The volume and curie content of wastes produced per gigawatt year are shown in Tables 2 (6) and 3 (7).

Table 2. Annual Radioactive Waste Generation Rates for 1000 MW_e Light Water Reactor

Spent Fuel Discharged	25	MTHM/YEAR
Reactor		
Experience	1300	cubic meters
Design Basis	420	cubic meters
Uranium Mills		
Tailings Solutions	250,000	MT
Tailings Solids	96,000	MT
UF_6 Conversion	35	cubic meters
Enrichment	1.5	cubic meters
Fuel Fabrication	20	cubic meters

The hazards presented by the wastes are shown in Figure 1 (8), as the amount of water necessary to decrease the concentration of wastes from one gigawatt year of electricity production to the maximum acceptable concentrations for public water supplies. These concentrations at equilibrium would lead to a yearly dose of 500 millirem for a person taking his entire drinking water supply from that source. We might note that for the first 100 years, the hazards from spent fuel are indistinguishable from the hazards of reprocessed wastes

TABLE 3. Estimated Activities of Selected Radionuclides in Wastes (Ci/MTHM) *

	Spent Fuel	Total Recycle
Volatile		
H -3	3.07 E + 2**	0
Kr -85	4.90 E + 3	0
C -14	1.08 E - 1	0
Uranium		
U -235	1.72 E - 1	2.00 E - 1
U -238	3.14 E - 1	3.17 E - 1
Plutonium		
Pu 239	3.31 E + 2	2.11 E 0
Pu 240	4.88 E + 2	1.58 E + 1
Pu 241	7.89 E + 4	9.28 E + 2
Particulates		
Sr -90	6.01 E + 1	5.38 E + 4
Tc 99	1.44 E + 2	1.44 E + 1
I 129	3.39 E - 2	3.74 E - 5
Cs 137	8.65 E + 4	8.71 E + 4
Ce 144	1.47 E + 2	1.42 E + 2
Am 241	1.75 E + 3	8.97 E + 2
Am 243	1.72 E + 1	8.09 E + 1
Cm 244	1.35 E + 3	1.04 E + 4

* PWR fuel irradiated 33000 MWD/MTHM at 30 MW /MTHM 10 year decay time
** E + 2 = 10^2

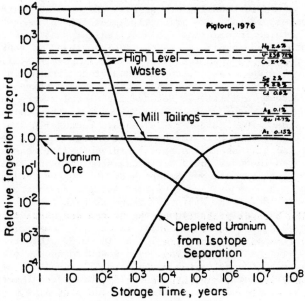

FIGURE 1. Relative Hazard of Toxic Substances

After that the spent fuel grows to be over 100 times as hazardous as the reprocessed wastes. The excess heat generated over 100,000 years indicates that spent fuel burial cannot be exactly the same as that of reprocessed waste. We also note that after about 600 years, the hazards from the mill tailings, though 4 orders of magnitude lower than the initial high level waste hazard, became indistinguishable from high level wastes and that after 20,000 years the depleted uranium hazards equal those from high level waste and that after 100,000 years the hazards from the mill tailings become less than the hazard from depleted uranium. We should compare the hazards of these wastes with those from which the uranium was extracted, and several common geological deposits. We can see that at approximately 600 years the reprocessed waste hazard is equal to that of the ore from which it came (primarily due to radium 226) and after 100 years and less than 600 years, high level waste is exceeded in hazard by average concentrations of ores of mercury, pitchblende, chromium, selenium, lead calcium, silver, barium and arsenic. We can also note that the lethal doses in nuclear wastes from an all-nuclear economy (decayed 100 years) is 3 orders of magnitude less than the number of lethal doses in the amounts in annual use of arsenic: 4, barium; 5 for ammonia and for hydrogen cyanide; 6 for phosgene; and 7 orders of magnitude lower than in the chlorine we use in everyday commerce (9).

Present plans call for all wastes to be segregated into solids or gases to be retained, and solids, liquids, and gases to be released to the environment. At the present time, the major dose to the public is from these gaseous wastes (10), though the APS study found this not to be the case (11).

III. PRESENT METHODS OF MANAGEMENT

How are we handling our wastes? High level wastes from commercial power programs are primarily spent fuel elements which are mostly stored in existing pools at reactor stations. A small amount of reprocessed commercial waste is also stored in tanks at the Nuclear Fuel Services facility. These tanks are similar to those used for defense wastes, cup and saucer configuration. To my knowledge, there have not been any leaks of radioactive materials from any of these facilities. Defense high level wastes have been stored in Richland, Washington, Savannah River, South Carolina, and Idaho Falls, Idaho in liquid form in tanks. Because it was felt that the wastes were too mobile in the liquid form, they are being converted to calcine at Idaho Falls by a fluidized bed process and the solids are stored in underground stainless steel bins on site.

At both Hanford and Savannah River, the wastes have been converted to a damp salt cake by removing the supernatent liquor and by additional drying at Hanford. Tanks at both Savannah River and Hanford have failed and thousands of gallons of high level liquid Wastes have leaked from the Hanford tanks. However, though 115,000 gallons leaked out of the 106T tank in 1973, other than tritium, the most mobile nuclide, Ru-106, moved downward about 16 meters, still 34 meters above the water table, and radially about 20 meters, and the least mobile nuclide detected, Cs-137, had moved 2 meters downward and a maximum of 5 meters laterally. No plutonium movement was detected. In July, 1977, ERDA noted that "further movement of the radioactivity from its present location is expected to be negligible" (12).

IV. PROPOSED METHODS OF MANAGEMENT

No decision has yet been reached on what to do with the high level wastes stored at the three Department of Energy plants nor with commercial wastes.

The proposed methods for handling high level wastes that have been recommended to the President are solidification and placement in mined repositories, in deep ocean sediments, in very deep drill holes, and in a mined cavity in a manner that leads to rock melting; partitioning of reprocessing wastes, transmutation of heavy radionuclides and geological disposal of fission products; and ejection into space. Emplacement into geological media is the first choice because it has been reasoned that, since geological media have been stable for millions of years, if the wastes could be emplaced without unduly disturbing the media, it would be safe for millions more years.

A systematic view of waste disposal facilities would include the waste form, canister material, overpack, the host rock, as well as the formations above and below the repository, ground water transport system including its pH, eH, dissolved solids, leach rate, ground water and its volume and utilization in the biosphere. We have shown in a sensitivity analysis of a crude mathematical model of the system, using reasonable bounds for the important parameters, a wide variety of geological media could provide suitable repositories for high level wastes under appropriate conditions (13).

The waste form could be the spent fuel or after reprocessing, glass, calcine, super calcine, cement, ceramics, metal matrix or chemical resynthesis into minerals that occur in nature. They would be thermodynamically stable under all conditions likely to occur in the repository, be in equilibrium with the minerals in the geologic media and have low solubility in the ground water. These minerals theoretically could be stable indefinitely, if they mimic minerals in nature which are stable. Examples of typical artificial minerals are monazite, Na-pollucite, feldspar, Movsodalite, hematite, apatite, nepheline, and I-sodalite (14).

After the wastes are solidified, they can be placed into canisters which will further retard the leaching of the waste form. For the estimated life-times, canisters could be made of mild steel, cast iron or Al alloys - 100 years; Fe and Pb alloys, 100 - 500 years; Ni, Cu, Ti, Zr alloys, 500 - 5000 years; noble metals, glass ceramics, or alumina > 5000 years (15). Outside the canisters there could be an overpack of clays, zeolites, iron oxides, redox controlling materials or vermiculite to absorb any released moisture and to retard the migration of any nuclide released.

The host geologic media would be chosen so that the repository would be at a depth sufficient to separate the repository from any surficial process or event that might cause a breach of the repository and of a size and shape sufficient to provide a buffer zone around the repository, within a structurally stable geologic block and not near a tectonic boundary, away from faults, abnormally high geothermal gradients, and areas of recent volcanic activity; and under stress conditions which would ensure the stability of the repository; in a hydrological and chemical regime which will prevent the fluid transport of radioactive material to the biosphere in hazardous amounts; and without valuable resources (16).

Most of the work done in this country and abroad for the disposal of high level wastes in geological media has been concerned with mined repositories in salt (NaCl). The only large scale disposal and experimental work has been in salt, with the Germans disposing of low level and intermediate level waste in their bedded salt mine at Asse. We have previously placed several millions curies in spent fuel elements into a bedded salt mine at Lyons, Kansas in Project Salt Vault (17), and left them there until the dosage to the salt was close to 10^9 rad and the temperature close to 200°C. Other geologic materials under intensive investigation are basalt at the Hanford, Washington, Department of Energy site; granite in Sweden with some United States participation; and lesser efforts are underway in shale, tuff and anhydrite.

These media differ in important respects in the properties which are essential for a suitable repository; natural moisture content, strength, thermal expansion, heat capacity, thermal conductivity, permeability, porosity, and solubility. Each medium has its own proponents: those at already-contaminated sites such as Hanford argue convincingly that, since the sites are already contaminated and the wastes must be disposed of someplace, why not there, if the geological media is suitable from a larger systems view? This would reduce handling, offsite transportation, radiation dosage, and probability of sabotage.

We also need to note the effects of the decay heat over time upon the temperature rise at the repository and the induced surface rise. Preliminary calculations do not take into account the subsidence that will occur after mining and backfill. Typical effects for basalt are shown in Table 4 (18).

TABLE 4. Temperature and Surface Rise for High Level Waste and Spent Fuel in a Basalt Repository [a]

Time, Years	High Level Wastes				Spent Fuel			
	Depths, Meters			Surface Rise, Meters	Depths, Meters			Surface Rise, Meters
	305	610	915		305	610	915	
	Temperature, °C				Temperature, °C			
100	38	151	59		38	193	59	0.8
500				0.6				0.15
1000	49	109	68	0.6	86	193	98	0.21
2000	57	93	76		105	199	124	
4000								
5000	61	84	82	0.04				0.26
7000					107	172	136	
10000								0.25
Geothermal Gradient	38	47	59		38	47	59	

[a] 77 watts/hectare

The maximum surface rise would occur at 7000 years for spent fuel and 3000 years for high level waste for the particular conditions outlined. Swedish scientists have recommended that the stresses be reduced by storing the wastes near the surface for 50 years, which will allow a reduction by a factor of about 60 of the heat generation rate from 120 days out of the reactor.

The effect of the choice of geologic media on the cost of waste management is negligible with the repository being only about 30% of waste disposal costs, which run from 0.4 to 1 mill/kwhr in mid-1978 dollars (19).

The time and cost to demonstrate the practicality of other disposal methods for high level wastes is far more uncertain then disposal in mined repositories. The solution to the problem of other radioactive wastes is technically easier than that for high level wastes.

V. CONCLUSIONS

Finally, if the nuclear waste problem is resolvable, why hasn't it been? Here, we come to an institutional failure. Waste treatment has been a stepchild, first out of necessity when priorities were to make bombs and afterwards, when the budgets were slashed, the question was always in terms of priority - are the wastes in tanks safe now? Are they safe for another year? The answer was always yes and so there they stayed. Now, with President Carter's anti-inflation stance, that inattention to waste is likely to continue. A firm national commitment is needed to resolve the issue by actually putting wastes into repositories and seeing whether it is as safe as its proponents claim.

We have radioactive wastes which are large in volume, toxic in nature, but with technology available to handle these wastes with the same degree of risk as we handle other toxic materials, though we have not yet demonstrated this technology.

We need to show, in the near term, when the hazard is the greatest that performance of the mined repositories conforms to our predictions. Given our present practice, near surface storage, which is obviously more hazardous, than burial in mined repositories, we need to proceed as expeditiously as possible and in a conservative manner to dispose of our radioactive wastes rather than continue to store them.

REFERENCES

1. (a) Council on Environmental Quality, "Environmental Quality," GPO, December, 1978, p. 167.
   (b) American Public Works Association, "Solid Waste Collection Practice," APWA, 1975, p. 21.
   (c) U.S. Department of Energy, "Report of Task Force for Review of Nuclear Waste Management," (Deutch Report), February, 1978, p. 127.
2. U.S. Nuclear Regulatory Commission, "Final Generic Environmental Statement on the Use of Recycle Plutonium in Mixed Oxide Fuel in Light Water Cooled Reactors," NUREG-0002, Vol. 1, p. s.-22.
3. (a) "Report to the President by the Interagency

Review Group on Nuclear Waste Management," TID-29442, NTIS, March, 1979, p. 11.
(b) MacNish, Charles F., and Lawhead, Harley F., "History of the Development of Use of the Great Lakes and Present Problems," in Proceedings of Great Lakes Water Resources Conference, American Society of Civil Engineers, 1968, p. 8.
(c) U.S. Department of Energy, "Management of Commercially Generated Radioactive Waste," DOE/EIS-0046D, April, 1979, p.A-4.

4. Reference 3 a, p. 11.
5. (a) Lieberman, Joseph A., et at, "High Level Waste Management," Testimony submitted to State of California Energy Resources Conservation and Development Commission, March 21, 1977, p. 109a.
(b) Office of Waste Isolation, Union Carbide Corporation, "Contribution to Draft Generic Environmental Impact Statement on Commercial Waste Management: Radioactive Waste Isolation in Geologic Formations," OWI GEIS Report Y/OWI/TM-44, April, 1978, p. 3-18.
6. Reference 3 a, p. D-6.
7. Reference 3 c.
8. Reference 9, p. VII,-14.
9. Cohen, Bernard L., "The Disposal of Radioactive Wastes from Fission Reactors," Scientific American, June, 1977, p. 30.
10. Reference 2, p. s-21.
11. Reference 9, 0. v-45.
12. Committee on Radioactive Waste Management, "Radioactive Wastes at the Hanford Reservation," National Academy of Sciences, 1978, p. 38.
13. Parker, F.L. and Andrew Ichel, unpublished manuscript, 1977.
14. Reference 3 c, p. 3.2.11.
15. Reference, p. 3.1.60.
16. Committee on Radioactive Waste Management, "Geological Criteria for Repositories for High-Level Radioactive Wastes," National Academy of Sciences, August 3, 1978.
17. Bradshaw, R. and McClain, W.C., "Project Salt Vault: A Demonstration of the Disposal of High-Activity Solidified Wastes in Underground Salt Mines," ORNL-4555, Oak Ridge National Laboratory, Oak Ridge, Tennessee, April, 1971, p. 336-338.
18. Office of Waste Isolation, Union Carbide Corporation, "Technical Support for GEIS: Radioactive Waste Isolation in Geologic Formations," OWI GEIS Report Y/OWI/TM-36/20, Oak Ridge, Tennessee, April, 1978, p. 7-40, 41, 44.
19. Reference p. 3.1.234.

ENVIRONMENTAL MONITORING AND DEEP OCEAN DISPOSAL OF PACKAGED RADIOACTIVE WASTE

N. T. Mitchell and A. Preston

INTRODUCTION

A basic tenet of the philosophy which underlies radioactive waste disposal control into the marine environment is that of environmental monitoring. Current monitoring principles have recently been discussed by Mitchell (1) with illustration of their application to controlled releases of radioactive effluents to estuarine and coastal waters of the U.K. This paper sets out to discuss environmental monitoring philosophy in the context of dumping of packaged waste in the deep ocean and how far it may be reasonable to apply it in practice.

THE AIMS AND OBJECTIVES OF ENVIRONMENTAL MONITORING:
SOME BASIC PRINCIPLES

Both the ICRP (2) and the IAEA (3) have devoted specific publications to the topic of environmental monitoring and U.K. practices are consistent with them. As with any operation associated with the management of radioactive waste, environmental monitoring should be in accord with the fundamental dose limitation principles of the ICRP and the principle of optimization is of particular importance in this respect. The IAEA in its recommendations to the London Dumping Convention (4) recognizes a need for environmental monitoring but adds an important rider 'to the extent feasible and meaningful'. The ICRP lists three objectives for environmental monitoring.
    (a) Assessment of the actual or potential exposure of man to radioactive materials or radiation present in his environment, or the estimation of the probable upper limits of such exposure.
    (b) Scientific investigation, sometimes related to the assessment of exposures, sometimes to other objectives.
    (c) Improved public relations.
To these the IAEA has added:
    (i) assessment of the adequacy of controls on the release of radioactive materials to the environment;
    (ii) demonstration of compliance with the applicable regulations, environmental standards, and other operational limits; and
    (iii) the possible detection of any long-term changes or trends in the environment resulting from the operation.

The estimation of radiation exposure

The fundamental objective of environmental monitoring, as the term is used in the U.K., is to facilitate or otherwise provide for the *direct* estimation of radiation exposure. Reflecting the emphasis of U.K. national waste disposal policy, which is primarily on limitation of radiation exposure to the public as opposed to that of effects on environmental resources, environmental monitoring has come to be associated mainly with measurements needed to assess doses to which human populations are exposed. The justification for this attitude is found in the realization that the potential risk to environmental resources is minor provided that exacting standards for control of public radiation exposure are met and maintained, i.e. those recommended by the ICRP.

Monitoring and research

There is still much confusion between monitoring and research, possibly because they are frequently practised as mutually supporting activities. The objective of estimation of radiation exposure has sometimes been met by utilization of data which have accrued from research programmes, replacing a need which would otherwise have had to be met by mounting monitoring programmes specifically for that purpose. Conversely, environmental monitoring programmes themselves have sometimes generated information of value to research and it is in this kind of situation where there may appear to be some confusion of purpose as between monitoring and research. Whilst in a few marine situations radiation exposure can be estimated from analysis of environmental materials, it is not possible to do this where contamination is below detection limits. In such situations monitoring and research have another interface, for mathematical modelling provides a means of indicating what the levels of radioactivity may be in the environment in relation to given rates of input and hence provides data for the evaluation of levels of dose. Research in support of this kind of modelling activity is especially important in connection with deep sea disposal of packaged radioactive waste.

Monitoring and control measures

As practised in the U.K., environmental monitoring has sometimes provided a means of checking adequacy of control measures, such as treatment plant or filtration systems, relating to the on-site management of radioactive waste. This has certainly been applied in some coastal situations related to the control of liquid radioactive waste discharges but, just as this is not always a practicable proposition there, it would be misleading to consider that such objectives could necessarily be met in regard to deep sea dumping.

Public information

A popular view in some circles is that monitoring should also be done on occasions for what is termed 'public relations purposes'. Whilst the need to provide the public with appropriate information and data is readily acknowledged and met, it is considered that the conduct of monitoring programmes for which there is no justifiable radiological or scientific need in relation to estimates of human radiation exposure is not a suitable objective. P.R. monitoring must not be allowed to become an end in itself and require the production of data for the sake of being able to state that some monitoring has been done. Such programmes would be misleading, and subject to severe criticism by the scientific community as a conscious effort to mislead the public and allay public concern at a cost out of proportion to the need. Data arising from programmes designed to fulfil the basic aim of providing a sound basis for estimates of radiation exposure should also be sufficient to answer public information needs.

ENVIRONMENTAL MONITORING IN PRACTICE

Compared with disposals of liquid wastes into coastal waters, disposals of packaged waste into the deep sea pose special problems for those involved in environmental monitoring. The capacity of the receiving environment, its remoteness from pathways back to man, coupled with the relative biological and physical unavailability of the radioactivity in the waste means that, curie-for-curie, there is must less chance of being able to measure activity in the critical materials from deep sea disposals. Whilst waste packages are designed to ensure that their active contents are at the very least delivered to the deep ocean bed it is clear that in general they will have a much longer life and in most cases the activity will be released only very slowly into the water.

The major disposals of liquid wastes to coastal waters are characterized by the ease with which environmental monitoring yields positive evidence of the disposal. In such labelled environments discharge rate can be correlated with radioactivity in environmental materials and from this with radiation exposure to man. In contrast, and whilst there is evidence to show that in at least certain circumstances activity from dumped waste may be detected on sediment very close by packages, it cannot be detected in pathways critical to man and monitoring along the lines of coastal disposals would not therefore be meaningful.

Faced with such a situation it is necessary to find an alternative to environmental monitoring, at least to an extent necessary to meet the fundamental objective of monitoring, that of assessing human radiation exposure. Situations where levels of activity attributable to a particular disposal are below limits of detection are not unique to deep sea disposal; it is typical of a majority of the disposals at coastal sites and two options are open to us.

The first option is to make an upper limit estimate based on a judgement of analytical detection limits. As such it provides only a very crude answer and whilst this will be sufficient to show that radiation exposure is within prescribed limits it has little scientific value and serves no other purpose.

The second option is to compute the dose by mathematical modelling. For a system on an oceanic scale this is a considerable undertaking and many of the factors involved are not known with any precision; neither are oceanic processes well understood. Nevertheless it is possible to model the system using pessimistic values of and data is readily acknowledged and met, it is considered that the conduct of monitoring programmes for which there is no justifiable radiological or scientific need in relation to estimates of human radiation exposure is not a suitable objective. P.R. monitoring must not be allowed to become an end in itself and require the production of data for the sake of being able to state that some monitoring has been done. Such programmes would be misleading, and subject to severe criticism by the scientific community as a conscious effort to mislead the public and allay public concern at a cost out of proportion to the need. Data arising from programmes designed to fulfil the basic aim of providing a sound basis for estimates of radiation exposure should also be sufficient to answer public information needs.

ENVIRONMENTAL MONITORING IN PRACTICE

Compared with disposals of liquid wastes into coastal waters, disposals of packaged waste into the deep sea pose special problems for those involved in environmental monitoring. The capacity of the receiving environment, its remoteness from pathways back to man, coupled with the relative biological and physical unavailability of the radioactivity in the waste means that, curie-for-curie, there is must less chance of being able to measure activity in the critical materials from deep sea disposals. Whilst waste packages are designed to ensure that their active contents are at the very least delivered to the deep ocean bed it is clear that in general they will have a much longer life and in most cases the activity will be released only very slowly into the water.

The major disposals of liquid wastes to coastal waters are characterized by the ease with which environmental monitoring yields positive evidence of the disposal. In such labelled environments discharge rate can be correlated with radioactivity in environmental materials and from this with radiation exposure to man. In contrast, and whilst there is evidence to show that in at least certain circumstances activity from dumped waste may be detected on sediment very close by packages, it cannot be detected in pathways critical to man and monitoring along the lines of coastal disposals would not therefore be meaningful.

Faced with such a situation it is necessary to find an alternative to environmental monitoring, at least to an extent necessary to meet the fundamental objective of monitoring, that of assessing human radiation exposure. Situations where levels of activity attributable to a particular disposal are below limits of detection are not unique to deep sea disposal; it is typical of a majority of the disposals at coastal sites and two options are open to us.

The first option is to make an upper limit estimate based on a judgement of analytical detection limits. As such it provides only a very crude answer and whilst this will be sufficient to show that radiation exposure is within prescribed limits it has little scientific value and serves no other purpose.

The second option is to compute the dose by mathematical modelling. For a system on an oceanic scale this is a considerable undertaking and many of the factors involved are not known with any precision; neither are oceanic processes well understood. Nevertheless it is possible to model the system using pessimistic values of

the necessary parameters, such that upper limit values are produced
which are more accurate than those derived from analytical detection
limits. The overall oceanographic/radiological model is divided into
several parts, viz, release of activity into the water, its dispersion and transport, and uptake into critical pathways. Simplifying
assumptions are made, for example that no removal by sediment occurs
to reduce the availability of activity to biological pathways, provided that they do not underestimate the dose received. Most of the
values of concentration factor needed are reasonably realistic, as
would be the consumption rates/occupancy factors used if the pathways were effective now. A fundamental problem to sea dumping
assessments is the long delay between dumping and the arrival of dose
to man. For the purpose of dose assessment the assumption is made of
prompt release after dumping has taken place, a maximizing assumption
which leads to exaggerating the resulting dose from the shorter-
lived radionuclides. For the present, it is only possible to calculate the dose at equilibrium from continued dumping over very long
periods of time and the results are therefore likely to be gross
overestimates of the true dose. Nevertheless, work is continuing, in
terms both of better mathematical models to predict dose to man and
research into oceanographic and biological transfer processes to provide better data and improve the accuracy of the models. Faced with
the inability of direct monitoring to provide data by which direct
estimates of dose to man can be made, resources can be used to
greater effect by devoting them to modelling and oceanic research.

REFERENCES

1. Mitchell, N. T. (1979): Monitoring radioactivity in the marine environment. In: Monitoring the Marine Environment, p. 153. Editor: David Nichols. Institute of Biology, London.
2. Anon. (1965): Principles of Environmental Monitoring Related to the Handling of Radioactive Materials, ICRP Publication 7. Pergamon Press, Oxford.
3. Anon. (1975): Objectives and Design of Environmental Monitoring Programmes for Radioactive Contaminants. Safety Series No. 41, International Atomic Energy Agency, Vienna.
4. Anon. (1978): Convention on the Prevention of Pollution by Dumping of Radioactive Wastes and Other Matter. The Definition Required by Annex I, Paragraph 6 to the Convention, and the Recommendations Required by Annex II, Section D. International Atomic Energy Agency, Vienna.

ABSORPTION OF DOSE RATE IN GREAT BITUMEN BLOCKS EXPERIMENTAL DETECTION

P. R. M. Patek

GENERAL

Development and widespread application of atomic energy involves a lot of problems, concerning the handling of increasing radioactive waste amounts. For storing and disposing of these big amounts of radioactive trashes several methods have been developed.

Concerning the fact that storing and disposing prices inside an activity group (medium level or low level waste) are nearly independent in a wide range from the activity of the immobilized waste, a volume reduction is desirable and necessary, regarding the principle of cost minimization.

Commonly LLW and MLW are reduced in volume by combustion (for burnable wastes) or by compaction. The residues of these volume reduction methods are conditioned partly by pouring over or mixing with cements or bitumen. Aime of these procedures is the immobilization of the radioactivity against leaching by ground water after a possible water contact. Both methods- cementation and bituminization - are showing advantages and disadvantages. Whereas cements are highly resistent against radiation damage, bitumen shows better properties against water leaching, for example.

Numerical calculations of dose rates occuring inside non artifical waste mixtures often showed great differences to the real values reached, especially regarding beta-doses. Using the bituminization technique as conditioning method, therefore different little "bitumen elements" will suffer a high dose, while others will remain nearly unirradiated. These effects are amplified when the amount of embedded waste in bitumen increases to a very high extent, what is reachable in principle by sedimentation technique (1,2).In this case the specific activity of the immobilised waste increases, the necessary storing place decreases. Presently a specific activity of 1 Ci/l (^{137}Cs + ^{90}Sr) is the limit for bitumen embedding techniques, accepted by many authorities. This value is on the very safe side of the activity limit, above which decomposition of bitumen is initiated, although there is some early work in higher activities enclosed inside bitumen (3). In any case the limits for loading waste into a bitumen matrix are defined by two parameters: chemical stability of the bitumen in a field of radiation, and heat generation and transfer out of the bitumen block.

Facing the goal of as high waste loading as possible inside a bitumen matrix we aimed to elaborate a method for detecting the irradiation affecting the bitumen matrix under actual conditions.

Absorption of Dose Rate in Great Bitumen Blocks

RADIATION SOURCE MATERIAL

To gain short exposition times during the experiments, on the one hand for getting the results after reasonable time, on the other to meet the endeavours in increasing the possible specific activity of the enclosed waste, we decided to use irradiated high temperature coated particle fuel as radiation source. To get reproducible conditions coated particles (cp) out of one run were taken. The showed specific activity of 75,46 mCi ^{137}Cs and 15,32 mCi ^{134}Cs and 90mCi ^{90}Sr per gramm particle material. The diameter of these particles was 800 μ in average, their weight 1,25 to 1,38 mg, the density 1,5g/cm^3. The bulk density was estimated with 0,6.

PERFORMANCE OF EXPERIMENTS

The experimental work was performed inside hot cells. In the first experiments we used film dosimeters as detecting device. The dosimeter badges were placed inside the hot cell and afterwards a weighed amount of cp was positioned overneath the dosimeter badge. The cp were fixed in a tub-shaped container on a thin Al-foil in a 2-3 mm layer (about 4 cp one upon another). After few irradiations we recognized that the possible exposure times were too short for reproducable handling operation with the remote controled equipment of a hot cell. The results of a 10 sec irradiation differed for $\pm$ 200 %. Therefore we changed over to TLD.

The arrangement of the experiments was the same like that with the film dosimeters. Although the exposition time was now long enough, great difficulties arose from the effect not to reach a sufficient plane layer of cp. All efforts to level the particle layer were successless. Also experiments with an amount of cp for a one-particle-layer only didn't succeded.

So we changed the philosophy of the experiments. We postulated, that not a little grain of waste is surrounded by bitumen, but a "differential element" of bitumen is enclosed by an infinite thick mass of waste.

In that case, the γ-intensity depends on the distance source - bitumen, so that a certain thickness of the particle layers enclosing the bitumen element, the γ-doserate will become constant. The ß-dose belongs mainly to the waste particles, situated next to the bitumen element, while the ß-irradiation of the outer grain-layers will be mainly absorbed by the inner waste layers, surrounding the bitumen element and will not affect the bitumen anymore.

For these experiments we used glass dosimeters, because we had no troubles with light exposition. Also the cleaning of the dosimeter surface before the evaluation was easier as by working with TLD.

To determine the gamma doserate of the used cp mixture we protected the ß-absorber of the glass dosimeters with a thin rubber layer and attached it inside a 250 ml beaker. Afterwards we poured the cp into it. Irradiation time was

10 minutes. Then the dosimeter was pulled out of the particles and brought out of the hot cell. The possible error in timing was about one second or less than 1 %.

Outside the hot cell the rubber was removed and after a conditioning time of one day the glass dosimeters were evaluated. The gamma doserate we found was about 1050 rad in ten minutes, with an accuracy better than $\pm$ 5 %.

For the determination of the ß + γ doserate we removed the ß-absorber from the dosimeters and irradiated them in direct contact with the fuel particles. For this purpose we filled half of the particle amount into the beaker, then we dropped the dosimeter into the glass too and immediately afterwards we poured the rest of the cp over the dosimeter. After an irradiation time of 5 min the whole content of the beaker was poured over a sieve, the fuel particles fell through and the dosimeter remained in the sieve. With a pincette we transferred the dosimeters into a little container and brought it out of the hot cell. The possible error in timing was about 3 to 6 seconds or 1-2 %. After a run of 10 irradiations all the dosimeters were decontaminated, dried and evaluated. The results were as accurate as that from the determination of the gamma doserate. The ß + γ dose we found, was about 4400 rad in ten minutes. That means that the ß-dose alone is about 3350 rad per 10 minutes.

NUMERICAL EVALUATION

In a speric shell of a thickness dr the activity is distributed homogeniously. We assume:

$$A_{Ci} = A_o \cdot V_b = A_o \cdot 4\pi r^2 \, dr \qquad (1)$$

A_o=specific activity (Ci/cm^3), V_b=volume of spheric shell

The radiation level in the centre of the sphere is equal and independent of the fact, that the activity is distributed in the spheric shell or concentrated at a certain point. The doserate in the centre of the sphere can be assumed with

$$D = \Gamma_\gamma \cdot \frac{A}{r^2} \qquad (2)$$

D=Doserate (R/h), γ=dose factor (R.cm^2.h^{-1}.Ci^{-1})
Considering the absorption inside the sphere we get

$$D = \Gamma_\gamma \cdot \frac{A}{r^2} \cdot e^{-\mu r} \cdot B \qquad (3)$$

μ=absorption coefficient, B=build up factor
In this model B=1, because the same amount of gammaquants are scattered out of the considered volume as can become scattered into it. So we can combine (1) and (3) to get

$$D = \Gamma_\gamma \frac{A_o 4\pi r^2 dr}{r^2} \cdot e^{-\mu r} = 4\pi \Gamma_\gamma A_o \cdot e^{-\mu r} \, dr \qquad (4)$$

Integration over the whole sphere leads to the doserate depending on the specific activity:

$$D = 4\pi \bar{\Gamma \gamma} A_o \cdot \int_0^r e^{-\mu r} dr \quad (5) \text{ or}$$

$$D = 4\pi \bar{\Gamma \gamma} A_o \frac{e^{-\mu r} - 1}{-\mu} = \frac{4\pi}{\mu} \bar{\Gamma \gamma} A_o (1 - e^{-\mu r}) \quad (6)$$

If there are several nuclides present, one has to sum over all activities:

$$D = 4\pi \sum_{i,j} \frac{\bar{\Gamma \gamma}(E_{i,j})}{\mu(E_{i,j})} \cdot A_i (1 - e^{-\mu(E_{i,j})r}) \quad (7)$$

When the diameter of this active sphere reaches infinite high values we get for $r \to \infty$ and (6)

$$\lim_{r \to \infty} D = \frac{4\pi \bar{\Gamma \gamma} A_o}{\mu} \quad (8)$$

With respect to the characteristics of the cp used and the numerical values for dose constants and absorption coefficients (4) the calculated value for the centre doserate of an infinitive great sphere is 37,5 kR/h. The doserate inside a sphere with a diameter of 4,2-4,4 cm can be calculated with 8457 to 8810 R/h. From this value one has to substract the volume of the dosimeter itself, corresponding with 2240 R/h.

The result of the numerical calculation with 6217 to 6570 R/h shows excellent agreement with the experimental results. Although it is to mention, that the uncertainties are of great influence on the calculated results, because of the exponential dependence.

FURTHER WORK

In case of embedding these used particles into bitumen in the highest possible volume loading, we will reach an absorbed dose of 10^8 rad in about 160 days. With such a bitumen fuelparticle-mixture we can simulate a 300 years storage in about half a year. To study the leaching behaviour of a bitumen-salt mixture with less active material we will dilute the fuel particles with salt or grafite powder and determine the doserate inside this mixture with the method described.

Finally we can say that this method enables us, to measure the doserate of all mixtures of active wastes we want to embed in several kinds of bitumen, so that we are able to determine the loading capacity of each certain sort of bitumen, according to the radiation damaging we want to reach.

REFERENCES

Lit. (1) J. Zeger et al;
 IAEA Symposion SM-207/55, Vienna 1976
Lit. (2) M. Sarontin et al;
 Nuclear engineering and design 34 (1975) 439-46
Lit. (3) K.P. Sachanoff et al;
 OECD/NEA Symposium SM-163/30, Paris 1972
Lit. (4) W. Comper; KFK-1615 (1972)

A REVIEW OF THE DISPOSAL OF MISCELLANEOUS RADIOACTIVE WASTES IN THE UNITED KINGDOM

B. Hookway

In the United Kingdom there are about 25 major nuclear establishments such as reprocessing plants and power reactors, which produce radioactive waste. There are about 1500 other establishments, such as hospitals, universities, research establishments and commercial firms which also produce radioactive waste. The problems of the first of these classes, the enormous research and development programmes being carried out to achieve acceptable disposal routes for the wastes at present stored and the effect of the wastes which are disposed of, are well documented elsewhere. This paper deals with the second of the classes, those which can perhaps conveniently be called "minor users" although some establishments in this class discharge more radioactive waste to the environment than a nuclear power station. The whole of the United Kingdom radioactive waste management policy has recently been reviewed by an expert group [1]; this present paper looks at current practices for waste disposal from the minor users and indicates where the expert group endorses these or recommends changes.

LEGISLATION AND STANDARDS

In the United Kingdom, disposals of radioactive waste are subject to the Radioactive Substances Act, 1960 (RSA 60). Waste can be disposed of only in accordance with the conditions and limitations contained either in an authorisation which is specific to the disposer or, in the case of certain very low level wastes, in one of a series of Exemption Orders made under the Act. The Act covers the disposal of radioactive waste of all forms, solid, liquid or gaseous; there are no lower limits on radioactivity except for the natural radionuclides at levels found in nature; it removes the control of radioactive waste disposal from local authorities and places it in the hands of central government. United Kingdom standards for radiological protection are based on the system of dose limitation recommended by ICRP. In the case of radioactive waste disposal this is achieved by a case by case approach, a practice endorsed by the expert group.

EXISTING DISPOSAL PRACTICE

The following are the practices whereby all the wastes from the minor users are disposed of: where appropriate these practices are also used for wastes from the major nuclear establishments.

(i) SOLID WASTES

Solid waste of activity less than $10^{-5} \mu Ci/g$ (0.4 Bq/g) is currently regarded as insignificant and it has not been the practice

Disposal of Miscellaneous Radioactive Wastes

for this to require authorisation. The expert group agreed with this and recommended it should be formalised in an Exemption Order.

Low Level Wastes in Domestic Refuse

Small amounts of solid radioactive waste are authorised for disposal with ordinary refuse. The limits applied for such "dustbin disposals" are 10 μCi (400 kBq) in 0.1m^3 and 1 μCi (40 kBq) per article. It is also usual to exclude alpha emitters and strontium-90 and to raise the first limit to 100 μCi (4000 kBq) in 0.1m^3 for the weak beta-emitters carbon-14 and tritium which are in common use. The expert group considered all the implications of this method of disposal and concluded that it represents no hazard. They therefore endorsed the practice.

Private Incineration

This method of disposal is useful for wastes which are unpleasant to handle; it also reduces the volume of waste requiring disposal. Separate authorisation is required for the disposal of the ash. The radionuclides and activities permitted for disposal this way are generally authorised on a case by case basis, taking all the local circumstances into account. However, it is usual to permit disposals of up to 100 μCi (4 MBq) a day of the commonly used tritium and carbon-14 without this detailed examination. The expert group find this practice acceptable.

Special Precautions Disposals

Where solid radioactive waste arises which is not suitable for dustbin disposal, disposal at a landfill tip is still permissible provided certain precautions are taken. Authorisations for such disposals specify the tip which is chosen after consideration of its management, its expected life, the probable subsequent use of the land, whether the tip is liable to catch fire, whether there is unauthorised salvage, drainage and any other special features. The two classes of waste suitable for such disposals are:

(a) **Packaged Wastes**: The limits and conditions normally imposed are:

  a. Waste shall be conveyed to the tip in a sealed, plain, unlabelled plastic or multilayer paper sack in a closed metal bin;
  b. At the tip, the sack shall be removed from the bin and placed either at the foot of the tipping face or in a hole dug for it and immediately covered with inactive refuse to a depth of 1.5m;
  c. No sack shall contain more than 100 μCi (4 MBq) of radio-nuclides of half-life greater than one year and one mCi (40 MBq) of others, except that in the case of tritium and carbon-14, up to 5 mCi (200 MBq) per sack is permitted.

(b) **Bulk Loads**: Radioactive waste consisting of relatively lightly contaminated rubble and soil frequently arises as a result of the demolition of premises in which work with radioactive substances has been performed. Typical examples are luminising works, gas mantle factories and ore-processing factories. Demolition and subsequent site decontamination often produce thousands of tonnes of lightly contaminated waste. When waste of this type is authorised for special precautions disposals, a limit of $10^{-4} \mu Ci/g$ (4 Bq/g) is placed on it. This corresponds to a surface dose rate about 10 times the background level; when buried to 1.5m it cannot be detected on the surface and if care is taken not to concentrate the waste in one part of the tip the dispersion reduces radiation virtually to background level and subsequent disturbance can create no hazard.

The expert group endorsed all these conditions and practices but recommended in addition that in the case of packaged wastes, radionuclides with half-lives greater than one year (except tritium and carbon-14) should be limited to $10 \mu Ci$ (400 KBq) per individual article. They also described this method of disposal as a valuable, and with the safeguards described, a radiologically sound method for the disposal of low-activity solid waste. They deplored the uninformed opposition to which these disposals are becoming increasingly subject and recommend that the duty already placed by RSA 60 on waste disposal authorities to accept radioactive waste should be extended to the operators of all landfill tips.

Disposal on Site

Disposal on the site on which the waste arises appears an alternative to special precautions burial on a landfill tip, but the expert group noted objections to this method and only recommended it provided certain conditions, including an assurance of ownership of the site for an appropriate period, could be met.

The National Disposal Service (NDS)

This is available for radioactive wastes not suitable for disposal by the means so far discussed. Generally, if the waste is bulky and within the authorised limits it goes for burial at Drigg, the site in Cumbria operated by British Nuclear Fuels Ltd, and individual items or small quantities go to the United Kingdom Atomic Energy Authority, Harwell where they are drummed for sea disposal. Full details of the operations at Drigg and Harwell are contained in "A Review of Cmnd 884", together with a description of the authorisation under which Drigg works and the international constraints on sea-dumping. In view of the fact that material from the minor users contributes only about 4% of the wastes disposed of by these routes, which have their main use in dealing with wastes arising from the nuclear fuel cycle, they are not considered further here.

(ii) LIQUID WASTES

Liquid radioactive discharges from the major nuclear establishments are, for the purpose of granting authorisations, evaluated individually against Government policy relating to disposals. The principles relating to the exposure of individuals and populations have been rigorously observed and discharge limits have been assessed quantitatively, often with high precision. But these sophisticated techniques, involving a knowledge of environmental pathways, habit surveys, members of critical groups etc, are rarely necessary for the discharges from minor users.

Drain Disposals

Disposal directly to the drains, without prior collection or storage in hold-up tanks, is the most convenient and radiologically safe method of disposal of relatively small amounts of low activity liquid radioactive waste. Authorisations are usually given in terms of activity per month with limits on individual radionuclides where this is necessary. A few Ci (several GBq) a month, more for tritium, have been authorised for some establishments; others are able to operate with much smaller limits. Hospitals discharging the excreta of patients who have been given therapeutic or diagnostic doses of radioactive substances are amongst the premises having the largest authorisations. The radioactive waste is diluted immediately with other waste waters and in most cases the average concentration in the effluent from the establishment is orders of magnitude below the permissible level for drinking water. There is no formal upper limit for the average concentration of radioactivity in liquid effluents. Each case is considered on its merits: taking account of the toxicity of the radionuclides discharged; the possibility that they may settle out in the sewerage system; their behaviour in the sewage treatment process and ultimately in the effluent, whether discharged to a soakaway, stream, river or the sea. The expert group endorsed these practices with the proviso that the authorising departments should continue to check their assessments by monitoring a few of the most important cases.

(iii) AIRBORNE DISCHARGES

In the case of the minor users, the authorisation contains a specified limit on the activity which may be discharged in a given period. The radionuclides and activities permitted for disposal this way are, as in the case of all authorisations, based on the need of the user to have a particular level of discharge and are granted on a case by case basis taking all the local circumstances into account. The expert group are satisfied that existing controls over emissions to atmosphere are adequate for the time being.

THE EFFECTS OF RADIOACTIVE WASTE DISPOSAL

The effects of all disposals are assessed by the appropriate Inspectorate before authorisations are given, premises are inspected to ensure compliance and any necessary environmental monitoring is carried out. It is therefore possible to say with confidence that the effects of disposals from minor users are insignificant.

REFERENCES

[1] A Review of Cmnd 884: "The Control of Radioactive Wastes". A Report by an expert group made to the Radioactive Waste Management Committee. Department of the Environment (1979).

Internal Contamination and Radiobiology

CHEMICAL PROTECTION AND SENSITIZATION TO IONIZING RADIATION MOLECULAR INVESTIGATIONS

R. Badiello

INTRODUCTION

Chemical radioprotection and radiosensitization are the phenomena induced by the presence of certain chemical compounds, which reduce or enhance respectively the effect of ionizing radiation on living organisms. Such substances are either naturally present or may be artificially introduced in the living cells. When these phenomena occur in complex biological systems, they are a synthesis of many processes of physical, chemical and biological nature. The study of the mecahnisms of chemical radioprotection and radiosensitization could also aim at a better understand - ing of how radiation acts on cells and tissues.

Chemical radioprotectors are interesting for the possible application in health protection of both professionally exposed workers and patients treated by radiation for diagnostic and therapeutic reasons. Although the initial enthusiasm has been not paid by the success of finding the "anti-radiation pill", the problem is so important that such studies are still up to date.

Chemical radiosensitization has boomed in the last years for its potential application in the radiotherapy of tumours since even a modest increase of radiosensitivity of neoplastic cells results in a better therapeutic treatment.

The main classes of radioprotective and radiosensitizing drugs include compounds with respectively reducing and oxidizing properties towards the radiation induced radicals derived from biological molecules. Both processes of radioprotection and radiosensitization occur by means of complicated mechanism, whose the very early stages correspond to very fast reactions. The mechanism of action of such sub - stances can be investigated by means of radiation chemical techniques, i.e. pulse radiolysis (1). Briefly pulse radiolysis uses a short intense pulse of radiation to induce the initial physical-chemical damage and fast recording technique (i.e. absorption kinetic spectrophotometry with oscillographic output) are used to investigate the short-lived chemical species produced, and to follow their subsequent pathway.

CHEMICAL RADIOPROTECTION.

Chemical radioprotection is still an important field of research in radiobiology at all levels of biological complexity. Different types of substances, including aliphatic alcohols, sulphur or selenium containing compounds etc., offer radioprotection "in vitro" and "in vivo". The most important class of radioprotective substances includes the sulphur-compounds. These compounds can act by different mechanisms: protection by mixed-disulfide formation protection by radical scavenging and protection by the hydrogen transfer mechanism. Of particular interest is the repair model by the hydrogen transfer mechanism, originally proposed by Alexander and Charlesby (2), and further applied to biological systems by Howard Flanders (3). According to this hypothesis, the radioprotection offered by -SH compounds is due to their ability to repair the radical damage of the target molecule by hydrogen donation from the -SH group. This reaction is in competition with the damage fixation produced by oxygen.

Adams and coll. (4,5) have directly observed by pulse radiolysis radical repair reactions such as:

$$\cdot CH_2OH + RSH \longrightarrow CH_3OH + RS^\cdot$$

In solution containing excess RSH, the radical anion $RSSR^-$ formed from $RS^\cdot$ in the equilibrium reaction:

$$RS^\cdot + RS^- \rightleftharpoons RSSR^-$$

absorbs strongly at 410 nm and can be used to monitor the reaction.

A great number of rate constant for such repair reactions have been determined and they are generally lower than rate constants for the reactions of the same radicals with oxygen. If the previous model is correct, one should think that the -SH compounds are present at relatively high concentration in the cell in order to show radioprotection. An important molecule ubiquitously present in the cells at relatively high concentration is the tripeptide glutathione.

Radiation chemical data obtained in this Laboratory (6, 7) show that the -SH group is responsable of all chemical events occurring in the molecule. Moreover the transfer of hydrogen atom from glutathione to carbon radicals has been demonstrated in model systems and appears to be the most probable mechanism of protection.

Repair of nucleic acid radicals by -SH compounds has not been observed directly by pulse radiolysis and it remains to be demonstrated directly that hydrogen transfer is a mechanism of protection in cellular systems.

CHEMICAL RADIOSENSITIZATION

The effect of oxygen in enhancing radiation damage in most biological systems, has been known for many years and fundamental and clinical work on both animal and human tumours has demonstrated the importance of anoxic regions in tumours as limiting factors in radiotherapy. ANy chemical agent which acts similarly to oxygen on the biological response to radiation is of potential value in radiotherapy. Nowdays some strongly electron-affinic compounds have been shown to be the most interesting radiosensitizers in view of the large number of investigations at fundamental level and for the promising pilot clinical studies still in progress (8).

They include quinones, dicarbonyl compounds, aromatic ketones, nitrofurans and nitroimidazoles. The radiosensitizing ability of these compounds is related to their electron-affinity and to their structure, in which the oxidizing property is due to the stabilizing effect of electron delocalization by resonance. Pulse radiolysis experiments have shown that these compounds are efficient oxidizing agents and can transfer electron rapidly and quantitatively from free radicals derived from different substrated including purines, pyrimidines, nucleic acids and amino acids. These experiments demonstrate that such sensitizers are more electron affinic than target molecules and give support to the electron-trapping model for the sensitization phenomenon proposed by Adams (9). The model suggests that, following direct ionization in the target molecule, thermalised electrons migrate to some electron trapping sites in the molecule. In the presence of electron affinic compounds, electron transfer reaction from the ionized molecule to the radiosensitizers could occur in competition with the internal charge recombination. The final result is an irreversible chemical damage to the critical molecule.

CONCLUSION

The purpose of this communication is to illustrate some significant examples of the application of radiation chemistry to chemical radioprotection and radiosensitization. The results demonstrate the important role played by molecular phenomena for the interpretazion of mechanism of chemical radioprotection and radiosensitization and for the development of more active substances. Much of the information concerning the involvement of fast processes in chemical radioprotection and radiosensitization, derives from studies of simple model chemical and cellular systems carried out with fast radiation chemical techniques. Even though the great complexity of "in vivo" systems excludes a unique explanation of chemical radioprotection and radiosensitization in molecular terms only, the contribution of fundamental radiation chemical studies is still of great importance.

REFERENCES

1. Adams G.E., Wardman P., (1977): in Free Radicals in Biology, Vol. III, Chapter II, Academic Press Inc, New York p.53.
2. Alexander P., Charlesby A. (1954): in Radiobiology Symposium, Butterworth's London, p. 49.
3. Howard-Flanders P. (1960): Nature 186,485.
4. Adams G.E., Mc Naughton G.S., Michael B.D. (1968): Trans. Faraday Soc. 64, 1902.
5. Adams G.E., Armstron R.G., Charlesby A., Michael B.D., Willson R.L. (1969): Trans.Faraday Soc. 65, 732.
6. Quintiliani M., Badiello R., Tamba M., Gorian G. (1976): Modification of Radionsensitivity of Biological Systems, Vienna, IAEA, p. 29.
7. Quintiliani M., Badiello R., Tamba M., Esfandi E., Gorin G. (1977): Int. J. Radiat. Biol. 32, 195.
8. Breccia A., Rimondi C., Adams G.E. Eds. (1979): Radiosensitizers of Hypoxic Cells, Elsevier/North-Holland Biomedical Press, Amsterdam.
9. Adams G.E. (1969): in Radiation Protection and Sensitization, Moroson J.L., Quintiliani M. Eds., Taylor and Francis Ltd, London, p. 3.

DYNAMICS OF Cs-137 DISTRIBUTION IN THE MUSCLE TISSUE OF SWINE BY SINGLE AND REPEATED CONTAMINATION

J. Begović, S. Stanković and R. Mitrović

Having in view that animal production is one of the essential links in the chain: environment-animal production-man, numerous investigations have been carried out on contamination pathways, resorption, distribution and elimination of radionuclides in dependence of contamination duration, biological and physico-chemical properties of radionuclides (1, 2).

Radionuclides enter the animal organism mostly through food. From the digestive tract by resorption through the blood they are deposited into organs and tissues.

Recently, we have undertaken systematic examinations on resorption, distribution and elimination of radioactive isotopes J-131, Sr-89,85 and Cs-137 applied to domestic animals as simulated fission mixtures. In this work is presented the level of deposited radioactive caesium-137 in muscle tissue of pigs and the rate of its elimination from the organism after repeated and single contamination. Muscle tissue was examined in particular owing to its great usability in human nutrition.

MATERIAL AND METHODS

Six month old pigs of "domestic white" breed were contaminated per os by radioactive cesium-137 (CsCl carrier free). Two groups of animals were repeatedly contaminated (for 7 consecutive days) with a total dose of 2,967 kBq (daily dose - 424 kBq) and a dose of 1,609 kBq (daily dose -215.5 kBq per animal).The animals were sacrificed by total bleeding under anesthesia on day 1, 7, 14, 28 and 60 after the contamination. For radioactivity measurings the samples of musculature were taken. Another two groups of animals were singly contaminated with the same isotope and in the same way, and were sacrificed on day 3 after the contamination. Radioactivity was measured in the same samples of the musculature as in previous groups.

The radioactivity was measured in a gamma "INTERTECHNIQUE" counter in the samples of musculus longisimus dorsi, musculus supraspinatus and musculus gluteus superficialis. In order to determine the Cs-137 elimination rate from the organism of the contaminated animals the radioactivity in the samples of urine and faeces was measured every day.

The results were statistically processed and represent the mean value of the six individual samples.

RESULTS

By considering the obtained results it was observed that Cs-137 is almost evenly distribution in the examined muscle tissue of the contaminated animals. The differences among individual muscles during the examined period in both repeatedly and singly contaminated animals are unsignificant. Slight differences among the samples of some muscles are observed only on the first day after the contamination when the highest radioactivity was detected: by the dose of 2,967 kBq: musc. supraspinatus: musc. longisimus dorsi: musc. gluteus -92.87+2.22 : 83.25+18.87 : 101.75+26.27, and by dose of 1,609 kBq by the same order: 50.32+9.32 : 39.59+5.00 : 44.77+5.33. The level of radioactivity by the dose of 2.967 kBq in muscle tissue is nearly twice as high when compared with the dose of 1.609 kBq Graph 1. The radioactivity in all muscles exponentially decreases in the function of time in both applied

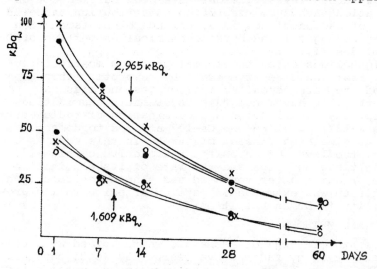

Graph. 1.- Concentration of Cs-137 in muscle tissue after the repeated contamination (in kBq/kg of wet sample) ●—● m. long.dorsi; o—o m. super; x—x m. gluteus.

Cs-137 doses. Intensity of decrease of radioactivity in muscle tissue is more expressed during the first 14 days in both applied doses than in the later experimental period, which is particularly expressed in a dose of 2,967 kBq.

The results of examination of the single contamination with two different doses are presented in Table 1. The level of radioactivity in muscle tissue of animals contaminated with a dose of 11.720 kBq is nearly twice as high when compared with the lower applied dose.

TABLE 1. Concentration of Cs-137 in muscle tissue on day 3 after single contamination (in kBq/kg of wet sample)

Musculus	Dose	
	5,772 kBq	11.729 kBq
m.supraspinatys	116.55±29.97	186.85±21.09
m.longisimus dosi	132.46±29.23	210.16±26.27
m.gluteus superficialis	121.73±5.18	219.78±31.82

Elimination of the radioactive Cs-137 from the organism of the contaminated animals is mostly performed by excretion through urine. Essential difference between the single and repeated contamination is found to be in elimination rate, i.e. in the effective time of half-life of excretion of the contaminant through the excrate (urine and faeces), Graph. 2. It is evident from the Graph that the effective half-life of excretion is between day 17 and 20.

By single contamination 50% of the applied Cs-137 is excreted by day 5 from the beginning of contamination (Tef. 1/2 is 5 days).

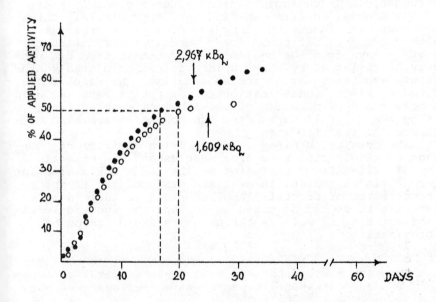

Graph.2.- Per cent of cumulatively excreted Cs-137 via urine and faeces by repeated contamination.

DISCUSSION

The analysis of the results presented shows that elimination of the deposited Cs-137 from muscle tissue of pigs, 6 months old, is not even. The elimination of this radionuclide from the muscle tissue is found to be more intensive between day 1 and 14 than later on (up to day 60). By the higher applied dose the constant of elimination rate up to day 14 was 1.50-1.71, and by the lower dose, for the same period, 0.83-1.50. Having in view the dynamics of the process of establishing equilibrium state during radionuclide deposition (1, 3) it could probably be explained the uneveness of elimination rate. The effective time of half-life of elimination of Cs-137 from muscle tissue, which was obtained for this age of pigs, was 14-15 days, while according to the data by Sirotkin (4), Buldakov and Moskalev it was 29.5, but the age of animals was not mentioned.

Burov (5) found that the deposit coefficient in two month old pigs was 7.9, while we found that in pigs of 6 months old by both applied doses was approximately the same deposition coefficient, i.e. 9.5 for the higher and 8.8 for the lower dose. In singly contaminated animals the difference in deposition coefficients was even lower, 11.4 for the higher and 11.2 for the lower dose.

Korneev et al. who established that Cs-137 and I-131 in pigs are excreted mostly via urine in contrast to the ruminants, in which these radionuclides are mostly excreted via faeces, confirm our findings. Approximately the same excretion intensity of Cs-137 from pig's organism (although doubly different doses were applied) during the first 17 days from the beginning of contamination can probably be explained by more rapid elimination of the contaminant immediately after its application. The results obtained by single contamination for excretion rate, as well as more rapid elimination of caesium from muscle tissue during the first 14 days after the contamination also contributed to this findings.

The results obtained as: deposition coefficient, constant of elimination rate from muscles and the effective time of half-life of excretion may be taken for evaluation of contaminant content in meat and its products if the concentration of perorally applied radionuclides is known, meanwhile it should be taken in account the age and species of animals, as well as the dynamics and duration of contamination.

REFERENCES

1. Корнеев, Н.В., Сироткин, Н.В. Корнеева (1977): Снижение радиоактивности в растениях и продуктах животноводства, Москва "Колос".

All other authors are cyted in the mentioned book.

TUMORIGENIC RESPONSES FROM SINGLE OR REPEATED INHALATION EXPOSURES TO RELATIVELY INSOLUBLE AEROSOLS OF ^{144}Ce

B. B. Boecker, F. F. Hahn, J. L. Mauderly and R. O. McClellan

People may inhale radioactive aerosols in (a) a single exposure from an accidental release, (b) repeated exposures in an occupational setting, or (c) chronically from an environmental exposure. It is normally assumed that the biological behavior and resulting long-term biological effects observed in a single exposure situation can be extrapolated to repeated or chronic exposure situations. The specific objective of this study is to compare the biological effects in Beagle dogs exposed by different sequences of repeated inhalation exposures to a relatively insoluble form of ^{144}Ce at dose levels known to produce tumorigenic responses in dogs exposed once to the same aerosol. After a single inhalation exposure, the dose rate to lung decreases with an effective half-life of about 175 d. For comparison, repeated inhalation exposure sequences were chosen that would (a) increase the dose rate to lung with each exposure, or (b) reestablish a given dose rate.

MATERIALS AND METHODS

Thirty-six Beagle dogs (14 to 18 months old) were given 13 brief (< 60 min.), nose-only inhalation exposures at 8-week intervals. Twenty-seven of them were exposed to ^{144}Ce in fused aluminosilicate particles (AMAD $\sim$ 1.8 µm, $\sigma_g \sim$ 1.6) and nine controls were exposed to non-radioactive fused aluminosilicate particles. The three exposure sequences to ^{144}Ce (nine dogs/group) were: repeated increase in lung burden of 2.5 µCi/kg body weight, reestablished lung burden of 9.0 µCi/kg body weight, and reestablished lung burden of 4.5 µCi/kg body weight. Post-exposure measurements included whole-body counting, physical examinations, radiography, hematology, clinical chemistry and pulmonary function. At death, necropsies were performed for gross and histopathologic evaluation and measurement of the levels of ^{144}Ce in different tissues.

RESULTS

Typical whole-body retention measurements are shown in Figure 1 for one dog from each exposure group. Each spike represents the increase in body burden due to an inhalation exposure. The difference in body burden immediately before and after each spike represents

---

This research was performed under U.S. Department of Energy Contract No. EY-76-C-04-1013 in facilities fully accredited by the American Association for the Accreditation of Laboratory Animals.

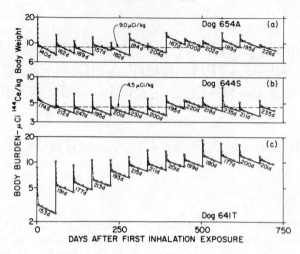

Figure 1. Whole-body counting data for 3 dogs repeatedly exposed at 56-day intervals to ^{144}Ce in fused aluminosilicate particles.

^{144}Ce deposited in the pulmonary region. Calculated pulmonary deposition varied considerably for a given dog. The overall mean lung deposition ± 1 s.d. was 26 ± 7% of the inhaled aerosol. For individual dogs, the values ranged from a low of 12 ± 4.1% to a high of 39 ± 10%.

Effective half-lives for retention in each 56-day interval between exposures are also shown in Figure 1. Group means and standard deviations are 193 ± 12, 205 ± 23, and 186 ± 19 days for the 9.0, 4.5, and 2.5 µCi/kg body weight groups, respectively.

The tissues from dog 655A, who died at 771 days after the initial inhalation exposure, were analyzed for their ^{144}Ce content. The body burden was divided among the lung (76%), liver (9.4%), skeleton (10%), tracheobronchial lymph nodes (1.4%), and other tissues (3.2%).

This study has been in progress for 5.8 years. During the first 2 years, the most pronounced biological effect was a decrease in circulating lymphocytes that occurred first in the 2.5 µCi/kg repeated and 9.0 µCi/kg reestablished groups and later in the 4.5 µCi/kg reestablished group. To date, 11 dogs have died as summarized in Table 1. The primary causes of death were radiation pneumonitis and pulmonary fibrosis (3), neoplasms (2), myelomalacia (1), autoimmune hemolytic anemia (2), bone marrow aplasia (1), parvovirus (1), and an anesthesia accident (1). Six neoplasms were noted, four in lung and one each in the spleen and tracheobronchial lymph nodes (TBLN). The four pulmonary tumors were all noted in dogs that had other primary causes of death.

All survivors in the 2.5 µCi/kg repeated and 9.0 µCi/kg reestablished groups are showing lymphopenia and radiographic signs of radiation penumonitis and pulmonary fibrosis. Alterations of gas exchange and lung mechanics are also becoming more apparent. Similar,

TABLE 1. Biological effects during first 5.8 years after first of 13 inhalation exposures to ^{144}Ce in fused aluminosilicate particles.

Lung Burden:	2.5 µCi/kg			9.0 µCi/kg			4.5 µCi/kg			C	
Dog Number:	644T	664C	645C	648S	648B	654A	665A	649U	655U	646B	648T
Pneumonitis/fibrosis	P				P	P					
Hemangiosarc., lung		X									
Sq. cell carc., lung	X										
Br. alv. carc., lung					X	X					
Hemangiosarc., spleen	P										
Hemangiosarc., TBLN									P		
Other (no tumors)			P	P			P	P		P	P

P = primary cause of death
X = other important observations at death

but less severe, changes of the same types are present in the 4.5 µCi/kg reestablished group.

DISCUSSION

This study provides a means of assessing variability in patterns of both dose and response. The pulmonary deposition data show that there was considerable variability among dogs (approximately 3X) as well as for any given dog. Such variability must be taken into account when assessing inhalation risks for a population (1).

The four pulmonary tumors seen to date in this study are plotted in Figure 2 as a function of the time of death and cumulative dose to lung. In a concurrent single exposure study, 15 pulmonary tumors were observed in 11 dogs during the first 5.8 yr (2). The rectangle drawn in Figure 2 illustrates the bounds of dose and time after initial exposure for these 15 tumors. All eight pulmonary hemangiosarcomas occurred in the dose-time region delineated by the horizontal and vertical arrows. The doses to lung received by dogs in the 2.5 µCi/kg repeated and 9.0 µCi/kg reestablished groups were within the dose range in which the 15 pulmonary tumors occurred in the singly exposed dogs. In spite of this, the first pulmonary tumor in the repeatedly exposed dogs occurred approximately 3 yr later than in the singly exposed dogs. Also in contrast to the results from the singly exposed dogs, the first pulmonary tumors in the repeatedly exposed dogs were not hemangiosarcomas. Another difference is that the only tumor seen to date in the 4.5 µCi/kg reestablished group was not in the lung but in the tracheobronchial lymph nodes.

It appears that the time of tumor occurrence and tumor type may relate to differences in patterns of dose rate to lung. In the single exposure study, the initial dose rates to lung in the 15 dogs with pulmonary tumors ranged from 150 to 320 rads/day. In the 2.5 µCi/kg repeated study, the average dose rate to lung was 19 rads/day

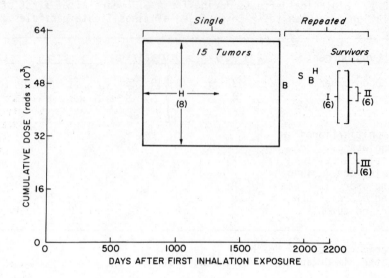

Figure 2. Schematic representation of the occurrence of pulmonary tumors. Tumor types are hemangiosarcoma (H), bronchioloalveolar carcinoma (B), and squamous cell carcinoma (S). Repeated exposure groups are (I) repeated 2.5 µCi/kg, (II) reestablished 9.0 µCi/kg, and (III) reestablished 4.5 µCi/kg.

after the first exposure and increased to 64 rads/day after the 13th exposure. In the 9.0 µCi/kg reestablished group, the dose rate was approximately 60 rads/day after each exposure. In all 3 studies, it has been assumed that the effective half-life of ^{144}Ce in the lung was 175 days after the exposures were completed.

Two major findings stand out in this continuing study. The first is the variability in deposition and retention seen among dogs. The second is the delay in occurrence of pulmonary tumors and the associated trend toward different types of tumors in the repeatedly exposed dogs as compared to singly exposed dogs. Such a comparison yields important information on how dose rate patterns can influence the resulting biological effects.

REFERENCES

1. Cuddihy, R. G., McClellan, R. O. and Griffith, W. C. (1979): Toxicol. and Appl. Pharmacol., 49, 179.
2. Hahn, F. F., Benjamin, S. A., Boecker, B. B., Hobbs, C. H., Jones, R. K., McClellan, R. O. and Snipes, M. B. (1977): In: Inhaled Particles IV, p. 625. Editor: W. H. Walton. Pergamon Press, Oxford.

RADIOBIOLOGICAL AND RADIOECOLOGICAL STUDIES WITH THE UNICELLULAR MARINE ALGAE
Acetabularia, Batophora and *Dunaliella*

S. Bonotto, A. Luttke, S. Strack, R. Kirchmann, D. Hoursiangou and S. Puiseux-Dao

Unicellular marine algae are particularly useful for investigating the effects of ionizing radiations on living organisms as well as for studying the radioactive contamination of the aquatic ecosystem (1,2). *Acetabularia* (*A. crenulata, A. mediterranea, A. peniculus*) and *Batophora* (*B. oerstedii*) are giant unicellular uninucleate green algae, containing several million cytoplasmic organelles (chloroplasts and mitochondria). *Dunaliella* (*D. bioculata*) is a flagellated microalga, belonging to the Volvocales. These algae are being used in our laboratories for biological, radiobiological and radioecological studies. Due to the development of nuclear facilities, a detailed knowledge of the effects of radiations and of the biological behaviour of radioactive substances in the biosphere is urgently needed. This paper deals with the biological and the biochemical effects of X-rays (*Acetabularia, Batophora*) and with the incorporation of ^{3}H (*Acetabularia, Dunaliella*).

MATERIALS AND METHODS

Most of the methods used have been previously reported (1,2,3). *Acetabularia* and *Batophora* cells were irradiated with increasing doses of X-rays (from 0 to 150 Kr) during their vegetative growth (stage 4) (4). Labeling experiments were performed with *Acetabularia* at stage 4 or with *Dunaliella* being in its stationary phase (about 2×10^6 cells/ml). Tritiated organic molecules, obtained from CIS Association, were added to the culture medium of the algae during various periods of time. Radioactivity was measured in a liquid scintillation spectrometer. The intracellular concentration of radioactivity was calculated on the basis of the cells' fresh weight. *Acetabularia* chloroplasts were observed with Nomarski interference optics. The starch content was determined with the Boehringer hexokinase test after hydrolysis of the storage material with α-amyloglucosidase.

RESULTS

Radiobiological studies.

a) Biological effects of X-rays on *Acetabularia mediterranea*. X-rays interfere with the morphogenesis of *A. mediterranea*. The formation of the reproductive cap and of cysts (gametangia) are strongly reduced only at relatively high doses (> 50 Kr). Several types of morphological anomalies are observed in irradiated cells : 1) loss of the whorls; 2) enlargement of the first order articles of the whorls; 3) development of a new stalk from a hair of the whorls or from a branch of the rhizoid; 4) enlargement of the apex; 5) alteration of the cap symmetry; 6) irregular growth of the caps' rays; 7) formation of irregularly shaped cysts in the caps' rays; 8) increase of cyst size; 9) cyst degeneration; 10) loss of or impaired compartition of the cytoplasm in the caps' rays prior to cyst formation. Light microscopical observations have shown the presence of elongated chloroplasts having large starch granules. Analyses performed with amyloglucosidase and hexokinase have revealed that starch accumulation in the chloroplasts increases with the radiation dose.

b) Biological effects of X-rays on *Acetabularia peniculus*. The morphogenetic processes of *A. peniculus* are affected by increasing doses of X-rays (Fig. 1A, B). Again, relatively high doses (> 50 Kr) are necessary to inhibit cap and cyst formation. Several types of morphological anomalies were found in irradiated cells : 1) loss of the apical whorl; 2) enlargement of the apical region of the stalk; 3) formation of irregularly shaped caps' rays; 4) reduction of the number and of the size of cap rays (see Fig. 1B); 5) increase of cyst size; 6) cyst degeneration; 7) absence of cyst formation in some caps' rays. Moreover, some irradiated cells turn dark green, suggesting a condensation of the cytoplasm and/or chlorophyll accumulation in the chloroplasts.

c) Biological effects of X-rays on *Batophora oerstedii*. In irradiated *B. oerstedii* cells the formation of the spherical compartments (sporangia), where later on cysts (gametangia) develop, is only delayed by the radiations (Fig. 1C, D). This finding shows that the morphological differentiation of *Batophora* is extremely radioresistant. X-rays, however, induce in *Batophora* several types of morphological anomalies; 1) reduction of the number of sporangia (see Fig. 1D); 2) formation of abnormal sporangia; 3) development of one or more sporangia along the first or second order articles of the whorls instead of at their tip; 4) enlargement of first order articles; 5) enlargement of the apical region of the stalk; 6) cyst formation in the first or second order articles of the whorls; 7) sporangia degeneration.

d) Biochemical effects of X-rays on *Acetabularia mediterranea*. Labeling experiments with thymidine-6-^{3}H, uridine-5-^{3}H and leucine-^{3}H have revealed that X-rays (50 Kr) provoke a strong reduction of DNA, RNA and protein synthesis in the chloroplasts. RNA synthesis was stimulated, however, for doses up to 25 Kr.

Radioecological studies.

a) Experiments with tritiated water (HTO). When *Acetabularia* cells (*A. crenulata, A. mediterranea*) are grown in the presence of HTO (0-5 µCi/ml), a significant amount of ^{3}H is incorporated in the total nucleic acid and protein fraction (2). However, ^{3}H supplied in the form of tritiated water is not accumulated by the algae.

b) Experiments with tritiated organic molecules. Since recent work suggested that ^{3}H may be accumulated when this element is bound to organic molecules (see ref. 3), we have studied the uptake of 10 different tritiated organic molecules by *Acetabularia mediterranea* and by *Dunaliella bioculata* : 1) thymidine-methyl-^{3}H; 2) adenine-2-^{3}H; 3) uridine-5-^{3}H; 4) L-leucine-4-^{3}H; 5) glycine-2-^{3}H; 6) L-arginine-3.4-^{3}H; 7) L-aspartic acid-2.3-^{3}H; 8) L-phenylalanine-

Figure 1. Morphological effects of X-rays on *Acetabularia* and *Batophora*. A : control cells of *A. peniculus* bearing a normal reproductive cap with 9 rays; B : 2 *A. peniculus* cells, 17 days after the irradiation with a dose of 150 Kr, showing a reduced number of caps' rays. C : Control cell of *B. oerstedii* showing the typical spherical sporangia, where later on cysts (gametangia) develop; D : *B. oerstedii* cell, 30 days after the irradiation with a dose of 50 Kr, having a reduced number of sporangia. Scale = 2 mm.

2.3-^{3}H; 9) D-glucose-1-^{3}H; 10) D-glucose-6-^{3}H. After a short incubation (30 min.), the intracellular concentration of the tritiated molecules can reach that of the external medium. However, *Acetabularia* accumulates adenine, arginine and glucose (respective concentration factors : 4.6; 5.1; 5.7), and *Dunaliella* is capable of concentrating adenine and leucine (respective concentration factors : 122.7; 11.4).

DISCUSSION

Our radiobiological studies show that the main morphogenetic processes of *A. mediterranea*, *A. peniculus* and *B. oerstedii* are affected by the radiations. Certainly, the sequence of events during the cells' developmental cycle is only realized under a well co-ordinated cooperation between the nucleus and the organelles. Most probably, X-rays interfere with this intergenomic co-operation, provoking different types of morphological anomalies. The radiations inhibit the syntheses of DNA, RNA and proteins in the chloroplasts of *Acetabularia* cells. Chloroplasts, which transform solar energy for the benefit of the cell, may play an important role for the realization of morphogenesis.

Experiments with tritiated water have revealed that *Acetabularia* cells are unable to concentrate ^{3}H. However, a significant amount of this radionuclide is incorporated into the genetic material of the cells (3). When organically bound ^{3}H is supplied to *Acetabularia* or to *Dunaliella*, a selective accumulation of some substances is observed. Our results contribute to a better understanding of the impact of radiations on living organisms and of the biological behaviour of ^{3}H in the aquatic system.

ACKNOWLEDGEMENT. We thank Mr. A. Bossus and Mr. G. Nuyts for assistance and Mrs. Jeannine Romeyer-Luyten for typewriting the text. The support of the Commission of the European Communities is acknowledged.

REFERENCES

1. Bonotto, S., D'Emilio, M.A. and Kirchmann, R. (1979) : In : Developmental Biology of *Acetabularia*, p. 49. Editors : S. Bonotto, V. Kefeli and S. Puiseux-Dao. Elsevier/North-Holland, Biomedical Press, Amsterdam-New York-Oxford.
2. Bonotto, S., Ndoite, I.O., Nuyts, G., Fagniart, E. and Kirchmann, R. (1977) : Current Topics in Radiation Research, 12, 115.
3. Strack, S., Bonotto, S. and Kirchmann, R. (1979) : Paper presented at the 14th European Marine Biology Symposium, Helgoland, September 23-29, 1979.
4. Bonotto, S. and Kirchmann, R. (1970) : Bull. Soc. Roy. Bot. Belgique, 103, 255.

PARA-HYDROXYBENZOIC ACID A HYPOXIC RADIOSENSITIZER IN BACTERIAL CELLS

N. Sade and G. P. Jacobs

The relative radioresistance of hypoxic cells present in some tumours is a serious limitation in attempts to increase the therapeutic ratio between tumour control and damage to normal tissue during radiotherapy. Hyperbaric oxygen, fast neutrons and π^- mesons are suggested methods of overcoming this problem. The use of chemical radiosensitizers which effectively act upon hypoxic cells is another approach (1). A major group of such hypoxic sensitizers resemble oxygen in their mode of action, and it has now been fairly well established that their efficiency as radiosensitizers, on a molar basis, is directly related to their electron-affinity (2-4). Sensitization in bacterial systems by p-nitroacetophenone (PNAP) (4-6) and its more soluble derivative NDPP (7,8) led to the testing of other nitro-aromatic compounds such as the nitroimidazoles. Two promising drugs to have emerged recently, metronidazole and misonidazole, have been shown to be effective both in vitro and in vivo, and, more recently, the results of tests with these drugs in patients have been encouraging (9,10). Both these compounds are of particular interest because of their clinical use as trichomonicides, with considerable pharmacological, toxicological and pharmacokinetic information available. It seems quite possible that they may prove to be of value in clinical radiotherapy.

The aim of the present investigation is to look for other compounds, whose medicinal use is established and which, by virtue of their electron affinity, might be anticipated to act as radiosensitizers. One such group of compounds are the esters of p-hydroxybenzoic acid which are used in many pharmaceutical formulations as antimicrobial preservatives. In the first instance, the effect of different concentrations of p-hydroxybenzoic acid on the radiation sensitivity of oxic and anoxic buffered suspensions of the bacterium Staphylococcus aureus has been examined.

MATERIALS AND METHODS

p-Hydroxybenzoic acid (PHBA) was supplied by Sigma Chemical Co. The test organism was Staphylococcus aureus Oxford (NCTC 8236) maintained on slopes of nutrient agar at 4°C. Suspensions were grown to log phase in nutrient broth, washed and resuspended in fresh 1/15M phosphate buffer saline (pH 7.0) at a concentration of approximately 10^8 cells per ml prior to irradiation. PHBA was incorporated into the buffer saline at suitable concentrations. Routinely the bacteria were in contact with the additive for at least 45 min prior to irradiation.

γ-irradiation was carried out at ambient temperature using either a M38-3 Gammator 6200Ci ^{137}Cs source or a Gammacell 220 24000 Ci ^{60}Co source. Irradiation vessels were glass vials of od 19.7mm sealed with gas-tight rubber closures. Using the ^{137}Cs source, arrangement of these vessels within the irradiation chamber permitted the simultaneous irradiation of 18 suspensions at an average dose rate of 1.75 krads per min. With the ^{60}Co source, 36 suspensions could be simultaneously irradiated at predetermined dose rates ranging from 5.1 to 7.4 krads per min. Deoxygenation was by bubbling oxygen-free N_2 (less than 3 ppm O_2) for 20 min through the suspensions immediately prior to irradiation. For maintenance of suspensions under aerated conditions O_2 was bubbled through the suspensions for 5 min.

Suspensions were irradiated for fixed time intervals such that at the maximum dose level tested at least two decades of inactivation were generally achieved. Following irradiation bacterial suspensions were appropriately diluted and four 0.05ml aliquots of each diluted suspension pipetted onto plates of nutrient agar. After overnight incubation at 37°C, the micro-colonies formed were scored. Each colony was taken as indicative of a single bacterium in the original suspension. Counts performed on unirradiated samples of test suspensions treated identically to those exposed to irradiation were used as control estimates of the number of viable cells in calculations of surviving fractions. Dose-ln survival curves were constructed from five experimental points. Variation in values of slopes of these curves (inactivation constants) with changes in test conditions have been used to demonstrate quantitatively changes in radiation sensitivity.

RESULTS

Typical dose survival curves for Staph. aureus suspensions irradiated in the presence of different concentrations of PHBA in anoxia are depicted in Figure 1. Included in this figure are the responses for suspensions irradiated in air and anoxia in the absence of PHBA. These curves are linear over the dose range tested. All our computed values of inactivation constants together with standard deviations have been plotted in Figure 2 as a function of PHBA concentration. The horizontal lines are the levels of sensitivity observed when cells are irradiated in the presence and absence of O_2 in buffer alone. The corresponding parallel dashed lines represent standard deviation about these values. Increasing concentrations of PHBA from 10^{-6} to 5×10^{-5}M in anoxic bacterial suspensions produced no change in the response characteristic of bacteria irradiated in anoxic buffer alone. However further increases in PHBA concentration to 6×10^{-3}M caused a marked increase in radiation sensitivity, with the maximal response very close to that for suspensions irradiated in the presence of oxygen. The testing of higher PHBA concentrations was confounded by problems of toxicity.

PHBA was also tested for radiosensitizing activity in the presence of O_2. These tests were carried out at concentrations of PHBA that cause significant sensitizing effect in anoxic suspensions. The responses obtained are very close to that characteristic of bacteria suspended in oxygenated buffer alone, clearly demonstrating that the enhancing effect of PHBA and oxygen are not additive and that the sensitizing action of PHBA operates within the oxygen effect.

Preliminary studies showed that PHBA was somewhat toxic to the bacterium at concentrations of and above 6×10^{-3}M. Experiments were done to ensure that PHBA at lower concentrations showed no toxicity

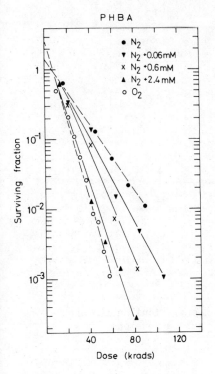

Figure 1. Survival data for hypoxic suspensions of Staph. aureus irradiated in the presence of different concentrations of PHBA.

towards the biological system. These control experiments demonstrated that the unirradiated PHBA had no effect on the viability of the unirradiated cells, that radiolysis products of PHBA at test concentrations had no effect on the viability of irradiated cells, and that these products carried over to the plating medium were without effect on the irradiated bacterium undergoing colony formation.

DISCUSSION

Our major finding is that PHBA acts as an efficient hypoxic radiosensitizer when tested against Staphylococcus aureus. The fact that the degree of sensitization is close to that observed in the presence of oxygen (DMF 2.5), suggests that PHBA may be simulating several of the previously proposed sensitizing actions of oxygen (11, 12). Only one such action may be as a result of its electron-affinity. Furthermore, the lack of sensitizing action in the presence of oxygen is evidence that this agent operates within the 'O_2 effect'.

Although the actual mechanism by which PHBA exerts its radiosensitizing action is unknown, several models have been proposed for the mechanism of sensitization by electron-affinic agents in general (for example, 3,13,14). The testing of PHBA in the presence of specific radical scavengers may help elucidate its mode of operation as a radiosensitizer. Whether the esters of PHBA, which are known to be less toxic than the parent compound, will likewise possess radiosensitizing properties remains to be investigated.

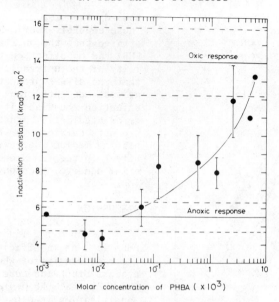

Figure 2. The effect of PHBA concentrations on the anoxic radiation response of <u>Staph. aureus</u> suspensions.

REFERENCES

1. Fowler, J.F. (1972): Clin. Radiol., 23, 257.
2. Adams, G.E. and Dewey, D.L. (1963): Biocem. Biophys. Res. Commun., 12, 473.
3. Adams, G.E. and Cooke, M.S. (1969): Int. J. Radiat. Biol. 15, 457.
4. Tallentire, A. and Jacobs, G.P. (1972): Ibid, 21, 205.
5. Adams, G.E., Asquith, J.C., Dewey, D.L., Foster, J.L., Michael, B.D. and Willson, R.L. (1971): Ibid, 19, 575.
6. Ewing, D. (1973): Ibid, 24, 505.
7. Jacobs, G.P. and Tallentire, A. (1976): Trans. Israel Nucl. Soc., 4, 125.
8. Adams, G.E., Asquith, J.C., Watts, M.E. and Smithen, C.E. (1972): Nature New Biol., 239, 23.
9. Deutsch, G., Foster, J.L., McFadzean, J.A. and Parnell, M. (1975): Br. J. Cancer, 31, 75.
10. Gray, A.J., Dische, S., Adams, G.E., Flockhart, I.R. and Foster, J.L. (1976): Clin. Radiol., 27, 151.
11. Tallentire, A. and Powers, E.L. (1963): Radiat. Res., 20, 270.
12. Ewing, D. and Powers, E.L. (1976): Science, 194, 1049.
13. Powers, E.L. (1972): Israel. J. Chem., 10, 1199.
14. Chapman, J.D., Reuvers, A.P., Borsa, J. and Greenstock, C.L. (1973): Radiat. Res. 56, 291.

A DOSIMETRIC MODEL FOR TISSUES OF THE HUMAN RESPIRATORY TRACT AT RISK FROM INHALED RADON AND THORON DAUGHTERS

A. C. James, J. Rosemary Greenhalgh and A. Birchall

1. INTRODUCTION

Epidemiological studies carried out amongst populations of underground miners since the 1950's, notably the follow-up of uranium miners in the United States and Czechoslovakia, have firmly established correlations between the risk of contracting bronchial cancer in mine environments and cumulative exposure to the short-lived daughters of radon-222 gas. The Working Level Month (WLM) has evolved as the measure of cumulative exposure best correlated with excess risk of lung cancer in each individual population studied. This WLM unit of exposure is equivalent to the total alpha energy potentially available for absorption in lung tissue as a consequence of inhaling for 1 month air containing any combination of the short-lived radon daughters that would release 1.3×10^5 MeV alpha energy per litre volume in decaying to ^{210}Pb (RaD).

We have attempted here to construct a model for dosimetry of bronchial epithelial tissues in the human that takes adequate account of the available knowledge of radon daughter aerosol characteristics, their deposition and subsequent redistribution by mucociliary action and absorption into the bloodstream and the location of sensitive cells in the respiratory tract. The model is applied to examine the influence of environmental and physiological factors on the relationship between the WLM unit of exposure and the pattern of dose absorbed by sensitive lung tissue.

2. DOSIMETRY OF BRONCHIAL STEM CELLS

Most lung cancer in uranium miners arises in the second to sixth bronchi with only about 10% occurring more distally. It is thought that the basal layer of bronchial stem cells in the upper airways constitutes the sensitive tissue.

Irradiation of basal cells from two principal sources of α-activity must be considered; (i) the Gel phase of mucus undergoing ciliary clearance and (ii) daughters diffusing through the mucosa in a concentration gradient.

We have weighted the contributions to basal cell dose from α-particles emitted from both mucus and mucosal tissue according to

This work was supported by CEC Contract 182-B10UK

our assumed probability distribution for epithelial thickness in each class of bronchi.[1]

3. COMPARTMENT MODEL OF RADON DAUGHTER RETENTION

In order to describe the different clearance rates for insoluble particles and ionic lead observed in ciliated mucosal tissue of laboratory animals[2], we have assumed that radon daughters deposited on the bronchial surface fractionate so that a proportion remain associated with the gel phase of mucus and are thus subject to mucociliary clearance, the complementary part transferring rapidly via the sol phase to mucosal tissue. The retention characteristics of lead ions in epithelial tissue are complex[3], but for dosimetric purposes, biological retention can be represented adequately by a single first order process with half-time of about 4 hours.

We have assumed a relationship between mean mucus velocity and airway calibre[4] and derived rate constants of mucociliary transport for the compartments of our model bronchial tree consistent with overall bronchial transit times ranging from 'fast' to 'slow' in the human.

4. CHARACTERISTICS OF RADON DAUGHTER AEROSOLS

The fraction of radon daughter activity remaining unattached to condensation nuclei in the inhaled atmosphere is recognised as a major variable influencing the dose absorbed by bronchial tissue. Under modern mining conditions, where diesel plant is used extensively, the concentration of airborne nuclei is higher than in mines operated during the 1950's to which epidemiological data relate[5] and the unattached fraction lower. Typical aerosol characteristics may be considered as follows:

	UNATTACHED FRACTION		ACTIVITY MEDIAN DIAMETER OF NUCLEI, μm ($\sigma_g = 2$)
	f_A	f_B	
RADON-222 DAUGHTERS			
Pre-1960 U-mine	0.1	0.01	0.1
Modern U-Mine	0.05	0.005	0.3
Indoors	0.2	0.2	0.08
RADON-220 DAUGHTERS			
Workplace	1.0	0.02	0.1

5. BRONCHIAL DEPOSITION

We have calculated the probability of depositing radon daughter aerosols in each bronchial airway assuming that bronchial anatomy can be represented by a regular dichotomous system of branching tubes under laminar flow conditions, and that 50-60% of inhaled unattached daughters are deposited in the nose but only 2-3% of nuclei. The probability of depositing unattached daughters in the larger bronchi is calculated to be about 2 orders of magnitude higher than that for nuclei. We conclude that deposition probability decreases as tidal volume and respiratory frequency increase except for the larger 0.3 μm median diameter aerosol in modern mines. For this size distribution a significantly different pattern of deposition is calculated because of the dominant contribution made by inertial impaction.

6. INFLUENCE OF BIOLOGICAL FACTORS ON DOSE PER WLM EXPOSURE

After bronchial dimensions and the distribution of epithelial thickness, the principal biological factors affecting dose absorbed by bronchial stem cells are the rates of mucociliary clearance, the fraction of deposited radon daughters absorbed by epithelial tissue and the retention characteristics of this absorbed fraction. For both radon-222 and radon-220 daughters variation of the absorbed fraction of deposited atoms over a probable range from 0-50% does not critically affect dose, except in the major bronchi where basal cells are mostly out of range of α-particles emitted from mucus. Biological retention half-time is, likewise not a critical factor, although variation of the rate of mucociliary clearance has a larger effect in redistributing dose between different regions of the bronchial tree, most markedly for deposited radon-220 daughters. Stem cell dose is calculated to increase through the larger airways, peaking in the airways most proximal to the bronchioles, largely because of the thin epithelium assumed in this region.

7. ENVIRONMENTAL AND PHYSIOLOGICAL VARIATION

The dose to critical cells in lobar and segmental bronchi of miners exposed to a pre-1960 uranium mine atmosphere is calculated to be about 7 mGy per WLM cumulative exposure (Fig. 1) at a moderate rate of physical work. Doubling minute volume would increase this dose to about 11 mGy. Variation of the unattached fraction of the inhaled daughters from zero to twice its typical value would change critical cell dose by approximately $\pm$ 40%. In a modern uranium mine environment unattached daughters are less important but doubling minute volume is expected to increase the dose to critical cells by about a factor 2.5.

Radon-222 daughters inhaled at rest by an adult man in a domestic indoor environment are calculated to deliver about 75% of the dose to critical cells per WLM exposure incurred under conditions of moderate physical activity in a pre-1960 uranium mine. Inhaled radon-220 daughters are expected to deliver about 50% of this reference dose.

These calculations imply that factors other than radiological dose have a dominant influence on the sensitivity of tissues of the respiratory tract to carcinogenesis, since bronchiolar epithelium distal to segmental bronchi was the most heavily irradiated tissue in exposed uranium miners, yet 90% of their bronchial cancer was observed in more proximal airways.

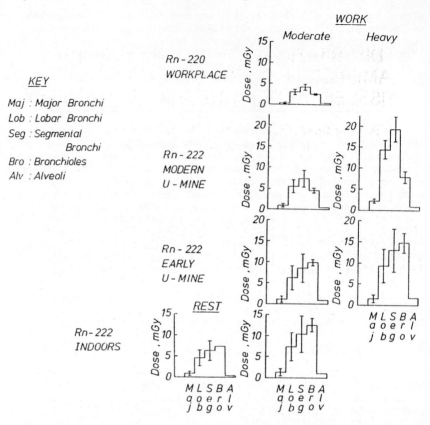

Figure 1. Comparative lung dosimetry for adult man exposed to 1 WLM of radon-222 or radon-220 daughters. Bars show the dispersion in the estimated dose per WLM exposure introduced by varying the free ion component of the inhaled aerosol from zero to twice its typical value.

REFERENCES

1. Gastineau, R.M., Walsh, P.J. and Underwood, N. Health Phys. 23 (6), 857 (1972).
2. Greenhalgh, J.R., James, A.C., Birchall, A. and Smith, H. in preparation.
3. Chamberlain, A.C., Heard, M.J., Little, P., Newton, D., Wells, A.C. and Wiffen, R.D. AERE - R9198 (1978).
4. Jacobi, W. Personal Communication.
5. George, A.C., Hinchliffe, L. and Sladowski, R. Am Ind. Hyg. Assoc. J. 36 484 (1975).

DISTRIBUTION OF PLUTONIUM AND AMERICIUM IN HUMAN AND ANIMAL TISSUES AFTER CHRONIC EXPOSURES

J. K. Miettinen, H. Mussalo, M. Hakanen, T. Jaakkola, M. Keinonen and P. Tähtinen

Distribution of plutonium in tissues of a rather homogenous group of fifty southern Finns has been elucidated. The subjects are healthy adults, ages 20 to 60, who died accidentally. Their intake of plutonium has been from fallout via inhalation; no occupational or significant dietary exposure seems possible.

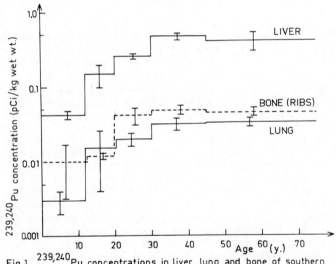

Fig.1. $^{239,240}Pu$ concentrations in liver, lung and bone of southern finns who died in 1976-1977.

Concentration in human liver was approximately 8 times higher than concentration in bones on the basis of wet weight (Fig. 1). Tissue concentrations in subjects over 20 years of age still reflect their high inhalatory intake during the 1960s. Whole liver contains 49, whole skeleton 40, muscles 8 per cent of the estimated total body burden, 1 pCi (1977-78) (Fig. 2).

Plutonium concentration in different bones: vertebrae, long bones, ribs, varied only little (Table 1). As seen in Table 1, Pu concentration in vertebrae is only 1.3 times that in other bones analyzed. Other authors have reported considerably bigger differences (1).

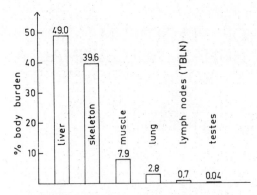

Fig.2. Distribution of plutonium in southern Finns in 1976-1979.

Table 1.
^{239,240}Pu in bone samples of southern Finns

	pCi/kg wet	pCi/kg ash wt.	
vertebrae	0.071±0.010	0.50±0.05	(10)[A]
ribs	0.054±0.008	0.31±0.05	(10)
femur shaft	0.055±0.006	0.12±0.01	(9)
femur condyles	0.05 ±0.03	0.26±0.14	(3)

A) Number of samples

Plutonium intake via inhalation during life time was calculated on the basis of analyses of air filters (Fig. 3A). Ratio of the determined liver concentration to the estimated total life time inhalation intake (Fig. 3B) indicated direct correlation (1).

A group of northern Finns, the reindeer-herding Lapps, also obtained the bulk of Pu via inhalation but in addition some plutonium in their diet, rich in reindeer liver. Samples of only a few individual have been obtained so far. The results on tissue-Pu do not differ significantly from those of southern Finns. Since their dietary intake of plutonium is fairly well known (males 40 pCi/a in 1967, 10 pCi/a in 1977), it can be calculated that on the basis of the absorbability stated by the ICRP ($3 \cdot 10^{-5}$ to 10^{-6}) their dietary plutonium retention is of the order of one per cent of the measured body burden. If the real absorbability would be higher, higher tissue concentrations could be expected, too. Although the Pu results of reindeer herders are not yet statistically conclusive, they provide a positive indication that the ICRP value is correct. Further analyses of this population will be valuable for corroboration of this ICRP value, so far mainly based on animal results.

Reindeer is an exceptional animal since it obtains large amounts of transuranium elements in its natural winter diet, which mainly consists of lichen (Pu-239,240: ca. 100.000 pCi/a in 1967, 15.000 pCi/a in 1976; Am-241 ca. 20.000 pCi/a in 1967, 3.500 pCi/a in 1976). Thus, reindeer provides an opportunity for direct determination with good accuracy of the gastrointestinal absorption of these transuranium nuclides from a natural diet.

Pu-239,240 and Am-241 were also analysed in elk (an animal resembling the American moose) because it is closely related to

Distribution of Plutonium and Americium

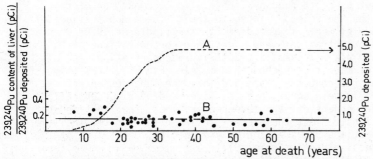

Fig. 3. A: Calculated life-time deposition of inhaled plutonium in total body of subjects who died in 1976-1977. B: The ratio of ^{239,240}Pu content in liver to the total lung deposition of plutonium during the life-time.

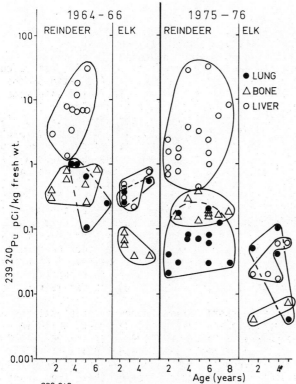

Fig. 4. ^{239,240}Pu in reindeer and elk liver, lung and bone during the periods of high (1964-1966) and low fallout (1975-1976).

reindeer but does not feed on lichen. Thus, comparison of reindeer and elk indicates the role of lichen to the tissue concentrations of these transuranium elements in reindeer. In the middle of 1960s when atmospheric Pu-concentrations were high, lung concentrations of Pu in reindeer and elk were similar, but liver concentrations in reindeer one order of magnitude higher than in elk (Fig. 4). Presently, when the atmospheric concentrations are low, the lung concentrations are still approximately equal, but the liver and bone concentrations of reindeer two orders of magnitude higher. Thus, the liver and bone concentrations reflect the dietary intake of plutonium. The liver of reindeer contains 50, skeleton 30 and muscles 10 per cent of the total body burden of plutonium (Fig. 5).

The Am-241 determinations of human tissues are still under study but in reindeer the Am-241 concentrations (on wet weight basis) are ca. 20 to 40 per cent of the Pu-239,240 concentrations in liver and lung, but 80 to 140 per cent in the trabecular bone of old animals (Table 2). Much of the Am-241 in reindeer tissues is due to ingrowth from Pu-241 in the animal. Accumulation of Am-241 in bone is much higher than of Pu-239,240. The purpose of this study is to establish wheather this situation is true also for the human bone.

REFERENCES

1. Mussalo, Helena, Jaakkola, T., Miettinen, J.K. and Laiho, K. (1980): Distribution of fallout plutonium in southern Finns, Health physics (In press).

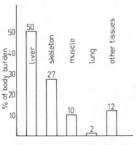

Fig. 5. Distribution of ^{239,240}Pu in reindeer in finnish lapland.

TABLE 2. ^{241}AM AND ^{239,240}Pu IN REINDEER IN FINNISH LAPLAND DURING 1974-1976. STANDARD DEVIATION OF THE RADIO-ASSAY (1σ) IS INDICATED.

		^{241}AM pCi/kg WET	^{239,240}Pu pCi/kg WET WT.	$\frac{^{241}AM}{^{239,240}Pu}$
LIVER				
REINDEER	I (o, 13A)	1.17±0.14	7.10±0.20	0.16±0.02
REINDEER	II (o, 4.5A)	0.19±0.05	0.5 ±0.2	0.39±0.18
REINDEER	III (o, 11A)	0.89±0.10	4.44±0.27	0.20±0.03
TRABECULAR BONE				
REINDEER	I	0.25±0.15	0.33±0.02	0.76±0.46
REINDEER	II	0.10±0.14	0.11±0.04	-
REINDEER	III	0.37±	0.27±0.08	1.37±0.46
LUNG				
REINDEER	I	0.06±0.02	0.13±0.03	0.47±0.15
REINDEER	II	0.09±0.05	0.35±0.04	0.26±0.13
REINDEER	III	-	0.20±0.14	-

ETUDE EXPERIMENTALE DES CANCERS INDUITS CHEZ LE RAT PAR DES PARTICULES A TRANSFERT LINEIQUE D'ENERGIE ELEVEE

Michèle E. Morin and J. E. Lafuma

Malgré l'hétérogénéité de répartition des radioéléments entre les différents organes et à l'intérieur de chaque organe après une inhalation de transuraniens émetteurs alpha chez le rat, on obtenait pour une même dose absorbée calculée (exprimée en nombre de particules émises par gramme d'organe) une fréquence identique de cancers dans le squelette et le poumon (réf.1). Pour une dose moyenne identique délivrée à l'organe, la dose délivrée à la cellule était différente pour l'os et le poumon. Dans le poumon, la dose était répartie d'une façon relativement homogène, alors qu'elle était très hétérogène dans l'os. Il en résultait que la cellule osseuse se trouvait recevoir en réalité une dose beaucoup plus élevée que la cellule pulmonaire. Les radiosensibilités de ces deux organes étant peu différentes, la cellule osseuse pouvait donc être moins sensible que la cellule du poumon à l'irradiation alpha.

Nous avons voulu vérifier ce phénomène pour un autre type d'irradiation, et nous avons soumis des rats mâles de race Sprague Dawley à une irradiation externe chronique par des neutrons, cette irradiation délivrant une dose beaucoup plus homogène à tous les organes de l'animal et le transfert linéique d'énergie était comparable dans les deux cas.

Les rats sont placés dans une cavité du réacteur Néréide situé à Fontenay aux Roses, conçue pour les expériences biologiques (réf.2). Les neutrons émis sont des neutrons de fission d'une énergie moyenne de 1 Mev. Nous avons, en plus, une composante gamma égale à 20% de la dose neutronique. En ce qui concernait les émetteurs alpha, les doses délivrées calculées à l'organe variaient de 3 à 3.000 rad et ne tenaient pas compte de l'hétérogénéité de distribution dans les organes; pour les neutrons, nous avons essayé d'obtenir des doses pouvant se comparer aux précédentes; nous avons choisi une gamme de doses mesurées à la peau variant de 1.5 à 800 rad et délivrées sur une durée variant de 1 jour à 6 semaines.

RESULTATS.

Tableaux 1 et 2.

Comme avec les transuraniens, la durée de vie des rats après irradiation neutronique dépend de la dose totale qui leur a été délivrée.

Pour le moment, nous n'avons de résultats définitifs avec les neutrons que pour 220 animaux irradiés à plus de 100 rad. Pour ces 220 rats, nous obtenons 221 cancers, soit en moyenne 1 cancer par animal.

TABLEAU 1. -Inhalation de transuraniens par des rats

Dose délivrée à l'organe (rad)	CANCERS %		
	Poumons	Squelette	Autres organes
0	1	0	0
3200	85	-	-
1000	54	-	-
750	-	40	-
350	26	22	-
150	10	10	-
110	-	5	-
35	-	2	-
25	-	-	31
3	-	-	10

TABLEAU 2. Irradiation de rats par des neutrons

Dose (rad)	CANCERS %				Durée de vie (jours)
	Poumon	Os	Peau	Autres organes	
0	0	1	1	8	650
800	10	15	10	20	239
460	15	2.5	23	60	365
240	22	8	37	43	455
150	20	6	25	51	485
50 ±	4	8	21	38	455
30 ±	6	3	6	42	460

± Expériences non terminées.

Les résultats donnés pour les doses de 50 et 30 rad sont des résultats qui ne sont pas définitifs puisque tous les animaux ne sont pas encore morts.

Les organes les plus sensibles à l'action des neutrons sont d'une part la peau et les tissus sous-cutanés et d'autre part les poumons.

Etude Experimentale des Cancers Induits Chez le Rat

Le rapport du nombre de cancers osseux au nombre de cancers pulmonaires obtenus après irradiation neutronique est égal à 0.4; pour des neutrons de 1 Mev le rapport des doses absorbées par l'os et le poumon est égal à 0.65 (réf.3). La radiosensibilité de ces organes est donc comparable à la précision de mesure et de calcul près. Les résultats obtenus avec les émetteurs alpha pouvaient faire croire à une sensibilité cellulaire différente, mais cette interprétation n'est pas compatible avec les résultats obtenus avec les neutrons; les différences de radiosensibilité obtenues ne sont pas assez grandes pour confirmer cette hypothèse.

Un phénomène particulier apparaît dans le cas de ces irradiations neutroniques, c'est la proportion beaucoup plus élevée d'une part de cancers de la peau et des tissus sous-cutanés (28%), d'autre part d'angiosarcomes de tous les tissus (8%). Si on compare ces résultats à ceux obtenus avec les émetteurs alpha où l'on avait 9% de cancers cutanés et 1,9% d'angiosarcomes, on constate que l'irradiation totale et homogène de ces deux systèmes provoque l'apparition d'un fort pourcentage de cancers dans ces tissus.

CONCLUSION.

Lorsque l'irradiation de l'organisme est totale, on voit apparaître une fréquence anormalement élevée de cancers du système vasculaire et du système de revêtement.

Les résulatts des irradiations neutroniques montrent que le fait de calculer la dose à l'organe pour les émetteurs alpha afin d'exprimer la relation dose-effet est une pratique correcte.

L'hypothèse que nous avions émise d'une radiosensibilité différente de la cellule osseuse et de la cellule pulmonaire n'est pas confirmée par ces nouvelles expériences.

REFERENCES.

1. MORIN M., MASSE R., NENOT J.C., METIVIER H., NOLIBE D., PONCY J.L. LAFUMA J.
   Etude expérimentale des différents effets observés après inhalation de radioéléments émetteurs alpha
   IV° Congrès International - I.R.P.A. - Paris, 24/30 Avril 1977
   Vol. 4, p. 1321/1328
2. PARMENTIER N., LAVIGNE B., CHEMTOB M., N'GUYEN V.D., SOULIER R.
   Paramètres microdosimétriques. Comparaison des valeurs mesurées directement et des valeurs calculées à partir du spectre des neutrons incidents.
   3° symposium sur la dosimétrie des neutrons en biologie et médecine. Neuherberg, Munich, 23-27 Mai 1977
3. I.C.R.U. Report n° 26 - 1977

PRESENT STATE OF RADIO-STRONTIUM DECORPORATION RESEARCH WITH CRYPTAND(222)

W. M. Müller

Strontium-90 appears in high percentages in reactor burn-up as well as in nuclear fall-out . In animal experiments it has been demonstrated numerously that this bone-seeking element provokes skeletal malignancy . Therefore it is not only potentially hazardous to workers of nuclear power plants and related industries, but as an environmental contaminant to every body .

Prevention of intestinal uptake , removal during circulation , and removal of bone deposits are the three possibilities of therpeutic onset for decorporating remedies after Sr-90 incorporation .

Cryptating agent (222) is presented here as a potent means to remove radioactive strontium during the circulation of the latter . With (222) the strongest decorporation effects ever reached on the radionuclides Sr-85, Ba-133, and Ra-224 were obtained by us in rats (1,2,3,4) . Our results were confirmed by Knajfl et al.(5) and Batsch et al.(6) . From recent experiments we extrapolate tentatively from rat to man , presenting here a probable treatment scheme, demonstrating the decorporation effect as function of the (222)-dose and the time interval between incorporation and treatment start .

Some toxicological aspects are also discussed.

MATERIALS AND METHODS

The basic parameter in decorporation experiments is the effectiveness quotient (EQ), which permits to judge on the pure decorporation effect of an agent . In our experiments this quotient is defined as

$$EQ = 100 \times \frac{\text{TBR of radionuclide in (222)-treated rats}}{\text{TBR of radionuclide in untreated rats}}$$

(TBR = Total Body Retention) . One obtains this quotient with high significance, using two groups of 5-6 rats in general of about 300 g body weight, which were injected i.v. or i.p. with about 37 KBq Sr-85,(Ba-133 and Ra-224). One of these groups received additionally either together with, or after a time delay of the incorporation of the radionuclide, a constant or varying doses of cryptand (222) either i.v. or i.p. . The TBR was measured with a one-channal gamma-spectrometer immediately after the incorporation and consecutively after 24 hours .

Two of the three questions , which had to be answered experimentally, are demanded by what we call Schubert - Catsch-Heller-Relationship (S-C-H) , valid within the re-

strictions for E and EQ given by Catsch (7) :
$$\frac{\log EQ_1}{\log EQ_2} = \frac{\log E_1}{\log E_2}; \log E = \log K_{Sr(222)} - \log K_{K(222)} + \log A$$
A comprises the dose of (222). The S-C-H rules decorporation involving the EQ as dependent of two main physico - chemical parameters, 1) the stability constant K_1 of cryptand (222) towards the radionuclide (8), and 2) the concentration e.g. the dose of (222) (9), e.g. $EQ=f(K_1)$ and $EQ=f(Dose)$. Third, the dependency of EQ from the starting point of treatment after incorporation of the nuclide e.g. EQ as a time-fuction $EQ=f(Time)$.

RESULTS

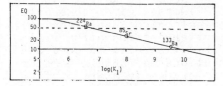 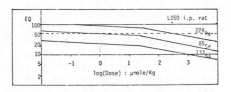

Graph 1 $EQ=f(K_1)$ Graph 2 $EQ=f(222-Dose)$

The decorporation effect rises, e.g. the EQ drops with a rising stability constant K_1 strictly obeying the (S-C-H) (Graph 1). It rises further with a rising dose, obeying also the S-C-H down to a dose of 25 µmole/Kg; further below it obeys again the S-C-H equations if one introduces instead of A, root of A (Graph 2)(9). The striking fact that EQ as time-function (Graph3)represents also a straight line in a log-log system is explained by the fact that decorporation with (222) follows the availability of strontium in the blood or extracellular space of rats(4). Graph 4, derived from the results of Graph 2 & 3, represents the EQ as function of the (222)-dose and a delayed treatment start ($EQ=f(Dose \& Time)$), which permited to extrapolate from rat to man(Table 1), using further : the Sr-retention in blood of rats (10) and men (11); and the following assumptions : 1) the extracellular space is as well reaction space as distribution space of (222) and Sr-85 ; 2) the Sr-retention in the extracellular space paralells the Sr-retention in the blood; 3) the blood volume of rats and men corresponds to 6% of the body weight; the extracellular space of rat and man corresponds to 16.6% of the body weight;4) the EQ-values obtained from experiments in rats do not essentially differ from those in men .

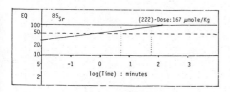 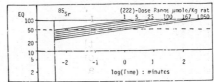

Graph 3 $EQ=f(Time)$ Graph 4 $EQ=f(Dose \& Time)$

DISCUSSION

Though, a significant decorporation effect (EQ) in man (Table 1) may obviously be obtained, even after a few days of strontium incorporation, it seems hard to obtain an EQ lower than 50% with a single and relatively high (222)-dose even after a few minutes of treatment start. Further, in man the decorporation of strontium seems to slow much more down with the reduction of the (222)-dose than with an increasing delay of treatment. Consequently high doses of (222) are necessary in order to obtain a sufficient decorporation. High doses of (222) may approach acute toxicity. (LD-50 = 292 μmole/Kg rat i.p.(12)). If several protracted small (222)-doses given consecutively, may avoid toxicity of a single high dose, eventually producing the same or even a better decorporation effect must still be demonstrated. Table 1 permits to conclude finally : With a (222)-treatment one may obtain a decorporation effect of 50% after a very early start of therapy, accopainied by an additional significant percentage of naturally excreted strontium, but it seems inavoidable, that a certain rest of the nuclide will be trapped by bone. Decorporation of radiobarium will be more beneficial because of a higher K_1 (Graph 1 & 2).

The therapeutic range of (222) deduced from Graph 2, varies from nuclide to nuclide along with its stability constant (8). For Sr-90-89-85 it equals 100, for Ba-140-133 even 146000 but for Ra-226-224 only 1.0 in the rat. If these ranges are large enough for man, even in the case of a steep mortality-dose function, cannot be answered at the moment.

If morphological and biochemical, e.g. enzymatical (5) disorders are limited to the lethal dose range of (222) or are reversible respectively not existent in the sub-lethal, therapeutic dose range, is not yet known.

Reversible side effects, such as impairment of protein- and DNA-synthesis (13) and urinary sodium/potassium retention (14) observed after single sub-lethal (222)-doses, may be accepted as those, because (222)--treatment will always be an acute-treatment but never a chronical one.

Future studies will teach us, if (222) becomes an acceptable agent for the decorporation of radio-strontium barium and perhaps radium during their circulation in the blood and extracellular space of man.

Present State of Radio-Strontium Decorporation Research

EXTRA-CELLUL. Sr-85 %	MINUTES		(222)-DOSE IN µ MOLE/Kg					
	RAT	MAN	167	100	50	25	5	1
5	450	-	-	-	-	-	-	-
10	110	-	-	-	-	-	-	-
13.8	60	5000	88	-	-	-	-	-
15	48	3300	86	96	-	-		
19.4	30	1050	82	91	-	-		
20	27	1000	80	90	100	-		
25	17	350	77	85	97	-		
30	12	130	73	82	93	-		
35	8.6	80	70	78	90	100		
40	6.6	50	68	76	87	98		
45	5.2	31	66	74	84	94		
45.5	5	26	66	74	84	94		
50	4.2	20	65	72	82	92		
55	3.3	12	63	70	80	90		
60	2.8	9	61	69	79	88		
65	2.4	6	60	67	77	87		
70	2.1	4.5	59	66	76	86		

EQ-VALUES

TABLE 1

REFERENCES

1. Müller, W.H. (1970): Naturwiss. 57, 248.
2. Müller, W.H. and Müller, W.A. (1974): Naturwiss. 61, 455.
3. Müller, W.H., Müller, W.A. and Linzner, U. (1977): Naturwiss. 64, 96.
4. Müller, W.H. (1971): V. Congr. Internat. Soc. Franc. Radio. Prot. Grenoble, p. 640.
5. Knajfl, J., Vondracek, V. and Neruda, U. (1976): 12th Ann. Meeting of Europ. Soc. Rad. Biol. Budapest.
6. Batsch, J., Geisler, J. and Szot, Z. (1978): Nukleonika, 23, 305.
7. Catsch, A.(1968): Dekorp. radioakt. und stab. Metallionen, p. 22, Verlag: Karl Thiemig K.G. München.
8. Müller, W.H. (1977): Strahlenther., 153, 570.
9. Müller, W.H., Friedrich, E. and Kollmer, W.E.(1978): 14th Ann. Meeting Europ. Soc. Rad. Biol., Jülich.
10. Baudot, P., Joque, M. and Robin, M.(1977): Toxicol. Appl. Pharmacol., 41, 2358.
11. Michon, G. and Maillard, M.J. (1961): Symposium Use Radioisotopes Animal Biol.& Med. Sci.,Mexico City.
12. I.C.R.P. Publication no. 20, p. 76 (1972).
13. Jackl, G., Personal communication.
14. Müller, W.H. and Beaumatin, J. (1976): Life Sciences 17, 1815.

SUSTAINED RELEASE OF RADIOPROTECTIVE AGENTS *IN VITRO*

J. Shani, S. Benita, A. Samuni and M. Donbrow

The major functional group of radioprotective agents contains a sulfur atom, either as a thiol (-SH) or in the oxidized state (-S-S-). These compounds protect against radiation as a result of their ability to trap primary free radicals formed via degradation radiolysis of water (1). The practical use of those radioprotectants is limited by their cumulative toxicity and rapid excretion and degradation. Moreover, for provision of adequate protection against irradiation effects, the concentration of the drug requires precise regulation, a difficult goal to achieve due to the rapidity of depletion on the one hand, and the low protective index on the other. Some protective doses and LD_{50} values for a highly effective radioprotective agent, cysteamine, in various mammalian species, are given in the following table (2):

Species	Route	Protective Dose (mg/kg)	LD_{50} (mg/kg)
Mouse	ip	75-150	260
Rat	ip	75-150	140
Dog	iv	75-110	
Rabbit	iv		150
Man	iv	150-400	

The radioprotective drugs chosen for the first phase of this study were cysteamine (β-mercaptoethylamine, $H_2N-CH_2-CH_2-SH$) and cysteine (β-mercaptoalanine, $H_2N-CH(COOH)-CH_2-SH$), the most effective agents explored in mammals. The objective of this research is to improve the efficacy of these radioprotectants by development of new pharmaceutical formulations rather than by synthesising of new agents. With the growing use of nuclear enrgy throughout the world, and the risk of accidental irradiation becoming a dangerous hazard, we considered it desirable to develop a sustained-release radioprotec= tive formulation. Prolonged release of the agent in the body at effective concentrations is aimed at avoiding increase in toxicity, and will provide a model for a novel approach to human protection against accidental irradiation. The ultimate goal of this study is then better control of the bioavailabilityof the drug in the serum, with increase in its protective index.

MATERIALS AND METHODS

Cysteamine and cysteine (Sigma) were dried, pulverised and mixed with stearic acid BP 1973 (sigma) and ethylcellulose (Hercules), in various proportions, in a mortar. Various concentrations of the active material, not exceeding 50%, were prepared in defined ratios of the matrix mixture. The mixture was than compressed into cylindrical tablets of 13.1 mm in diameter and a mean of 4.0cm^2 surface area (13.6 mm height) in a vacuum KBr die, with a laboratory press, at various pressure values (3), as given previosly (4). Release of the drug was measured spectrophotometrically (Unicam, model SP-1805) at 240 nm, using a rotating basket dissolution apparatus as described in USP XIX. The dissolution medium (phosphate beffer pH 7.4) was kept under nitrogen. Exactly one liter of the buffer, previously heated and maintained at 37±0.3°C, was used for each experiment. The basket was immersed in the buffer and rotated at 100 rpm. Experiments were carried out for 8-16 hours, and the concentration of the unoxidized drug was monitored continuously, using a 10 mm flow-cell fed by a peristaltic pump (Buchler, model 2-6100), at flow rate of 60 ml.min^{-1}. The tablets were wrapped in aluminium foil immediately after their compression, and kept at room temperature pending assay. Experimental results obtained spectrophotometrically were computerized, using a programme to fit Higuchi equation treatment described as follows: 10-20 points from each spectrophotometric run were fed into a PDP-15 computer and plotted against the square root of t. The parabolic UV absorption curve, which demonstrates the accumulated released drug when plotted on a linear scale of t, is converted into a linear correlation when plotted against the square root of t, following the Higuchi law. For these calculation the molar extinction coefficients for cysteine (427 M^{-1}.cm^{-1}) and cysteamine (563 M^{-1}.cm^{-1}) were determined experimentally under nitrogen.

The mechanisms for the sustained-release of drugs dispersed uniformly in hydrophobic matrices were proposed by Higuchi in 1963 (5). He developed equation suitable both for drugs partially soluble in the matrix, and for drugs insoluble in the matrix but soluble in the external medium. In the latter type of system the drug particles are released by leaching action of the penetrating solvent. The Higuchi square-root law, which has been validated in numerous cases (6-8), takes the following form:

$$Q = \sqrt{\frac{D \cdot \varepsilon}{\tau} (2A - \varepsilon \cdot C_s) C_s \cdot t}$$

where Q = the amount of drug released at time t per unit exposed area
 ε = the porosity of the matrix
 τ = the tortuosity of the matrix
 D = the dffusibility of the drug in the dissolution medium
 A = the total amount of drug present in the matrix / unit volume
 C_s = the solubility of the drug in the dissolution medium

With all other parameters fixed, the release rate is linearily related to the square root of t. Release rate from spherical pellets by this mechanism does not follow a first-order relationship, as the time required to release 50% of the drug from a matrix is expected to be approximately 10% of the time required to dissolve the last trace of the solid drug phase in the center of the pellet. It was shown (9) that dissolution of a soluble drug at high concentration from an insoluble matrix follows the Higuchi square-root equation except during an initial lag phase and a terminal diffusion phase. The main assumptions made are that the two-dimentional cross-sectional porosity has the same mean as the volumetric porosity, and that the dissolving substance is sufficiently dilute not to affect the porosity. A graphic demonstration of a tablet before (left) and during (right) the drug-dissolution process is shown in the following figure, where the lined areas represent the solvent liquid, and the bold lines - the drug (9):

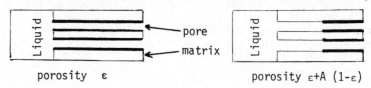

porosity ε porosity $\varepsilon + A (1-\varepsilon)$

RESULTS AND DISCUSSION

A very distinct correlation was obtained in this study between the stearic acid / ethylcellulose ratios and the release rate of 20% cysteine, pressed at 4 tons, from the tablets. The mean release rate at five increasing ratios (expressed as percentage of stearic acid in the matrix mixture) is shown in the following table:

Stearic acid (% in matrix mixture)	Release rate of 20% cysteine
25	4.90
33	4.52
50	3.77
56	3.02
75	2.86

Tablets of 100% stearic acid as matrix liquified during their preparation and could not be manufactured. The significant <u>decrease</u> in rate of cysteine release with <u>increasing</u> the concentration of stearic acid was calculated, and obtained graphically by the computer. In general, the release rate of 20% cysteamine from similar matrix mixtures were much higher than those of cysteine, with 17.20% for 0% stearic acid (i.e. pure ethylcellulose) and 7.10% for 50% stearic acid (i.e. 1:1 ratio). The release rate of cysteamine also <u>decreases</u> with <u>increasing</u> stearic acid concentration in the matrix, and experiments with 90-100% of stearic acid are being carried out currently.

The relation between the concentration of the active drug and its rate of release was also investigated. Preliminary results suggest that the rate <u>increases</u> with <u>increasing</u> percentage of stearic acid. Further studies are needed in order to validate this preliminary finding.

One other factor that has been investigated in this study was the effect on the release rate of the pressure applied to the tablet during its formation. Here, preliminary findings suggest that with 20% cysteamine as the active material, a higher formation pressure results in quicker drug release, while with 20% cysteine no such irregularity was noticed. Although both components of the matrix are hydrophobic, stearic acid is much more hydrophobic than ethylcellulose, due to its waxy character. This character is probably the reason for the decrease in the porosity of the matrix.

Sustained Release of Radioprotective Agents *in Vitro*

Studies of the release of the radioprotectants from their matrices in simulated gastro-intestinal juices are currently being carried out. The results of this work indicate that a correlation might be anticipated between the in vitro profile of release and the in vivo efficacy of the radioprotective agent, due to the sustained-release of the drug.

This work was supported in part by the International Atomic Energy Agency (IAEA), Vienna.

REFERENCES

1. Bacq, Z.M. (1975): Sulfur-containing Radioprotective Agents. Perganon Press.
2. Doull, J., Plzak, V. & Brois, S.J. (1962): A survey of Compounds for Radiation Protection. USAF Aerospace Medical Division, Texas.
3. Parrott, E.L. & Sharma, V.K. (1967): J. Pharm. Sci. 56,1341.
4. Donbrow, M. & Friedman, M. (1975): J. Pharm. Pharmacol. 27,633.
5. Higuchi, T. (1963): J. Pharm. Sci. 52, 1145.
6. Desai, S.J., Singh, P., Simonelli, A.P. & Higuchi, W.I. (1966): J. Pharm. Sci. 55, 1224.
7. Desai, S.J., Singh, P., Simonelli, A.P. & Higuchi, W.I. (1966): J. Pharm. Sci. 55, 1230.
8. Desai, S.J., Singh, P, Simonelli, A.P. & Higuchi, W.I. (1966): J. Pharm. Sci.,55, 1235.
9. Fessi, H., Marty, J.P., Puisieux, F. & Carstensen, J.T. (1978): Intern. J. Pharmac., 1, 265.

INVESTIGATION OF THE SOLUBILITY OF YELLOWCAKE IN THE LUNG OF URANIUM MILL YELLOWCAKE WORKERS BY ASSAY FOR URANIUM IN URINE AND *IN VIVO* PHOTON MEASUREMENTS OF INTERNALLY DEPOSITED URANIUM COMPOUNDS

H. P. Spitz, B. Robinson, D. R. Fisher and K. R. Heid

INTRODUCTION

Evaluation of occupational inhalation exposure to uranium is routinely performed using the results of bioassay measurements for uranium in excreta and in vivo scintillation measurements for uranium deposited in the respiratory system of potentially exposed workers (1). Uranium toxicity is similar to other heavy metals, such as arsenic, lead, and mercury, and differs essentially in that uranium is an alpha particle-emitting radioactive material. Therefore, in order to establish an effective worker protection program at a uranium mill, both nephrotoxicity and the radiological hazard to the lung must be considered for the overall assessment of worker exposure.

Recent studies of the solubility of uranium in simulated lung fluid have demonstrated that yellowcake, the generic name of the end-product material from the uranium milling process, does not have a unique solubility classification (2). The solubility of yellowcake has been shown to depend upon the specific chemical separation process employed at the mill (3). That is to say, yellowcake solubility is influenced by the chemical form of the material, the method of chemical extraction of the uranium from the ore, and the temperature at which the yellowcake is calcined or dried. To further exacerbate the lack of a unique solubility classification for yellowcake, each mill will vary its own chemical separation process, depending upon the ore quality, and produce a yellowcake material with slightly different physiochemical characteristics including solubility.

The lack of a specific solubility classification for yellowcake impacts the worker protection program at its most fundamental level. It prohibits development of a simple method for interpreting bioassay and in vivo measurements for uranium which are performed to assess the occupational exposure of workers who come into contact with yellowcake. It is due to the question of the solubility of yellowcake in the human lung that in vivo scintillation measurements for uranium in the respiratory system must be performed as an adjunct to the assay for uranium in excreta.

METHOD

The United States Nuclear Regulatory Commission, Division of Safeguards, Fuelcycle, and Environmental Research is sponsoring a study at the Pacific Northwest Laboratory (PNL) to assess the solubility of yellowcake in the human lung as determined from measurements of workers who were accidentally exposed on the job at a uranium mill. Extensive evaluation of accidental human yellowcake exposures, using

Investigation of the Solubility of Yellowcake

sequential measurements of uranium in excreta and in the respiratory system, should provide data to characterize the solubility of the inhaled material.

Analysis of excreta samples from uranium mill workers routinely indicate the presence of microgram amounts of uranium resulting, presumably, from low-level, chronic exposure at the mill. It is not possible to absolutely identify the source of uranium in these samples as arising from an actual worker inhalation exposure or from external contamination, such as dust in the mill environment.

The first phase of this study is to determine baseline uranium measurements in excreta samples from yellowcake and other mill workers. Intercomparisons of routine assay procedures at the mills and the methods employed at PNL are performed to relate records of past measurements to the analysis of the current program. Urine and feces are obtained from workers at many different uranium mills. Management personnel from several uranium mills in Wyoming, Washington, Colorado and New Mexico have been very cooperative with this voluntary program and have enthusiastically supported the study. Their assistance in reporting potential overexposure incidents is paramount to the success of the second part of this study, i. e., measuring yellowcake solubility in the lungs of an overexposed worker.

In order to alleviate the potential for external contamination during sample collection and assay volunteer workers are transported to PNL for collection of excreta samples under more controlled circumstances than are available at the mill. The travel arrangements also provide a period of time for recently inhaled soluble airborne particulate to be eliminated prior to initiating sequential sample collections (4). In other words, this phase of the research is an attempt to collect excreta samples for uranium assay in order to characterize any chronic exposure that may exist at the mills.

Measurement of any insoluble uranium material deposited in the respiratory system is performed by simultaneously detecting photon emissions from ^{235}U, ^{234}Th, and x-rays from uranium, protactinium, and thorium (5). Two dual-crystal NaI(Tl)/CsI(Tl) scintillation detectors are placed on the anterior thorax of the worker while he lays prone in a shielded room at the laboratory whole body counting unit. An overabundance of the $\simeq$16 keV x-rays in an in vivo measurement for uranium in the lung indicates that some fraction of the material is located on the surface of the worker. Evaluation of the internal uranium deposition must be adjusted to eliminate the influence of surface contamination on the worker. Prior in vivo measurements for uranium deposited in the lungs of uranium mill workers have been performed by another researcher without special regard for the presence of external contamination (6). Although, in that study, workers were required to shower and wear coveralls, the presence of uranium bearing dust and imbedded soil on the skin could not have been entirely eliminated by washing.

Once baseline reference values are established for uranium in excreta and in the lungs of uranium mill workers, measurements will be performed with an overexposed yellowcake worker. Uranium excretion from the acutely overexposed yellowcake worker will be compared to baseline values determined from "unexposed" mill workers. The temporal relationship between uranium in urine and feces and the change in the lung burden of the overexposed worker will provide a basis for describing the solubility of the inhaled material.

It is unlikely that any one of the volunteers who are measured during the initial phase of the program will later become contaminated with yellowcake as a result of an accidental overexposure at the mill. For this reason a representative number of volunteers from several mills must be measured according to the aforementioned protocol. The millworkers will typically spend three to five days at PNL and need only collect excreta and be available for in vivo measurements each day.

The schedule for the overexposed worker will require at least two weeks time at the laboratory for initial sample collections. Immediately following the inhalation incident the worker will be transported by air to PNL so that the metabolism and translocation properties of the rapidly soluble, class D material can be studied. Excreta samples will be collected for several weeks after this initial period. Follow-up in vivo measurements for uranium in the lung will be performed in order to identify any insoluble uranium material remaining in the lung following the early clearance.

DISCUSSION

A unique application (7) (8) of the dual-crystal detector to the in vivo measurement for uranium in the lung provides a mechanism to distinguish external uranium contamination from that actually deposited in the lung tissue. The thin (3 mm) NaI(Tl) scintillator in the dual crystal detector can identify the presence of uranium on the skin by measuring the $\simeq$ 16 keV x-rays from U, Pa, and Th. Using this same scintillator, the 63 keV and 93 keV ^{234}Th photons indicate the presence of ^{238}U. The thicker (5 cm) CsI(Tl) scintillator simultaneously detects the ^{235}U photons at 186 keV.

In order to take advantage of this technique, a surrogate thorax structure (phantom) is fabricated with a known amount of yellowcake uranium material deposited within the lungs. Two detectors are placed in contact with the surface of the anterior thorax, one detector centered over each lung. A standard ratio of the counts in the x-ray region to the number of counts in the ^{234}Th region is determined from a measure of the phantom. A similar ratio is calculated from an in vivo measurement of a potentially exposed subject. This ratio is first modified for chest wall attenuation and then compared to the surrogate thorax phantom. Detection of an abundance of 16 keV x-rays with the exposed subject indicates that skin contamination is present. The subject in vivo measurement is corrected to account for surface contamination so that the amount of uranium in the lung can be determined.

SUMMARY

Experience has shown that uranium has a low order of chemical toxicity in man and that other heavy metals, such as lead, arsenic, and mercury would produce severe injury at the same levels of exposure (9). However, without a precise knowledge of the solubility of yellowcake in the human lung its hazard to man from an inhalation exposure cannot be adequately determined.

Experiments to measure the solubility of yellowcake in simulated lung fluid indicate that the material exhibits some properties of each classification, i. e., D, W, and Y. This research study attempts

Investigation of the Solubility of Yellowcake

to measure the solubility of yellowcake in the human lung from analysis of the metabolic and translocation characteristics of the material in a worker who has been exposed at a uranium mill. The elimination of uranium from the body will be measured in urine and fecal samples collected from the exposed worker. In vivo scintillation measurements for uranium in the lung will be performed to determine the clearance of the material from the lung.

REFERENCES

1. Alexander, R. E., (1974): "Applications of Bioassay of Uranium", U. S. Nuclear Regulatory Commission, WASH-1251.
2. Edison, A. F. and Mewhinney, J. A., (1978): "In Vitro Dissolution of Uranium Product Samples from Four Uranium Mills", Lovelace Biomedical and Environmental Research Institute, NUREG/CR-0414.
3. Kalkwarf, D. R., (1979): Solubility Classification of Airborne Products from Uranium Ores and Tailings Piles", Pacific Northwest Laboratory-Battelle Memorial Institute,NUREG/CR--530.
4. Lippman, M., (1958): "Correlation of Urine Data and Medical Findings with Environmental Exposure to Uranium Compounds", Symposium on Occupational Health Experience and Practices in the Uranium Industry. U. S. Atomic Energy Commission, HASL-58, 103.
5. Cohen, N., (1978): "In Vivo Measurements of Bone-Seeking Radionuclides", New York University Medical Center Progress Report to U. S. Department of Energy. COO-4326-1.
6. Helgeson , G. L., (1979): In Vivo Counting at Selected Uranium Mills", Final Report to the U. S. Nuclear Regulatory Commission, NUREG/CR-0841.
7. Cohen, N. Spitz, H. B., and Wrenn, M. E., (1977): "Estimation of Skeletal Burden of 'Bone-Seeking' Radionuclides in Man from In Vivo Scintillation Measurements of the Head", Health Physics 33, 431.
8. Shapiro, E. G., and Anderson, A. L., (1974): "Dual Energy Analysis Using Phoswich Scintillation Detectors for Low-Level in vivo Counting", IEEE NS-21, 201/
9. Eisenbud, M. E., and Quigley, J. A., (1956): Industrial Hygiene of Uranium Processing", A. M. A. Arch. Ind. Health 14, 12.

EXCRETION OF ORGANIC AND INORGANIC TRITIATED COMPOUNDS IN COW'S MILK AFTER INGESTION OF TRITIUM OXIDE

J. Van den Hoek, G. B. Gerber and R. Kirchmann

INTRODUCTION

The evaluation of the transfer of tritium in the environment is complicated by the fact that tritium can be incorporated into a variety of organic molecules whose metabolic behavior may differ greatly from that of tritium oxide (2, 3, 6, 7). An important link in the transfer of tritium to man is the secretion of tritiated molecules in milk (4, 5, 8, 10) because this has its impact on infants, the most sensitive part of the population. The secretion in milk of tritium as water and organic molecules (casein and lipids) was, therefore, studied after giving cows tritiated water to drink for a period of 25 days.

MATERIALS AND METHODS

Two lactating cows (weighing about 560 kg) were given drinking water containing 18.9 mCi/l tritium oxide for a period of 25 days (a tritium intake of 871.3 µCi/day, all values given being the means for both cows). The diet consisted of a hay supplemented with a concentrate consumption of food and water (46.6 l/day) as well as production of milk (20.8 l/day), urine (21.9 l/day) and faeces (32 kg/day) were determined daily. Radioactivity in water, dry matter, lactose, lipids and casein of milk was determined during this "loading" period and during a "decay period" of 75 days after application had been terminated. The different constituents were isolated as follows : milk water by distillation, whole dry matter by lyophilization, lactose by cristallization in ethanol, casein by precipitation with acid and fats by the Rose-Gottlieb procedure. The radioactivity was determined by liquid scintillation counting after the material had been combusted, except for water and fats which were counted directly. The data were fitted to appropriate functions for the loading and decay period using a nonlinear regression procedure. The calculations were carried out separately for turnover rates and turnover times to obtain standard errors directly. Statistical weighing according to a Poisson distribution was required for the data of the decay period to obtain the second exponential term. No such weighing was needed for the loading period since the range of the activities measured is relatively small. Analysis of variance indicated a good fit for all functions shown. The loading phase is best described by a single term integrated exponential function with a time delay.

$$A = A^\infty [1 - e^{-\alpha[t - T_d]}]$$

where A^∞ the activity reached after very long times of application, α the turnover rate (days^{-1}) and T_d the time delay (days).
The behavior during the decay period can be approached by a two term exponential function :

$$A = A_1 e^{-\alpha_1 t} + A_2 e^{-\alpha_2 t}$$

where A_1 and A_2 the fractions of activity for each turnover component and α_1 and α_2 the respective turnover rates. The formulas for turnover times T (days) are obtained by replacing α with t.

RESULTS

Table 1 presents the parameters and their standard errors for tritium oxide, casein and lipid during the loading period. One recognizes that secretion starts after a short (0.5 day), but significant

Table 1: Parameters ± S.E. for "Loading phase" $\left[A = A^\infty \left[1 - e^{-\alpha[t - Td]}\right]\right]$

Compound	A^∞ [pCi/g]	α [days^{-1}]	T delay [days]	Tσ [turnover time] [days]
Water	15.680 ± 49	0.2095 ± 0.00397	0.624 ± 0.057	4.773 ± 0.0904
Lipids	5865 ± 41	0.1478 ± 0.0049	1.08 ± 0.11	6.764 ± 0.225
Casein	2927 ± 25	0.1756 ± 0.0728	0.545 ± 0.105	5.68 ± 0.496

delay; for lipids this delay is about one day. Turnover of all three compounds is short, corresponding to an half life time of about 5 days; it is somewhat longer for lipids than for tritium oxide or casein. No second component can be distinguished during this phase but this is expected since the statistical variability exceeds by far the contribution of such a small fraction. The half life found appears relatively short if compared to that of other nonlactating great mammals (1, 5, 8), but is readily explained by the large water turnover of lactating cows. From the known specific activity of the drinking water and the tritium content of lipid and casein (11% and 7.6% respectively), one can calculate the dilution of the ingested water during metabolism. Thus about 83% of the milk water secreted is found to originate from drinking water; the rest comes from water in food and from that formed in metabolism. About 38% of the hydrogen in dry matter, 30% of that in lipids, 48% of that in lactose and 22% of that in casein are derived from ingested water or from labeled recycling organic molecules. These values are compatible with observations by others (2, 6, 7).

The parameters of the decay phase presented in table 2 indicate that the principal components of water, casein and lipids have about

Table 2: Parameters ± S.E. for "decay phase" $\left[A = A_1 e^{-\alpha_1 t} + A_2 e^{-\alpha_2 t}\right]$

Compound	$A_1 [pCi/g]$	$A_2 [pCi/g]$	Time constants		Turnover times	
			$\alpha_1 [days^{-1}]$	$\alpha_2 [days^{-1}]$	$T_0^1 [days]$	$T_0^2 [days]$
Water	16.539 ± 50	7.41 ± 2.15	0.2073 ± 0.00047	0.0172 ± 0.00525	4.822 ± 0.011	58.2 ± 17.8
Lipids	6 845 ± 105	26.6 ± 7.11	0.1834 ± 0.0026	0.00335 ± .00469	5.451 ± 0.071	29.9 ± 41.8
Casein	2994 ± 41	246 ± 47	0.1889 ± 0.0045	0.0413 ± 0.0056	5.29 ± 0.12	24.2 ± 3.3

the same turnover rates as during the loading phase. In addition, small fractions (about 0.04% for tritium oxide, 0.4% for lipids and 8% for casein) display much longer half lifes. The half life seems longest for lipids, but its is error marge. It should be noted that the fractions A_1 and A_2 presented in table 2 are those appearing after a loading period of 25 days; the long lived component would be much less important after a single application (table 3).

The data shown allow to calculate how much activity of each compounds and each metabolic component would be excreted after a single and after a continuous application of tritium oxide when integrated over infinite times (table 3). One recognizes that after single application tritium is mainly excreted as tritium water of rapid turnover, organic molecules participate only to about 3.9% and all slow metabolic components represent together about 0.2% of the total activity excreted. It should be noted, however, that when the milk is ingested by man the organic molecules may give rise to a much higher fraction of tritiated molecules with long life span than does tritium water. After continuous application, molecules with long half life, particularly lipids become more important and may represent somewhat more than 4% of the total; the values for lipids remain, however, subject to a rather large error and require further study.

	Fraction (pCi / g or %)		Unique application		(% of total) Continuous application	
	Rapid Component	Slow Component	Rapid Component	Slow Component	Rapid Component	Slow Component
Water %	$2.92 \cdot 10^6$ 96.7 %	11.4 $3.97 \cdot 10^{-4}$ %	96	0.0045	92.3	0.052
Lipids %	67083 2.2 %	28.9 $9.57 \cdot 10^{-4}$ %	2.49	0.058	2.71	3.48
Casein %	32918 1.09 %	691 $2.38 \cdot 10^{-4}$ %	1.19	0.114	0.85	0.55

Table 3 - Metabolic components after single application and integrated activities after single and continuous application.

REFERENCES

1. Black, A.L., Baker, N.F., Bartley, J.C., Chapman, T.E. and Philips, R.W. (1964) : Science, 144, 876.
2. Commerford, S.L., Carsten, A.L. and Cronkite, E.P. (1977) : Rad. Res.,72, 333.
3. Hatch, F.F. and Mazrimas, J.A. (1972) : Rad. Res.,50, 339.
4. Kirchmann, R., Van den Hoek, J. and Lafontaine,A. (1971) : Health Physics, 21, 61.
5. Mullen, A.L., Moghissi, A.A., Wawerna, J.C., Mitchell, B.A. and Bretthauer, E.W. (1976) : EPA-600/3-77-076.
6. NCRP Report 62 - Tritium in the Environment, Washington 1979.
7. Pietrzak-Flis, Z., Radwan, I. and Indeka, L. (1978) : Rad. Res., 76, 420.
8. Potter, G.D., Wattuone, G.M. and Mc Intyre (1972) : Health Physics, 22, 405.
9. Van den Hoek, J. and Kirchmann, R. (1971) : EUR 4800 1121.
10. Van den Hoek, J., Kirchmann, R. and Juan, N.B. (1979) : In : Behaviour of Tritium in the Environment, p. 433, IAEA, Vienna.

Supported by the contract n° 235-76-7 BIO F of the radiation protection program of the Commission of the European Communities (Publication n° 1643).

CORRELATION BETWEEN CONCENTRATIONS OF ^{210}Pb IN THE BIOLOGIC SAMPLES FROM MINERS AND INDIVIDUAL LEVELS OF EXPOSURE TO SHORT LIVED RADON-222 DAUGHTER PRODUCTS

H. I. M. Weissbuch, M. D. Grădinaru and G. T. Mihail

The ^{210}Pb - ^{210}Po skeletal burden may be correlated with specific activities of these elements in biological samples, especially blood and hair levels (1,3). There exists thus the possibility to state some quantitative relations between these parameters and the exposure to inhaled radon daughters, the dose delivered to the bronchial tissue as well as the expected respiratory cancer among miners (4,6).

Compared to radiation exposures evaluated from short lived radon daughters content of mine air, the skeletal levels of lead-210 represents a potentially more accurate measure of miners' exposure. By determining the skeletal reservoir of lead-210 it becomes possible to evaluate the individual cumulated exposures and to make some estimates of prior mine exposures.

MATERIALS AND METHODS

The epidemiological investigation performed by the authors was based on the methodological principle of estimating the miners' exposure both by determining the mine air contamination and the lead-210 blood levels.

As it was expected that cumulated exposure would be estimated from skeletal lead-210, only when steady-state equilibrium was attained between lead-210 bone content in the body fluids, the investigation was carried out only on men who had been exposed over periods of 10 years or more, when miners may be considered to be close to equilibrium (9/10),(2).

The study group included eighty miners and as controls forty adult males, living in the same residence aria as the miners but never being exposed occupationaly to radon daughters.

From the wet-ashed blood samples both ^{210}Bi and ^{210}Po were quantitatively plated on disks. The recovery yield was typically 95 - 8%. The activity of the electroplated elements has been measured by quantitative autoradiography of tracks, using I.F.A.-E.N.1 nuclear emulssion.

The basical hypothesis adopted in the calculation considered that the power function relationship is a statistical property of any biologic sufficiently complex system that has evolved to a state of dynamic equilibrium (7). It is accepted that this kind of mathematical relation characterizes the different biochemical and metabolic mechanisms that involve lead-210 between the actual exposure and the time of sampling (4).

As the power function will be valid only if all lead-210 activity reaching the skeleton is supplied by inhaled radon daughters, the lead-210 body burden originating from other sources was evaluated and suitable corrections had been made (5).

Both $\overline{WL}$ values calculated from blood ^{210}Pb (4) and data about exposure conditions obtained from occupational histories were utilised in estimating individual cumulated exposures (CWLM).

Simultaneously with the bioassay study, working levels (WL) were determined by dynamic determinations of radon daughter concentrations in the mines, where the investigated group had been working.

For a more accurate quantification of the risk the equilibrium grade between Ra A:Ra B:Ra C, the free atoms fraction and the granulometric distribution of aerosols have been considered.

Rolle's modified method (8) was utilised for routine control of the risk factors in the underground work places with noxious radioactivity.

RESULTS AND DISCUSSION

Pair values of the individual exposures resulted from radon decay products inhalation determined by the two methods are plotted in figure 1.

The good agreement between the WLM values obtained from excess lead-210 in bone and those derived from direct measurements of radon daughters in the mine atmosphere emphasizes that the skeletal reservoir of lead-210 may be an extent of the individual exposure to short lived radon daughter products and that the miner's organism behaves as its own dosemeter.

The WLM values resulting from the biological assay are in most cases higher, revealing more important exposures prior to our making the determinations. It may be due either to increased radon emanations or to worse work conditions.

It must be noted that the two value series are not completely independent of each other, as both resort to the occupational histories of miners.

On the whole it may be concluded that the cumulated exposure estimated from the lead-210 body burden expresses more accurately the actual risk, as it is an individual indicator that takes into account the feature

PAIR VALUES OF THE INDIVIDUAL EXPOSURES

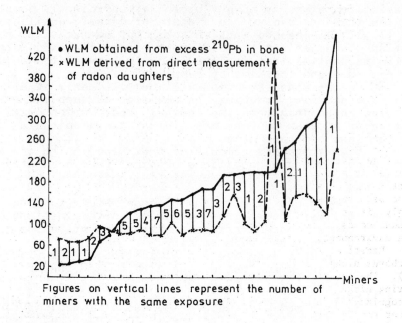

Figure 1

of the organism, the specific of the worker's profession and integrates the aleatory changes in radon daughter concentrations in both their temporal and work place dynamics.

REFERENCES

1. Black,C.S.,Archer,V.E.,Dixon,W.C. and Saccomanno,G. (1968) :Health Physics,14,81.
2. Blair,H.A. (1969) :In :Radiation-Induced Cancer,p.203, I.A.E.A.,Vienna,1969,S.T.I./B.U.B./228.
3. Blanchard,R.L.,Archer,V.E. and Saccomanno,G.(1969) : Health Physics,16,585.
4. Gotchy,R.L. and Schiager,K.J. (1969) :Health Physics,17, 199.
5. Holzman,R.B. (1970) :Health Physics,18,103.
6. Lundin,F.E.,Jr.,Wagoner,J.K. and Archer,V.E. (1971) :In: Radon daughter exposure and respiratory cancer quantitative and temporal aspects,p.62,N.J.O.S.H.-N.J.E.H.S. Joint Monograph.No.1.
7. Marshall,J.H. (1963) :A.N.L.-6646,Argonne National Laboratory.
8. Rolle,R.(1972) :Health Physics,22,233.

DISTRIBUTION AND CLEARANCE OF INHALED $^{63}NiCl_2$ BY RATS

P. L. Ziemer and S. M. Carvalho

The radionuclide ^{63}Ni has been reported to occur in biological samples from the Bikini and Eniwetok Atolls, and in other environmental samples as well (1). Possible release of this nuclide has also been suggested for advanced gas-cooled reactors (2). Accordingly, it is important to characterize the inhalation of ^{63}Ni compounds so as to assess dose commitments associated with releases to the environment and subsequent human exposures.

Recent attention has also been focused on increases in levels of airborne nickel (stable) resulting from the burning of fossil fuels (3). The hazards of exposure to airborne nickel and its compounds during nickel refining and other industrial operations have been recognized for many decades. Health effects include allergic and other sensitivity responses, systemic poisoning, pneumoconiosis, and respiratory tract and other cancers (4). One nickel compound widely used in the nickel industry is nickel chloride. Animal studies have linked exposure to $NiCl_2$ to effects such as lung changes and increases in mucus secretion (5), inhibition of insulin release (6), and alteration of ciliary activity (7). With recent findings (5,7,8) of effects at lower and lower doses of stable nickel, it is important to establish distribution patterns of nickel in animals and man.

Most distribution and excretion studies done with ^{63}Ni in rats and mice have dealt with relatively large amounts of nickel (>100 µg per animal) administered intratracheally (6), intraperitoneally (9), intravenously (10,11), and orally (12). The distribution and excretion of nickel following administration of small amounts (<1.5 µg) administered intravenously have also been reported (13).

This research was designed to study lung deposition and clearance of ^{63}Ni administered as an aerosol to rats. Concentrations of nickel as a function of time were also determined for eight organ systems and blood.

MATERIALS AND METHODS

An inhalation chamber constructed for previous studies (14,15) was used to expose male rats (Sprague-Dawley descendents) to a $^{63}NiCl_2$ aerosol. A compressed air nebulizer* generated the aerosol from an aqueous solution of nickel chloride whose specific activity was 3.44 mCi/mg of the compound at a concentration of 2.14 mg/ml. The aerosol was introduced into the exposure chamber which contained six exposure ports, five of which were used to expose the rats (nose only) and one which served as a sampling position for a Mercer-type

---

*Retec X-70/N, Retec Development Lab., Portland, Oregon, U.S.A.

7-stage round-jet cascade impactor. The average aerosol concentration in the inhalation chamber for the exposures was 9.28 µCi/l (2.7 µg Ni/l). At this concentration, the average amount of nickel deposited in each rat was 2.04 µg with an activity of 7 µCi.

Twenty rats (160 to 180 g size) were each exposed to the aerosol for 15 minutes. The animals were randomly assigned in groups of 5 to four exposure runs. Within each group, the rats were further randomly assigned to one of five sacrifice times, namely, 35 minutes, one day, 3, 7, and 21 days after exposure.

A mild anesthetic was used to sedate the rats so that they would remain in place during the exposure period. For the level of anesthetic used, the average respiration rates during sedation were observed to be the same as the normal sleeping respiratory rates.

After exposure, the rats were removed from the chamber and their heads were washed to remove external contamination. The rats were kept in individual metabolism cages after exposure until sacrifice so that urine and feces could be collected. At the appropriated sacrifice times, the rats were anesthetized and 2 blood samples of 0.1 ml each were taken from the inferior vena cava. The animals were then perfused with normal saline solution, causing death, and their organs were removed. The organs were then homogenized in preparation for assay. Two samples (50-100 mg each) were taken from each of the homogenized organs and acid-digested for liquid scintillation counting.

Each carcass was frozen and then homogenized in a blender with 250 ml of water. Two samples (0.2 to 0.5 ml each) were taken from the mixture and acid digested for liquid scintillation counting. Likewise, ground feces samples and blood were prepared by acid digestion for counting. Urine (two 0.1-ml samples in each case) was placed directly in the scintillation solutions for counting.

The liquid scintillation counter* was set so that all samples were counted with no more than a 5% counting error (at the 95% confidence level). The counting efficiency of each sample was determined by internal standardization. Tissue samples from a control rat (unexposed to the aerosol) were used to obtain the background count for each type of sample analyzed.

Cascade impactor sampling times were 5 minutes for each run. The seven stages and collectors, on which glass cover slips had been placed for particle collection, were disassembled after each run and counted for ^{63}Ni activity. The impactor stages had been previously calibrated using monodisperse particles of known diameters. The activity mean aerodynamic diameter (AMAD) of the $NiCl_2$ particles for all the runs was found to be 0.56 µm with a geometric standard deviation of 1.83.

RESULTS

For each rat, the total activity was determined for 14 organs and body components. The results were expressed in terms of percentage of the initial body burden as determined by the total activity recovered within each rat. Table 1 is a summary of the concentration

---

*ISOCAP/300, Model 6868, Searle Analytic, Des Plaines, IL, U.S.A.

Distribution and Clearance of Inhaled $^{63}NiCl_2$

of ^{63}Ni in nine organs and in blood as a function of sacrifice time. Table 2 gives the total activity (expressed as a percentage of initial body burden) in the organs (combined), the blood, the carcass, the GI tract, and the total excreted, for the various sacrifice times.

Total urinary excretion by day one was 26.2% of the initial body burden, while 38.5% was measured in the feces on day one. After 7 days, 95% of the initial activity had been excreted. Very little of the excretion beyond day one was via the feces, the average fecal activity being just 42.1% for the full 21 day period.

Lung retention, expressed as a percentage of the initial body burden, is shown for the various sacrifice times in Fig. 1. The data can be fitted with a two component exponential curve of the form:

$$LR = 19.06\ e^{-0.942\ t} \pm 20.84\ e^{-0.253\ t}$$

where LR is the percentage of the initial body burden which is retained in the lungs and t is the time, in days, following the inhalation period.

Fig. 2 is the whole-body retention curve showing the fraction of the activity remaining in the rats at each sacrifice time. The whole body retention is described by the two-component exponential equation:

$$BR = 81.54\ e^{-1.179t} + 20.83\ e^{-0.190t}$$

where BR represents the percentage of the initial body burden remaining in the body at time, t, days following the inhalation.

Table 1. Ni Distribution Per Unit Weight at Timed Intervals After Inhalation Exposure.*

Organs	0†	1	3	7	21
rt	$2.5 \times 10^{-2} \pm 5.9 \times 10^{-3}$	$6.9 \times 10^{-3} \pm 1.7 \times 10^{-3}$	$2.8 \times 10^{-3} \pm 3 \times 10^{-4}$	$2.8 \times 10^{-3} \pm 9 \times 10^{-4}$	$5 \times 10^{-4} \pm 3 \times 10^{-4}$
nt Lung	$3.7 \times 10^{1} \pm 7.5$	$2.4 \times 10^{1} \pm 7.9$	9.5 ± 1.6	2.6 ± 1.1	$7.3 \times 10^{-2} \pm 1.6 \times 10^{-2}$
t Lung	$4.7 \times 10^{1} \pm 1.6 \times 10^{1}$	$2.5 \times 10^{1} \pm 8.7$	$1.0 \times 10^{1} \pm 2.3$	$3.0 \pm 8.7 \times 10^{-1}$	$7.8 \times 10^{-2} \pm 1.6 \times 10^{-2}$
een	$1.2 \times 10^{-2} \pm 3.5 \times 10^{-3}$	$7.4 \times 10^{-3} \pm 1.3 \times 10^{-3}$	$6.1 \times 10^{-3} \pm 1.0 \times 10^{-3}$	$5.1 \times 10^{-3} \pm 1.7 \times 10^{-3}$	$1.1 \times 10^{-3} \pm 3 \times 10^{-4}$
neys	$4.2 \times 10^{-1} \pm 1.0 \times 10^{-1}$	$1.4 \times 10^{-1} \pm 2.7 \times 10^{-2}$	$8.8 \times 10^{-2} \pm 1.0 \times 10^{-2}$	$4.3 \times 10^{-2} \pm 2.3 \times 10^{-2}$	$5.5 \times 10^{-3} \pm 1.3 \times 10^{-3}$
tes	$1.2 \times 10^{-2} \pm 2.2 \times 10^{-3}$	$6.0 \times 10^{-3} \pm 4 \times 10^{-4}$	$3.7 \times 10^{-3} \pm 4 \times 10^{-4}$	$2.8 \times 10^{-3} \pm 8 \times 10^{-4}$	nd‡
er	$6.6 \times 10^{-3} \pm 1.4 \times 10^{-3}$	$3.7 \times 10^{-3} \pm 4 \times 10^{-4}$	$3.0 \times 10^{-3} \pm 4 \times 10^{-4}$	$2.3 \times 10^{-3} \pm 8 \times 10^{-4}$	nd‡
n	$4.3 \times 10^{-2} \pm 1.3 \times 10^{-2}$	$1.0 \times 10^{-2} \pm 8 \times 10^{-4}$	$9.3 \times 10^{-3} \pm 3.3 \times 10^{-3}$	$4.7 \times 10^{-3} \pm 1.5 \times 10^{-3}$	$7 \times 10^{-4} \pm 5 \times 10^{-4}$
e	$2.4 \times 10^{-2} \pm 5.9 \times 10^{-3}$	$8.3 \times 10^{-3} \pm 7 \times 10^{-4}$	$6.5 \times 10^{-3} \pm 1.1 \times 10^{-3}$	$4.7 \times 10^{-3} \pm 1.4 \times 10^{-3}$	$1.0 \times 10^{-3} \pm 2 \times 10^{-4}$
d§	$6.6 \times 10^{-2} \pm 2.6 \times 10^{-2}$	$1.5 \times 10^{-2} \pm 2.6 \times 10^{-3}$	$3.3 \times 10^{-3} \pm 1.0 \times 10^{-3}$	$3 \times 10^{-4} \pm 3 \times 10^{-4}$	nd‡

n percent/g ± SE, 4 rats/ST, 2 samples/organ.

ning 35 min after exposure period ended.

detectable.

n percent/ml ± SE.

Table 2. Ni Distribution at Timed Intervals Following Inhalation Exposure.*
Mean Percent ± SE

ST	Organs	Blood[†]	Carcass[‡]	GI Tract	Excreted
35 min	$3.9 \times 10^1 \pm 6.4$	$7.0 \times 10^{-1} \pm 2.4 \times 10^{-1}$	$5.5 \times 10^1 \pm 1.2 \times 10^1$	5.4 ± 3.9	...
1 day	$2.4 \times 10^1 \pm 5.6$	$1.7 \times 10^{-1} \pm 3.0 \times 10^{-2}$	$5.0 \pm 6.9 \times 10^{-1}$	5.5 ± 2.2	$6.5 \times 10^1 \pm 7.5$
3 days	9.6 ± 1.2	$3.7 \times 10^{-2} \pm 1.2 \times 10^{-2}$	$2.6 \pm 4.7 \times 10^{-1}$	$4.4 \times 10^{-1} \pm 5.0 \times 10^{-2}$	$8.7 \times 10^1 \pm 1.0 \times 10^1$
7 days	$3.7 \pm 9.5 \times 10^{-1}$	$3.6 \times 10^{-3} \pm 3.6 \times 10^{-3}$	$1.2 \pm 4.0 \times 10^{-1}$	$1.2 \times 10^{-1} \pm 5.6 \times 10^{-2}$	$9.5 \times 10^1 \pm 2.3 \times 10^1$
21 days	$1.2 \times 10^{-1} \pm 1.2 \times 10^{-2}$	nd[§]	$2.4 \times 10^{-1} \pm 1.7 \times 10^{-1}$	$4.0 \times 10^{-2} \pm 7.6 \times 10^{-3}$	$9.9 \times 10^1 \pm 1.0 \times 10^1$

*4 rats/group, 2 samples per organ type.
[†]Blood volume equation from Donaldson (1924).
[‡]Skin and bone (left femur and skin area) samples included.
[§]Not detectable.

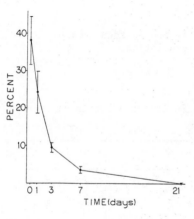

Fig. 1. ^{63}Ni present in the rat lungs at timed intervals following inhalation of ^{63}NiCl$_2$ (expressed as percent of initial body activity).

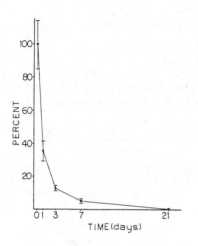

Fig. 2. ^{63}Ni present in the rat body at timed intervals following inhalation of ^{63}NiCl$_2$ (expressed as percent of the initial body activity).

DISCUSSION

The results of this study are in general agreement with a previous study (16) involving the intratracheal instillation of $^{63}NiCl_2$ in male rats. The results also agree with other studies with respect to relative amounts of nickel deposited in the kidneys, regardless of the route of administration. Levels of ^{63}Ni measured in the liver were lower than those reported by others (6,9).

It is important to note that even after 21 days following exposure, the spleen, bone, skin, and heart showed traces of the nuclide, and the lungs and kidneys showed relatively high levels. Also, slow nickel removal, with 64% of the initial (35 min) lung burden still present after one day, indicates that nickel may be incorporated in the lungs as suggested by Charles et al. (17), in existent "high affinity binding sites".

The route of early ^{63}Ni elimination was not primarily urinary as was the case in a previous intratracheal study (17). On the contrary, fecal excretion averaged 38.5% of the initial body burden on the first day following exposure to the aerosol, while urinary excretion was 26.2%.

REFERENCES

1. Beasley, T.M. and Held, E.E. (1969): Science 164, 1161.
2. Dutton, J.W.R. and Harvey, B.R. (1967): Water Res. 1, 744.
3. Abernathy, R.F. and Gibson, F.H. (1975): Rare Elements in Coal, Bureau of Mines Information Circular No. 8163, U.S. Dept. of the Interior, Bureau of Mines, Washington.
4. National Institute for Occupational Safety and Health (1977): Occupational Exposure to Inorganic Nickel. U.S. Dept. of Health, Education, and Welfare, Washington, D.C.
5. Bingham, E., Barkley, W., Zerwas, M., Stemmer, K. and Taylor P. (1972): Arch. Environ. Health 25, 406.
6. Clary, J.J. (1975): Toxicol. Appl. Pharmacol. 31, 55.
7. Adalis, D., Gardner, D. and Miller, F.J. (1978): Am. Rev. Respir. Dis. 118, 347.
8. Adkins, B. and Gardner, D.E. (1976): Proc. Am. Soc. Microbiol. 18, B.45.
9. Wase, A.W., Goss, D.M. and Boyd, M.J. (1954): Arch. Biochem. 51, 1.
10. Onkelinx, C., Becker, J. and Sunderman, F.W., Jr. (1973): Res. Commun. Chem. Path. Pharmacol. 6, 663.
11. Parker, K. and Sunderman, F.W., Jr. (1974): Res. Commun. Chem. Path. Pharmacol. 7, 755.
12. Ho, W. and Furst, A. (1973): Proc. West. Pharmacol. Soc. 16, 245, as cited by NIOSH, 1977.
13. Smith, J.C. and Hackley, B. (1968): J. Nutrition 95, 541.
14. Johnson, R.F., Jr. and Ziemer, P.L. (1971): Health Physics 20, 187.
15. Oberg, S.G. (1976): Ph.D. Thesis, Purdue University, W. Lafayette, IN.
16. Carvalho, S.M. (1979): Ph.D. Thesis, Purdue University, W. Lafayette, IN.
17. Charles, J.M., Williams, S.J. and Menzel, D.B. (1978): Toxicol. Appl. Pharmacol. 45, 302.

Fate of Radionuclides in the Environment

STUDIES OF AGE-DEPENDENT STRONTIUM METABOLISM WITH APPLICATION TO FALLOUT DATA*

S. R. Bernard and C. W. Nestor, Jr

The model for relating organ burden to dietary intake I can be written in integral equation form as

$$R_c(t) = R_c(t=0)\exp[-\lambda_e t] + \int_0^t f_w I \exp[-\lambda_e(t-\tau)]\,d\tau,$$

where t is the age, λ_e is the effective bone release constant = 0.089 yr^{-1} (Bernard [1]), R_c^e is the chronic bone burden for dietary intake of $f_w I$ units of ^{90}Sr per year; f_w is the fractional uptake from diet to bone. To set this up for computation involving fallout data we take t_B as the calendar year of birth and write

$$dR_c/dt = f_w(t-t_B)I(t) - \lambda_e R_c, \quad 0 \leq t-t_B \leq 20.$$

The British intake of the ^{90}Sr data in Papworth and Vennart [2] have been plotted in Fig. 1. These data can be expressed as

$$I(t) = \begin{cases} 0, & t<1953 \\ B_1 \exp[\beta_1(t-1953)], & 1953 \leq t \leq 1964 \\ B_2 \exp[-\beta_2(t-1964)], & t > 1964 . \end{cases}$$

Here B_1 and B_2 are approximately 256 and 1.10 × 10^4 pCi per year respectively, and β_1 and β_2 are 0.34 and 0.29 yr^{-1}, respectively. We take the burden at birth from Bernard [1] as

$$R_c(t_B) = C_1 I(t_B), \quad C_1 \sim 0.0096.$$

From the Kulp model (Bernard [1]),

$$f_w(t-t_B) = \begin{cases} A[1-C\sin\alpha(t-t_B)], & 0 \leq t-t_B \leq 20 \\ A, & t-t_B > 20, \end{cases}$$

where $C = 1/2$, $\alpha = \pi/10$ yr^{-1}, $A = 0.7$.

A plot of f_w as a function of age is given in Fig. 2. Here, as would be seen in Bernard [1], the reason the f_w is periodic is because it can be shown from the Kulp model that $f_w = K(d/dt\, C_a + (\lambda_e - \lambda_r)C_a)$, where C_a is the amount of calcium in the bone and λ_r is the radiological decay constant of ^{90}Sr, K is the bone-diet observed ratio -- the OR factor of Comar [3]. The bone-Ca data of Mitchell [4] fitted with a 4th degree polynomial is depicted in Fig. 3 as skeletal calcium weight vs age. Substitution of the data into the equation for f_w and plotting of the resultant information illustrates that the simple sine fit of the data renders a fair approximation of the curve shown in Fig. 2.

Denoting the present calendar year as t_p and the present age (Θ_p) as t_p-t_B then

$$dR_c/d\Theta = f_w(\Theta)I(\Theta+t_B) - \lambda_e R_c,$$

---

*Research sponsored by the Office of Health and Environmental Research, U.S. Department of Energy under contract W-7405-eng-26 with the Union Carbide Corporation.

$$R_c(\Theta_p) = R_c(\Theta_p=0)e^{-\lambda_e \Theta_p} + e^{-\lambda_e \Theta_p} \int_0^{\Theta_p} e^{\lambda_e \Theta} f_w(\Theta) I(\Theta + t_p - \Theta_p) d\Theta.$$

One has to take into consideration the fact that the intake is not continuous in time. Because of this, the integration must be taken in a piecewise fashion. To do this we set

$$R_c(\Theta_p) = R_c(\Theta_p=0) \exp(-\lambda_e \Theta_p) + A \exp(-\lambda_e \Theta)(B_1 Z_1 + B_2 Z_2).$$

The Z_1 and Z_2 parameters are of similar functional form, namely

$$Z = e^{-\gamma_1 X} \left[\int_{\Phi_1}^{\Phi_2} \exp(\gamma_2 \Theta) d\Theta - C \int_{\Phi_3}^{\Phi_4} \exp(\gamma_3 \Theta) \sin(\alpha \Theta) d\Theta \right].$$

For t_p less than 1964, $Z_2 = 0$ and the Z_1 parameter is conditional on Θ_p as noted below:

Θ_p	γ_1	$\gamma_2 = \gamma_3$	X	Φ_1	Φ_2	Φ_3	Φ_4
≤ 20	β_1	$\gamma_e + \beta_1$	$\Theta_p + 1953 - t_p$	Θ_L	Θ_p	Θ_p	20
>20	β_1	$\lambda_e + \beta_1$	$\Theta_p + 1953 - t_p$	Θ_L	Θ_p	Θ_L	20

where $\Theta_L = \mathrm{MAX}(\Theta_p + 1953 - t_p, 0)$. For t_p greater than 1964 the expressions are conditional on both t_B and Θ_p. Letting $\Theta_1 = \Theta_p + 1964 - t_p$, then for $t_B > 1964$, $Z_1 = 0$ and Z_2 is conditional on Θ_p as noted below:

Θ_p	γ_1	γ_2	γ_3	X	Φ_1	Φ_2	Φ_3	Φ_4
≤ 20	β_1	$\lambda_e - \beta_2$	$\gamma_e + \beta_1$	Θ_1	0	Θ_p	0	Θ_1
>20	β_1	$\lambda_e - \beta_2$	$\lambda_e + \beta_1$	Θ_1	0	Θ_p	0	20

For $t_B < 1964$ both Z_1 and Z_2 are conditional on Θ_1 as below:

Z_1:

Θ_1	γ_1	$\gamma_2 = \gamma_3$	X	Φ_1	Φ_2	Φ_3	Φ_4
≤ 20	β_1	$\lambda_e + \beta_1$	$\Theta_p + 1953 - t_p$	Θ_L	Θ_1	Θ_L	Θ_1
>0	β_1	$\lambda_e + \beta_1$	$\Theta_L + 1953 - t_p$	Θ_L	Θ_1	Θ_L	20

Z_2:

Θ_1	γ_1	$\gamma_2 = \gamma_3$	X	Φ_1	Φ_2	Φ_3	Φ_4
≤ 20	β_2	$\lambda_e - \beta_2$	Θ_1	Θ_1	Θ_p	Θ_1	Θ_p
>20	β_2	$\lambda_e - \beta_2$	Θ_1	Θ_1	Θ_p	Θ_1	20

Note that if the lower limit of integration exceeds the upper limit the integral is omitted. One of us (CWN) used the ORNL computers to make the necessary computations. Figs. 4, 5, 6, and 7 show plots of the British bone data (dots) and the smooth curves that are the computations from the above equations.

As can be seen in the early periods, 1957-60, and 1964, the computed curve tends to underestimate the data, while in 1968 there is an overestimate in the young members of the population. Papworth's and Vennart's model [2] yields a closer approximation to the data. In their model the fractional uptake is approximately constant, independent of age and the bone release constant is a *2nd* degree decreasing function of age. One way to use their model to estimate burdens for the intakes shown is to use tabulated data of the normalized Gaussian probability distribution function. It is not easily seen how to obtain closed form solutions from their model, at least by these authors. Mention is made that the calculated burden is probably not significantly different from the measured burden if the standard deviation is

Studies of Age-Dependent Strontium Metabolism

of the order of the mean. For the case of a single input into a compartment this is true [5]. It is also true for a catenary system of compartments. Also there is a need for error bars to be placed on the experimental data to get adequate comparison with the calculated curves. Further work in this area will concentrate on that path.

REFERENCES

1. Bernard, S.R., (1965): An Age Dependent Model for ^{90}Sr Metabolism, ORNL-3849, p. 195.
2. Papworth, D., and Vennart, J., (1973): Phy. Med. Biol. 18, p.169.
3. Comar, C., (1967): Strontium Metabolism, J.M.A. Lenihan, J.F. Lontit and and J.H. Martin, Acad. Press, p. 17.
4. Mitchell, H.H., et al., (1945): J. Biol. Chem., 158, p. 625.
5. Bernard, S.R., (1977): An Urn Model Study of Variability in a Compartment, Bull. Math. Biol. 39, p. 463.

Figure 1.

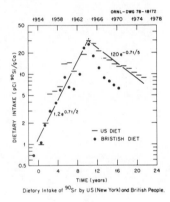

Dietary Intake of ^{90}Sr by US (New York) and British People.

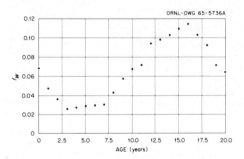

Figure 2. Plot of fraction of Sr retained in bone versus age.

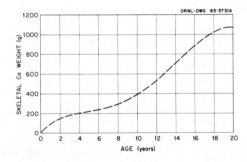

Figure 3. Mitchell's skeletal calcium curve.

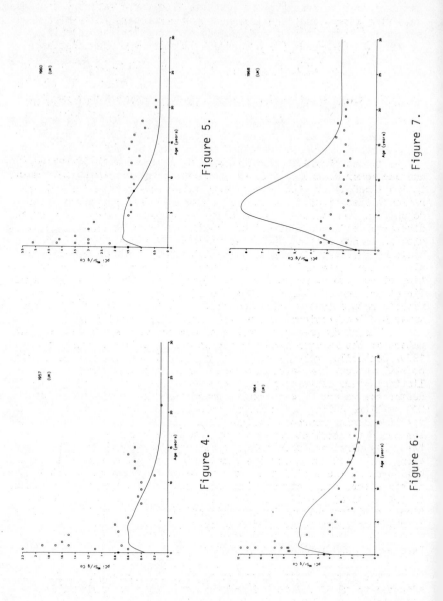

Figures 4, 5, 6, & 7. Observed concentration (data points) and theoretically predicted curve of ^{90}Sr in British subjects of various ages for the year indicated.

FATE OF MAJOR RADIONUCLIDES IN THE LIQUID WASTES RELEASED TO COASTAL WATERS*

I. S. Bhat, P. C. Verma, R. S. Iyer and S. Chandramouli

In the liquid radwaste released to coastal waters from the nuclear power stations, fission products - radiocesium and radioiodine - and activation product - ^{60}Co are reported as the critical radionuclides (1,2). Behaviour of these nuclides in the sea water depends on the ionic or colloidal state of radionuclide. Part of the discharged radionuclides get absorbed as soon as it comes in contact with suspended silt of sea water and slowly settle to the bottom. Radionuclides are biologically accumulated by marine organisms and also by the algae and ultimately reach human body through consumption of sea foods. In case of sea foods like crabs and prawns significant portion of radionuclides are in the non-edible part which reduces naturally the intake by man. The behaviours of ^{60}Co in sea water has been reviewed by Fukai & Murray (3) in detail.

The study presented here has been carried out in the coastal waters at the Tarapur Atomic Power Station (TAPS) operating from 1969. Controlled release of low level liquid wastes are made to the coastal waters. The water movement studies (1) have shown the oscillating nature of Tarapur near shore water with slow mixing with deeper sea waters. The coastal waters have silt varying from 50 to 200 mg/litre during non-monsoon period, but it goes as high as 1000 mg/litre during monsoon. At TAPS the liquid effluent is injected to the condenser coolant sea water at the outfall where it gets throughly mixed and then this diluted effluent flows out in open canal along the coast and joins the tidal waters. The activity discharged builds up in the oscillating coastal water body and gets distributed in sea water, silt, algae, fish and other foods. The chemical state of the critical nuclides in sea water, silt absorption and desorption and biological uptake of radionuclides by the marine organisms in the near shore region are described in this paper.

FATE OF THE NUCLIDES AT THE STATION OUTFALL

The liquid waste before release to the condenser coolant is adjusted to pH of 7.5 to 8.5 and the sea waters receiving this effluent has pH of 8.0. Dialysis experiments were carried out on the radwaste sample and also on the sea water mixed radwaste using cellulose tube membrane of 4.8 millimicron pore size. In case of $^{134+137}Cs$ and ^{131}I in radwaste nearly 100% passed through the dialyser membrane but in case of ^{60}Co only 15 to 20% passed through the membrane. Filtered sea water after spiking with radwaste and dialysis showed almost same result for ^{137}Cs and ^{131}I. Only 5 to 10% of ^{60}Co in filtered sea water passed through the dialyser. Thus, I and Cs which are in ionic state in sea water pass through the membrane but Co

---

*Work done under IAEA Research Contract No.1639

being hydrolysed completely at pH 8, more than 90% is likely to be in hydrous oxide particles of size more than the pore diameter (4.8 millimicron) of the dialyser. In case of cobalt Fukai & Murray (3) have reported negligible formation of cobalt hydroxide in sea water contrary to our observation.

IMMEDIATE SILT ABSORPTION BY SUSPENDED SILT

Liquid radwaste (10 ml) of known composition was mixed with fresh sea water (2 litres) stirred for a minute and filtered through millipore filter paper. The silt with adsorbed radionuclides was retained on the paper which was quantitatively counted by gamma spectrometry. The sea water silt content was 82 mg/litre during the experiments. Observed instantaneous silt adsorptions were ^{137}Cs 1.5 to 5%, ^{131}I 1 to 3% and ^{60}Co 80 to 99.5%.

EXCHANGE OF SILT ADSORBED RADIONUCLIDES

Samples of silt deposited at the bottom of discharge canal and nearby coastal areas were studied for the exchange of activity of the radiocesium and radiocobalt adsorbed on them. Silt at Tarapur coastal area contains about 85% of particles of size 50 microns or smaller. Coarse and fine silt having adsorbed ^{60}Co and ^{137}Cs were leached with 1 M ammonium acetate. The total activity on silt and in ammonium acetate leach were determined by gamma spectrometry. It was observed that only 0.5% of ^{60}Co was leached from fine silt and about 5% was leached from coarse silt by amm.acetate. 30 to 40% of total ^{137}Cs was leached from both coarse and fine silt by amm.acetate. Thus, cobalt adsorbed on silt is not easily available for exchange reaction where as cesium is labile to a significant extent. The exchangeability is seen decreasing with particle size. Organisms get their food through deposited silt and the labile activity in the silt would be easily available for biological uptake.

ACCUMULATION OF RADIONUCLIDES IN MARINE ALGAE

In the coastal environment around the radwaste release point, there is an abundant growth of algae (sea weeds) during the non-monsoon months. From the average radionuclide contents of sea water and algae, the concentration factors calculated for the three varieties of weeds are given in Table 1.

TABLE 1. Radionuclide concentration factors observed for Marine algae in the Tarapur environment.

Species	Concentration factor = $\frac{pCi/kg \text{ (algae)}}{pCi/litre \text{ sea water}}$ Range (average) for the nuclides		
	^{131}I	$^{134+137}Cs$	^{60}Co
1. Sargassum	354 − 2295 (1098)	76 − 195 (115)	235 − 920 (567)
2. Ulva Lactuca	77 − 226 (153)	3.6−33.5 (20)	65 − 292 (175)
3. Enteromorpha	20 − 436 (192)	51 − 502 (275)	73 − 358 (215)

Fukai & Murray (3) have reported the concentration factor for marine algae as 2000 from the analysis of stable cobalt. The (CF) obtained here from ^{60}Co concentrations is 4 to 10 times less.

ACCUMULATION IN ORGANISMS AND SEA FOODS

Crabs, prawns/shrimps, shell fish, onchidium (gastrapod) and small fishes are the main coastal organisms caught in the Tarapur near shore region. The edible soft tissues and non-edible shells and scales were tested separately. Significant amount of radionuclide accumulations on the non-edible portions were observed. The percentage of radionuclide in the edible tissue portion of the organisms are given in Table 2.

TABLE 2. Percentage of radionuclide activity in the edible portion of the coastal sea food.

Sea food variety	Percent of edible tissue to the total wt.	Percentage activity in the edible tissue of sea food		
		^{131}I	^{137}Cs	^{60}Co
1. Shrimps	38.0	12.0	78.0	40.0
2. Prawns	60.0	8.0	47.0	35.0
3. Crabs	81.0	70.0	79.3	78.2
4. Shell fish	11.0	Not analysed	81.0	85.3

Significant amount of activity present in the non-edible portion is a favourable factor decreasing the human intake through sea food.

CONCENTRATION FACTORS FOR SOFT TISSUES OF ORGANISMS

Radionuclide concentration factors (CF) in the sea foods help in radiation exposure evaluation. In the near shore environment of Tarapur Power Station, the CFs for the radionuclides in the coastal organisms were determined under natural conditions and the results obtained are shown in Table 3.

TABLE 3. Concentration factors of radionuclides in marine organisms in the near shore environment.

Organisms	Concentration factor = $\frac{pCi/kg}{pCi/litre}$		
	^{131}I	^{137}Cs	^{60}Co
1. Prawns	11 to 68	6 to 41	10.5
2. Onchidium	13.2	22.5	1.36×10^3
3. Oysters	29.7	26.7	40.0
4. Bombay Duck	11.2	8.5	15.5
5. Crabs	31 to 93	8 to 51	47 to 186

The near shore fishes taking up radiocobalt from coastal water medium have concentration factors ranging from 10 to 50 only, compared to 10^2 to 10^3 reported (3,4) for marine environment from the study of stable nuclides.

CONCLUSION

When ^{131}I, ^{137}Cs and ^{60}Co are released to coastal waters, ^{60}Co gets almost completely and irreversibly adsorbed on the silt, ^{131}I and ^{137}Cs get reversibly adsorbed to a fractional extent.

All the three nuclides are picked up by the crustaceans and benthos to a higher extent than fishes in the near shore region. The radionuclides considered reach the population mainly through sea food items.

The concentration factors observed for these radionuclides in coastal waters are low compared to general CFs reported (4) for sea waters. This may be due to the fact that near shore waters have about 10 to 50 times more inactive trace element content compared to off-shore waters.

ACKNOWLEDGEMENTS

The authors take pleasure in acknowledging their indebtness to S/shri S.D.Soman and P.R.Kamath for suggestions and discussions and approval of publication.

REFERENCES

1. Kamath P.R., Bhat I.S., Ganguly A.K. : Environmental behaviour of discharged radioactive effluents at Tarapur Atomic Power Station : IAEA/SM/146/58, 1970.
2. Bhat I.S. et al : Population exposure evaluation by environmental measurements and whole body counting in the environment of nuclear installations. : IAEA/SM/184/6, 1974.
3. Fukai R. and Murray C.N. : Environmental behaviour of radiocobalt and Radiosilver released from Nuclear Power Station into aquatic systems. : IAEA/SM/172/42, 1973.
4. Polikarpov G.G. : Radioecology of Aquatic Organisms. Reinholds, New York, 1966.

ASSESSMENT OF ^{210}Po EXPOSURE FOR THE ITALIAN POPULATION

G. F. Clemente, A. Renzetti, G. Santori and F. Breuer

The transfer to man of the normal environmental levels of the naturally occurring ^{210}Pb and of its descendant ^{210}Po is considered to contribute a large fraction of the natural radiation dose to man from internally deposited radionuclides (8). While ^{210}Pb being a beta emitter, contributes a small fraction of the entire dose due to the ^{210}Pb series, the alpha emitter ^{210}Po can be taken as the largest concurrent to the total natural internal dose to the skeleton of members of the general population (8).
^{210}Po has by itself a short effective half-life in the human body (6), which reduces its ability to accumulate in tissue, while the parent nuclide ^{210}Pb due to both its long physical half-life and metabolic behaviour in man, accumulates in bone(6). However, the ^{210}Po formed from ^{210}Pb in bone, it appears to remain in the skeleton in equilibrium with its parent (6). Therefore the knowledge of the pathways from environment to man and of the metabolic properties of ^{210}Pb and ^{210}Po is needed to assess the natural radiation dose due to the ^{210}Po burden in the members of the public. The dietary intake of ^{210}Pb is usually considered the main source of the ^{210}Pb - ^{210}Po internal burden in the general population (4, 7-8).
Increases of ^{210}Pb - ^{210}Po skeletal burden have been reported (1, 3, 6) in subjects exposed in uranium mines and spas to high levels of radon and daughter which are the progenitor of the ^{210}Pb series.
In this paper are summarized the experimental data needed to evaluate the ^{210}Po internal burden and the relative dose given to members of the general italian population and to those groups of subjects which are exposed in non uranium mines and spas to high radon and daughter air concentration.

MATERIALS AND METHODS

The ^{210}Po content, reported to the sampling time by correcting for decay and ingrowth from ^{210}Pb, has been measured in the following items: - About 40 samples of complete daily diet collected in six different regions of Italy and referred to adult members of the public.
- About 40 samples of urinary excretion collected from adult members of the public both non-smokers and smokers.

- More than 150 samples of urinary excretion collected from subjects working in spas and in non-uranium mines.
- About 20 samples of tooth and bone collected at various ages from members of the public.

The ^{210}Pb content, reported to the sampling time, was measured in most samples. The ^{210}Pb analysis has not been performed in those samples where the long time elapsed between collection and analysis was sufficient to give equilibrium conditions between the components of the ^{210}Pb series at the time of the analysis. The analytical procedure employed to determine ^{210}Pb and ^{210}Po in the various samples considered has been described in detail elsewhere (4). The method is based on a coprecipitation of polonium with manganese dioxide after a wet ashing of the sample. The ^{210}Po plated on silver disk is then measured by solid state detector spectrometry. The chemical yield of any analysis is tested by using ^{208}Po as an internal standard. The ^{210}Pb was evaluated from the ^{210}Po ingrowth in the sample by repeating the ^{210}Po analysis some months after the initial analysis.

RESULTS AND DISCUSSIONS

In table 1 are summarized the data referred to the main components of the metabolic balance of ^{210}Pb - ^{210}Po, evaluated on the basis of the ^{210}Pb - ^{210}Po content measured in the analyzed samples.

TABLE 1. Summary of the components of the ^{210}Pb - ^{210}Po metabolic balance in adult members of the general italian population.

	^{210}Pb	^{210}Po
Daily intake (Bq/day)		
Ingestion (Diet+Water)	0.11	0.11
Inhalation (non smoker)	0.006	0.0007
Daily excretion (Bq/day)		
Feces	0.095	0.095
Urine	0.015	0.015
Total Body Burden (Bq)	26	23
Skeletal Burden (Bq)	20	16

The daily inhalation has been calculated by measuring the ^{210}Pb ^{210}Po concentration in some air samples. The daily fecal excretion has been evaluated by applying the recently reported value (7) of the (daily fecal excretion) / (daily urinary excretion) ratio to the urinary excretion measured in our subjects. The ^{210}Pb - ^{210}Po concentrations measured in teeth have been found not significantly different from those in bone samples, the mean ^{210}Po concentration value ± S.E. being 3.2 ±0.5 mBq/g. The data of table 1 should be considered entirely indicative of the ^{210}Pb - ^{210}Po intake, excretion and internal burden of the non-smoker adult members of the general italian population, owing to the large individual variability of the ^{210}Pb - ^{210}Po excretion rates at natural levels (5). Furthermore the daily dietary intake of ^{210}Pb - ^{210}Po has been found quite variable as a function of many factors (4). The highest ^{210}Pb - ^{210}Po daily dietary intake (0.3 Bq/day) being measured in diet samples collected in an area with high natural background. The mean ^{210}Pb - ^{210}Po daily intake measured in Italy is in good agreement with the data reported (4,8) for other european countries, while is higher than in some extraeuropean countries (4,8).

A mathematical model has been developed on the basis of the data given in table 1 to describe the metabolic behaviour in man of the systemic ^{210}Po and it should be considered valid for any introduction of ^{210}Po into the systemic compartment of the human body. A detailed description of the model, reporting the retention function in soft tissue and bone and the urinary excretion function for ^{210}Po, is given in (2). These functions have been used to evaluate the ^{210}Pb - ^{210}Po body burden of the subjects exposed to high levels of radon and daughter on the basis of their ^{210}Pb - ^{210}Po daily excretion. The model has been satisfactorily tested on ^{210}Pb - ^{210}Po intake and excretion data reported by other authors (7).

In table 2 are summarized the ^{210}Pb - ^{210}Po internal burden and the resulting dose rate to the skeleton of four groups of subjects selected according to the origin of the ^{210}Pb - ^{210}Po exposure.

TABLE 2. Body burden and yearly dose rate for adult members of the italian population according to the origin of exposure.

| Population Group | Body burden (Bq) | | | | Yearly dose rate to the skeleton (mGy/year) |
| | Skeleton | | Soft Tissue | | |
	^{210}Pb	^{210}Po	^{210}Pb	^{210}Po	
General population non smoker	20	16	6	7	0.07
General population smoker	31	20	9	11	0.10
Non-uranium miners	150	110	30	25	0.60

Spa employees	480	370	110	90	2

The smoker subjects have a ^{210}Pb - ^{210}Po urinary excretion significantly higher than the non smokers, thus confirming the importance of smoking among the various natural sources of ^{210}Pb ^{210}Po contamination in man. A detailed discussion of the data reported in table 2, of their significance and of the dosimetric evaluations is given in (4). The mean dose rate to the human skeleton, due to the ^{210}Po bone content, has been assumed to be equal to the mean ^{210}Po dose rate received by the cells of the cortical bone (4,8). The yearly dose rate reported in table 2 for the adult members (smoker and non smoker) of the general italian population is in good agreement with similar data reported by the Unscear (8). The dose rate referred to spa workers and non-uranium miners should be considered purely indicative of the highest levels of exposure to radon and daughter in those environments. The ^{210}Pb-^{210}Po skeletal burden due to the radon and daughter exposure resulted to have large individual variability as a function of many factors which may seriously affect the lung retention and clearance of the radon daughter which are decaying to ^{210}Pb inside the lung (3,4). Nevertheless the reported data permit to establish the relevance of the ^{210}Po dose to the skeleton of subjects which are currently exposed or have been exposed in the past to high levels of radon and daughter in various environmental conditions.

REFERENCES

1. Blanchard, R.L., Kaufmann, E.L., Ide H.M. (1973): Health Phys. 25, 129.
2. Breuer, F., Clemente, G.F., (1979) : In: Radon Monitoring, p. 239, NEA-OECD, Paris.
3. Clemente, G.F., Renzetti, A., Santori, G. (1979): Envir. Res. 18, 120.
4. Clemente, G.F., Renzetti, A., Santori, G., and Bachvarova, A.K. (1979): CNEN Report RT/PROT (79) 6.
5. Holtzman, R.B., Spencer, H., Ilcewicz, F.H. and Kramer, L. (1977): In: Proc of the 10th Midyear Topical Symp. of Health Phys. Society, p. 245, Reusselaer Polytechnic Institute, Troy, N.Y.
6. Parfenov, Y.D. (1974): At. Energy Rev. 12, 75.
7. Spencer, H., Holtzman, R.B., Kramer, L., Ilcewicz, F.M. (1977): Radiation Res. 69,166.
8. Unscear (1977): Sources and Effects of Ionizing Radiation, p.35, United Nations, New York.

REPRESENTATION GEOGRAPHIQUE DE DIVERSES DONNEES DANS UNE GRILLE EUROPEENNE

A. Garnier, A. Sauvé and C. Madelmont

RESUME

On décrit un système de représentation uniforme permettant de collationner de manière homogène diverses données : population, données physiques, ressources agricoles, informations sanitaires.. dans les mailles de dimensions appropriées d'une grille européenne. Divers programmes ont été mis au point pour l'utilisation de ces données en vue de l'évaluation des doses collectives.

INTRODUCTION

Pour l'évaluation de l'équivalent de dose collective consécutive à la libération d'effluents radioactifs dans l'environnement, il est nécessaire de connaître la répartition géographique de la population et des productions agricoles dans les régions susceptibles d'être contaminées, afin d'évaluer les quantités de polluants qui, directement ou indirectement, pourraient être inhalées ou ingérées. A l'échelle européenne, on a établi un système défini par des coordonnées géographiques, permettant d'obtenir un ensemble homogène d'informations à partir de données initiales diversement présentées selon les pays - les unes (populations) se rapportant à des régions géographiquement définies, mais dans des systèmes différents de coordonnées - les autres (statistiques agricoles) se référant à des divisions administratives.

EXPOSE ET DISCUSSION

Définition du système (figure 1)

La zone considérée est délimitée environ par les parallèles 37° N et 60° N et par les méridiens 10° W et 20° E de Greenwich, mais le système retenu permet de l'étendre en latitude et en longitude. En effet, la zone est partagée en fuseaux de 1,5°, de part et d'autre du méridien de Greenwich ; le long des méridiens, on a porté des subdivisions variables, de façon à obtenir des mailles de surface équivalente se déformant progressivement du Nord au Sud. L'espacement des parallèles a été calculé de façon que chaque maille, quelle que soit sa position, ait une superficie de 10 000 km2 qui peut être subdivisée par interpolation linéaire, celle-ci étant facilitée par la transformation des coordonnées dans le système centésimal (grades). Ce maillage est directement utilisé dans la transformation des données agricoles disponibles au niveau de divisions administratives (proportionnalité pour les productions et cheptel avec les surfaces) C'est un maillage cent fois plus fin qui est utilisé pour répartir les effectifs des populations.

Le code de repérage des mailles est simple et très souple. Il utilise un double indiçage, en latitude à partir du parallèle 49° N, en longitude à partir du méridien de Greenwich. Chaque indice peut être positif (vers le Nord ou l'Est) ou négatif (vers le Sud ou l'Ouest) et comporte deux ou trois chiffres selon la finesse de la maille. L'extension du code à une grille plus étendue ou plus fine est donc aisée.

Répartition géographique de la population [1]

Les données proviennent des recensements nationaux, effectués selon les pays entre 1970 et 1975. Elles sont présentées soit par unités administratives (généralement les communes), soit dans une grille à mailles régulières (de 10 km x 10 km ou de 1 km x 1 km selon les cas).

La localisation des centres des communes ou des mailles dans le système de coordonnées géographiques indispensable pour unifier la présentation des données a exigé un gros travail, soit sur le plan cartographique, soit sur le plan informatique.

Plusieurs programmes informatiques ont été mis au point pour associer les données provenant des recensements de population et celles relatives aux coordonnées des centres des agglomérations ou des bornes de maillage. On a notamment classé, par ordre croissant des coordonnées géographiques, les données population des différents pays, ce qui permet, par exemple, de calculer la densité de population dans chaque maille en confondant les nationalités (tableau 1).

TABLEAU 1. Population dans la maille (5, 1) = 2679368

Ab. Or.	50	51	52	53	54	55	56	57	58	59
19	8105	4401	7802	7185	108060	1759	5466	9521	9218	16736
18	10504	2352	7627	23363	6924	6690	5474	1482	45191	5420
17	89107	6832	7324	13545	3357	6400	11655	15321	10360	2217
16	19545	10164	8621	12117	10347	17174	13550	18748	13092	19410
15	31065	3933	10123	11051	31207	25996	29573	30440	15534	18623
14	71057	34629	3825	4846	40744	28408	60344	96812	33521	16224
13	107472	47471	5210	9513	24417	102305	63491	106512	47662	45768
12	30694	27525	1648	7240	27656	80100	192464	41145	18192	10159
11	40162	149796	4917	3756	35958	49890	27544	44782	6992	3644
10	16341	14317	4838	10726	9302	5656	11380	8831	7658	10730

Acquisition et transformation des données agricoles [2][3]

Les données agricoles de base (moyenne des années 1970 à 1973) sont obtenues à partir d'un fichier donnant des caractéristiques de production végétale (surface, rendement, production des effectifs animaux, et ce pour des divisions administratives de base équivalentes d'un pays à l'autre (quelques milliers de km^2)). Les données manquantes sont estimées en fonction de surfaces cultivées pour les végétaux et d'effectifs animaux (produits laitiers et viande) et de données plus générales disponibles au niveau national pour les pays

Representation Geographique de Diverses Donnees dans une Grille Europeenne

à faible étendue et au niveau régional pour les autres. Un exemple de résultats figure au tableau 2.

TABLEAU 2. Production de lait de consommation (milliers de tonnes, années 1970-1973) dans la zone MNPQ (figure 1)

Ab. Or.	-2	-1	1	2	3	4	5	6	7	8
+4	417,1	227,3	108,6	6,4	90,0	382,3	338,4	166,2	111,4	22,4
+3	605,6	260,0	154,5	89,5	213,7	208,8	169,1	118,7	90,4	1,1
+2	45,9	11,6	14,6	210,0	253,2	241,8	160,3	104,4	107,6	111,3
+1	10,6	37,9	83,1	148,5	133,0	89,1	113,2	86,0	132,7	162,5
-1	57,1	89,0	53,9	30,3	20,2	52,5	106,1	141,7	193,5	295,3

DISCUSSION

. Les dimensions des mailles utilisées paraissent bien adaptées aux possibilités d'obtention des données et à leur utilisation pour les évaluations à l'échelle régionale en ce qui concerne tant les populations que les productions agricoles. Il va de soi que l'étude approfondie de l'environnement d'un site particulier requiert des données plus détaillées .
. La grille de données concernant les productions agricoles sera améliorée dans les mises à jour ultérieures (tous les cinq ans environ) : en s'efforçant d'acquérir plus de données au niveau des divisions administratives de base, en précisant, dans certains cas, les productions destinées à l'homme ou à l'animal (cas des pommes de terre, des céréales), en effectuant des corrections de répartition utilisant le maillage fin avec pour les zones côtières (la présence de population) et pour les zones montagneuses (la prise en compte de l'altitude).
. D'autres programmes permettant de produire les données dans des configurations différentes, telles que cercle ou secteur, selon le mode de présentation des autres données du problème.

CONCLUSION

Ce système homogène d'informations, établi à l'échelle européenne, et encore susceptible d'améliorations, est directement utilisable pour les évaluations d'impact radiologique des installations nucléaires [4] .

Remerciements

Nous remercions très vivement Monsieur le Professeur Kormoss du Collège d'Europe (Brugge) et l'Institut Géographique National (France) ainsi que l'Institut Royal de Groningen.

REFERENCE

1. Garnier A., Sauvé A., Etude géographique de la population européenne en vue de l'évaluation des conséquences radiologiques des rejets d'effluents radioactifs (Rapport CEA à paraître).
2. Eurostat
3. Fichier de l'Institut Royal de Groningen.
4. CCE-NRPB-CEA, Méthodologie pour l'évaluation des conséquences radiologiques des rejets d'effluents radioactifs, 1979.

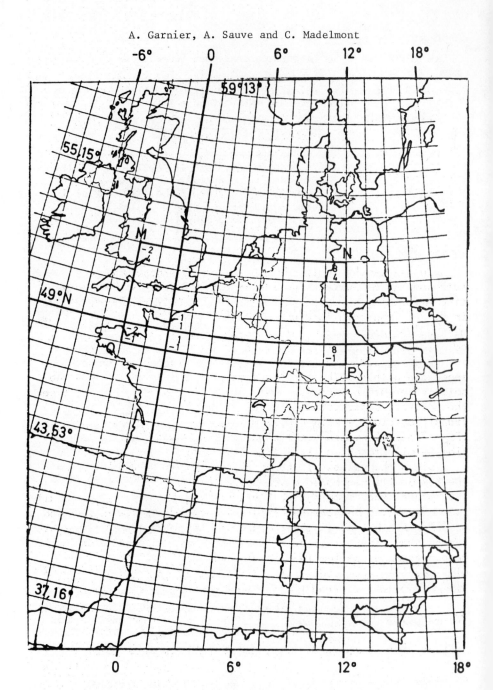

Figure 1 : Définition du système de grille européenne

STUDIES OF THE TRANSFER FACTORS OF SR90 AND CS137 IN THE FOOD-CHAIN SOIL-PLANT-MILK

K. Heine and A. Wiechen

To assess the radiation exposure of the population in the surroundings of nuclear industrial installations with radioactive effluents, calculations of the radionuclide concentrations in the environmental mediums, based on the quantities discharged, must be performed during planning and operation. For these calculations it is necessary to know the transfer factors describing the radionuclide transfer in the foodchains for the ingestion pathway. The plant uptake of a radionuclide from the soil is defined by the transfer factor TF_{SP} as the ratio between the activity concentrations in plant and soil. For the transition plant - milk of the milk pathway the transfer factor TF_{PM} of a radionuclide is the ratio of the activity concentrattion in milk to the daily intake rate of an animal.

Since the values of the transfer factors recommended for the F.R.G. (1) are based on data from the international literature, they remain to be verified for the ecological conditions prevailing in Germany. This holds especially for Sr 90 and Cs 137 which belong to the relevant radionuclides of the ingestion pathway. Therefore, transfer factors of Sr 90 and Cs 137 were determined by low-level measurements of soil, plant and milk samples. The studies were performed in the surroundings of the site of the planned nuclear waste center at Gorleben.

MATERIALS AND METHODS

In 1978, samples of soil, plant and milk were taken at several locations. The pastures under study had either podsolic or loamy soils. Podsol is a soil type with high contents of sand which is common in the F.R.G. The loamy soils are found in the lowland of the river Elbe. The sampling depth in the soil was 10 cm. The following parameters were determined in soil samples: pH - value (0,1 nKCl), exchangeable Ca and K (Mehlich), glow loss (400 $^\circ$C ashing), concentration of K and Cs 137 (γ-spectrometry) and Sr 90 concentration after HCl extraction (β-spectrometry of Y 90).

Sampling of plant was performed twice during the vegetation period on the same pastures, the soil samples had been taken from. Determination of Sr 90 after fusion of the ashes with Na_2CO_3 and K_2CO_3 and of Cs 137 was performed in a similar manner as for the soil samples.

During the outdoor season raw milk samples were collected from 5 farms under study in about monthly intervals. Further, bulk milk of two different collection areas was studied. After drying and ashing of the milk samples Sr 90 and Cs 137 were determined.

RESULTS

Transfer factors TF_{SP} soil-plant were calculated based upon the results obtained from the Sr 90 and Cs 137 determinations in the soil and plant samples. A survey is given in table 1 where the transfer factors are listed according to increasing glow loss of the soil samples. The values of the Sr 90 transfer factor are negatively correlated with the glow loss at the 1 % level. This tendency is due to the positive correlation between the Sr 90 concentration of the soil samples and the glow loss that is significant even at the 0,1 % level.

TABLE 1. Transfer factors TF_{SP} soil-plant for pastures

Pasture No.	Soil type	Soil glow-loss (%)	TF_{SP}* Sr 90	Cs 137
1	Podsolic	2,7	0,97	0,086
2	Loamy	6,0	0,46	0,038
3	Podsolic	6,3	0,41	0,108
4	Loamy	7,0	0,40	0,045
5	Podsolic	8,3	0,35	0,094
6	Podsolic	8,3	0,29	0,055
7	Podsolic	12,5	0,27	0,051
8	Loamy	16,2	0,10	0,018

* pCi/kg plant wet weight: pCi/kg dry soil

The glow loss is approximately equal to the organic matter content in the soil. Hence, the cation-exchange capacity and the sorption of Sr 90 at the available exchange sites in soil increases with increasing organic matter content. This may be the cause of the reduction of the Sr 90 uptake by plants. The Sr 90-transfer factors are correlated with the available Ca

Studies of the Transfer Factors of Sr 90 and Cs 137

content in soil at the 5 % level of significance. No correlations with the other soil parameters were observed.

The values of the Cs 137-transfer factor TF_{SP} were higher in podsol than in loamy soils. Very low values were obtained for pasture No. 8 which is situated in the floodplain of the river Elbe. There was a correlation between the transfer factors and the soil parameters only for whole potassium, which was negatively correlated with the values.

Transfer factors TF_{PM} were determined for the transition plant-milk assuming a daily intake for cows of 11 kg dry matter. A survey of the values with the reference to the respective pastures is given in table 2.

TABLE 2. Transfer factors TF_{PM} plant-milk

	Associated Pasture No.	$TF_{PM} \cdot 10^{3*}$			
		Sr 90		Cs 137	
		Range	Mean	Range	Mean
Area 1	1;2;8	1,3 - 2,1	1,5	7,0 - 13	9,7
Area 2	2;8	1,1 - 2,4	1,8	6,5 - 11	8,6
Farm 1	3	1,3 - 2,7	1,8	8,5 - 57	32
Farm 2	4	1,4 - 1,7	1,5	2,3 - 5,5	3,8
Farm 3	5	0,7 - 3,2	1,3	6,7 - 11	9,4
Farm 4	6	1,5 - 2,3	1,9	5,3 - 19	11
Farm 5	7	1,4 - 2,0	1,7	5,4 - 30	14

* pCi/l : pCi/daily intake

The mean values of the Sr 90 factors agree well for the different locations. On the other hand, the Cs 137-transfer factors TF_{PM} were found to vary markedly which was due to the great variations of Cs 137 concentrations in the milk. As it has been the case with the transition soil-plant, the highest values were found at farms with sandy soils. Throughout the outdoor season, the lowest factors were determined for farm 2 situated in the lowland of the river Elbe. As shown by sorption studies of different soils using Cs 134 tracer, this may be caused by sorption of Cs 137 on soil particles ingested by the cows together with the feed (2).

DISCUSSION

The determined transfer factors TF_{SP} for the transition soil-plant were found to be dependent on both the soil type and the soil composition which is in agreement with the literature (3, 4). The values obtained for Sr 90 and Cs 137 were of the same order of magnitude than those determined by lysimeter experiments with podsol of Gorleben (5). Since no discrimination between root uptake and direct contamination of plants by fall out is possible in field studies of Sr 90 and Cs 137, the determined values can be considered the upper limit of the transfer factors TF_{SP}.

The data of the Sr 90-transfer factor TF_{PM} with a mean value of 0,0016 for the transition plant-milk agree well with the value of 0,0014 estimated from the literature (6). Considering the given ecological conditions the recommended factor 0,0008 (1) should be doubled.

The determined range of the Cs 137-transfer factor TF_{PM} between 0,0023 and 0,057 is generally higher than the value 0,0071 calculated from the literature (6). The mean value measured at 2 farms during the outdoor season has been found to be higher than the recommended value of 0,012 (1). From the results obtained so far, it is not possible to decide definitely whether the recommended Cs 137-transfer factor TF_{PM} can be verified for the ecological conditions of the surroundings of Gorleben. The studies will be continued for several years in an attempt to support the present data.

REFERENCES

1. Gemeinsames Ministerialblatt (1979): Allgemeine Berechnungsgrundlage für die Strahlenexposition bei radioaktiven Ableitungen mit der Abluft oder in Oberflächengewässer. No. 21, 369, Bonn, F.R.G.
2. Heine, K., Wiechen, A. (1979): Milchwissenschaft, 34, 275.
3. Francis, C.W. (1978): Radiostrontium Movement in Soils and Uptake by Plants. Technical Information Center, US Department of Energy, TID-27564.
4. Dahlman, R.C., Francis, C.W., Tamura, T. (1974): Radiocesium Cycling in Vegetation and Soil. Proc. Symp. on Mineral Cycling in Southeastern Environment. Savannah River Laboratory.
5. Führ, F. (1979): Die Aufnahme von Radionukliden aus dem Boden. Conference Radioecology. Deutsches Atomforum, Bonn, F.R.G.
6. Ng, Y.C., Colsher, C.S., Thompson, S.E. (1978): Transfer Coefficients for Terrestrial Foodchain - Their Derivation and Limitations. 12th Annual Conference of the German-Swiss Fachverband Strahlenschutz, Norderney. F.R.G.

STUDIES OF CONCENTRATION AND TRANSFER FACTORS OF NATURAL AND ARTIFICIAL ACTINIDE ELEMENTS IN A MARINE ENVIRONMENT

E. Holm, B. R. Persson and S. Mattsson

Natural (thorium and uranium) and artificial (plutonium and americium) actinides have been studied in a marine environment at the Swedish south west coast.

The actinides have reached the water from fall-out, run off from land, leaching, resuspension or by in situ build up. The partition of these elements between water, sediment and organisms was studied. Special concern was given to plutonium and americium.
Such investigations play an important role for predicting the fate of actinides released into the sea by dumping or wastes from nuclear industry. Their fate will depend on type of source and the physical and chemical parameters.

The brown algae Fucus vesiculosus and Fucus serratus was shown to be excellent bioindicators for actinide elements. Other investigations have frequently used mussels as bioindicator for transuranium elements.

MATERIALS AND METHODS

Fucus vesiculosus and F. serratus have been collected at the Swedish south west coast occationally during 1967-1976 (1). After 1976 a more regular collection has taken place.
During 1978 and 1979 several watersamples and sediment cores were taken.
The actinides were separated radiochemically by procedures published elsewhere (2). As radiochemical yielddeterminants ^{242}Pu, ^{243}Am, ^{232}Th and ^{229}Th were used. The alpha activity of samples electro deposited onto stainless steeldiscs were measured with surface barrier detectors for 1-20 days.

RESULTS AND DISCUSSION

The area studied is situated in an area around 56°N, 13°E. It could briefly be characterized as very shallow with strong currents, high resuspension and high organic particulate in the water.

ACTINIDES IN WATER

In table 1 the results for water samples are given. All values refer to unfiltered water. We find lower plutonium and americium concentrations compared to most other areas such as the Atlantic and the Mediterranean (3, 4). Our results are in agreement with investigations by Murray et. al. (5).
Bottomwater (19 m) shows higher plutonium and americium concentrations than surface water. The uraniumconcentration could be repre-

sentative for this brackish water (salinity 7-10 o/oo at surface). We can not observe any influence of transuranics from european reprocessing facilities although this has been done for ^{134}Cs (6).

We have not yet investigated the partition between particulate and soluble fraction of the actinides. We believe the transuranium elements are predominantly in a particulate form due to the high particulate load in this region.

Table 1. Actinides in water and Fucus

	^{228}Th	^{230}Th	^{232}Th	^{234}U	^{235}U	^{238}U	$^{239+240}$Pu	^{241}Am
Water (µBq/kg)								
Surface	450	270	70	18100	650	15900	14	3
Bottom (19m)							40	7
Fucus (mBq/kg dry)	5000	400	90	12700	450	11100	190	53
CF (dry)	11000	1300	1300	700	700	700	13500	18000

ACTINIDES IN SEDIMENT

The results for sediment support the theory of rapid sedimentation for the transuranium elements in this area. The concentrations in surface sediments are a factor of 3-4 higher than for other regions contaminated from global fall-out. This in combination with the low concentrations in water gives a dry sediment/water activity concentration ratio for plutonium as high as 250 000.

The results varied much depending on site of collection, but in table 2 are given the average values for 6 representative cores. The integrated area content for these cores was estimated to 72 Bq/m^2 and 20 Bq/m^2 for $^{239+240}$Pu and ^{241}Am respectively. This is much higher than expected from integrated fall-out at this latitude (48 and 13 Bq/m^2) indicating significant run off from land.

Table 2. Plutonium and americium in sediment cores. (Bq/kg dry)

Depth (cm)	$^{239+240}$Pu	^{241}Am	^{241}Am/$^{239+240}$Pu
0-2	80	18	0.20
2-4	79	17	0.23
4-10	12	4	0.32

Americium and plutonium show different vertical distributions. The activity ratio Am/Pu increase with depth although this ratio agrees with integrated fall-out in the integrated cores. This difference is explained by that americium has mainly been formed in situ by the decay of ^{241}Pu($T_{\frac{1}{2}}$=14.2 a). Plutonium deeper down is older and shows then higher Am/Pu ratio. We do not believe that this is an effect of higher penetration for americium. Of the integrated area content for water and sediment only 0.3 % of plutonium and americium is present in the watercolumn. This is in large contrast to the situation for example in the Mediterranean (7).

ACTINIDES IN ALGAE

The results for actinides in Fucus can be seen in Table 1 and Fig. 1. All values refer to the date of collection. The measured ^{241}Am was corrected to the date of collection by using ^{241}Pu/$^{239+240}$Pu activity ratios measured in global fall-out (8). The $^{239+240}$Pu concentration in Fucus has decreased during the period. Such a decrease is expected due to that the main fall-out delivery ocurred shortly after 1962.

^{241}Am on the other hand shows an increase during the same period. This increase must be related to the in situ build up of ^{241}Am ($T_{\frac{1}{2}}$= 433 a) from ^{241}Pu($T_{\frac{1}{2}}$=14.2 a) in the environment.

Most americium values fall below the expected ones from Am/Pu activity-ratios in integrated fall-out.

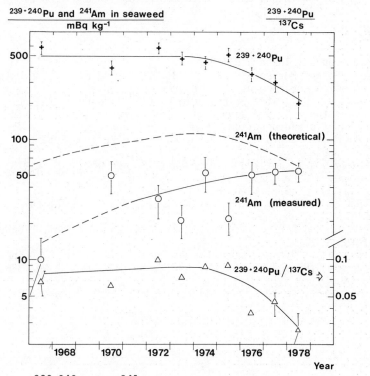

Figure 1. $^{239+240}$Pu and ^{241}Am activity concentrations in Fucus during 1967-1978. The Pu/Cs activity ratio is also displayed.

Our results indicate that the concentration factor for americium from water to Fucus is higher than for plutonium. The processes involved to reach water and organisms for in situ produced americium are different to those for plutonium.

The ^{137}Cs concentration in Fucus has been rather constant during 1967-1978 (1), resulting in a decreasing Pu/Cs activity ratio from 0.1 to 0.02 during the period.

By using ^{137}Cs values from Aarkrog et al (9) the $^{239+240}$Pu/^{137}Cs activity ratio in water in 1978 was about 0.002-0.003. This illustrates the much higher uptake of plutonium than cesium by Fucus.

Recently Hodge et.al.(10) showed that uptake of actinides in particulate form by coastal marine organisms was significant. Our ^{234}U/^{238}U activity ratio in Fucus is about 1.14. The soluble element uranium shows the lowest concentration factor. Plutonium and americium which are expected to be predominantly in the particulate form, in this area, show the highest concentration factors. Our results indicate an organic particulate sorbtion of actinides by Fucus.

Of the thoriumisotopes ^{228}Th differs remarkably from the others. The lifespan for the algae is not long enough to explain this high CF for ^{228}Th by in vivo build up from ^{228}Ra. This might indicate that ^{228}Th is brought into a more bioavailable form through the decay - chain ^{228}Ra-^{228}Ac.

REFERENCES

1. Mattsson S,: To be published.
2. Holm E. and Persson R.B.R. (1978): In: Proceedings, The Natural Radiation Environment III, in press.
3. Fukai R., Ballestra S. and Holm E. (1976): Nature, 264, p. 739.
4. Livingston H.D. and Bowen V.T. (1976): Environmental toxicity of aquatic radionuclides: Models and Mechanisms, p. 107. Editor: M.W. Miller and J.N. Stannard. Ann Arbor Science Publ. Inc. Ann Arbor, USA.
5. Murray C.N., Kautsky H., Hoppenheit M. and Domain M. (1978): Nature, 276, p. 225.
6. Aarkrog A. (1979): Personal communication.
7. Fukai R., Holm E. and Ballestra S. (1979): Oceanologica Acta, 2, p. 129.
8. Holm E. and Persson R.B.R. (1979): Rad. and Environm. Biophys., 15, p. 261.
9. Aarkrog A., Bøtter Jensen L., Dahlgaard H., Hansen H., Lippert J., Nielsen S.P. and Nilsson K. (1978): In: Risø report No. 386, p. 58.
10. Hodge V.F., Koide M. and Goldberg E.D. (1979): Nature, 277, p. 206.

^{226}Ra- AND ^{222}Rn-CONTENT OF DRINKING WATER

J. Kiefer, A. Wicke, F. Glaum and J. Porstendörfer

^{226}Ra- and ^{222}Rn-concentrations were measured in drinking water in parts of the Federal Republic of Germany. 254 samples were analysed for ^{226}Ra, 57 for ^{222}Rn. The average contents were found to be $4,4 \times 10^{-3}$ Bq/l (0.2 pCi/l) for ^{226}Ra and 7,4 Bq/l (0.2 nCi/l) for ^{222}Rn with maximum values of 0.11 Bq/l (3 pCi/l) and 43 Bq/l (1.2 nCi/l), respectively. For both radionuclides there was a log-normal distribution of concentrations. Statistical analysis revealed no correlation between ^{226}Ra and ^{222}Rn, both for processed drinking and raw water. It is shown that there is a significant influence of the processing method on the final content of radioactivity. The average yearly doses to the critical organs are estimated to be about 0.003 mSv/a for ^{226}Ra (bone lining) and about 70 µSv/a for ^{222}Rn (stomach).

INTRODUCTION

Radium and its daughter products constitute an important part of natural environmental radiation exposure. Since ingestion forms a major pathway - apart from inhalation for Radon - for internal irradiation, the measurement of radioactivity in drinking water is relevant to assess the contribution of these environmental radiation hazards. Although a considerable body of information is already available it is generally felt that more data are desirable (1). We started therefore a survey of ^{226}Ra- and ^{222}Rn-concentrations in a part of central Germany to obtain also a better insight into the distribution of the measured parameters.

MATERIAL AND METHODS

Scope:

Random samples of drinking water (257 for ^{226}Ra and 57 for ^{226}Ra and ^{222}Rn) were taken in Hessia, a state of the Federal Republic of Germany during 1977 - 1979. In some instances the waterpath was followed from the well to the final consumer.

Measurement

^{226}Ra and initially ^{222}Rn were measured using Lucas szintillation chambers as already described (2). Later ^{222}Rn was determined by means of liquid szintillation in a Beckman LS 8000 according to Pritchard and Gesell (3).

RESULTS

Figure 1 shows the sum distributions of the measured values on a probit scale with a logarithmic abscissa. The approximately straight line indicates a log-normal distribution. The mean values are 0.2 pCi/l for ^{226}Ra and 0.2 nCi/l for ^{222}Rn. In 50 samples both nuclides had been measured. Statistical analysis of these data showed no correlation, the correlation coefficient was 0.12 (p < 0.9). Since it is known that water processing strongly influences the radionuclide distribution (see below) the same analysis was carried out with unprocessed well water (22 samples). The correlation coefficient was 0.28, which is again too small to accept a statistical correlation.

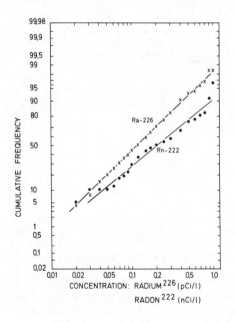

Figure 1.
Sum distribution of ^{226}Ra- and ^{222}Rn-concentrations

In some places it was possible to follow the radionuclide content from the well to the final consumer. Two examples are shown in figure 2A and 2B: In the first one the processing consisted of an aeration stage with subsequent fast filtration through a bed containing oxydizing substances to precipitate Fe- and Mn-contamination.

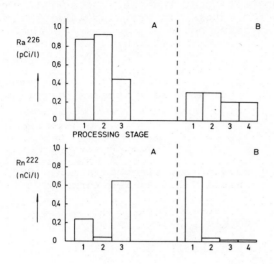

Figure 2. Influence of water processing on radionuclide content: A: Processing using rapid filtration with oxydizing substances; B: slow filtration; stages: 1. Well water; 2: after aeration; 3: after filtration; upper panel: ^{226}Ra; lower panel: ^{222}Rn

It is seen that the ^{226}Ra-content drops to about 50% during filtration, presumably due to Ra-binding in the filter bed. ^{222}Rn is lost drastically by aeration, but its concentration rises to rather high levels after filtration. We suggest that this is caused by Ra-accumulation in the filter bed. If a slow filtration process is used (figure 2) the decrease in ^{226}Ra is smaller and there is no increase in ^{222}Rn after the last stage. It is clear from this comparison that the processing method influences the final concentrations in drinking water.

DISCUSSION

We have measured the ^{226}Ra- and ^{222}Rn-content in drinking water in a part of Germany. The average values found agree reasonably with those reported for other areas (4). If we assume a daily consumption of 1.2 l drinking water the mean dietary intake is 0.14 pCi/l. According to UNSCEAR 77 (1) this would lead to a bone activity of 0.9 pCi/kg^{-1}. From this we estimate a yearly dose equivalent of about 15 µ Sv/a (1.5 mrem/a). For ^{222}Rn only unboiled drinking water has to be taken into account (about 0.15 l/d) which exerts its action mainly on the stomach. The mean somatically significant dose equivalent to this

organ is then estimated to be 70 µ Sv/a (7 mrem/a) (5).
Our results demonstrate that the yearly dose-rate expected from ^{226}Ra and ^{222}Rn intake via drinking water is low compared to other radiation sources.

Acknowledgement

This study was supported by a grant from the German Federal Ministry of Internal Affairs ("Bundesministerium des Innern")

REFERENCES

1. Sources and effects of ionizing radiation, 1977 report (1977): United Nations, New York

2. Scheibel, H.G., Porstendörfer, J. and Wicke, A.(1979): Nucl. Instr. Meth. 165, 345

3. Pritchard, H.M. and Gesell, T.F. (1977): Health Physics 33, 577

4. Aurand, K., Gans, I. and Rühle, H. (1974) in: Die natürliche Strahlenexposition des Menschen (eds. K. Aurand, H. Bücker, O. Hug, W. Jacobi, A. Kaul, H. Muth, W. Pohlit, W. Stahlhofen), Georg Thieme Verlag, Stuttgart, p. 29

5. Aurand, K., Gans, I., Wollenhaupt, H., personal communication

MOBILITY AND RETENTION OF ^{60}Co IN SOILS IN COASTAL AREAS

Y. Mahara and A. Kudo

The operation of more than 20 commercial nuclear reactors, at coastal sites in Japan during the past several years, has produced a tremendous amount of radioactive wastes which have saturated the waste storage facilities at the reactor sites. Because of present Japanese safety regulations, the wastes cannot be disposed of permanently underground nor on the ocean floor. Two methods of disposal (either into the ocean floor or underground) have been much discussed, but still more information is needed on the movements of radionuclides in the environments. This is definitely so for the movements of radionuclides underground, especially in underground water-flow systems. The lack of information includes data on and the mechanisms of the chemical and/or biological transformation of radioactive materials during the movements under various geochemical conditions.

Among the operating wastes from electricity generating nuclear stations, radioactive cobalt (^{60}Co) is a major radionuclide in quantity and as a radiological hazard. It has a relatively long half life (5.3 years) and emits two gamma rays with high energy (1.17 Mev and 1.33 Mev). Furthermore, cobalt is one of the essential elements to sustain life in animals and plants (for example, as Vitamin B_{12}) and hence is easily accumulated by various organisms, if available to them (1,2).

This paper summarizes the results of our previous investigation and reports some recent findings of ^{60}Co interactions in seawater-sediment systems, especially pertaining to the difference in the mobility of ^{60}Co under aerobic and anaerobic conditions. Furthermore, the difference in the mobility of ^{60}Co in seawater and in freshwater was investigated under the influence of various environmental factors. Details of the investigation will be reported elsewhere (3).

MATERIALS AND METHODS

A series of laboratory experiments was conducted simulating some of the underground environments at coastal areas. The interaction of ^{60}Co with various sediments in both freshwater and seawater systems was observed for a considerable period under either aerobic or anaerobic conditions. The degree of ^{60}Co mobility, after it had interacted with water and sediments, was determined by measuring radioactivity in the water phase and in the sediment phase.

The reversible (and irreversible) nature of ^{60}Co interaction between water and sediments was evaluated by a quick change of the surrounding environments. The ^{60}Co, for example, after having interacted under anaerobic conditions for the first 30 days, was suddenly introduced into aerobic conditions for the next 30 days. On the other hand, the ^{60}Co, kept under an aerobic condition for the first 30 days, was moved into a condition of anaerobic for a further 30 days. To achieve the new anaerobic and aerobic conditions, nitrogen and oxygen gases were bubbled respectively into the systems.

To evaluate some of the physical and chemical characteristics of the highly mobile ^{60}Co produced in a seawater-sediment system, the ^{60}Co was separated from the rest of the system by means of a dialysis membrane (24 Å pore size), a chemical extraction (dithizone-benzene), and a sorption (various sediments). Details of the physical and chemical characteristic of the sediments used were reported elsewhere (4-6).

RESULTS AND DISCUSSION

The magnitude of ^{60}Co mobility under four different environmental conditions is illustrated in Fig. 1. The mobility in the system of the sand sediments and freshwater under aerobic condition was used as a base line for a comparison. The mobility of ^{60}Co increased at pH values between 6 and 8 in the following order: (1) in freshwater under aerobic conditions, (2) in seawater under aerobic conditions, (3) in freshwater under anaerobic conditions, and (4) in seawater under anaerobic conditions. It is seen that the ^{60}Co in the system of seawater under anaerobic conditions is by far the most mobile.

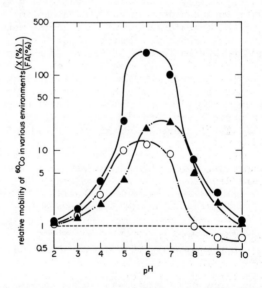

Fig. 1. Magnitude of ^{60}Co mobility in various environments: ▲---▲ X = freshwater under anaerobic, ●—● X = seawater under anaerobic, ○--○ X = seawater under aerobic, - - - - FA = freshwater under aerobic.)

Mobility and Retention of ^{60}Co

The irreversible nature of ^{60}Co derived from the changes in the redox environments was observed. In other words, the initial conditions such as aerobic or anaerobic conditions in the sediment and water systems where ^{60}Co was initially discharged seemed to determine the magnitude of ^{60}Co mobility. In Fig. 2, characteristic variation of ^{60}Co mobility was shown when incubating anaerobic conditions were suddenly converted into aerobic, and vice versa. The following results were observed in a series of these experiments; (i) if ^{60}Co initially had a high mobility under anaerobic conditions, it could have the high mobility even if surrounding environments were changed drastically, and (ii) if ^{60}Co initially had a low mobility under aerobic condtions, it would keep the low original mobility even if surrounding conditions were changed.

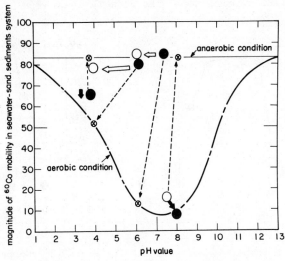

Fig. 2. Mobility of ^{60}Co, in the system of seawater and the sand sediments, with respect to changes in pH. (Incubating condition was changed suddenly from aerobic to anaerobic, or vice versa.) The following five symbols (○,●,⊗,⇨,➡) indicate respectively: the magnitude of ^{60}Co mobility under aerobic conditions: mobility under an anaerobic condition; mobility of reversible ^{60}Co expected by changes in the incubating condition; direction of change from anaerobic to aerobic conditions; and direction of change from aerobic to anaerobic conditions.

A highly mobile form of ^{60}Co produced in the seawater system under anaerobic conditions could pass through a dialysis membrane freely (more than 99%). This result indicated that the mobile form of ^{60}Co consisted of a mixture of ionic cobalt, cobalt compounds with low molecular weight, and cobalt adsorbed (or absorbed) on the surface of very fine particulates whose diameter is less than 24 Å.

A comparison was made between ionic form of cobalt ($CoCl_2$) and the cobalt which was highly mobile and could pass through a dialysis membrane freely. The purpose of this comparison was to define some of the characteristics of the highly mobile cobalt produced in the system of seawater under anaerobic conditions. Figure 3A shows a result of dithizone-benzene extraction for two forms of the cobalt at various pH values. The highly mobile cobalt was not extracted at all by the chelating agent with an organic solvent, though almost all of the ionic form was extracted at pH values between 4 and 8.

There is a clear difference between two forms of the cobalt in sorption to a fresh sand sediment. Again, the highly mobile cobalt was not reactive to the sediments, though the ionic form was sorbed nearly 100% by the sediments at pH values between 5.5 and 7, Fig. 3B. Unfortunately, the precise nature of the highly mobile cobalt has not yet been determined, but the result of these experiments suggest that it is stable and moves freely with the water in the flow systems of underground environments at the coastal areas.

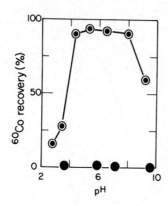

Fig. 3-A. Comparison between recovery of mobile ^{60}Co in seawater under anaerobic conditions by dithizone-benzene and recovery of ionic ^{60}Co.

Fig. 3-B. Comparison between sorption of mobile ^{60}Co in seawater under anaerobic conditions on the sand sediments and sorption of ionic ^{60}Co.

(● ; ^{60}Co in seawater under anaerobic, ◉ ; ionic ^{60}Co)

The authors thank the staff members of the National Research Council of Canada for their contribution to this work. Further acknowledgement is due to Dr. D.R. Champ, Chalk River Nuclear Laboratories for his valuable suggestions and to Dr. S. Senshu, Central Research Institute of Electric Power Industries, who made this visiting research possible for Y.M. N.R.C.C. 17873.

REFERENCES

1. Robertson, J.B. (1979): IAEA-SM-243, #152.
2. Morgan, K.Z. et al. (1964): Health Physics, 10, 151.
3. Mahara, Y. and Kudo, A. (in preparation).
4. Kudo, A. and Hart, J.S. (1974): J. Environmental Quality, 3, 273.
5. Kudo, A. et al. (1975): Canadian J. Earth Sciences, 12, 1036.
6. Kudo, A. et al. (1978): Progress in Water Technology, 10, 329.

RADIOECOLOGICAL STUDIES OF ACTIVATION PRODUCTS RELEASED FROM A NUCLEAR POWER PLANT INTO THE MARINE ENVIRONMENT

S. Mattsson, M. Nilsson and E. Holm

In order to study the transport of radionuclides released under controlled conditions from the nuclear power industry it is convenient to use bioindicators which concentrate the substances since analytical procedures for seawater are very time-consuming. We have found the brown algae Fucus vesiculosus and F. serratus to be very suitable bioindicators for such studies (3,5). It is also of importance to study the activity concentration in algae because they are used in different kinds of food.

The purpose of this work was to study the distribution along the Swedish westcoast of radionuclides released from the Barsebäck nuclear power plant, to compare the activity concentration in Fucus with that of other algae and crustaceans in the area and to try to study the ^{60}Co-concentration in seawater and sediment profiles.

MATERIAL AND METHODS

Samples of Fucus vesiculosus, F. serratus, Ascophyllum nodosum and Cladophora glomerata, firmly rooted on the bottom, were collected from water depths between 0.5 and 1 m along the coast. Samples of the crustaceans Idothea, mainly I. baltica and I. viridis, as well as Gammarus found on the Fucus plants were sorted out for analysis. A limited number of sediment profiles were also taken. After drying in air at room temperature for 2-3 days the samples were ground and packed in plastic containers of 5 or 180 ml volume. The measurements were carried out using Ge(Li)-detectors of volume 46-100 cm^3 with counting times between 5 and 100 hours. The detector efficiencies were determined carefully using samples of different densities containing accurately known activities of ^{152}Eu, ^{57}Co and ^{22}Na. After measurement the samples were dried at 105° C for 24 hours to get the reference weight.

Water samples with volumes between 80 and 170 liters were taken at the Fucus fields 2.9 km NNE of the power plant. ^{57}Co was added as a yield determinator and NaOH was then added to pH=12. The precipitate ($\simeq$ 5 l) was the next day disolved in HCl and coprecipitated with 200 mg Fe as carrier. The pH was adjusted to 12 with ammonia. This precipitation was centrifuged and dried before measurement.

RESULTS AND DISCUSSION

Activity concentration in Fucus

The time variation of the concentration of activation products in Fucus well reflects the increase in the discharged activity during the overhaul periods (3). The local and distant spreads of radionuclides from the power plant were investigated earlier at 8 stations along the Swedish westcoast (3,5) and showed a distance dependence of the activity concentration in Fucus which could be described by a power function of form:

$$C(z) = \alpha \cdot z^{-\beta}$$

where $C(z)$ is the activity concentration at distance z and α and β are constants. The value of β was found to be around 1.4 for the radionuclides ^{60}Co, ^{58}Co, ^{65}Zn and ^{54}Mn, somewhat lower, 0.8 for $^{110}Ag^m$ and still lower for ^{131}I (3,5). The results of our extended sampling now at 15 stations is given for ^{60}Co in Fig 1 and for ^{60}Co, ^{58}Co, ^{65}Zn, ^{54}Mn, $^{110}Ag^m$ and ^{57}Co in Fig 2. No significant deviation from the above mentioned β-values is recorded. Some of the sampling stations recently used have been chosen in bays. The important parameter regarding the activity concentration in Fucus seems to be the shortest distance in water between the power plant and the sampling station, regardless of the costal configuration. Up to now all our results indicate a very efficient mixture of surface water in the area.

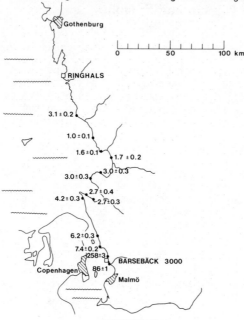

Figure 1.
Map over the Swedish westcoast showing the nuclear power plants BARSEBÄCK and RINGHALS and the sampling stations. The ^{60}Co activity concentration (Bq/kg dry weight) of Fucus in August 1979 is also given.

Co-60 in water and sediment

The ^{60}Co activity concentration in unfiltered seawater 2.9 km NNW of the power plant was determined to 5.4 mBq/l (1978-06-15), 4.1 and 5.1 mBq/l (1978-09-22) and 1.3 mBq/l (1979-04-11). Our results, indicate that the concentration factor for Fucus (dry weight) is as high as $(2\pm 1)\cdot 10^5$ corresponding to a wet weight concentration factor of $(4\pm 2)\cdot 10^4$. This value is considerably higher than those computed by Fukai and Murray (1) or measured in aquariums by Nakahara et al (4). It is interesting to note that the concentration factors for actinides in this area have also been found to be unusually high (2). Measurable concentrations of ^{60}Co have been found in sediments from sampling stations NW of the power plant 1.7-2.5 km off-shore. The total area content of ^{60}Co was found to be 190 Bq/m² at 4.1 km from the plant, 76 Bq/m² at 6.4 and 560 Bq/m² at 11 km. Around 70% of the

^{60}Co activity was found in the upper 25 mm of the sediment layer and the rest in the next 25 mm.

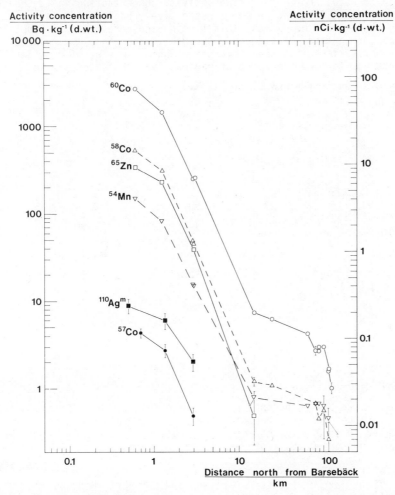

Figure 2. Variation in activity concentration in Fucus by distance northwards from BARSEBÄCK in August 1979.

Comparative studies of activity concentration in different algae and crustaceans.

The activity concentration in some algae and crustaceans relative to that of Fucus is given in Fig 3. All concentrations are given on dry weight basis. It is interesting to note the differences in activity concentration in Idothea and Gammarus living side by side in Fucus plants. For the cobalt isotopes the concentration in Idothea is a factor of 4 higher than in Gammarus. This may indicate a difference in their feeding-habits. The highest activity concentration of ^{54}Mn, ^{60,58,57}Co, ^{65}Zn and ^{131}I are found in Fucus while crustaceans for these nuclides show an activity concentration of 4-40% compared to Fucus. For ^{110}Ag^m, the activity concentration in Idothea and Gammarus is a factor of 9 and 6 respectively higher than in the Fucus plants in which they lives. This finding indicates an active uptake of silver in these crustaceans. Because Idothea and Gammarus are eaten by various fish species, these findings may be of considerable importance for the transport of radioactive substances to man.

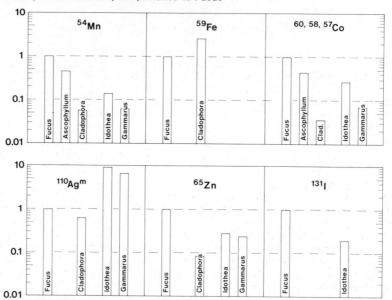

Figure 3. Relative content, per dry weight unit, in different algae and crustaceans of radionuclides released from BARSEBÄCK in August-November 1979. Dry weight/wet weight ratios: Fucus-0.20, Ascophyllum-0.2, Cladophora-0.05, Idothea-0.19 and Gammarus-0.18.

ACKNOWLEDGEMENT

Thanks are due to Lena Carlsson and Lars-Eric Persson for valuable help with collection and identification of samples.

REFERENCES

1. Fukai, R., and Murray, C.N., (1973): In: Environmental Behavior of Radionuclides Released in the Nuclear Industry. IAEA, Vienna, p.217.
2. Holm, E., Persson, B., and Mattsson, S. This conference.
3. Mattsson, M., Finck, R., and Nilsson, M.,(1980): Environmental Pollution, (In Press).
4. Nakahara, M., Koyanagi, T., and Saiki, M. (1975): In: Impact of Nuclear Release into the Aquatic Environment, IAEA, Vienna, p. 301.
5. Nilsson, M., and Mattsson, S., (1980): In: Proceedings, 3rd NEA Sem. Marine Radioecology, Tokyo, Japan, 1st-5th Oct.,1979 (In press)

MONITORING OF ENVIRONMENTAL RADON-222 IN SELECTED AREAS OF TAIWAN PROVINCE OF THE REPUBLIC OF CHINA

T.-Y. C. Mei, C.-F. Wu and P.-S. Weng

Radon-222 and its daughters in air may enter human body by inhalation and cause radiation hazard to the epithelial basal cells of the respiratory tract. It may induce lung cancer (1). Therefore, measurement of ^{222}Rn and its daughters in air is an important program in environmental monitoring.

The experimental method used in this work is to measure ^{222}Rn activity concentration in selected areas of Taiwan by sampling air through a filter paper and detecting the gross α counts of the filter paper with an α scintillation counter. The experimental results obtained from sampling in a radioactive waste plant, a storage room for uranium, a natural cave, a room using natural gas, etc., are presented. Continuous sampling of environment at the Tsing-Hua open-pool reactor site for a period of six months is included. The annual dose equivalent of ^{222}Rn at the reactor site is estimated.

EXPERIMENTAL

The experimental setup were simply an air sampler and a ZnS(Ag) scintillation counter. The Whatman No.41 filters were used in air sampler. After a sampling time of 5 min, the filter was α counted with the ZnS(Ag) scintillation counter which was calibrated against an ^{241}Am α standard source. The ratios of ^{222}Rn concentration to its daughters' were assumed to be 10:9:6:4 for RaA, RaB, RaC, respectively (2, 3). A simplified equation can thus be obtained as below:

$$Q = \frac{R}{c \, \eta \, \varepsilon \, v} \qquad (1)$$

where

- Q is the concentration of ^{222}Rn in air (Bq/m^3)
- R is the counts from 2 to 7 min after the end of the sampling
- c is a coefficient determined by the decay constants of ^{222}Rn and daughters (0.975 x 10^3)
- η is the sampling efficiency (η = 1 in this experiment) (4)
- ε is the counting efficiency (ε = 8.25%)

v = is the air sampling rate (0.85 m^3/min).

RESULTS AND DISCUSSION

The ^{222}Rn concentration in air of some selected areas in Taiwan are summarized in Table 1.

Table 1. Rn-222 concentrations in air of some selected area in Taiwan

Location	Concentration (Bq/m^3)	Sampling Date
Radioactive waste plant	3.15 ∿ 13.99	1978.7.7
Uranium storage room	3.96 ∿ 11.95	1978.9.21
A room 30 m from uranium storage room	3.55 ∿ 10.73	1978.9.21
Room using natural gas	3.63 ∿ 8.55	1978.10.18
Room not using natural gas	3.81 ∿ 8.34	1978.10.18
Natural cave (inside)	13.83 ∿ 16.69	1978.11.7
Natural cave (outside)	1.93	1978.11.7

The results of continuous sampling of environment at the Tsing-Hua open-pool reactor site for a period of six months are shown in Fig.1, and Fig.2 shows a typical daily variation in ^{222}Rn concentration.

According to the BEIR Report (5), the mean dose from ^{222}Rn to the epithelial basal cells is 0.01 Gy/WLM. As mentioned before, the concentration ratio of ^{222}Rn to its daughters is 10:9:6:4, then the ^{222}Rn concentration of 1 Bq/m^3 is equal to 1.5×10^{-4} WL. If the quality factor is assumed to be 20, and a multiplier factor of 1.5 is included because of the clean air, then the annual dose equivalent from 1 Bq/m^3 of ^{222}Rn in air to the epithelial basal cells could amount to 2.3 mSv/yr.

From the data shown in Fig.2, the annual dose equivalent of ^{222}Rn can be calculated, and it is equal to 29.6 mSv/yr (indoor) and 13.8 mSv/yr (outdoor).

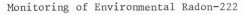

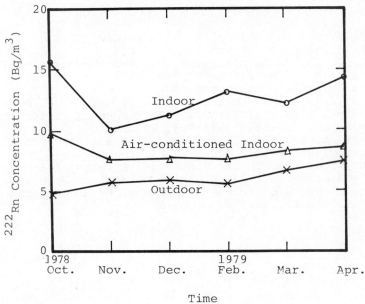

Figure 1

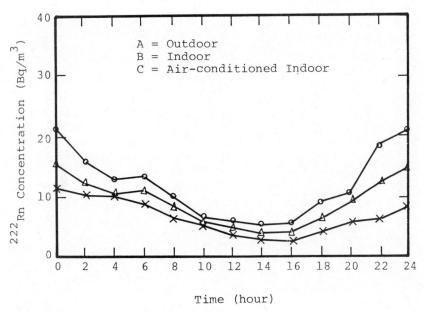

Figure 2

REFERENCES

1. Archer, V. E., Wagoner, J. K., and Lundin, F. E., (1973) *Health Phys.* 25, 351.
2. Gesell, T. F., Johnson, Jr., R. H., and Bernhardt, D. E., (1977) U.S. Environmental Protection Agency Report, EPA-520/1-75-002, Washington, D.C.
3. Johnson, Jr., R. H., Bernkardt, D. E., Nelson, N. S., and Calley, H. W., (1973) U.S. Environmental Protection Agency Report, EPA-520/1-73-004, Washington, D.C.
4. Lockhart, L. B., *et al.*, (1964) U.S. Naval Research Laboratory, NRL Report 6054, Washington, D.C.
5. Advisory Committee on the Biological Effects of Ionizing Radiations, (1972) Division of Medical Sciences, National Academy of Sciences, National Research Council, BEIR Report, Washington, D.C.

DISTRIBUTION OF URANIUM, ^{226}Ra, ^{210}Pb AND ^{210}Po IN THE ECOLOGICAL CYCLE IN MOUNTAIN REGIONS OF CENTRAL YUGOSLAVIA

Z. Miloševic, E. Horšić, R. Kljajić and A. Bauman

INTRODUCTION

The distribution of the uranium decay series was investigated in the central parts of Yugoslavia. The characteristic of region are mountains 1000-1500 m high with unexploited and partially unexplored forests, extensive cattle breeding and cheese and milk production. The soil has neve been cultivated and the grass on the pastures grew without human interference.

RESULTS

Four nuclides of the uranium decay series were investigated: ^{210}Pb, ^{210}Po, ^{226}Ra and uranium, either for their longer $T_{1/2}$, or radiotoxicity. Samples of beef (meat and bones), milk, cheese, grass and soil were analyzed to round off a vital part of the ecological cycle.

TABLE 1. CONCENTRATION OF ^{210}Pb, ^{210}Po, ^{226}Ra AND URANIUM IN BEEF (MEAT AND BONES)

Location No	^{210}Pb (mBq/kg)		^{210}Po (mBq/kg)	
	meat	bone	meat	bone
1.	18.50	3082,10	37.00	827.32
2.	99.53	777.00	21.46	555.00
3.	15.91	1914.49	12.95	1202.50
4.	227.92	869.50	46.25	1295.00
5.	124.69	1058.20	12.95	1047.10
6.	0	3903.50	18.50	8221.40
7.	134.16	670.44	190.92	1417.10
8.	45.88	3136.86	64.75	111.00

Location No	^{226}Ra (mBq/kg)		URANIUM (μg/kg)	
	meat	bone	meat	bone
1.	167.98	827.32	0.57	5.98
2.	4.43	235.69	0.60	0
3.	16.65	966.44	0.11	2.36
4.	79.92	994.56	0	16.75
5.	12.72	740.00	0.16	24.80
6.	109.15	1733.08	9.25	0.47
7.	122.10	1169.20	0	23.28
8.	79.55	617.90	0	9.22

The contamination of beef bones (Tb.1) with ^{210}Pb, ^{210}Po and ^{226}Ra was on the average 10 to 100 times higher than in meat. While bones were boiled for bouillon only up to 1,0% of ^{210}Pb and 13,8% of ^{210}Po passed over into the solution. The transfer of contamination thus becames negligible. The level of meat contamination with ^{210}Pb and $^{210}Po_2$ was of the same order of magnitude as found in literature (1). The ^{226}Ra results for meat were slightly higher (2).

^{226}Ra in dairy products (Tb.2) was higher than in USA where the average value (2) amounts to 10 mBq/kg.

Taking nito consideration a daily intake of soup, meat and dairy products in the whole region, the daily amount of ^{210}Pb, ^{210}Po and ^{226}Ra would only occasionally surpass the values obtained in other countries. The concentration of uranium in meat varies from 0 - 9,25 µg/kg. Supposing that all the beef on this location had the highest uranium concentration the uranium intake in a balanced diet, would be high if 2 ug/kg of total diet is taken as the average intake of uranium for unexposed population. The same applies for uranium milk samples (3).

TABLE 2. CONCENTRATION OF ^{210}Pb, ^{210}Po, ^{226}Ra AND URANIUM IN MILK AND CHEESE

Location No	^{210}Pb (mBq/kg)		^{210}Po (mBq/kg)	
	milk	cheese	milk	cheese
1.	5.92	283.79	18.50	99.90
2.	88.43	173.16	69.19	49.21
3.	18.13	2104.93	18.50	166.50
4.	89.17	0	26.27	55.50
5.	25.53	284.16	52.17	141.34
6.	52.91	227.18	9.25	81.40
7.	23.31	309.69	50.69	122.11
8.	50.32	173.16	9.25	77.70

Location No	^{226}Ra (mBq/kg)		URANIUM (µg/kg)	
	milk	cheese	milk	cheese
1. 1.	22.94	0	0.60	0
2.	15.17	0	0	0
3.	15.17	82.88	0	0.60
4.	51.06	104.71	0.36	0.87
5.	2.22	91.39	0	0.68
6.	28.12	0	0.62	0
7.	87.32	97.31	1.60	0
8.	37.74	55,50	0,67	0.34

Distribution of Uranium

TABLE 3. CONCENTRATION OF ^{210}Pb, ^{210}Po, ^{226}Ra AND URANIUM IN GRASS

Location No	^{210}Pb (mBq/kg)	^{210}Po (mBq/kg)	^{226}Ra (mBq/kg)	URANIUM (µg/kg)
1.	841.75	555.00	132.83	0.77
2.	3784.36	240.50	480.63	11.88
3.	15932.22	0	721.13	0.97
4.	370.37	222.00	385.91	5.09
5.	11939.30	351.50	392.94	7.96
6.	3241.94	277.50	4329.00	5.46
7.	3592.70	462.50	1951.75	0.48
8.	14936.16	610.50	235.69	8.85

Higher concentrations of ^{210}Pb in soil are followed by higher concentrations in grass on 3 locations. The same is valid for ^{210}Po for locations 5, 7 and 8. ^{226}Ra and uranium do not follow this pattern. The amount of ^{210}Pb, ^{210}Po and ^{226}Ra in soil coresponds to podzolic soils found in different parts of the world (1). The concentration of ^{210}Pb and ^{210}Po diminish proportionally to the soil depth. ^{226}Ra and uranium do not show any distinctive differences of concentration between soil layers.

TABLE 4. CONCENTRATION OF ^{210}Pb, ^{210}Po ^{226}Ra AND URANIUM IN PODZOLIC SOILS OF CENTRAL YUGOSLAVIA

Location No	210Pb (Bq/kg)	^{210}Po (Bq/kg)	^{226}Ra (Bq/kg)	URINIUM (ppm)
1.	32.92	43.14	43.73	0.50
2.	36.78	58.09	56.16	0.57
3.	153.18	75.63	61.31	0.64
4.	31.82	64.19	75.74	1.45
5.	122.56	127.28	34.41	1.17
6.	56.39	87.69	76.55	1.72
7.	87.84	108.67	40.55	0.91.
8.	108.78	277.13	55.13	1.46

soil layer: 0 - 5 cm

TABLE 4. (cont.)

Location No	^{210}Pb (Bq/kg)	^{210}Po (Bq/kg)	^{226}Ra (Bq/kg)	URANIUM (ppm)
1.	36.92	39.99	43.73	0.20
2.	36.04	35.48	71.30	0.92
3.	50.69	49.73	62.27	0.80
4.	26.31	52.06	106.56	1.17
5.	83.77	83.62	49.58	1.08
6.	53.42	43.66	71.00	1.64
7.	43.92	272.69	67.30	1.49
8.	272.69	348.54	84.36	0.95

soil layer: 5 - 10 Cm

Location No	^{210}Pb (Bq/kg)	^{210}Po (Bq/kg)	^{226}Ra (Bq/kg)	URANIUM (ppm)
L 1.	23.30	34.70	26.89	1.78
2.	31.06	33.74	66.34	0.86
3.	43.58	35.89	16.52	0.92
4.	26.53	43.63	93.98	0.22
5.	78.55	72.52	65.86	1.10
6.	35.75	66.93	70.70	1.60
7.	34.95	75.48	65.86	1.64
8.	66.93	241.98	110.63	0.59

soil layer: 10 - 15 cm

DISCUSSION

The distribution of uranium, ^{226}Ra, ^{210}Pb and ^{210}Po in an ecologically unpolluted environment is not different from regions in other parts of the world which are under cultivation.

REFERENCES

1. Ermolaeva-Mahovskaya A.R. Litver B.Y.(1978): Pb-210 y Po-210 v byosphere. Atomizdat, Moskva.

2. Eisenbud M. (1973): Environmental Radioactivity, Academic Press, Brace Jovanović , New York, NY.

3. Magno P.J. Groul P.R. Apidionihis J.C. (1970): Health Physic, 18 383.

RA-226 COLLECTIVE DOSIMETRY FOR SURFACE WATERS IN THE URANIUM MINING REGION OF POÇOS DE CALDAS*

A. S. Paschoa, G. M. Sigaud, E. C. Montenegro and G. B. Baptista

A monitoring survey of the ^{226}Ra concentrations in river waters in the vicinity of the uranium mining, and future milling facilities, in the Poços de Caldas region started in January 1977. Results of this survey have been published elsewhere (1). Dosimetric and environmental models are used in the present work to calculate the Annual Collective Dose Equivalent (ACDE) to the populations potentially exposed. The ACDE for the whole-body, bone, gastro-intestinal tract - lower large intestine (GI tract-LLI), kidneys, and liver, via the pathways of drinking water and ingestion of food, grown in irrigated fields, are presented as a function of the ^{226}Ra concentrations in the surface waters of the Poços de Caldas region.

POPULATION DISTRIBUTION

Most of the population of the Poços de Caldas region are concentrated in cities located near the quasi-circular border of the plateau, which is represented by a dashed line in Figure 1. The central region of the plateau looks uninhabited at a permanent basis, except for the resident farm workers of the rural properties extant in the internal area of the plateau. The economical growth of the Poços de Caldas plateau, allied with the generalized use of motor vehicles in the region, made the local population much more mobile today than in 1970. Besides that, most cities of the plateau are touristic resorts with large seasonal fluctuations in the temporary population.

The Poços de Caldas plateau was divided in areas defined by concentric subregions having the common central point in the mining site of Campo do Cercado, as shown in Figure 1. The subregions are denoted by their radial border limits followed by the distance from the center to the outer ring. The populations, P_i, of the subregions i are shown in Table 1.

The approximate populational distribution of the Poços de Caldas plateau given in Table 1 is based upon the results of the 1970 census, which is the last census available. The city of Poços de Caldas had about 5.8×10^4 inhabitants in 1970, over 95% living within the city limits. The 1980 brazilian census has just started, but new populational data will not be available before two years from now. However, preliminary data collected, unofficially, indicated that the population of the municipality of Poços de Caldas might have exceeded 1.0×10^5 inhabitants sometime along the last ten years. Today the population distribution in the plateau is such that 90% of the entire population

---

* This work was supported by the International Atomic Energy Agency, Comissão Nacional de Energia Nuclear, Financiadora de Estudos e Projetos, and Conselho Nacional de Desenvolvimento Científico e Tecnológico.

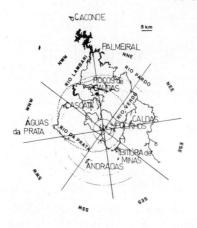
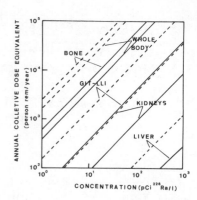

Figure 1. Hydrographic chart and subregions of the P.C. plateau.

Figure 2. ACDEs for the populations of the Poços de Caldas plateau (drinking water) and subregions NWN20 and NWN25 (food ingestion) as a function of ^{226}Ra concentrations in waters.

reside in urban areas, while 10% can be found in rural areas. As a consequence of the facts mentioned above, the populational data shown in Table 1 do not correspond to the actual population of the region at any given time today, but shall only be interpreted as an indication of the percentual distribution that one can expect to find in the region, if the seasonal variations and other fluctuations in the population of the Poços de Caldas plateau are discounted. Collective dosimetry for ^{226}Ra concentrations in the surface waters of the region were undertaken with full awareness of these and other shortcomings.

DOSIMETRIC MODEL

The ACDE, $\dot{S}_j$, for the population of the Poços de Caldas plateau was calculated based upon the formulae recommended by the International Commission on Radiological Protection (ICRP). In particular, the ACDE to an organ j, in person-rem/year, was calculated from the per caput annual dose equivalent, $\dot{H}_j$, in rem/year, for an organ j of a generic individual of a population, assuming that P_i persons in the subgroup i of the total population were submitted to the same dose equivalent (2). However, to evaluate the annual absorbed dose, $\dot{R}_j$, in mrad/year, to an organ j, the dose factors, $\dot{D}_j$, were derived from the exponential models recommended by ICRP (3). The dosimetric data of Tables 1 and/or Figure 2 reflect the following simplifying assumptions: (i) Urban and rural populations are considered together for the calculational purpose of regarding a population, P_i, characteristic of a particular subregion i; (ii) Temporal and spatial variations in the populations P_i of each subregion i were not taken into account; (iii) The maximum ^{226}Ra concentrations as listed in Table 1 for each subregion were considered as indicators for the constant values adopted to allow estimates of the annual absorbed doses, $\dot{R}_j$, to an organ j of a generic individual of

TABLE 1. Populational distribution, ^{226}Ra concentration in water, and ACDEs for the Poços de Caldas plateau

Subregion	City or village	Population * P_i (×10³)	%	Maximum ^{226}Ra concentration pCi/ℓ	Collective annual dose equivalent (person rem/yr)				
					Whole body	Bone	Liver	Kidneys	GI-tract-LLI
NWN 20[†] NWN 25	Poços de Caldas	58.0	57.5	0.6	1.3×10²	1.9×10²	9.5×10⁻²	2.7	1.2×10
WNW 25	Águas da Prata e Cascata	4.1	4.1	29	4.3×10²	6.6×10²	3.3×10⁻¹	9.3	4.0×10
SSW 15 SSW 20	Andradas	17.4	17.2	≤ 0.2	≤ 6.4	≤ 9.7	≤ 4.7×10⁻³	≤ 1.4×10⁻¹	≤ 5.9×10⁻¹
SES 15	Ibitiúra de Minas	2.7	2.7	≤ 0.2	≤ 2.2	≤ 3.4	≤ 1.6×10⁻³	≤ 4.7×10⁻²	≤ 2.0×10⁻¹
ESE 20	Santa Rita de Caldas	7.0	6.9	≤ 0.2	≤ 1.3×10	≤ 1.9×10	≤ 9.4×10⁻³	≤ 2.7×10⁻¹	≤ 1.2
NEE 10 NEE 15	Caldas e Pocinhos	8.7	8.6	≤ 0.2	≤ 2.0	≤ 3.0	≤ 1.5×10⁻³	≤ 4.2×10⁻²	≤ 1.8×10⁻¹
NNE 20	Santana de Caldas	3.0	3.0	≤ 0.2	≤ 5.2	≤ 7.8	≤ 3.8×10⁻³	≤ 1.1×10⁻¹	≤ 4.8×10⁻¹
				TOTAL	≤ 5.9×10²	≤ 8.9×10²	≤ 4.5×10⁻¹	≤ 1.3×10	≤ 5.5×10

* According to the most recent data available.

† NWN 20 means subregion limited by radii defining, respectively, the directions NW and N, and between concentric rings distant 15 and 20 km from the Campo do Cercado uranium mine.

the population P_i; (iv) The ^{226}Ra concentrations in food products grown in irrigated fields were estimated based upon local data, from the 1970 census, on agricultural production, and site specific environmental models; (v) The usage factors for drinking water and food consumption, used in the calculations, were essentially those of the Reference Man (4) for Latin American inhabitants. This assumption is made based upon the fact that there are wide variations in usage factors for regions of fast changing parameters and non-uniform distribution of wealth, so the Reference Man values can be considered suitable for the present dosimetric calculations, because of the uncertainties associated to alternative values; (vi) The ACDEs were calculated by assuming no ^{222}Rn loss by gaseous diffusion after ^{226}Ra uptake by the human body.

The above simplifying assumptions are not in complete agreement with the recommendations of ICRP26 (2). However, one should bear in mind when applying the ICRP26 recommendations to a particular region of a developing country, that the demographic distribution as well as food habits may experience significant fluctuations under certain conditions like, for example, fast population growth and local economical development. As a consequence, care must be exercised when using the estimates of collective dose equivalent, from data of a fast developing region, as the basis for decision-making processes as suggested by the ICRP26 (2), since dramatic changes may occur in the calculational parameters within relative short periods of time. Accordingly, ACDEs values of Table 1, like 590 person-rem/year for the population of the Poços de Caldas plateau, are not to be used in decision-making processes.

Figure 2 shows graphs of the annual collective dose equivalents to the whole-body and selected organs as a function of the ^{226}Ra concentrations in water via the pathway of drinking water for the population of the Poços de Caldas plateau, and via the pathway of food ingestion for the populations of the subregions NWN20 and NWN25, by far the most populated subregions. These estimates were made as a contribution for future comparisons, although other parameters used in the dosimetric calculations may also change considerably along the future.

CONCLUDING REMARKS

Paragraph 22 of ICRP26 (2) is very carefully worded to warn about the complexity of the relationship between the distribution of dose equivalent in an exposed population and the assessment of detriment. Furthermore the careful wording of ICRP26, can be found notably also in paragraphs 219, 221 and 232 which are bound to prevent misuses of the concept of collective dose equivalent to make dubious quantitative appraisals of detriment associated with practices, however decisions are likely to be made based upon such appraisals.

The present work intends to be a contribution to the understanding of the shortcomings involved in calculating collective dose equivalent due to the potential enhancement of ^{226}Ra concentrations in the surface waters near the uranium mining region of Poços de Caldas.

General observations and tentative conclusions are listed as follows:

1. The ACDEs for selected populational subgroups as well as for the total population living in the Poços de Caldas plateau have been calculated based upon simplifying assumptions. These ACDEs, rather than the dose-equivalent commitments or the collective dose-equivalent commitments for the predicted time of operation of the uranium mine of the Poços de Caldas region were calculated, because there were several intrinsic uncertainties in the parameters available for the calculations.

2. The fluctuations expected to occur in the data on populational distribution, irrigational practices, agricultural production, and food consumption in developing regions, like the Poços de Caldas plateau, makes the quantitative assessment of the collective dose-equivalent commitments meaningless, unless reliable long range predictions can be made on the varying parameters to enable time integration.

3. Linear models can be used to estimate the ACDE as a function of the ^{226}Ra concentration, based on parameters which may be valid at a particular time, but the actual collective dose equivalent commitment is difficult to predict.

4. Dosimetric models based upon site specific environmental models are helpful to estimate collective dose equivalents to populations from a particular practice, but extreme care should be exercised by competent national authorities when using such estimates in decision-making processes.

5. Paragraphs 22, 219, 221 and 232 of ICRP26 (2) should be taken into full account when estimating collective dose equivalents.

REFERENCES

1. Paschoa, A.S., Baptista,G.B., Montenegro, E.C., Miranda, A.C., Sigaud, G.M. (1979): In: Low-Level Radioactive Waste Management. EPA-520/3-79-002: 337.
2. ICRP-(1977). Publication 26. Annals of the ICRP, Vol.1: 1-53.
3. ICRP-(1960). Publication 2. Health Physics 3, 1-380.
4. ICRP-(1975). Publication 23. Pergamon Press, Oxford: 1-480.

THE RADIUM CONTAMINATION IN THE SOUTHERN BLACK FOREST

H. Schüttelkopf and H. Kiefer

The high natural radium contamination prevailing in the Southern Black Forest was used to assess the degree of contamination of the environment, the mechanisms of radium transport to man, as well as the radiation impact on the population from natural Ra-226.

The Ra-226 concentration in the environmental air amounted to values between 0.16 and 1.2 fCi/m^3. The mean value was 0.5 ± 0.2 fCi/m^3. Considering the Ra-226 concentration in the soil, this value corresponds to about 150 µg of $dust/m^3$.

The Ra-226 concentration in the drinking water was measured between March and December 1978. The mean value is 0.31 ± 0.03 pCi/l. The Ra-226 contamination found in the Krunkelbach, was likewise measured continuously during the same period. The mean value of the measured concentration was 0.5 ± 0.3 pCi/l. 23 water samples were taken from the Feldberger Alb brook, the most important tributary of the Krunkelbach brook; the average concentration was 0.19 ± 0.02 pCi of Ra-226/l.

In the Krunkelbach valley the Ra-226 contents were determined of 22 samples taken from several brooks. The concentrations found were < 0.03 to 2.6 pCi/l. In 23 water samples collected in the immediate and more distant vicinity of that valley the Ra-226 contents were determined. The results ranged from 0.03 to 1550 pCi/l. Generally speaking it can be stated that Ra-226 concentrations above 1 pCi are very rare, that the Ra-226 concentrations in a brook or river are maximum at its spring, and that high concentrations are observed when on account of extended dry periods the water stays longer in the soil or when in the winter season the salt concentration in the residual liquid increases by freezing of most of the water.

More than 100 soil samples were examined for Ra-226. In the topmost 20 cm soil layer the mean value was 2.9 pCi/g.

A great number of sediment samples from several brooks and rivers were measured to determine their Ra-226 content. The mean value found in the valley of the Menzenschwander Alb was 3.2 ± 1.0 pCi/g. In other sediments a mean value of 1.1 ± 0.1 pCi Ra-226/g was measured. The Ra-226 concentrations in trouts were measured. The average concentration was 50 pCi Ra-226/kg of fresh weight. The maximum value was 211 pCi Ra-226/kg.

Milk samples were taken regularly from two farms and analyzed for their Ra-226 content. The results measured for one farm are plotted in Fig. 1. The Ra-226 concentrations scatter by more than one order of magnitude and take an average value of 11 pCi Ra-226/l. For the second farm a mean value of 7 pCi/l was measured. Random samples collected at other farms confirmed these results.

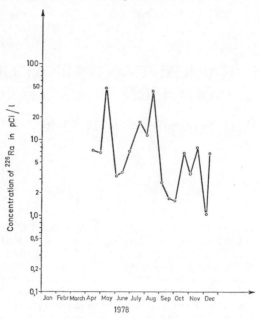

Fig.1: ^{226}Ra concentration in the milk from farm A, Menzenschwand.

Foodstuffs produced in the Southern Black Forest were measured for their Ra-226 content. Besides the milk and fish samples already mentioned, wheat, barley, oat, potatoes, salad, cabbage, beans, kohlrabis, blueberries, beef, entrails, venison, eggs, mushrooms and beer were investigated. A summary of values taken from the literature, compared with the radium contents found in the Southern Black Forest, is given in Table 1. The Ra-226 contents of corn, potatoes and milk are surprisingly high as compared with the published data. A more accurate examination is presently performed in order to find out which part of the corn has been contaminated with Ra-226.

Since the Ra-226 transfer from the soil to the grass is very significant for the contamination of milk, a great number of grass and hay samples were examined for Ra-226. Moreover, in search of a bioindicator for Ra-226, quite a number of wild plants were measured to determine their Ra-226 content. The average concentration of the grass and hay samples found was 0.30 ± 0.04 pCi/g of dry substance. For the other wild plants values were found of 0.7 to 22 pCi Ra-226/g of dry substance. A clear bioindicator for Ra-226 was not identified.

Since water and fish samples, grass and milk samples, soil and grass samples as well as soil and plant samples were always taken jointly, it was possible to calculate many transfer factors. The following transfer factors were determined: The transfer factor for fish / water was 28 ± 12. In the guts, on the one part, and in the meat and heads + bones, on the other part, 1/3 each of Ra-226 was contained. The ratio of water to sediment concentration was calculated for a settling tank to be 11×10^{-5} and for the brooks and rivers

Radium Contamination in the Southern Black Forest

26×10^{-5}. The grass to milk transfer factor indicates the percentage of Ra-226 contained in one liter of milk consumed daily. A value of 0.3 ± 0.1 % per liter was calculated from the measured grass and milk concentrations. The milk to grass transfer factor is 0.20 ± 0.05. The transfer factor calculated for grass to soil was 0.027 ± 0.005. Slightly higher values were determined for the wild plants. For the milk to soil transfer factor the value 0.005 ± 0.002 can be calculated from the data indicated. The transfer factors calculated for the individual foodstuffs vary from 0.003 for the white of egg up to 0.2 for corn.

Sampling Material	^{226}Ra concentration in pCi/kg	
	Menzenschwand	[1]
Soil:	1200 - 1500	150 - 3100
Water:		
river and lake water	0.03 - 2.5	0.002 - 62
ground and spring water	0.1 - 1549	0.001 - 237800
dringking water	0.11 - 0.57	0.005 - 50
Foodstuffs:		
potatoes	30 - 40	0.8 - 2.8
corn, flour	20 - 240	1.9 - 2.8
meat	2	0.01 - 1.1
milk	0.3 - 48	0.3
vegetables	5 - 170	0.5 - 3.8
fish	1.4 - 211	5.1
eggs	80	3.1 - 6.1
entrails	10 - 200	0.1

Table 1: ^{226}Ra concentration in environmental samples taken in and around Menzenschwand, as compared with values taken from the literature [1].

Taking into account the average habits of consumption in Germany the maximum possible intake per annum of radium was calculated for the population living in the region under investigation. It amounts to 7.1 nCi/a if one assumes that the total demand for foodstuffs is satisfied by local produces. This value is higher by the factor 12 than the annual Ra-226 ingestion value of 580 pCi permitted by the German Radiation Protection Ordinance. The body burden corresponding to this annual intake should amount to 7.4 nCi of Ra-226. To verify this conclusion, 28 members of the local population were measured in a body counter to determine the amount of Ra-226 incorporated. All 28 measuring values were below the detection limit of about 7.5 nCi of Ra-226. If one takes these 28 measuring values as part of a random sample, the mean value and its error can be calculated for the Ra-226 body burden received by the whole group from the measured values, taking into account the standard deviations according to the counting statistics. The value so determined was 0.3 ± 0.7 nCi of Ra-226. A comparison of the most probable value of 0.3 nCi of Ra-226 with the calculated maximum value of 7.4 nCi of Ra-226 for the body burden yields that either only 4-5 % of the foodstuffs consumed by the group examined can stem from local production or that the radium transport from the gastro-intestinal tract into the blood according to the ICRP model [2] is overestimated.

To assess exactly the body burden of the inhabitants of the Southern Black Forest, the Ra-226 contained in the teeth of people living there is presently examined.

REFERENCE

[1] K. Aurand et al., "Die natürliche Strahlenexposition des Menschen", Georg Thieme-Verlag, Stuttgart, 1974

[2] Report of Committee II on Permissible Dose for Internal Radiation, International Commission on Radiological Protection, ICRP Publ. 2, Pergamon Press, London, 1959

EVALUATION OF SMALL SCALE LABORATORY AND POT EXPERIMENTS TO DETERMINE REALISTIC TRANSFER FACTORS FOR THE RADIONUCLIDES ^{90}Sr, ^{137}Cs, ^{60}Co and ^{54}Mn

W. Steffens, F. Führ and W. Mittelstaedt

Much of the information on uptake of radionuclides from soils for determination of transfer factors have been obtained from laboratory experiments with prepared soils or soils contaminated by nuclear weapon tests. These results may not be valid for estimation of transfer factors in the field. On the other hand, it is nearly impossible to conduct field experiments to determine the transfer of radionuclides for every site and condition. We have started lysimeter experiments in a controlled experimental field to study the root uptake of ^{90}Sr, ^{137}Cs, ^{60}Co, and ^{54}Mn under outdoor conditions. (Figure 1).

Since this type of experiments are time-consuming and expensive, and can only be conducted at locations which permit the use of radionuclides under outdoor conditions, we have also set up experiments parallel to those in the outdoor lysimeter using the Kick-Brauckmann experimental pots (3) under greenhouse and using the Neubauer cups under growth chamber conditions. The results obtained from the three types of experiments are compared.

Figure 1. Test units used: 1 m^2 lysimeter filled with topsoil (Ap-horizon), 0,25 m^2 lysimeter with undisturbed soil profile, Kick-Brauckmann pot (8 kg topsoil), and Neubauer cup (400 g of topsoil).

MATERIALS AND METHODS

Two soils were used: a podzolic soil (spodosol) and a degraded loess soil (alfisol). Their properties and details about the lysimeter experiment were presented by Steffens et al. (4). For the pot and cup experiments the soils were sieved (<5 mm and <2 mm, respectively), fertilized (120 mg N, 120 mg P_2O_5, 200 mg K_2O/kg soil), treated with 10 μCi ^{90}Sr $(NO_3)_2$ or $^{54}MnCl_2$ or 5 μCi $^{137}CsCl$ or $^{60}CoCl_2$/kg soil, and thoroughly mixed. The pots in 4 replicates containing 8 kg soil each were watered to 65 % of the water holding capacity of the soil and placed in greenhouse under natural light conditions. The cups in 4 replicates and containing 400 g soil each were placed in the growth chamber maintained at 12-hour day/night photoperiod at 23 °/16 °C and 65/85 % relative humidity. The soils in the lysimeters were equilibrated for 8 months and in the pots and cups for 2 weeks before the initiation of the experiments. Plants were grown to ripeness in the lysimeters and pots, but were harvested from the cups 4 weeks after sowing or planting. The plants were dried at 105 °C. ^{90}Sr was measured according to Cerenkov via the decay product ^{90}Y, ^{137}Cs, ^{60}Co, and ^{54}Mn on a surface Ge(Li) detector. Additional details are listed by Führ (1). The transfer factors are expressed as the quotient

$$TF_{SP} = \frac{dpm/g \text{ plant fresh weight}}{dpm/g \text{ dry soil (application)}}$$

The transfer factors from the pot and cup experiments were expressed in relative values based on those from the lysimeter experiments equal to 1.

RESULTS

The results (Table 1) demonstrate that the transfer factors for ^{90}Sr, ^{137}Cs, ^{60}Co, and ^{54}Mn obtained under the specific conditions of the small scale Neubauer cup experiment differed greatly from those obtained from the outdoor lysimeter experiment. These differences may be mainly due to the short equilibration period, daily water addition (repeated desorption), plant density, short vegetation period, and higher transpiration rates in the cup experiment.

In the pot experiment the transfer factors for ^{90}Sr, ^{137}Cs, and ^{54}Mn showed less deviation especially in crops grown on podzolic soil. On the average they are 1.5 to 2 times higher than the lysimeter values, with a tendency to higher deviation in the root crops. In the case of ^{60}Co, transfer factors were found to be similar in potatoe leaves and tuber on podzolic soil, 2 to 3 times higher in sugar beets on loess soil, but up to 23 times higher in barley and salad. Yields and to a smaller extent some of the nutrient contents were generally higher in plants grown in pots than those from field trials (2), probably due to higher ferti-

Small Scale Laboratory and Pot Experiments

Table 1: Relative transfer factors of pot and cup experiments compared with transfer factors from outdoor lysimeter experiments.

Plants		Transfer factors[1]	Relative transfer factors[2] pot	cup
Sr-90: podzolic soil				
Barley	straw	2,08	1,6	0,5
	grain	0,17	1,5	-
Potatoes	leaf	0,79	1,3	-
	tuber	0,014	3,7	-
Salad		0,46	2,1	1,7
Sr-90: loess soil				
Barley	straw	1,91	1,2	0,25
	grain	0,09	1,8	-
Sugar beets	leaf	0,21	2,9	3,5
	beet	0,29	2,2	-
Salad		0,34	1,4	3,5
Cs-137: podzolic soil				
Barley	straw	0,081	2,0	5,9
	grain	0,039	1,4	-
Potatoes	leaf	0,057	1,9	-
	tuber	0,046	2,1	-
Salad		0,018	2,2	16,1
Cs-137: loess soil				
Barley	leaf	0,0051	-	5,9
	grain	0,0023	-	-
Sugar beet	leaf	0,0087	5,2	7,9
	beet	0,0037	1,7	-
Salad		0,0036	0,8	0,8
Co-60: podzolic soil				
Barley	straw	0,018	17,8	6,8
	grain	0,013	23,1	-
Potatoes	leaf	0,32	1,0	-
	tuber	0,082	0,9	-
Salad		0,0074	15,0	13,5
Co-60: loess soil				
Barley	straw	0,0088	7,2	0,5
	grain	0,0059	8,1	-
Sugar beet	leaf	0,012	3,1	3,0
	beet	0,0099	2,1	-
Salad		0,0037	6,2	1,7

[1] Obtained from lysimeter experiments
[2] Based on lysimeter data equal to 1

Continued to next page

Plants		Transfer factors[1]	Relative transfer factors[2]	
			pot	cup
Mn-54: podzolic soil				
Barley	straw	3,40	2,1	0,2
	grain	1,00	1,5	-
Potatoes	leaf	2,00	1,5	-
	tuber	0,14	1,0	-
Salad		0,48	4,2	2,1
Mn-54: loess soil				
Barley	straw	1,30	1,5	0,1
	grain	0,31	1,0	-
Sugar beet	leaf	0,15	8,7	1,9
	beet	0,08	3,9	-
Salad		0,28	2,4	0,8

[1] Obtained from lysimeter experiments
[2] Based on lysimeter data equal to 1

lizer application rates, less competition for light and water, higher root mass per soil unit, and hence higher interception. Therefore transfer factors obtained in pot experiments can only be applicable to a limited extent to field conditions. Factors dominant in influencing the transfer factors in pot experiments may include soil volume, root density and root/shoot ratio, water supply, and fertilizer application rate.

In order to make use of the advantages of pot experiments outlined earlier, these experiments will be continued for two more vegetation periods to reduce the magnitude of variation in transfer factors for some plants on a given soil. The following plants will also be included: wheat, alfalfa, grass, bush beans, carrots, and radish.

REFERENCES

1. Führ, F. (1979): In: "Radioökologie", Tagungsbericht der Fachtagung Radioökologie, Bonn 2./2. October 1979 (in press ISBN 3-8027-2121-7)

2. Große-Brauckmann, E. (1973): Z.Pflanzenernaehr.Bodenkd. 134, 102-107

3. Große-Brauckmann, E. (1977): Z.Pflanzenernaehr.Bodenkd. 140, 617-626

4. Steffens, W., Mittelstaedt, W., and Führ, F. (1980): In: 5th Int. Congr. of IRPA, Jerusalem/Israel, 9.-14. March 1980

THE TRANSFER OF Sr-90, Cs-137, Co-60 and Mn-54 FROM SOILS TO PLANTS RESULTS FROM LYSIMETER EXPERIMENTS

W. Steffens, W. Mittelstaedt and F. Führ

Predicting irradiation of man from food intake using computer simulation models (1) the magnitude of transfer of radionuclides from soil to plant known as transfer factor is of importance. This transfer factor is influenced by a number of environmental parameters such as climate, plant species, soil properties, concentration of the stable and radioactive isotope in the soil etc. (2). Therefore, results from laboratory studies or from experiments using radioactively contaminated soils of non-agricultural origin may not be validly applicable from one region to another.

In our investigations outdoor lysimeters containing 2 typical German soils were used to evaluate the transfer of Sr-90, Cs-137, Co-60 and Mn-54 from the soil to a variety of crop plants.

MATERIALS AND METHODS

Soils: A parabrown earth form the Eschweiler region and a podzol (sandy soil) from the Gorleben area being very different in their properties (Tab. 1).

Table 1: Chemical and physical properties of the soils.

Origin	Eschweiler	Gorleben
Soil type	Parabrown earth	Podzol
Horizon	Ap	Ap
pH ($CaCl_2$)	5,9	4,7
Org. C %	1,4	1,1
Total N %	0,1	0,1
Clay %	12,0	2,6
Silt %	28,4	2,9
Fine sand %	58,3	34,3
Coarse sand %	1,4	60,2
T-value	11,2	6,2
S-value	12,8	2,4
Ca (meq/100g)	11,8	2,0
K (meq/100g)	0,8	0,2

Lysimeter:
1) 0,25 m^2 surface undisturbed soil profile
2) 1 m^2 surface, uniformly mixed surface soil (3)

Treatment: Carrier-free Cs-137 Co-60, Mn-54 in chloride and Sr-90 in nitrate form mixed with
a) 0 - 1 cm soil layer to simulate an accidental contamination, 77 and 80 µCi/kg dry soil
b) 0 - 20 cm soil layer to simulate a 50-years contamination, 5,8 and 6,0 µCi/kg dry soil.

Plants: Grass and alfalfa pasture, sugar beets, winter wheat and summer barley grown in a rotation with bush beans, carrots, radish, lettuce and clover as intermediate crops. The fertilization corresponded to the amounts usual in practice.

Analysis: Radioactivity measuring was done in a well type Ge(Li)-detector and in a liquid scintillation spectrometer (Sr-90).

RESULTS AND DISCUSSION

The transfer factors for Sr-90 determined in pasture grass, grass hay, and alfalfa grown on the podzol soil contaminated in the 0 - 1 cm layer were similar or up to 3 times higher than those for the parabrown earth soil, but the transfer factors for Cs-137 were 3 - 73 times higher on podzol soil (Table 2). At least in pasture grass, the transfer of both radionuclides was lower than that published previously (3) due to longer time of equilibration of Sr and Cs in the soil. The tendency of the transfer factors in the consecutive cuts was contrary to the expected decrease of transfer.

Table 2: Transfer factors (TFSP) of Sr-90 and Cs-137 for plants grown on parabrown earth and podzol soils after contamination of the 0 - 1 cm soil layer.

Plants	Dry matter %	TFSP x 10^{-1} Sr-90 Parabr. earth	Podzol	Dry matter %	TFSP x 10^{-1} Cs-137 Parabr. earth	Podzol
Grass 1. cut	11,3	4,3	4,0	12,8	0,05	0,60
2. cut	12,0	2,6	2,9	11,0	0,06	0,53
3. cut	26,8	7,0	7,7	21,3	0,08	0,51
4. cut	16,2	4,6	10,7	29,8	0,12	0,99
Grass hay 1. cut	21,0	3,7	5,7	16,5	0,01	0,73
2. cut	24,3	7,5	12,5	24,4	0,03	0,75
Alfalfa 1. cut	26,2	4,1	13,0	20,4	0,02	0,54
2. cut	29,7	30,0	41,0	26,0	0,10	0,31

Soil sampling 0 - 10 cm deep, time intervals between the cuts: grass 28 days, grass hay and alfalfa 56 days.

Similary, for winter wheat, summer barley, radish and lettuce grown on soil contaminated in the 0 - 20 cm layer (Table 3 and 4), the transfer factors for Sr-90 were only up to 2 times higher, but for Cs-137 5-59 times, for Co-60 2-27 times, and for Mn-54 1,5-17 times higher for the podzol soil than for the parabrown earth soil. The differences between the 2 soils might be due essentially to the lower sorption capacity and base saturation in the podzol soil.

In all plants and on both soils the transfer factors for Sr-90 and Mn-54 were higher by an order of 1-3 than those for Co-60 and Cs-137 (Table 2-5).

The Transfer of Sr-90, Cs-137, Co-60 and Mn-54

Table 3: Transfer factors (TFSP) of Sr-90 and Mn-54 for plants grown on parabrown earth and podzol soils after contamination of th 0 - 20 cm soil layer.

Plants		Parabrown earth			Podzol		
		D.M. %	TFSP Sr-90	Mn-54	D.M. %	TFSP Sr-90	Mn-54
Winter wheat	straw	95,0	1,35	0,53	96,0	2,77	9,16
	grain	91,0	0,05	0,35	90,0	0,14	3,41
Summer barley	straw	92,0	1,91	1,33	92,0	2,09	3,40
	grain	95,0	0,09	0,31	95,0	0,17	1,00
Radish	leaf	8,5	1,20	0,20	8,5	1,77	0,46
	beet	5,8	0,13	0,02	5,7	0,10	0,03
Lettuce		8,6	0,34	0,28	8,6	0,46	0,48

Table 4: Transfer factors (TFSP) of Cs-137 and Co-60 for plants grown on parabrown earth and podzol soils after contamination of the 0 - 20 cm soil layer.

Plants		Parabrown earth			Podzol		
		D.M. %	TFSP x 10^{-2} Cs-137	Co-60	D.M. %	TFSP x 10^{-2} Cs-137	Co-60
Winter wheat	straw	95,0	0,46	1,00	96,0	11,1	26,1
	grain	91,0	0,10	0,74	90,0	5,9	19,8
Summer barley	straw	92,0	0,51	0,88	92,0	8,1	1,8
	grain	95,0	0,23	0,59	95,0	3,9	1,8
Radish	leaf	7,3	0,37	0,56	9,0	2,6	1,4
	beet	5,3	0,04	0,25	5,3	0,4	0,5
Lettuce		7,7	0,36	0,37	5,7	1,8	0,7

Table 5: Transfer factors (TFSP) of Sr-90, Mn-54, Cs-137, and Co-60 for plants grown on parabrown earth and podzol soils after contamination of the 0 - 20 cm soil layer.

Plants		D.M. %	TFSP x 10^{-1} Sr-90	Mn-54	D.M. %	TFSP x 10^{-1} Cs-137	Co-60
Sugar beet	leaf	13,7	2,06	1,49	16,1	0,09	0,12
	beet	26,8	2,62	0,79	26,8	0,04	0,10
Potatoes tuber	leaf	9,2	7,85	19,66	10,7	0,57	3,23
	peel	21,4	0,67	1,32	22,0	0,64	1,05
	flesh	21,2	0,14	1,37	21,7	0,47	0,82

Sugar beets on parabrown earth, potatoes on podzol soil

Because of the higher dry matter content in grain crops, higher transfer factors for all nuclides were found in dicotyledons than in monocotyledons, Lower transfers were found in generative and storage plant parts (grains and tubers) than in straw and leaves except for sugar beets and Sr-90 (Table 3-5).

Comparing with the calculated transfer factors for these radionuclides suggested in the provisional Radioecology Regulatory Guide (1), the transfer factors for plants grown on parabrown earth soil were similar. For plants grown on podzol soil, however, they exceeded the suggested values for Sr-90, Co-60, Mn-54, and Cs-137 by factors up to 20, 30, 65, and 82, respectively.

REFERENCES

1. Der Bundesminister des Inneren (1979): RS II 2-515603/2

2. Biesold, H. et al. (1978): In: 12. Jahrestagung Fachverband Strahlenschutz, Norderney

3. Führ, F. (1979): In: "Radioökologie", Tagungsbericht der Fachtagung Radioökologie, Bonn 2./2. Oktober 1979 (in press, ISBN 3-8027-2121-7)

THE ASSESSMENT OF RADON AND ITS DAUGHTERS IN NORTH SEA GAS USED IN THE UNITED KINGDOM

B. T. Wilkins

There has been considerable interest in the radon content of natural gas in North America for some years, and activity concentrations of up to 1450 pCi l^{-1} at STP have been measured at wellheads(1). In contrast, reported measurements of radioactivity in North Sea natural gas streams have only appeared relatively recently(2). The National Radiological Protection Board began the project described in the present paper in 1977 at the request of the British Gas Corporation. The objectives were to measure the activity concentrations of radon-222, lead-210 and polonium-210 in those gas streams coming ashore in the UK, and to assess the possible exposure of the public that might result from the use of the gas in the home.

MEASUREMENTS

Natural gas is routinely filtered soon after coming ashore to remove particulate material of greater than 1µm aerodynamic diameter before being distributed on the national grid system. The maximum activity concentrations that could reach domestic appliances were therefore assessed by measurements made by sampling after filtration, but before distribution. The radon-222 content was measured directly using 3 litre chambers designed within NRPB(3). The rate of decay was followed and found to be consistent with the 3.8 day half-life of radon-222. The lead-210 and polonium-210 activity concentrations were determined by passing a known amount of gas through concentrated nitric acid, an effective scrubbing agent for metals. A high pressure bubbling system was employed, which had been developed by the British Gas Corporation. This permitted a representative sample of 1m^3 of the gas at STP to be scrubbed in a reasonable time. Four samples of gas from each stream were separately scrubbed so that duplicate analyses could be performed for each radionuclide. Lead was selectively stripped from the acid solution by a two-stage solvent extraction, the chemical recovery being estimated gravimetrically. After a suitable ingrowth period, the more energetic beta particles from the bismuth-210 daughter were counted using a coincidence-shielded gas-flow proportional counter; the lead-210 content of the sample was inferred from the results. Polonium-210 was determined by electrodeposition and alpha spectrometry using polonium-208 as a tracer. The results available at the time of writing for all the gas lines currently on stream are shown in Table 1.

TABLE 1. Radon-222, polonium-210 and lead-210 activity concentrations of North Sea Gas streams supplying the U.K.

Stream	Radon-222, $Bq\,m^{-3}$	Polonium-210, $mBq\,m^{-3}$	Lead-210, $mBq\,m^{-3}$
A	31	2.5	< 17.4
		2.1	< 15.5
B	35	14.1	21.5
		17.0	< 8.1
C	33	4.1	22.2
		3.3	< 7.0
D	27	< 1.5	21.5
		< 1.9	< 7.4
E	36	< 5.6	< 12.2
		< 2.2	< 8.9
F	39	< 2.6	< 7.8
		< 2.2	< 7.0
G	11	9.3	19.6
		10.0	11.5

Note that since 1Bq corresponds approximately to 27 pCi, an activity concentration of 37 Bq m^{-3} is equivalent to about 1 pCi l^{-1}.

The limits of detection are determined both by the sensitivity of instrumentation and by the chemical recovery. Consequently, the limit of detection is particular to each sample. Typical relative standard deviations for these results are 20% for radon and 25% for polonium and lead, based on counting statistics alone.

ASSESSMENT

If the cautious assumption is made that these activity concentrations persist until the gas is combusted in the home, then the maximum possible exposure of individual members of the public may be calculated using a simple model. As a result of filtration, any particulate matter remaining in the gas stream is well within the so-called respirable range.

If there is a steady input rate of activity A into a room which has a ventilation rate of λ air changes per hour, then the total activity present at time t is given by the equation

$$N_t = \frac{A}{\lambda}(1 - e^{-\lambda t}) \quad \ldots\ldots(1)$$

As t increases, N_t tends towards an equilibrium value N_e where

$$N_e = \frac{A}{\lambda} \quad \ldots\ldots(2)$$

A typical room might have a volume of some 30m^3 and a ventilation rate of 1 air change per hour. A typical gas burner used continuously would consume approximately 0.3m^3 of gas per hour. Under these circumstances the equilibrium activity concentration C_e in Bq m^{-3} is given by

$$C_e = \frac{N_e}{30} = \frac{0.3 \, a}{30} = a \cdot 10^{-2} \qquad \ldots\ldots(3)$$

Here a is the activity concentration in the gas in Bq m^{-3}.

RESULTS

Using the highest measured activity concentrations in the gas, the highest concentrations in air for polonium-210 and lead-210 for the conditions assumed are those shown in Table 2. These may be compared to the Derived Air Concentrations for these nuclides appropriate to members of the public which are taken to be one tenth of those for occupationally exposed workers (4).

TABLE 2. Comparison of maximum possible air concentrations with appropriate derived limits

Nuclide	Air concentration, Bq m^{-3}	Derived air concentration, Bqm^{-3}
Pb-210	$2.2 \cdot 10^{-4}$	$3 \cdot 10^{-1}$
Po-210	$1.7 \cdot 10^{-4}$	$9 \cdot 10^{-1}$

In the case of exposure to radon-222, its short-lived daughters contribute significantly to the dose and calculations are made in terms of the Working Level (WL). The highest calculated exposure will be obtained if it is assumed that the radon daughters are in full equilibrium in the gas at the time of supply to the house. By definition, an activity concentration of 3.7 kBq m^{-3} (100 pCi l^{-1}) of radon-222 in equilibrium with its daughters corresponds to 1 WL. Applying equation (3), the highest radon concentration measured in the gas then corresponds to $1.05 \cdot 10^{-4}$ WL in the room. The unit of radon daughter exposure is the Working Level Month (WLM), which corresponds to an exposure to 1 WL for one working month (170h). One year of continuous exposure (8760h) to $1.05 \cdot 10^{-4}$ WL will therefore correspond to $5.4 \cdot 10^{-3}$ WLM in a year. This can be compared to what might be regarded as the maximum permissible exposure for a member of the public, namely 0.4 WLM in a year, (i.e. one tenth of that for occupationally exposed workers) and to an annual exposure of 0.16 WLM from natural background radiation in the U.K. (5).

In summary, the activity concentrations in the home have been calculated using cautious assumptions. No allowances have been made for dilution with gas from other sources (which is presently not significant) deposition of lead or polonium in pipework and, in the case of radon, the time lapse between coming ashore and combustion. It is also known that ventilation rates are usually increased during cooking perhaps by as much as an order of magnitude. Nevertheless the activity concentrations in the house calculated here are two to three orders of magnitude lower than the appropriate limits for members of the public.

It is therefore concluded that no significant radiation exposure of the public results from the distribution of natural gas from the fields in the North Sea that supply the United Kingdom. Nonetheless the activity concentrations of these radionuclides are being periodically monitored, and measurements will be carried out on any new fields that come on stream.

The co-operation of the British Gas Corporation is acknowledged, for permission to publish these results and for their substantial contributions to the sampling programme.

REFERENCES

1. UNSCEAR Sources and effects of ionising radiation. United Nations Scientific Committee on the Effects of Atomic Radiation 1977. Report to the General Assembly, New York, UN (1977).
2. Heide, H.B. v.d., Beens, H. and de Monchy, A.R. Ecotoxicol. and Environ. Safety 1 (1), 49, (1977).
3. Roberts, P.B. and Davies, B.L. A transistorized radon measuring equipment. J. Scient. Instrumentation 43, 32-35 (1966).
4. Annual Limits of Intake of Radionuclides for Workers - Adams, N., Hunt, B.W. and Reissland, J.A. NRPB R-82 (1978).
5. Taylor, F.E. and Webb, G.A.M., (1978), Radiation Exposure of the UK population, NRPB R-77.

Radiological and Industrial Hygiene

EPIDEMIOLOGY, OCCUPATIONAL HYGIENE AND HEALTH PHYSICS

J. A. Bonnell

As a practising Occupational Health Physician of over thirty years standing, I have always believed and have said on many occasions, that occupational exposure to ionising radiations should be regarded in the same way as any other occupational hazard. Unfortunately, because the nuclear age was born in a spectacular way at the end of World War II, ionising radiation is imbued emotionally with almost magical and evil qualities. Half a century before the explosions at Hiroshima and Nagasaki the discoveries of Roentgen and Marie Curie were hailed as major advances. These discoveries have universally benefited mankind and I feel that we should keep emphasising this, rather than the association with the bomb. The emotional impact of releases of radioactive material can be totally disproportional to their real effect; this again was well illustrated in the United States when the incident at the Three Mile Island plant caused widespread anxiety, despite the fact that its effect in terms of physical detriment to the environment and the local population was nil. At the same time in Florida 6 people died, 34 were hospitalised and 2,500 were evacuated following a release of chlorine after a freight train accident. This hardly merited mention even though it was the second such accident in the U.S.A. in 48 hours.

I am not advocating a relaxation of standards, but we do need to keep in perspective the relative risk of nuclear power in comparison with other sources of power and other industrial processes. My task this afternoon is to bring together some views and thoughts about occupational hygiene and health physics, or radiation hygiene which, I think, is a better term. Health physics is really a specialized part of occupational hygiene, but because health physics developed behind a powerful security screen in most countries, there was little if any contact or exchange between occupational hygienists and health physicists in the early days.

The major difference between occupational hygiene and health physics is the high level of concern of radiation protection for the general as well as the occupational environment. This is due partly to the all pervading existence of radiation exposure in our daily lives from natural background and also from routine medical investigative procedures. The long term effects of ionising radiation are limited to an increase in the incidence of cancer and the possibility of an increase in the incidence of hereditary disease.

It is apparent that the principles of protection are identical, viz. the control of exposure to acceptable limits and the monitoring of the exposed individuals to ensure the efficiency of the safety procedures adopted. Whenever possible alternative less hazardous materials are substituted e.g. man-made mineral fibres for asbestos and tritium based luminising paint for radium based luminisers.

For the purpose of the conference I propose to confine my remarks to the working environment, with only passing reference to the general environment since this will probably be dealt with at a number of the other sessions.

Occupational hygiene is concerned with the assessment of the relationship between the health of occupational groups and their exposure to chemical, physical and biological agents; the study of dose response relationships and the promulgation of dose limits referred to as threshold limit values (TLV). TLV's are defined as time weighted average air concentrations which if not exceeded over an 8 hour period can be expected not to cause unacceptable or permanent adverse effects on the health of the majority of persons exposed. Leaving aside acute effects the philosophy adopted for the protection of workers exposed to such chemical and physical agents are the same as for radiation protection. In many instances the relationship between exposure and disease is specific e.g. lead poisoning; cadmium poisoning or pneumoconiosis. The problem of understanding the dose response relationship so far as the induction of cancer by chemical carcinogens is concerned is similar to that of the induction of cancer by ionising radiation.

For the purpose of discussion I propose to consider asbestos and bladder carcinogens with ionising radiations, since a number of points of interest will become evident.

ASBESTOS

Asbestos is a generic term for all fibrous silicates. Exposure to asbestos dust in sufficiently large quantities will give rise to asbestosis. A disease characterised by progressive fibrosis of the lungs. In the U.K. 60-70% of persons diagnosed as suffering from asbestosis die from bronchogenic carcinoma (lung cancer), this high incidence is almost certainly associated with concurrent cigarette smoking.

Asbestos is also associated with a rare tumour of the pleura and peritoneum called mesothelioma; mesotheliomata are highly malignant and invariably fatal. This cancer is associated with exposure to croccidolite (blue asbestos) and to a lesser extent amosite (brown asbestos). Mesothelioma may follow transient exposure to asbestos and there is frequently a latent interval of 40 years or more.

Mesothelioma of the pleura is rare in the absence of documented asbestos exposure, the association of disease with exposure to a specific substance eases the problem of deciding whether or not an exposure is significant. Unfortunately the extent of the exposure and in particular the 'dose' to the affected tissue cannot even be estimated. Whilst blue asbestos is no longer imported into the U.K. other forms of asbestos continue to be used in some specific situations, e.g. in brake linings for automobile brakes or lift hoists where there is no alternative to asbestos. Discussions of acceptable risk and cost-benefit analysis are therefore just as important as those concerned with radiation risks. These risk assessments are also valid for general public exposures.

CHEMICAL CARCINOGENESIS OF THE BLADDER

A number of chemical substances used in the dyestuffs and rubber industry, α and β naphthylamine and benzidine have the capacity to cause cancer of the bladder even following exposure to low concentrations of the chemicals and with an accompanying long latent interval. The recognition of the disease and its association with exposure to these chemical substances merits further consideration because of similarities to the problem of assessing the significance of low level radiation exposure.

Carcinoma of the bladder has been increasing in frequency in the U.K. over the last thirty years.

In common with many cancers, bladder cancer shows a marked regional variation within the national pattern in the U.K. The figures show that deaths from bladder cancer in U.K. males were increased from 2,000 in 1958 to 2608 in 1968, an annual increase of 30% in 10 years. Within this increasing 'background' of bladder cancer, epidemiological studies in England and in the U.S.A. were able to demonstrate that appreciable numbers of those who died from bladder cancer had an occupational history that warranted further investigation (Veys 1973).

The recognition of the association of bladder tumours with occupational exposure to specific chemicals in a situation when cancer of the bladder was increasing in frequency demonstrate the important role of epidemiological analysis of exposed populations. I believe this emphasizes the need for obtaining basic epidemiological information about occupationally exposed groups of radiation workers. There is no other way of testing the validity of the current dose limits propounded by ICRP.

J. A. Bonnell

IONISING RADIATION

The problem of ionising radiation is that of assessing the association of radiation dose with the development of malignant disease and possible harm to the descendants of those persons exposed. The mechanism of cancer production by asbestos (mesothelioma) and bladder cancer from the chemicals mentioned, appear to be similar to that of ionising radiation, i.e. the probability of the development of the disease is associated with the level of exposure. A "stochastic" effect since once the disease process has started it is totally independent of the "dose" of carcinogen.

Unlike the problems associated with the occupational hygiene of the situations discussed above, workers are exposed to radiation the whole time, from natural background, from potassium 40, from medical procedures and from the small additions due to world wide fall-out from bomb testing in the atmosphere.

If we now consider the significance of continued exposure to low level radiation i.e. at doses of about the ICRP recommended limit of 5 rem per annum, and how this limit was arrived at, there are a number of immediate and obvious differences between radiation dose limits and TLV's.

The ICRP dose limit assumes a linear dose response relationship derived from studies of populations exposed to large single doses of radiation at high dose rate. From these data risk factors have been calculated relating the probability of cancer arising in individual tissues with exposure to units of radiation dose in those tissues. This has been possible because reasonably accurate estimates of the acute dose is possible, unlike any retrospective study of asbestos exposure described previously. It is immediately apparent that these risk factors are probably pessimistic since no account can be taken of the modifying effects of dose rate or dose fractionation.

Radiation exposure is also assumed to have no threshold effect, but in these dose response relationships no account is taken of the effect of the level of dose on the latent period between exposure and the appearance of any disease. It could well be that each rem of dose has an effect, but if it takes 200 years to develop the relationship becomes rather academic.

There is one other factor in radiation control which I would like to mention and that is the sensitivity of the measuring instruments and their specificity. Because accurate measurement of a few molecules of radio active substances or photons of energy can be made it is a great temptation to ascribe some effect to these measurable quantities.

DISCUSSION

Since exposure to ionising radiation causes an increase in the incidence of cancer, and the radiation induced cancer is indistinguishable from cancer due to natural or other causes, a number of obvious practical difficulties present themselves.

It is generally believed that some 70% of all cancers are environmental in origin, but less than 4% are due to occupational causes. Where a specific disease is caused by a particular substance such as mesothelioma and asbestos exposure then it can be assumed with some confidence that if exposure to asbestos can be established then cause and effect is apparent.

Again, using experience gained from studying the effects of asbestos, when over 60% of people suffering from asbestosis develop lung cancer, if any individual asbestotic develops this disease, it is reasonable to assume a direct relationship. The fact that the individual has contributed to the disease by smoking cigarettes is an additional consideration.

The level of risk which is acceptable will I feel sure be widely discussed at this conference, but what should be done about the radiation worker who develops cancer, but who has not accrued a dose in excess of the dose limits. Is it necessary to accept that all cancers in an occupationally exposed group of radiation workers are related to radiation exposure? Should ICRP risk factors be used to determine the relationship?

It is vital to establish the validity of the ICRP dose limits and it is only by the study of occupationally exposed workers that this can be achieved.

If an excess of cancer is demonstrated in a group of workers whose exposure to radiation is known and is within the ICRP dose limits, then morally any individual suffering from cancer should be accepted as having a radiation induced disease. Care should be taken to ensure that only radiation workers who are exposed to radiation are included in the group. By the same token if no excess cancers are demonstrable in such a group no individual suffering from cancer should be accepted as a case of radiation induced cancer.

Such an approach has the added merit of being the accepted practice in the assessment of the incidence of cancer in fields other than radiation, and also of course for assessing the acceptability of other occupational hazards such as the dust concentration for the control of pneumoconiosis.

It is apparent that radiation protection practices have influenced occupational hygiene by emphasising the need for control of gaseous and liquid effluents to the general environment, by studies on lung aerodynamics which hitherto had always been concerned with mass inhalation of concentrations of dust in the study of pneumoconiosis rather than the behaviour of dust particles in the lung.

It is appropriate for those concerned with radiation protection to look at the practice of their colleagues in the safety and hygiene fields. The significance of environmental measurements is tested by epidemiological studies of the exposed population. I would therefore suggest that good epidemiological data are required for assessing the relationship of low level radiation exposure and possible effects on the health of those exposed. It is apparent from the studies of bladder cancer in the rubber and chemical industries that an excess of a particular cancer can be recognised even in the presence of an increasing incidence of bladder cancer generally. Attempts are currently under way in the U.K. to assess radiation exposure levels where the large employing organisations are co-operating with the National Radiological Protection Board to establish a national register for radiation workers.

The desirability of such a requirement was discussed at the IAEA Symposium on The Late Biological Effects of Radiation in March 1978 but there is a danger of attempting to collect too much information.

At the end of the day the significance of environmental measurement in relation to the health of an exposed population must be assessed. It is not enough to calculate artificial relationships based upon mathematical extrapolations, it is vital to obtain good epidemiological studies. It is not necessary to wait 20 years or more. If the present dose limits are unacceptably high then this will become evident in a much shorter period of time during the course of a properly designed epidemiological study.

In summary, it is apparent that radiation protection practices have contributed a great deal to the practice of occupational medicine and occupational hygiene. This is particularly the case in the development of philosophies of protection and in stimulating accurate studies in a number of biological systems. It is suggested that it is now time for those of us who have responsibilities for the radiation protection of workers to accept the desirability for accurate epidemiological assessment of persons exposed at or below the recommended dose limits for radiation.

REFERENCE

Veys, C.A., (1973) : Proceedings of the Symposium on the Assessment of Exposure and Risk, Society of Occupational Medicine, London (July 1973)

- - - - - -

N.B. This paper represents the personal views of the author.

AN ANALYTICAL APPROACH TO THE COMPARISON OF CHEMICAL AND RADIATION HAZARDS TO MAN

H. P. Leenhouts, K. H. Chadwick and A. Cebulska-Wasilewska[1]

INTRODUCTION

The main hazards arising from the exposure to ionizing radiation are the so called late effects; the induction of cancer and of hereditary defects. However, there is an increasing awareness that many other environmental agents such as UVR and chemicals are mutagenic and potentially capable of causing cancer and hereditary defects. Thus, the effects caused by radiation are not unique and consequently it is illogical to place radiation in an exceptional position with respect to the protection of man. A comparison between radiation and other mutagenic agents depends on the availability of comparative analytical and assessment techniques. In this paper we present an analytical model, based on radiation biological concepts at the molecular level which permits an analysis of the effects of other agents and also predicts that a synergistic interaction between two different mutagenic agents can occur at the molecular level.

ANALYSIS

Figure 1 presents schematically the molecular mechanisms assumed to be responsible for biological effects such as cell death, aberrations and mutations. The molecular theory (1,2,3) assumes that radiation induced DNA double strand breaks are the crucial radio-biological lesions. The figure shows that double strand breaks have a linear - quadratic dose relationship, that UVR or a mutagenic chemical causing single strand lesions will have a quadratic exposure relationship and that the combined action of radiation and agent gives an additional contribution of double strand lesions arising from the interaction between a single strand break and a single strand lesion. The total number of lesions is

$$N_T = \alpha D + \beta D^2 + \eta X D + \varepsilon X^2 \qquad (1)$$

This predicts that for radiation curves the α coefficient remains constant but combined treatment increases the linear coefficient ($\alpha' = \alpha + \eta X$). For agent curves the ε coefficient remains constant but combined treatment leads to a linear coefficient (ηD).

Figure 2 presents a series of radiation survival curves after a UVR pre-dose and a series of UVR survival curves after a radiation pre-dose. The analysis has been made according to equation (1), all radiation curves have the same β coefficient, α increases with increasing UVR pre-dose, all UVR curves have the same coefficient, the linear coefficient increases with radiation pre-dose.

Figure 3 presents the combination of radiation and BUdR on cell survival and chromosomal aberrations. Again the change in shape of the

---

[1]. A. Cebulska-Wasilewska was a visiting Fellow from the Institute of Nuclear Physics, Krakow. Poland and was supported by the Dutch International Agricultural Centre.
This publication is contribution no. 1646 of the Biology Division of the Commission of the European Communities and is also supported by the Dutch Ministry of Agriculture and Fisheries.

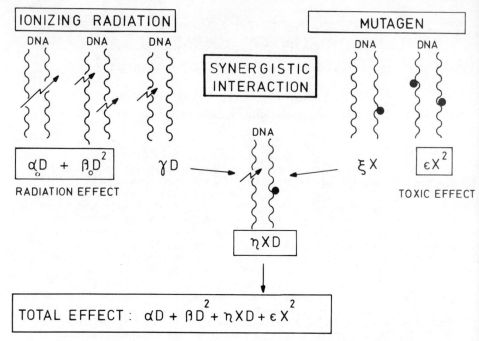

Figure 1. Schematic representation of the action of ionizing radiation, other mutagenic agents and the synergistic interaction between the two agents.

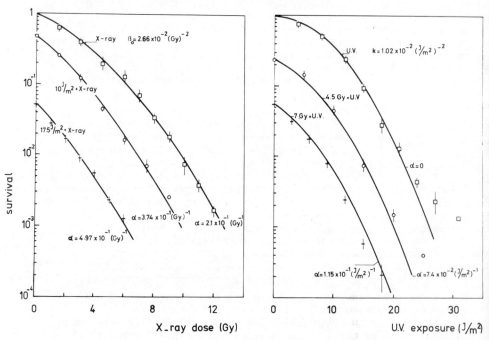

Figure 2. Survival of Chinese hamster cells for combined treatments of X-rays and UVR (4) analysed according to equation (1).

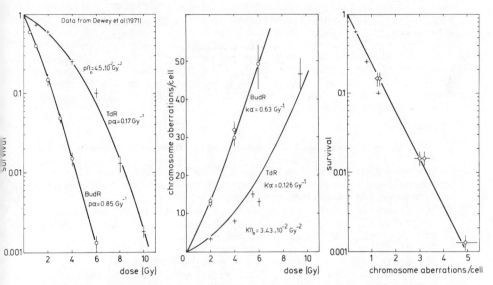

Figure 3. Survival and chromosomal aberrations induced by radiation in synchronised Chinese hamster cells with and without BUdR (5) analysed according to equation (1).

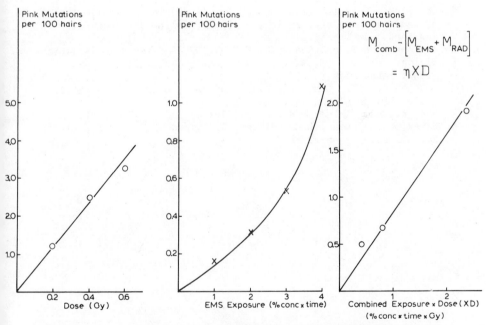

Figure 4. The induction of pink mutations in <u>Tradescantia</u> stamen hairs by X-rays and EMS (ethyl methane sulphonate) and the contribution of mutations arising from the interaction between the EMS and the X-rays as a function of the product of chemical exposure and radiation dose.

curves is reflected in a change in the linear coefficient only. The last part of the figure demonstrates that the change in survival is paralleled by the change in aberrations.

Figure 4 illustrates the effect of EMS and radiation on the induction of pink mutations in the stamen hairs of Tradescantia. The last part of this figure reveals that the difference between the combined treatment and the sum of the separate treatments is proportional with the product of chemical exposure and radiation dose (i.e. ηXD) as expected from equation (1).

CONCLUSIONS

1. The analysis shows that the synergistic interaction occurs at the molecular level in the DNA.
2. The synergism can be demonstrated in cell survival, chromosomal aberrations and somatic mutations and can therefore be expected for cancer and hereditary defects.
3. The synergistic interaction leads to an increase in the linear component of the radiation effect which is critical at low doses and important for radiological protection.
4. The model implies that a synergistic interaction between two different mutagenic chemicals can also be anticipated.
5. The model provides an analytical vehicle which can be used to compare the effects of radiation and other mutagenic agents at the mechanistic level.
6. If we are concerned to protect man from the increasing mutagenic, and thus carcinogenic load, an integral protection philosophy is essential.

REFERENCES

1. Chadwick, K.H., and Leenhouts, H.P., (1980) The Molecular Theory of Radiation Biology. Springer Verlag, Berlin-Göttingen-Heidelberg-New York. (In Press).
2. Leenhouts, H.P. and Chadwick, K.H., (1978) Advances in Radiation Biology Vol 7, p 55. Editors: J.T.Lett and H.I.Adler. Academic Press, London-New York.
3. Leenhouts, H.P., Chadwick, K.H., and Deen, D.F., (1980) Int.J. Radiat. Biol. (In Press).
4. Han, A., and Elkind, M.M., (1977) Int.J. Radiat. Biol. 31, 275.
5. Dewey, W.C., Stone, L.E., Miller, H.H., and Giblak, R.E., (1971) Radiat. Res. 47, 672.

TECHNOLOGY TRANSFER FROM NUCLEAR AND RADIOLOGICAL TO INDUSTRIAL SAFETY

Y. G. Gonen

The term technology transfer usually means the supply of equipment, information and techniques from a more developed supplier country to a less developed recipient country, together with an educational, human developmental package. In the ideal case, these enable the recipient to use the acquired technology independently and to develop it further.

Technology transfer in a broader sense includes situations in which the transfer occurs between previously not interconnected fields, in which the subject of the transfer is the philosophical, scientific, technical or managerial approach to a set of problems or in which the educational package is replaced by transit of individuals or groups from one activity to another. There are many such examples.

During the last decade another technology transfer process has developed, from the nuclear and radiological to the occupational safety and hygiene fields.

The responsible governmental organs in many developed countries realized that their efforts result in significantly lower levels of safety and health at work than desirable, reasonably achievable and socially acceptable.

The recommendations of governmental review committees on the subject (1) and the legislative actions taken upon their acceptance (2) show the marked influence of approaches well known and widely practised in nuclear and radiological safety, recognizing that the level of occupational safety achieved in these activities - by using specially developed approaches - is significantly better than in comparable non-nuclear ones.

In August 1978 the Minister of Labour and Social Affairs of Israel appointed a committee - under the chairmanship of the author - to review the state of safety at work in Israel. The committee had a broad mandate but rather limited time to present its recommendations. These were submitted to the Minister in March 1979. The final report (3) was published in August 1979 in Hebrew.

The following basic findings of the committee may well have general relevance, as similar conditions existed, exist or may exist in many developed and other countries.

(1) The organs responsible for the control of safety at work are fragmented and scattered between different authorities. The level of interaction between these organs is low and there is no integral approach to safety at any given facility. Coordination between these organs is ineffective and hazards of different nature are handled independently by the different authorities.

(2) The major responsibility for controlling safety at work is that of the work safety inspectorate. It acts on outdated and inadequate legislation, with limited manpower, and is unable to cope with the growing complexity of the industry, with the increasing concentrations of potential energy, hazardous materials at single locations and

mainly with the less obvious and apparent nature of some modern hazards. Actual and potential off-site effects (on populations or on the ecology) are unfortunately outside the jurisdiction of the inspectorate.

(3) The penalties in the law and the actual sentences imposed by the Courts for safety violations and for maintaining unsafe conditions in facilities are not an effective deterrent.

(4) Present legal, taxation and insurance practices enable employers to transfer the burden of economic losses due to unsafe operational conditions including accidents to the State or in general to the public. There is no economic incentive to employers to invest in the safety of their operations.

(5) There is no suitable data base for priority/policy decisions on matters of safety. Courses of action are chosen on intuitive basis. The marginal investment in risk reduction - where it exists - varies by orders of magnitude between different industries and work places both in monetary terms and in inspection effort.

(6) R&D projects are carried out without the establishment of a general plan, as no such plan or priority system could be developed in the absence of a suitable data base. There is no reliable system for the transfer of R&D results into general practice.

(7) There is no systematic and timely approach to hazards of stochastic nature and there is no systematic control of them. The information base on these hazards is insufficient.

(8) Improvements are made on a patchwork basis, usually following severe accidents or the detection of grossly unsafe situations.

(9) Training opportunities are scarce and generally non-specific in quantity and relevance.

(10) In general, there is no emergency planning - not at the off-site level nor at the on-site, intervention level.

(11) The professional level of many engaged in safety related activities (in the work places and in the active governmental organs) needs improvement. This is especially valid with respect to safety personnel at the facility level.

(12) Workers' participation is ineffective due to reasons similar to those already mentioned, little or no influence and deterrent, lack of professional ability and information.

Some of the specific findings of the committee, regarding the Israeli situation were:

(1) the rate of reportable work accidents (absence of three or more days from work) is approximately 80 per year per 1000 employed persons in all types of employment and about twice this rate among the employed in construction and industry.

(2) the rate of fatal accidents is approximately 0.2 per year per 1000 employed (of which ~ 50% are traffic accidents on the way to or from work - considered as work accidents for compensation).

(3) the rate of accidents causing permanent disability above 20% is approximately 0.7 per year per 1000 employed and of those causing temporary disability, about four times higher.

(4) the estimated financial loss to the economy due to accidents at work is in the range of $300-500 million, i.e., ~ 3 ÷ 5% of the GNP. The public expenditure on safety is less than one percent of this sum.

(5) there has been no trend of improvement during the past 5-10 years.

Technology Transfer

The major recommendations of the committee were:

I. AT THE FACILITY LEVEL

(1) Approach the safety of work places on integrative basis. Aim for safety as a built-in aspect of every activity, not an extra, add-on type one.

For new facilities the consideration of the desirability of an obligatory safety assessment was recommended, as a condition for a licence. For existing facilities the recommendations call for written safety policy statements and improvement programmes based on a thorough review of the existing hazards.

Both types of document should contain a description of the potential hazards of the operation and the appropriate countermeasures employed.

(2) Encouragement of worker-management co-operation in facility safety committees required by law on a more objective and professional basis, with the help of outside specialists if needed. All the potential hazards should be openly discussed - the workers should be properly informed about the risks they are exposed to.

(3) Safety should be part of management's responsibility at all levels. Duties and responsibilities and the authority of the different managerial levels with regard to safety should be clearly defined.

II. AT THE NATIONAL LEVEL

(4) Reorganization of the functions presently dispersed between the different authorities, ministries, etc., into one coherent unit, preferably in a unified national occupational safety, hygiene and health service.

(5) Establishment of research coordination and engineering development units within the unified service. This will also contribute towards the increase of the professional capability of the service as a whole.

(6) Establishment of coordinating functions within the unified service with other, presently not-interacting services fully or marginally relevant to safety at work.

(7) Encouragement of voluntary organizations in the field (professional societies, unions, associations of employers, etc.).

IN SPECIFIC FIELDS

(8) Modernization and updating of the legislation
   (a) to reflect the recommended changes;
   (b) to re-establish the deterrent force of the law by severely increasing the penalties on safety violations and by speeding up Court procedures; and empowering the safety inspectorate to impose administrative fines.
   (c) to widen the coverage of hazardous situations by regulations, including the promulgation of satisfactory, technically acceptable solutions. (Codes of practice).

(9) Establishment of inspection priorities according to the hazard level, safety record and safety arrangements in the inspected facilities. Introduction of obligatory investigation of every serious work accident.

(10) Establishment of a satisfactory data base and of a specialized safety information centre.

(11) Improvements in safety-related R&D management, in occupational hygiene and health control, in training activities at all levels, according to a priority system to be developed.

(12) Encouragement of research and publication on the economic aspects of work safety. Promotion of a cost benefit approach as a tool for priority assessment and investment evaluations.

As said above many of the recommendations, including part of those mentioned above, are based on practices which evolved and are used in the nuclear field. Basically the report calls for the introduction of a similarly organized, systematic approach to the different industrial hazards.

REFERENCES

1. Safety and Health at Work. Report of the Committee 1970-1972 (Robens report) H.M.S.O.
2. Health and Safety at Work Act, 1974. H.M.S.O.
3. Report of the Review Committee on Safety at Work in Israel (1979) Ministry of Labour and Social Affairs (in Hebrew)

RADIATION PROTECTION PRINCIPLES APPLIED TO CONVENTIONAL INDUSTRIES PRODUCING DELETERIOUS ENVIRONMENTAL EFFECTS

J. Tadmor

Comparison of the radiation protection standards for the population-at-large with the ambient standards for conventional pollutants, reveals differences in the principles upon which the standards are based.

The most important factors considered in establishing the radiation protection standards for populations-at-large (as well as for specific sections of the population) are as follows:

1. Somatic effects, influencing the exposed person himself
2. Genetic effects, influencing the descendants
3. Effects on specific tissues and organs of the human body (leading to the concept of critical organ)
4. Additivity of effects, considering the simultaneous effects on several tissues and/or the whole body, and the total risk to all irradiated tissues
5. Sensitivity of exposed person, e.g. children
6. Stochastic health effects, i.e. the probability of the occurrence of an effect, as a function of dose
7. Non-stochastic health effects, i.e. the severity of an effect, as a function of dose
8. Quantitative acceptable risk (e.g., expressed as deaths per person per year) based on risks experienced or acceptable in other human activities
9. Size of population exposed, expressed as the product of the exposure dose and the number of persons exposed, assuming a linear dose-effect relationship
10. Cost-benefit and ALARA (as low as reasonably achievable) considerations, taking into account technical, economic and social factors in limiting the exposure.

With the exception of the non-stochastic health effects, none of the other factors considered in establishing the radiation protection standards, are taken into account (certainly not explicitly) in spelling out the ambient standards for conventional pollutants, such as SO_2 and NO_x.

These and other differences result in more relaxed ambient standards for conventional pollutants, in comparison with the radiation standards, as illustrated by the ratios between the standards and the natural, medically perceivable and lethal levels (Table 1). The differences are in the range of orders of magnitude. The consequence of the severity of the limitation of exposure to radiation, as compared with conventional pollutants, is the penalization of the nuclear industry due to the increased cost of its safety measures.

TABLE 1. Comparison of different levels of exposure to radiation, SO_2, and NO_2

	SO_2^* (ppm)	NO_2^* (ppm)	Radiation (mR/day)
Level			
Background	0.0002	0.01	0.35 (130mR/y)
Maximum permissible exposure	0.03	0.05	0.03 (8mR/y)
Medically perceivable effect**	0.02	2	2×10^4 (20R)
Lethal**	0.5	500	5×10^5 (500R)
Ratios of levels			
Maximum permissible/background	150	5	~0.1
Medically perceivable effect/ maximum permissible	0.6	40	~6×10^5
Lethal/maximum permissible	17	10^4	~10^7

* with particulates
**one-day level

The fact that deleterious exposures are not restricted to the same extent in different human activities, appears to cause a misuse of public resources. Considering that there is a limit to the public economic means available, societal expenditures for reducing risks should be spread, as much as possible, over all human activities to get the maximum return from investments. Indeed, the law of diminishing returns indicates that the return on investments to decrease the marginal hazards of an activity is insignificant as compared with the return on initial expenditures which diminish the hazards substantially. The nuclear industry is already at the stage where additional expenditure brings only marginal returns, while many conventional industries are at the initial stage of safety expenditures.

The greater safety cost imposed on nuclear power plants, as compared with conventional power plants, may result in the substitution of a hazard worse than the radiation hazard due to the release of SO_2 and other harmful pollutants from conventional power plants.

It is proposed that, to diminish the hazards to the public uniformly and effectively and also to get an optimum return on the safety investments made by the public, radiation protection principles should be used as prototypes for pollutants having harmful environmental effects. It is also proposed that radiation health physicists should be active in the application of these principles of population protection.

The application of one of the principles of radiation protection, that of limiting the integrated population exposure (expressed as person x rem), is illustrated here for a conventional pollutant, such as SO_2. A study of the atmospheric release of SO_2 under different conditions is analyzed, to emphasize the importance of considering the size of the exposed population.

Assume that the only requirement concerning the release of SO_2 from any installation is to keep the ambient air concentration at the fence of the installation below the half-hour standard of 0.3 ppm, adopted in Israel. Assume also that the density of the population, uniformly distributed around the installation is 400 persons/km^2.

With these assumptions, the ambient SO_2 concentration (ppm) and the total integrated population concentration (person x ppm) were calculated for a distance up to 80 km from the source for two release cases: a) ground-level release (Table 2) and b) release from a height of 200 m (Table 3).

TABLE 2. Integrated SO_2 population concentration for a ground level release. Assumptions: 1) concentration of SO_2 at the fence of the plant (1 km from the source) = 0.3 ppm, 2) deposition velocity of SO_2 = 1 cm/sec, 3) population density = 400 persons/km^2, 4) average atmospheric conditions

Distance (km)	Population	Concentration (ppm)	Integrated concentration (person x ppm)
0-2	4×10^3	1.2×10^{-1}	500
2-3	8×10^3	5×10^{-2}	400
3-5	1.8×10^4	2×10^{-2}	300
5-8	3.3×10^4	7×10^{-3}	200
8-16	3×10^5	5×10^{-3}	1,500
16-30	7×10^5	5×10^{-4}	350
30-50	1.5×10^6	2.5×10^{-4}	400
50-65	2×10^6	2×10^{-4}	400
65-80	2.5×10^6	10^{-4}	200
			~4,000

The classical Gaussian plume formula (1) was used in these calculations, assuming average atmospheric conditions, and the integrated population concentration was calculated by multiplying the number of persons in concentric rings at various distances around the installation by the concentration calculated for the middle of the ring.

Assuming that the ambient concentration at the fence of the installation is at the level of the half-hour ambient standard, it was found, as expected, that the ambient concentration at any distance further from the source is below this concentration, for both the ground-level and 200 m height releases.

However, there is a very significant difference between the integrated population concentrations in the two aforementioned release cases. Assuming a deposition velocity of 1 cm/sec, the integrated population concentration for a ground-level release is about 4,000 person x ppm, while for a release at a height of 200 m, it is about 138,000 person x ppm.

It should be stressed again that in both release cases, the ambient cocentrations are below the standard. However, the difference by a factor of up to about 35 in the integrated population concentrations indicates that the integrated population concentration for conventional pollutants should also be limited as in the case of radiation protection, in addition to the limitation of the ambient concentration.

TABLE 3. Integrated SO_2 population concentration for an elevated release. Assumptions same as in Table 1, except for height of release which is assumed to be 200 m.

Distance (km)	Population	Concentration (ppm)	Integrated concentration (person x ppm)
0-2	4×10^3	0.3	1,200
2-3	8×10^3	0.3	2,400
3-5	1.8×10^4	0.25	4,500
5-8	3.3×10^4	0.25	8,200
8-16	3×10^5	0.07	21,000
16-30	7×10^5	0.04	28,000
30-50	1.5×10^6	0.02	30,000
50-65	2×10^6	0.01	20,000
65-80	2.5×10^6	7×10^{-3}	17,500
			~132,800

The importance of applying to conventional pollutants, the other principles and factors considered in establishing the radiation protection standards, could be similarly demonstrated.

REFERENCES

1. Gifford, F.A. Jr., (1960): Nuclear Safety, 2 (2), 56.

A HEALTH AND RESEARCH ORGANIZATION TO MEET COMPLEX NEEDS OF DEVELOPING ENERGY TECHNOLOGIES

R. V. Griffith

INTRODUCTION

The rapid development of technology, particularly in the area of energy research and development, brings with it increasingly more complex and sophisticated health and safety problems. The complex safety requirements also bring the need for greater interaction between health and safety disciplines that have historically functioned quite independently. The Lawrence Livermore Laboratory employs approximately 7000 people in various areas of energy technology development and nuclear research.

All of the LLL safety functions (except the Medical Department) are assigned to the Hazards Control Department. This department has responsibility for radiation safety, industrial hygiene, industrial safety, fire safety, and explosive safety. The safety program is managed through a system of program oriented field teams. Each team has a leader and a cadre of health and safety technicians. The resources necessary to make the field teams functional are provided from the health and safety discipline areas listed above. For example, the field team responsible for the general area of chemistry would have representatives from industrial hygiene, radiation safety, fire safety, etc. assigned to work together, insuring a multidisciplinary approach to solution of safety problems in that area

We, as many other moderate or large size laboratories, have found that operational safety questions frequently arise that cannot be answered or solved through available data, techniques or instrumentation. These questions can only be resolved through in-house cost effective safety technology development. At LLL, this development is done through a single division within the Hazards Control Department, that responds to health and safety technology problems in radiation science, fire science, industrial hygiene, and general safety.

SPECIAL PROJECTS DIVISION

The Special Projects Division was formed because it was recognized early in the formation of the Hazards Control Department that development of health and safety technology was sufficiently important to require a separate dedicated effort. Special Projects is a 27-person safety technology development organization representing approximately 10% of the Hazards Control Department work force. The primary mission of the Division is to provide health and safety technology development required by the operational safety program at the Laboratory. Division responsibilities include:
- Development of equipment or instrumentation
- Development of techniques for health and safety
- Studies to provide information needed by management or operational safety professionals

- Measurements of safety related parameters of Laboratory facilities, equipment, or materials
- Calculations or calculational studies.

The Special Projects Division is organized into three functional groups - Radiation Science, Safety Science, and Fire Science. These groups share common office and laboratory spaces, and interact with each other on a regular basis. This organization factors multi-disciplinary attention to general safety problems; i.e., a skill generally required for one class of health and safety discipline is available for solutions to problems that occur in others. For example, aerosol physics and filtration are areas that present problems for each of the three of the groups in our Division. Rather than hiring an aerosol physicist for each group, we are able to share the talents of one or two individuals who meet our technological needs. This is also true in areas such as instrumentation, chemistry, computer science, and others. Table 1 shows the current skills inventory in the Special Projects Division. This inventory meets the current needs of our Department and Laboratory. It is intended as an example, and may not apply to other facilities.

Table 1.

SPECIAL PROJECTS SKILLS INVENTORY

Combustion Science	X-Ray Fluorescence Analysis
Analytical Chemistry	Radiation Spectrometry
Risk Analysis	Health Physics Instrumentation
Industrial Hygiene	Industrial Hygiene Instrumenation
Aerosol Physics	Radiation Physics
Health Physics	
Solid State Dosimetry	

The <u>Radiation Science Group</u> (RSG) develops instrumentation and techniques for radiation protection and measurement. The group provides technical support to the Radiation Safety Program at the Laboratory, particularly in the areas of personnel dosimetry development, applied health physics, calibration and standards development, and the Laboratory environmental monitoring program. Expertise developed in radiation safety technology is often valuable to other Laboratory programs. Consequently, this group provides radiation instrumentation/measurement support to particular programmatic efforts at the Laboratory.

The Safety Science Group (SSG) is responsible for solution of problems in the fields of occupational health and general safety. The SSG obtains data and develops specialized equipment to support safety discipline needs in respiratory protection, work place monitoring, improved air cleaning and monitoring techniques, and evaluation of protective clothing. Recently the SSG has undertaken the additional responsibility of technical support to the Laboratory Waste Disposal and Decontamination facility efforts.

A Health and Research Organization

The Fire Science Group (FSG) develops solutions to problems faced by the Fire Safety Program at LLL. The FSG also uses its expertise to answer fire safety related questions for the Department of Energy, in general. The technical scope includes: materials testing for small-scale flammability tests to full-scale enclosure fires, development of unique modes and mechanisms for fire extinguishants; tests and analysis of fire detection concepts of hardware, analysis of physical and chemical properties of smoke aerosols; studies of fire retardant application to natural and synthetic materials; and parametric analysis of the interaction between fire management systems and the fire hazards potential of experimental facilities. The FSG maintains a full-scale fire test enclosure that is equipped to simulate fires in laboratory environments with typical fuels that might be found in the laboratory. It also has a fine chemical analytical laboratory including a sophisticated gas chromatograph-mass spectrometer for analysis of fire decomposition products.

PROJECT DEVELOPMENT

A key to the success of an organization such as the Special Projects Division is the need for close, working interaction with the operational health and safety staff of the department. Special Projects personnel are encouraged to interact with their operational counterparts daily in a candid and personal way. Some of the best technology development in the safety field comes from such interaction. A project may be conceived either by a Special Projects Senior Investigator, an operational client or both. Once formulated, projects must be approved by the client and Special Projects Division Leaders. Projects are reviewed formally once each quarter through a system of status reports. The progress of each project is monitored by the Steering Committee. The Steering Committee is composed of the Special Projects Group Leader, the Special Projects Division Leader, and the Division Leader and Group Leader in the counterpart Operational Health and Safety Division in Hazards Control. These steering committees meet regularly to review the status of ongoing projects. It is important for successful completion of a project that the senior investigator and client review the project frequently.

Depending on the complexity and time requirements on any given project, a Senior Investigator, who is generally a safety professional with an advanced degree, will have from one to four projects assigned at a given time. A Senior Investigator may also be called upon occasionally to provide programmatic support to research and develop problems arising from outside the Hazards Control Department, although that is not part of their prime mission.

PROJECT EXAMPLES

A major strength of the Special Projects Division is its multidisciplinary structure - the ability to combine a wide spectrum of skills to achieve successful project completion. Examples of the potential for interdisciplinary accomplishment include testing the fire resistance of neutron shielded storage containers; on-line measurement of toxic metal concentrations in Laboratory sewage effluent using special X-ray fluorescence and radiation detection techniques; and collaboration between the Safety Science and the Fire Science Groups to develop filtration and sampling techniques for dense smokes.

Other examples such as these could be cited to demonstrate the interaction between the disciplines, but perhaps the greatest benefit to the Laboratory is derived from non-specific daily interactions and discussions both within the Division and the Department. Without the unified structure that the Special Projects Division enjoys, interdisciplinary communications would be much less frequent and productive. An additional advantage to the organization and the Department is the ability to share resources among groups rather than having to duplicate resources. Single high quality laboratories for testing, analysis, dosimetry, etc. provide a maximum flexibility at a minimum cost.

SUMMARY

At the Lawrence Livermore Laboratory, a unique safety technology organization has been established that is especially geared to respond to interdisciplinary health and safety questions in response to rapidly growing energy technology problems. This concept can be adopted by smaller organizations at a more modest cost, and still maintains the efficiency, flexibility, and technical rigor that are needed more and more in support of any industry health and safety problem. The separation of the technology development role from the operational safety organization allows the operational safety specialists to spend more time upgrading the occupational health and safety program but yet provides the opportunity for interchange with health and safety technology development specialists. In fact, a personnel assignment flow between an operational health and safety organization and a special technology development organization provides a mechanism for upgrading the overall safety capability and program provided by a given industrial or major laboratory.

ACKNOWLEDGEMENTS

We would like to recognize Mr. Seymour Block, Dr. Joseph Tinney, Dr. Charles Prevo, and Dr. Thomas Crites who as the previous leaders of the Special Projects Division, did so much to develop its character and contributions. We also want to thank the operational staff of the Hazards Control Department for their support, understanding, and encouragement.

*Review of Activities
of International Organizations*

DEVELOPMENT AND TRENDS IN RADIOLOGICAL PROTECTION AND THE NEA PROGRAMME IN THIS FIELD

O. Ilari and E. Wallauschek

INTRODUCTION

The Nuclear Energy Agency (NEA) was initially created as a European organisation (ENEA) in 1957, but it assumed its present broader configuration in 1972 with the entrance of Japan among participating countries, successively followed by Australia, Canada, and the United States. A major task of NEA is to encourage harmonization of governments' regulatory policies and practices and to promote the exchange of information and the co-ordination of research and development in the field of radiological health and safety.

The work of the Agency is carried out under the authority of the OECD Council by the Steering Committee for Nuclear Energy which is assisted in its work by a number of specialised committees. In particular, the Committee on Radiation Protection and Public Health, which had originally been set up in 1958 as the Health and Safety Sub-Committee, is responsible for the Agency's activities concerned with radiological protection and related environmental problems. Its activities include the review and discussion of national radiation protection policies and practices, the review of progress of radiation protection philosophy and the interpretation of the ICRP Recommendations, the study of the means of their conversion into practical applications, including the establishment of radiological protection standards as well as the preparation of technical studies and state-of-the-art reviews on specific problems.

THE EARLY DEVELOPMENTS

Since the beginning, radiological protection became a major part of the Agency's programme. The areas of concern for the national and international radiation protection communities changed significantly during the last 25 years, and the Agency programme evolved with them. During the 1950s and the early '60s, the attention of public authorities and radiation protection specialists was primarily focused on two major questions.

One was the need to elaborate radiological protection regulations. Many countries, at that time, had not yet worked out these measures in detail, and still lacked the necessary experience to implement them. Moreover, it was soon realised that these regulations should have been as uniform as possible throughout countries in order to avoid international trade and public opinion repercussions resulting from inadequacy or excessive severity of individual national regulations. This need for an international co-ordination was soon appreciated by the

Steering Committee, and the first task of the newly-created Health and Safety Sub-Committee was the preparation of basic norms for protection against radiation. These norms were adopted in 1959 for use in Member countries as one of the bases for their forthcoming legislations. One specific concern in preparing these norms, in line with the Agency's vocation, was to transfer the ICRP policy and conceptual language into more practical terms, more easily applicable to a regulatory context. The Agency did not limit its effort to the preparation of basic norms, but it continued to keep under constant review the developments and trends of radiation protection policy and the ICRP recommendations. In this context, the basic norms were subjected to revision in 1963 and 1968. A major revision is also currently under way in order to take into account the new principles set down in ICRP Publication 26. This revision presents a significant difference from the previous ones; the latter were in fact carried out independently by NEA, though in coordination with other international organisations. In the last few years, however, it was realised that it was in the interest of Member countries to have a single series of international recommendations on radiation protection norms. Therefore, arrangements were made in 1977 with IAEA, WHO and ILO to prepare a joint revision of the relevant norms and to publish a unified set of standards applicable by all international organisations. The publication of the joint revised basic norms is expected by 1981.

The second major concern in the period of the 1950s and early 60s was the public health risks associated with the radioactive fall-out from nuclear explosions, and the methods and techniques for the measurement of low levels of activity in environmental matrices. Also in this field, the Agency played an active role. This was the setting up, in 1959, of a system for the centralised collection, comparison and dissemination of information and data resulting from the network of environmental radioactivity measurement stations located in Europe. This mechanism for an oriented exchange of information was operated for about 10 years; then, with the dramatic decrease in environmental radioactivity levels in the years after 1962, interest in an international exchange of information declined and the system was discontinued. Also, the Agency deployed a limited effort towards standardisation of sampling and measurement methods used in the different countries.

The concern of national authorities at the increasing levels of environmental contamination due to fall-out, and the growing awareness of the risk of nuclear accidents possibly involving bordering countries, induced NEA to set up, in 1961, an international system of supervision and emergency warning in the case of an increase in environmental radioactivity in one Member country. The technical features of this system, the scope of which was based only on the results from airborne radioactivity measuring stations, were of course not such as to generate a really timely and effective warning mechanism. However, those features were in line with the technical means and knowledge available at that time and, in any event, the NEA system had the undeniable merit of supplying a pre-arranged framework for quick contacts and consultations between national authorities in case of any problems concerning radiological protection of the public. It should also be appreciated that this system, with its obvious limitations, was probably the first international attempt to set up forms of co-operation between countries in the field of nuclear emergencies. The system was operated for several years, and successively, although it was never officially abrogated, it was practically abandoned in the last few years. In any event, the seed of international co-operation had been sown, and since then better and technically more adequate emergency warning and co-operation systems have been set up between bordering countries.

THE MATURITY OF THE AGENCY'S PROGRAMME

The 1960s saw the suspension of significant nuclear tests in the atmosphere, and the parallel impetuous development of nuclear energy and its various applications. In this changing scenario, other issues assumed a growing emphasis. These were the problems and techniques involved in the protection of workers in nuclear establishments, and the methods and instruments for the health physics surveillance and dosimetry. In these fields NEA did not play a leading role, apart from organising a number of seminars and symposia on health physics and dosimetry matters. In the same period, however, NEA showed a particular sensitivity and promptness in tackling some other problem areas which were at that time only beginning to concern public health authorities.

The ICRP Publication 7, in 1965, introduced a significant quality step into the criteria for the environmental monitoring around nuclear facilities. Concepts such as the critical groups of population and critical parameters of the environment were the basis for a more rational and cost-effective approach to monitoring, and paved the way for the development of the analytical models which are now an essential tool for the assessment of the environmental impact of any nuclear facility or waste management operation. Since 1961, the NEA Health and Safety Sub-Committee realised the importance of this subject, and focused its attention on the study of the environmental behaviour of radioactive wastes in the marine environment, and the associated radiological risks. A number of studies were therefore carried out between 1962 and 1964 on the oceanographic and radiological aspects of the presence of radioactive materials and the discharge of radioactive wastes into the North Sea. These preparatory studies culminated with the preparation of a joint theoretical study of the radiological capacity of the North Sea for the discharge of radioactive wastes. This was probably one of the first examples of a radiological risk assessment applied to a large ecosystem and, we suppose, it was the first to result from a joint international effort. This specialisation of NEA expertise on the problems associated with waste disposal into the sea constituted the basis for the increasing involvement of the Agency in the study and, successively, the implementation of operations involving waste disposal into the ocean under international surveillance. In this field, the Agency had, and still conserves in the framework of the Multilateral Consultation and Surveillance Mechanism for Sea Dumping of Radioactive Waste established in 1977, the role of organising an international co-operation, including radiological risk assessments and radiological protection inspection during the disposal operations, on a practice followed by several countries and which might otherwise be carried out without proper guarantees and information to all countries.

The other problem area in which NEA showed a prompt adaptation to the forthcoming requirements of Member countries was the spreading application of small quantities of radioactive materials in consumer goods of any sort. At that time the concern was focused on the individual risk from radiation exposure and the protection of individuals. Only a few, far-sighted experts were anticipating the far-reaching importance that the collective radiation detriment would take on in the following years and the resulting potential public health problems to be raised by the exposure of large groups of people to small levels of radiation. It is therefore to be ascribed to the merit of NEA the vigorous programme, carried out from 1962 to 1978, of studies and guidelines on the radiation protection problems raised by public exposure to consumer goods containing radioactivity and by natural, but artificially enhanced, radiation exposures. The NEA action in this field was in line with its institutional goal of transferring the philosophical language of the ICRP recommendations and the administrative/regulatory language of the basic radiation protection norms into terms more suitable for direct application to specific situations.

This activity started in 1962 with the preparation of radiation protection standards for radioluminous timepieces (published in 1967) and continued very actively with the publication (between 1970 and 1977) of standards or guidelines concerning the design, construction and use of radioisotopic power generators, gaseous tritium light devices, particle accelerators, cardiac pacemakers, smoke detectors. It is important to note that several of the above guides and standards were developed in collaboration with the other international organisations and some of them were also adopted by IAEA as their own standards.

A particularly interesting case of a guide in this

area was that entitled "Basic approach for safety analysis and control of products containing radionuclides and available to the general public" published by NEA in 1970. While the other guides were intended for use not only by regulatory authorities, but also by designers and utilisers, the latter was exclusively addressed to regulatory authorities. It contains the general principles of the analysis that licensing authorities should carry out before granting authorisations, as well as the technical bases for this analysis and the control procedures to be set up.

At this point, it is worth noting that, besides the basic radiation protection norms, the field of consumer goods is the only one where NEA embarked on an activity of international standards and guides. In the other fields, also in harmony with a general co-operation and co-ordination agreement signed with the IAEA in September 1960, the Agency focused its activity on the study and catalysation of co-operative efforts in advanced areas still presenting open problems.

The activity of NEA in the area of consumer goods and natural radiation was not limited to the publication of standards and guides, but also a number of studies were carried out. It is sufficient to mention the studies on radiation protection problems of lightning conductors, use of depleted uranium as ballast in aircraft, radioactivity of building materials, airborne natural radioactivity, etc. Not all of these studies were widely disseminated through formal publication. A few of them, in fact, remained, for various reasons, at the stage of reports for internal use by the national authorities. But even in this limited distribution, they represented a useful reference contribution to the development of knowledge and radiation protection criteria in Member countries.

THE PRESENT EVOLUTION AND FUTURE TRENDS

In the last ten years nuclear energy, which had previously shown promising development, began to face growing opposition as well as increasing economic and political difficulties. One of the concerns at the basis of this crisis was a deep and sometimes exasperated attention to the environmental impact of nuclear energy. This called for great efforts to be focused on the treatment and retention of radioactive effluents, as well as on the methods for assessment of the environmental behaviour and radiological impact of these effluents and for their monitoring. But the attention of the experts and the public opinion were progressively concentrating on the question of the long-term management of radioactive wastes. This immediately began to raise the problems of the assessment and criteria for acceptance of the associated long-term radiological risks. In this context, a particular international resonance was given to the publication by NEA, in 1977, of a state-of-the-art report, known as the Polvani report, on the concepts and strategies for the management of radioactive wastes.

During the 1970s, NEA rapidly became sensitive to these new requirements and its programme was progressively adapted by the Committee on Radiation Protection and Public Health to cope with the new challenges. In particular, the Committee reoriented the emphasis of the programme towards the problems of radiation protection and environmental impact of nuclear fuel cycle facilities, with special attention to the front-end (uranium mining and milling) and the back-end (waste management) of the fuel cycle. Its growing involvement in the problems associated with the nuclear fuel cycle obliged the Agency to concentrate its limited resources in those areas where a priority effort appeared warranted. Therefore, while increasing efforts were devoted to sensitive issues in the nuclear fuel cycle, the activities in areas not directly connected with the fuel cycle, such as the consumer goods and other low-level sources of radiation detriment, were progressively reduced to minimal levels.

The turning point of this evolving policy was in 1976. In that year, for the first time, the major part of the Agency's programme was devoted to radiation protection in the nuclear fuel cycle. Here again, the Agency fulfilled its vocation for tackling new problems in the moment of their formation and catalysing international interest and efforts towards their solution. In the last few years, in fact, the areas which called for attention were the transfer into practical terms of the new ICRP recommendations of Publication 26, the protection of the public and future generations against the long-term sources of radiation such as the uranium mill tailings and high level transuranic wastes, and the protection of workers in selected occupations with relatively high risks, such as uranium mining and maintenance work in nuclear power plants. Most of the new lines of activity started in 1976 were in fact focusing on selected issues in the abovementioned areas. In order to put into a proper perspective the radiological impact of the different stages of the nuclear fuel cycle and the problem areas requiring a selective effort, the first action of NEA was to sponsor a study on the relative radiological significance of all potential sources of human exposure to radiations. The report on this study, prepared by Sir Edward Pochin, was published in 1976.

In the field of the long-term impact of the nuclear fuel cycle, the criterion adopted was to select a few radionuclides of particular significance as potential long-term sources of human irradiation, and to study in detail their behaviour and their potential risk for present and future generations. A first study was focused on plutonium and other transuranics: this was of a scientific nature and its objective was to bring together the salient facts about the biological and environmental behaviour of these nuclides in order to assist in the appreciation of their potential risk. The result of this study is a comprehensive state-of-the-art report, publication of which is planned by mid-1980.

Another case of a specific study in this general area was an analysis of the radiological significance of four long-lived radionuclides arising, as airborne effluents, from the operations of the nuclear fuel cycle, namely ^{3}H, ^{14}C, ^{85}Kr and ^{129}I. The particular interest of this study lies in the fact that it is one of the first examples, perhaps the first from an international group of experts, of an attempt to demonstrate a practical application of the ICRP optimisation principle. The study, in fact, assesses the radiation detriment and costs associated with different retention technologies and combines them into a differential cost-benefit analysis, concluding with recommendations as to the policies to be followed for an optimised management of the above nuclides. In spite of a certain delay due to its industrial policy implications, it is hoped that the final report will be published during 1980.

Another issue of concern, both from the occupational exposure and environmental protection viewpoints, was the expansion in several countries of uranium mining and milling activities. The involvement of NEA in this field began in 1976 with the organisation of exchanges of information on personal dosimetry and area monitoring in uranium mines. The international debate progressed during the last three years with other meetings and seminars, within the framework of NEA, and permitted the clarification of the nature of the problems and identification of key subjects requiring a priority attention or effort. On this basis, the Agency was able to establish, in 1979, a consolidated programme on the long-term radiation protection and waste management aspects of the uranium mill tailings. These were, in fact, identified as a form of radioactive waste of low specific activity but having significant long-term implications due to their enormous quantities and very long half-life of the radionuclides involved. This three-year programme is presently starting and includes items such as the formulation of radiological protection principles and criteria for application to the long-term management of tailings, based on the ICRP system of dose limitation, as well as the study of environmental models for the assessment of the impact of the release of contaminants from the tailings.

Another area where the abovementioned international debate led to the establishment of a stimulating programme was the parallel subject of radon, and its daughters dosimetry and monitoring for workers and for environmental surveillance in connection with uranium mining and with mill tailings management. A sizeable effort is presently going to be devoted to studies on the analytical models for radon dosimetry, the influence of factors and parameters affecting the dose to lung from radon and its daughters, the principles and methods for measurement and monitoring of radon and daughters.

During the last few years, increasing improvements have been made in the safety of nuclear plants against accidents and the treatment and containment of radioactive wastes. However, these achievements were partly obtained at the cost of increasing radiation exposure to workers. Therefore, the concern of experts and authorities is now focusing on the trends of occupational exposure in nuclear plants, and no doubt this will continue to be a major issue during the next few years until adequate solutions are found and implemented. The achievement of this goal should be pursued applying the ICRP principle of optimisation. For this purpose, an adequate data base should be assembled, from which the criteria for optimising occupational exposure in the design and operation might be derived. In order to contribute to the above data base, NEA, in co-operation with IAEA, launched in 1978 a study on this subject based on an international enquiry aimed at collecting information on the levels and trends of occupational exposure in nuclear facilities and helping to identify critical groups of workers, critical operations and critical equipment in the plants, as a basis for the study of design and operational procedures improvements. This enquiry supplied a large amount of valuable data and the resulting study is very well advanced. Publication is envisaged for 1980.

Only the most important components of the current NEA programme have been briefly described here, but several other activities are also carried out or planned. These include studies and other forms of international co-operation in fields such as the development of practical examples of application of the ICRP optimisation principle, the preparation of state-of-the-art reports on the radiological and environmental protection aspects of the nuclear fuel cycle facilities, the development of radiation protection criteria for the geologic disposal of high level radioactive wastes, the study of radiation protection problems of the decommissioning of nuclear plants, the international review of nuclear emergency planning criteria and reference levels. This last item, which was already the object of a specialist meeting organised by NEA in 1976, has seen a renewed interest since the Three Mile Island accident in 1979 and is presently being considered by the Agency with a view to launching a programme of international co-operation in this area.

We have tried to give an overview of the evolution of the radiation protection problems and concerns in the last 25 years, in connection with the effort of NEA to assist Member countries in finding a timely and efficient solution to their national problems through a harmonised gathering of efforts and contributions at the international level. This effort has frequently been successful in spite of the obvious difficulties of debating certain sensitive matters at the international level, and we are confident that NEA will continue to be a useful contributor to international co-operation for the constant improvement of radiological protection.

THE RADIATION PROTECTION PROGRAMME ACTIVITIES OF THE WORLD HEALTH ORGANIZATION

E. Komarov and M. J. Suess

A number of World Health Assembly and the Executive Board resolutions have stressed the World Health Organization (WHO) responsibilities at the international level in respect of protection from radiation hazards, in collaboration with other interested organizations and societies in the radiation field, and in particular the IAEA, UNSCEAR, ICRP, IRPA etc. They have also called "the attention of Member States and Associate Members to the responsibility of their national health authorities in the protection of the population from radiation hazards" and emphasized inter alia the role of WHO in "encouraging and assisting" those authorities "to accept their major role in the public health aspects of radiation from all sources".

RADIATION PROTECTION STANDARDS AND GUIDELINES

The IAEA, WHO, ILO and OECD(NEA) are the four international organizations which have statutory obligations and responsibilities in the field of radiation protection standards. These organizations are now engaged in revising the Basic Safety Standards for Radiation Protection (IAEA Safety Series No. 9, 1967 edition) to implement new ICRP recommendation 26.

A topical seminar on the practical implications of the ICRP recommendations was held in March 1979, by the above-mentioned four organizations, to discuss the practical problems of the implementation of these recommendations. The joint IAEA/WHO Code of Practice on the Basic Requirements for Personnel Monitoring was revised and will be published in 1980. In August 1978 a joint WHO/IAEA/ILO meeting was held to discuss a code of practice on Radiation Protection in Mining and Milling of Radioactive Ores. The resulting document will be published in 1980.

PUBLIC HEALTH ASPECTS OF NUCLEAR POWER

The rapid development of nuclear power in developed countries since the early seventies and, especially, the plans of a number of developing countries to offset the high costs of oil and other fuels through the generation of nuclear energy, will pose new public health problems. Of particular concern is the obvious fact that the operation of nuclear power reactors requires an even higher level of training and competence in health and safety fields than has been considered adequate in highly developed countries a year ago. The strong diversity of opinions between the advocates of nuclear energy as a necessary alternative to power from fossil fuels and their opponents who wish to avoid nuclear power completely, makes it not only more difficult, but also more important for WHO as the lead international organization in health problems, to present a balanced view of health detriment of nuclear power and its alternatives, and other applications of radiation.

Special attention has been given in the WHO programme to the environmental and health aspects of nuclear energy production. In the joint IAEA/WHO publication Nuclear Power and the Environment, published in 1972, the main principle and philosophy of public health implications of nuclear energy production has been presented, and a statement was made that the nuclear power industry could operate safely for the general public and environment when all technical and control measures are taken and properly executed. To further advise public health authorities on this subject, a report on Health Implications of Nuclear Power Production was published by the WHO European Office in 1977. In the same series of activities, the European Office has held a meeting on the health aspects of transuranium elements, and will organize a meeting on the health implication of the handling of high-level radioactive waste.

The emergence of nuclear power as a significant component of energy systems in developing countries has created the need for extensive training of personnel, particularly of those to hold posts of responsibility in the various aspects of a nuclear programme at all stages of planning and implementation. In these countries, the scarcity of resources and the other pressing priorities of public health will render the task of public health authorities, even though limited to a few key ones, particularly important and at the same time difficult.

To assist these public health authorities, WHO is planning an interregional training course on public health aspects of national nuclear power programmes.

Radiation accidents are of major concern to national and international authorities in relation to nuclear power development and the increasing use of radiation in industry, agriculture and medicine. To cover this aspect, appropriate activities have been developed in close collaboration with the IAEA, ILO and FAO. In 1978, the IAEA/WHO manual "Early Treatment in Radiation Accidents" and a report on the Treatment of Incorporated Transuranium Elements have been published.

WHO has been running an extensive study on biological indicators of radiation injury, including the use of chromosome aberrations. Investigations performed in various countries have indicated the usefulness of this method for biological evaluation of radiation dose, by scoring the chromosome aberrations in lymphocytes of accidentally-irradiated persons.

The question of diagnosis and treatment of internal and external accidental exposure of persons has been dealt with extensively in terms of manuals and scientific meetings. Moreover, WHO is playing a prominent role in this connection because of its traditional competence and responsibilities in preventive and curative medical domaines.

To extend the service to member states WHO are now planning to establish three WHO collaborating centres on human radiopathology to serve in actual cases of human radiation injuries:
- one in Paris (Institute Curie, Department of Radiation Protection) for Member States in the African, Eastern Mediterranean and European Regions
- one in Oak Ridge University Medical Research Center; for the Americas
- one centre for Western Pacific and South East Asian Regions.

The terms of reference of centres would be:
  (i) to serve as focal points for advice and possible medical care in cases of human radiation injuries;
  (ii) to facilitate when necessary the establishment of a network of equipment and specialized staff in human radiopathology;
  (iii) to assist in the establishment of medical emergency plans for the event of large-scale radiation accidents;
  (iv) to develop and carry out coordinated studies on human radiopathology and epidemiological studies that may be appropriate;
  (v) to assist in the preparation of relevant documents and guidelines.

In addition, revision of the IAEA/FAO/ILO/UNDRO document on the mutual emergency assistance in radiation accidents is in preparation to provide information on the assistance that Member States might be able to make available at the request of another country.

SURVEILLANCE AND CONTROL OF ENVIRONMENTAL RADIOACTIVITY, ASSESSMENT OF POPULATION EXPOSURE AND HEALTH EFFECTS

Periodical reports prepared by 4 WHO collaborating centres in cooperation with 27 laboratories in 19 countries, are being provided to the Regions and Member States, which include the world data on radioactivity in air, water and food. Intercomparison of measurements of radioactivity in samples of milk and bones were maintained by collaborating centres, to improve the accuracy and comparability of data from various countries. Information on strontium-90 in human bones obtained from tropical area countries were made available to UNSCEAR and participating countries.

A WHO European working group has considered the problem of the acceptable levels of radionuclides in drinking water for the revision of WHO International Standards for Drinking Water. The report of the working group makes few significant changes in comparison with the previously published WHO International Standards. A significant departure from previous practice is the explicit statement that "Where these levels are exceeded, it is recommended that the competent authorities be required to decide what further action, if any, is necessary".

The radiological analysis of fresh water is dealt with in two chapters of a three-volume handbook on "Examination of Water for Pollution Control", sponsored by the WHO European Office and due for publication. Radiation exposure from naturally occuring Radon as one of the pollutants of Indoor Air has been discussed by a working group in Bilthoven in April 1979.

In the assessment and control of health risks from radiation exposure, the Organization also plans to study the exposure of the population due to radiation-emitting consumer products, technologically enhanced natural radiation, building materials etc.

WHO will assist the public health authorities in keeping the public currently informed on the likely health consequences of various uses of radiation. Promotion of education and training of medical personnel in radiation hygiene and protection is an important task of WHO in establishing contacts with populations in order to assure adequate health protection.

NONIONIZING RADIATION

The very high potential for cancer and genetic damage from ionizing radiation in comparison to a high tolerance for less energetic forms of radiation have led most health officials to disregard non-ionizing radiation (NIR). However, NIR, which ranges from ultraviolet through visible light, infrared and microwaves to radiofrequency radiation, as well as electomagnetic fields, is the result of a fast developing technology with a growing potential for affecting public health. To draw the attention of public health to this problem, in 1972 a special chapter on non-ionizing radiation was included in WHO monograph "Health hazards of the human environment"; and an International symposium on biological effects and health hazards of microwave radiation was jointly organised by WHO and the health authorities of the USA and Poland in 1973, with the participation of an IRPA representative. The WHO European Office has organized a series of meetings on various aspects of NIR which will result in the publication of a manual on NIR protection. The WHO/UNEP/IRPA environmental health criteria document on ultraviolet radiation was published in 1979, and the document on microwave and radiofrequency radiation has been presented to the publisher.

New environmental health criteria documents on lasers and ultrasound are at present in preparation, as joint activities with IRPA.

— : —

WHO resources for all radiation protection activities are limited but considerable support has been received from WHO collaborating centres and national authorities, e.g. Belgium's and the Federal Republic of Germany's contributions.

Most of these programmes are being carried out in close cooperation with other international organizations and bodies such as the International Atomic Energy Agency, the International Labour Organisation, the United Nations Scientific Committee on the Effects of Atomic Radiation, the International Commission on Radiological Protection and the International Radiation Protection Association.

REVIEW OF ACTIVITIES OF INTERNATIONAL ORGANIZATIONS THAT COOPERATE WITH IRPA

K. Z. Morgan

As we near the close of this Fifth International Congress of IRPA, I would like to point out that maybe it was no accident that the first meeting of IRPA was held in the eternal city of Rome, the home of the religion of the majority of our members who are Christians. With many of us, health physics is like a second religion where, instead of one God, we have one goal, viz. to reduce the harmful effects of radiation while enhancing its benefits.

I opened our first congress and in the years that followed other presidents were Doctors Marley, Polvani and Palmiter and now, for the next term, Dr Nooteboom-Beeckman will provide us with leadership. Like the children of Israel, and the trials and tribulations of Moses in search of the promised land, some of our great leaders have faithfully and fearlessly lead us out of the wilderness and into an unknown land of promise if we properly sort out the good from the evil, and then they left us to move ahead in our quests for the truth while they walked through the valley of the shadow of death. I respectfully mention the names of a few of these - Rolf Sievert, Paul Bonet-Maury, Jim Hart, Lia Forti, Walter Snyder and Yehuda Feige.

This meeting, so ably organized by Dr Tuvya Shlesinger and his co-workers in Jerusalem, the birthplace of three great religions - Judaism, Christianity and Mohammadism - symbolizes our working together as individuals but members of many radiation protection societies and as individuals of strong opinions and often with conflicting convictions. We do not see eye-to-eye on how best to reach this promised land. For example, I and many other members of IRPA believe the cancer risk of low level radiation exposure is far greater than do others whose views were espoused by Dr Lauriston Taylor in his Sievert Lecture delivered earlier in this Congress. But as the three religions in this great city of Jerusalem, we will continue to work together in accomplishing our goal of reducing unnecessary radiation exposure and look forward to our next meeting with the Fachverband fur Strahlenschutz in 1984 in Berlin, which is near where Protestantism had its beginning.

As we impatiently push ahead, progress of IRPA seems so slow, but when we glance back over our shoulders, we realize we have made some important advances, only to mention a few: 25 affiliated societies; 5 cooperating international agencies; leadership in standards setting in the field of non-ionizing radiation; our first woman president, Mrs Dr Nooteboom-Beeckman, elected at this Congress. We expect

that before our next congress the Iran Health Physics Society will be an
affiliated IRPA member. Iran is predominantly a country of Mohammedan faith,
the third great religion of Jerusalem.

I do not wish to leave the impression that you must belong to one of these
religions to be a health physicist, but one must have the cooperative spirit,
the zeal and a religious-like fervor for good, hard, honest work, to be a leader
in this profession. Like our religion, our health physics profession demands the
highest level of honesty, a continuous unselfish striving for the well-being of
others, and demands a responsibility to children yet to be born, and an
unshakeable faith in the future.

*Reports of
International Committees*

SUMMARY ACCOUNT ON THE ACTIVITIES OF THE IRPA/INTERNATIONAL NIR COMMITTEE

H. Jammet

Since its setting up in 1977, the Committee has developed its activities in two directions: on the one hand, its efforts have been directed to extending its membership and setting up its internal organization; on the other hand, it has undertaken to carry out scientific activities according to the objective which was assigned to it.

ORGANIZATION OF THE COMMITTEE

In the first place it seemed reasonable to extend the Committee membership up to 12 members. These are selected chiefly on the basis of their recognized activity in the different fields of the NIR spectrum, and of their expertise in appropriate disciplines (health physics, biology, physics, medicine and engineering) without forgetting the need for appropriate representation of the different scientific trends. At the present time, the Committee membership amounts to 9, no answer having been received from three other invited experts. Thus, the Committee membership now is as follows: H.Jammet (Chairman, France), B.F.M.Bosnjakovic (Netherlands), P.Czerski (Poland), M.Faber (Denmark), G.Kossof (Australia), M.H.Repacholi (Canada), D.H.Sliney (USA), J.C.Villforth (USA), G.M.Wilkening (USA).

Because of the very different physical, biological and technical problems associated with the different types of NIR, it would have been best to study these various aspects for each kind of NIR in a specialized Subcommittee, as previously proposed by the Study Group. However, such a distribution of the work would have required more important financial means. For the time being, therefore, the Committee has considered that it should adopt a more flexible way of working. This consists of distributing specific topics among the Committee members, around which small working groups will be set up, either within the Committee itself, or in co-operation with other organizations.

SCIENTIFIC ACTIVITIES AND WORKING PROGRAMME OF THE COMMITTEE

In order to prepare appropriate recommendations on protection against NIR, the necessary first step consists of collecting and analyzing all available data on the interaction of the different NIR with living matter, their biological effects, exposure sources and levels, the possible means of measurement, the quantities and units used, as well as the protection standards and regulations which are already applied or foreseen.

The health criteria documents for the different NIR are prepared by WHO/IRPA joint task groups under the chairmanship of a Committee member. The joint work has been in progress since the end of 1977 and the Committee has actively contributed to it.

The first document which deals with ultraviolet radiation was issued in December 1979. The document on microwave and radio frequency radiation is near completion and will be published towards mid-1980, or in the early fall. With regard to lasers and to ultrasound, work is under way and it is anticipated that the documents will be completed in 1983. Finally, it is also anticipated that the document relating to extremely low frequencies (ELF) will be ready in 1983.

The health criteria thus defined will serve as a basis for the Committee to determine, in each case, the fundamental and operational limits best suited to the protection of workers and of the general public. Meanwhile, however, preliminary reports dealing with this topic will be prepared for discussion at the next Committee meeting in 1980. As a matter of fact, the Committee hopes that between 1981 and 1983 it will be able to make some recommendations relating to the different NIR. Furthermore, as it has often been deplored that there is lack of general policy of protection against NIR, which could serve as a basis for a coherent system of exposure limitation in the field of NIR, the Committee intends to prepare a document on general principles of protection against NIR.

Finally, the Committee considered that some guidelines for operational protection and codes of practice relating to the different NIR should be established. From a review of the quantities and units proposed for NIR by different organizations and mainly the International Commission on Radiological Units and Measurements, considering the evident need for harmonization in this field, the Committee decided to prepare a report on this topic, which it hopes to complete in 1983.

RELATIONS WITH INTERNATIONAL ORGANIZATIONS

As mentioned before, the Committee first established working relations with the WHO as early as November 1977. The WHO is interested in all aspects relating to health, with respect to the workers as well as to the general public, including also environmental pollution.

The International Labour Organization, the activities of which are focussed on the protection of workers, listed NIR among its priorities a few years ago. The ILO attended the Committee's first consultations with the WHO and expressed the wish to be informed on the work of the Committee on protection standards for workers.

According to a former agreement between the IRPA and the URSI (International Radioscientific Union), the Committee is co-operating with URSI's Committee A Working Group on measurements related to the interaction of electromagnetic field with biological systems.

In preliminary formal contacts, the ICRU has expressed its willingness to co-operate with the Committee on the quantities and units to be used for NIR. Furthermore, liaisons are established with various international organizations in which certain Committee members assume some responsibilities.

FINANCIAL ASPECTS

In order to achieve its working program, the Committee considered it essential to be able to meet each year, for 5 to 7 days, to discuss the draft reports prepared by Committee members or working groups. Therefore Committee members have asked

Summary Account

to inform the Executive Council and the General Assembly about their concern with respect to the Committee's financial means.

In the past, the parent institutions of a few members did agree to support their travel. Furthermore, some meetings of the Committee, which was then smaller, were held concomitantly with a task group meeting, which meant at no cost to the Committee.

However, in the future the Committee's own activities should prevail over the joint activities with the WHO. The cost of an annual meeting would be too heavy for IRPA. In this case, there would be a need to obtain financial support from other sources through the IRPA Executive Council. Therefore, the Committee would be very grateful for any suggestion concerning this matter.

I hope that this short account has given you an idea of the importance of the three-year programme of the IRPA/INIRC and of the means that it requires.

So far, IRPA's NIR Committee has passed a period of "running in". The Committee does hope, now, to be in a position to produce tangible results for the benefit of both IRPA's membership and the communities working for the different fields in NIR.

ACTIVITES DU COMITE INTERNATIONAL DE L'IRPA ET PROBLEMES POSES PAR LA PROTECTION CONTRE LES RAYONNEMENTS NON IONISANTS

H. Jammet

Le Comité International sur les Rayonnements Non-Ionisants (CIRNI) a été créé, comme vous le savez, par l'IRPA en Avril 1977 à la suite du rapport établi par le Groupe d'Etudes que le Conseil Exécutif avait chargé d'étudier la situation dans le domaine de la protection contre les rayonnements non-ionisants (RNI).

Les principaux objectifs fixés au Comité furent :

- d'élaborer des documents de base et des recommandations qui seraient acceptées sur le plan international;
- de prospecter, avec d'autres organisations internationales si nécessaire, les voies et les moyens permettant de faire progresser la protection contre les RNI et de collaborer, en particulier, avec l'Organisation Mondiale de la Santé pour l'élaboration de critères fondamentaux de protection.

Le Comité se mit immédiatement au travail et, malgré des moyens financiers extrêmement réduits, arriva à tenir au moins une réunion par an.

Activites du Comite International de l'IRPA

Les trois années qui viennent de s'écouler depuis sa création ont été consacrées par le Comité d'une part à des tâches d'organisation, d'autre part à déterminer le champ de ses activités, fixer des priorités compte tenu de l'importance des problèmes posés par les différents types de RNI, rassembler les données de base et participer à l'élaboration des premiers critères sanitaires pour les rayonnements considérés comme prioritaires.

CHAMP D'ACTION DU COMITE

Il convenait tout d'abord de définir exactement le domaine d'action du Comité.

D'une manière générale, le terme de rayonnements non-ionisants se réfère à tous les types de rayonnements qui, lors de leur interaction avec la matière, ne peuvent céder une énergie suffisante pour produire une ionisation. Le Comité a décidé de prendre pour ligne de démarcation entre les rayonnements ionisants et les rayonnements non-ionisants une énergie des photons égale à 10 eV, ce qui correspond à une longueur d'onde d'environ 10^{-7}m. Les rayonnements non-ionisants comprennent tous les rayonnements électromagnétiques ayant une longueur d'onde égale ou supérieure à 10^{-7}m, c'est-à-dire : le rayonnement ultraviolet, la lumière visible, le rayonnement infrarouge et les ondes utilisées dans les télécommunications depuis les microondes jusqu'aux ondes radio les plus longues. A des fins de protection, le domaine des RNI est généralement étendu aux champs électrostatiques et magnétostatiques. D'autre part, tout en excluant de son domaine de préoccupation les ondes sonores dont les effets biologiques et les moyens de protection sont connus depuis longtemps, le Comité, suivant en cela les recommandations du groupe d'études, a inclus dans son champ d'action les ultrasons dont les problèmes de protection, très similaires à ceux des rayonnements électromagnétiques, n'ont été pris spécifiquement en charge, jusqu'à présent, par aucun organisme international.

ACTIVITES DU COMITE

La détermination de limites d'exposition qui soient acceptables sur un plan international, principal objectif du Comité, exige tout d'abord une analyse approfondie de toutes les données physiques, biologiques et réglementaires existant pour chaque type de rayonnement considéré ainsi qu'une bonne connaissance des sources et des niveaux d'exposition rencontrés en pratique.

Le but d'un tel travail est de dégager les bases scientifiques et techniques sur lesquelles pourront être fondées des limites d'exposition valables et réalistes. De même que dans le domaine des rayonnements ionisants, il s'agit là d'une tâche extrêmement lourde, d'autant plus que, comme nous allons le voir, les problèmes tant physiques que biologiques posés par les différents RNI sont loin d'être résolus.

En premier lieu, il fallait donc collecter les données. Dans un souci d'efficacité et suivant les recommandations du Conseil Exécutif et de l'Assemblée générale de l'IRPA, des contacts furent établis avec l'Organisation Mondiale de la Santé qui avait entrepris un programme de recherches en ce domaine et créé deux centres d'études sur les rayonnements non-ionisants respectivement aux USA et en Pologne. Ainsi, dès la fin de 1977, un accord de coopération pour la préparation de documents sur les critères sanitaires relatifs aux différents RNI put être établi entre l'IRPA et l'OMS dans le cadre du programme pour l'environnement financé par les Nations Unies.

Ce travail est actuellement en cours. Le premier document sur le rayonnement ultraviolet a été publié en décembre 1979. Le document sur les radiofréquences est dans son stade final de préparation, celui sur les lasers vient d'être entrepris et sera suivi de près par le document sur les ultrasons.

PROBLEMES POSES PAR LA PROTECTION CONTRE LES RNI

Nous allons maintenant passer rapidement en revue les principaux problèmes que posent sur le plan de la protection les rayonnements non-ionisants.

INTERACTION DES RNI AVEC LA MATIERE

Encore plus mal connu que pour les rayonnements ionisants, le mécanisme d'interaction des rayonnements non-ionisants et de la matière vivante est extrêmement complexe. Comme pour tous les rayonnements électromagnétiques, la profondeur de pénétration des RNI dépend en premier lieu de l'énergie des photons incidents. Cependant, contrairement à ce qui se passe avec les rayonnements X et γ où la profondeur de pénétration croît lorsque l'énergie des photons augmente, ici, en règle générale, depuis l'infrarouge jusqu'aux ondes radio les plus longues, la profondeur de pénétration a tendance à croître lorsque la fréquence et l'énergie des photons diminuent. Toutefois elle est extrêmement influencée par la nature du tissu rencontré. Lorsque les rayonnements électromagnétiques passent tout d'abord de l'air dans le milieu biologique, puis d'une couche de tissu à l'autre, ils peuvent être réfléchis, réfractés, transmis ou absorbés selon les constituants et la structure de la matière biologique et selon la fréquence (ou longueur d'onde) des rayonnements. Des phénomènes de diffusion et de résonance interfèrent avec les lois simples d'absorption de l'énergie et de sa conversion en chaleur.

Ainsi, dans le cas de l'ultraviolet la pénétration est limitée à l'épiderme et aux couches superficielles du derme. L'absorption dépendra des constituants du tissu, protéines et ADN étant des absorbeurs particulièrement importants. Tout le monde connaît, par exemple, l'importance pour la protection contre les coups de soleil de la présence, dans la couche cornée extérieure de la peau, d'un pigment appelé mélanine, protéine macromoléculaire complexe qui a la propriété d'absorber fortement la lumière et le rayonnement ultraviolet. Cependant, en raison de la complexité des phénomènes produits, il n'a pas été possible jusqu'à présent d'élaborer un modèle théorique quantitatif rendant compte de façon satisfaisante de la pénétration et de la réflexion des rayonnements optiques dans la peau. Dans l'oeil, le rayonnement ultraviolet est absorbé au niveau de la cornée et du cristallin où il peut interférer avec la synthèse de certaines protéines, et entraîner des modifications chromatiques. Il a, d'autre part, été clairement démontré que des modifications cataractogènes pouvaient être induites dans le cristallin.

Lorsque les rayonnements optiques se présentent sous forme d'un faisceau cohérent et monochromatique, les lasers, les mécanismes généraux sont de trois types : thermiques, mécaniques et électromagnétiques. Les effets thermiques sont les plus importants et les mieux connus, sauf en ce qui concerne la distribution de l'énergie cédée au niveau des différents tissus le long de la trajectoire du rayonnement. On sait encore peu de chose, par contre, sur les effets mécaniques dus aux variations de pression liées à la cession très rapide de l'énergie et sur les modifications du champ électromagnétique, difficiles à analyser.

Dans le cas des microondes et des ondes courtes, l'action des champs électriques dépend étroitement des propriétés diélectriques et de la conductivité des milieux biologiques. La cession d'énergie se traduit surtout par la production de chaleur, mais l'existence de structures moléculaires complexes et variées peut aboutir à des dépôts d'énergie répartis de façon non-uniforme et à des gradients de température importants. Des considérations théoriques et certains résultats expérimentaux montrent cependant que la conversion en chaleur n'est pas le seul mode de cession d'énergie. Des interactions au niveau de certaines molécules biologiques complexes donnent lieu à divers phénomènes qui peuvent être la cause de perturbations dans les fonctions plus ou moins importantes des structures moléculaires. De nombreuses hypothèses ont été faites sur les mécanismes primaires de ces interactions qui sont encore loin d'être parfaitement connus.

Pour les ultrasons, trois principaux types de phénomènes jouent un rôle dans l'interaction avec le milieu vivant : l'absorption, la réflexion et la cavitation. L'absorption correspond au transfert de l'énergie mécanique portée par le faisceau aux édifices moléculaires. La majeure partie de cette énergie est convertie en chaleur. L'absorption a lieu essentiellement au niveau des macromolécules et des interfaces des tissus. Le coefficient d'absorption varie avec la fréquence des ultrasons (puisqu'il est proportionnel au carré de cette fréquence) et avec la nature du milieu traversé. Des pénomènes de réflexion des ultrasons apparaissent aux interfaces, lorsque les tissus présentent des différences de densité (coefficient de réflexion à l'interface os/cerveau = 0,6, muscle/graisse = 0,08, foie/sang = 0,03). C'est le phénomène de réflexion qui est utilisé en échographie. Lorsque les interfaces sont en mouvement (vaisseau/sang, par exemple), il y a apparition d'un effet Döppler, c'est-à-dire variation de fréquence des ondes se réfléchissant sur une cible en mouvement. Cet effet a des applications diagnostiques importantes, en particulier lorsqu'il s'agit de mesurer la vitesse d'écoulement du sang par exemple. Le processus d'atténuation par cavitation se produit lui au niveau des liquides biologiques. Ces derniers contiennent des bulles microscopiques ou submicroscopiques (gaz du sang, vapeur d'eau qui, sous l'action de l'onde ultrasonore, tendent à grossir jusqu'à une taille déterminée par la fréquence, la pression, les amortissements thermiques et visqueux. Les bulles, une fois formées, peuvent soit rester stables et se comporter comme des cavités résonnantes dont les vibrations peuvent provoquer des modifications dans le tissu biologique, soit s'effondrer brusquement en entraînant des ruptures dans les structures environnantes. Il semble cependant que ce phénomène ne se produise que dans un domaine de fréquence déterminé que l'on évite avec les fréquences actuellement utilisées en médecine.

On voit donc que, pour l'ensemble des RNI, des études nombreuses et approfondies sont encore nécessaires pour éclaircir aussi bien qualitativement que quantitativement les mécanismes primaires d'interaction et notamment leur relation avec longueur d'onde ou la fréquence.

METROLOGIE DES RNI

Pour le physicien de santé accoutumé au contrôle des rayonnements ionisants, il est tentant de chercher à transposer les concepts, les méthodes voire les appareils de la dosimétrie des rayonnements ionisants à celle des RNI.

En fait cette transposition, si elle se révèle souvent très utile, est aussi très limitée.

Utile, par exemple, en incitant les photobiologistes et les radiobiologistes à unifier leurs définitions des grandeurs caractérisant un champ de rayonnements ionisants ou non-ionisants. Ainsi, en 1975, certaines grandeurs avaient été proposées à l'issue d'une réunion groupant des représentants de la Commission Internationale de l'Eclairage, de l'Association Internationale de Photobiologie, de l'Union Radioscientifique Internationale et de la Commission Internationale des Unités et Mesures Radiologiques au Bureau International des Poids et Mesures à Sèvres (France). Si actuellement l'agrément est loin d'être unanime sur ces définitions, le besoin d'unification est néanmoins évident.

Cette transposition est également très limitée par le fait même que la notion de dose absorbée qui désigne une grandeur généralement mesurable dans le cas des rayonnements ionisants n'a, le plus souvent, pas son équivalent dans le cas des RNI où elle ne peut généralement être que le résultat d'évaluations indirectes très sophistiquées.

Dans tous les domaines des rayonnements non-ionisants nous allons retrouver cette dualité : des champs de rayonnement relativement bien mesurés en l'absence de l'individu irradié, mais une grande difficulté d'évaluer l'énergie absorbée par celui-ci lorsqu'il pénètre dans ce même champ, surtout si cette pénétration se fait dans des conditions imprévues, hors d'un laboratoire spécialisé dans ce genre de mesure!

Rayonnements ultraviolets

Dans la gamme actinique (200-315 nm) qui correspond aux rayonnements responsables des principaux effets biologiques nocifs toutes les longueurs d'onde n'ont pas la même efficacité : la mesure d'énergie radiante doit être accompagnée d'une détermination du spectre de façon à pondérer les fluences monoénergétiques par un coefficient d'efficacité relative S très analogue à l'EBR des rayonnements ionisants. Dans ce cas, c'est le rayonnement de 270 nm de longueur d'onde qui sert de référence pour définir l'efficacité unitaire.

Les appareils de mesure du champ de rayonnement sont soit des débitmètres comprenant un détecteur photoélectrique (photodiode ou photomultiplicateur) associé à un filtre qui en modifie la réponse spectrale en fonction de l'efficacité S_λ, soit des intégrateurs tels qu'un dosimètre chimique ayant autant que faire se peut une réponse spectrale voisine de celle de l'érythème cutané.

La dosimétrie individuelle des rayonnements ultraviolets est sans doute le seul domaine des RNI qui puisse bénéficier, dans un proche avenir, des retombées de la dosimétrie des rayonnements ionisants. En effet, un certain nombre de dosimètres thermoluminescents ou photoluminescents présentent une sensibilité spécifique aux ultraviolets.

Rayonnements de haute fréquence et microondes

Dans ce domaine la mesure se heurte au problème de la perturbation apportée dans le champ de rayonnement par la personne irradiée, par celle qui effectue la mesure ou par l'appareil de mesure lui-même.

C'est pourquoi les constructeurs s'efforcent de faire des sondes non seulement isotropes mais aussi peu réfléchissantes que possible, allant jusqu'à supprimer les conducteurs métalliques pour les remplacer par des fibres optiques.

Dès qu'il s'agit de mesurer l'énergie absorbée par l'organisme exposé, la difficulté s'accroît du fait que sa forme, sa position dans le champ, la composition des tissus modifient l'absorption des rayonnements.

On a ainsi été amené à introduire la notion de TAS : taux d'absorption spécifique (specific absorption rate, SAR), qui exprime la puissance absorbée dans l'organisme par unité de masse pour un débit de fluence énergétique (densité de puissance) donné.

Le TAS est généralement déterminé par calorimétrie en mesurant l'élévation de température dans les tissus. Des relations empiriques permettent de déterminer le TAS en fonction du débit de fluence énergétique en tenant compte de la taille et du poids du sujet irradié. Pour illustrer la différence que cette notion implique avec la dosimétrie des rayonnements ionisants, considérons par exemple un rayonnement de 10 MHz dont le débit de fluence énergétique est 10 mW.cm^{-2} : un rat (TAS $\sim 4.10^{-3}$) y absorbera 4 mW/kg, un homme (TAS $\sim 0,02$) 20 mW/kg pour la même exposition, alors que dans un même faisceau de γ ils auraient reçu sensiblement la même dose.

Inutile d'ajouter que les déplacements d'un individu dans le champ de rayonnement modifient considérablement l'énergie absorbée.

C'est pourquoi, il serait extrêmement utile de disposer d'un dosimètre intégrateur individuel qui jouerait le même rôle que le film dosimètre pour les rayonnements ionisants. Des efforts ont été faits en ce sens. Cependant, un tel dosimètre, même s'il n'était pas trop perturbé comme on peut le craindre, par le moindre gradient thermique, ne peut donner qu'une indication très grossière de la dose biologiquement significative : en effet le corps d'un individu exposé à des microondes peut présenter des focalisations internes aboutissant à une distribution hétérogène de la dose, même si l'exposition est parfaitement homogène. L'utilisation de méthodes de thermographie infrarouge qui pourrait éventuellement donner une image de cette distribution est actuellement étudiée sur fantôme.

Il est donc très difficile de savoir comment un champ incident va se distribuer à l'intérieur d'un être vivant et les progrès qui restent à faire en dosimétrie sont l'un des points cruciaux notamment pour la protection des travailleurs.

Ultrasons

Des appareils de laboratoire utilisant différents types de capteurs peuvent être utilisés pour la mesure de l'énergie rayonnée par les émetteurs industriels ou médicaux auxquels les personnes peuvent être exposées.

Des appareils de contrôle tels que des hydrophones piézoélectriques miniaturisés sont maintenant commercialisés.

Si l'on a donc quelques moyens de mesurer une intensité ultrasonore, on n'en a pratiquement aucun pour enregistrer l'énergie absorbée par un individu.

Un effort dans ce sens a cependant été fait par la mise au point d'un capteur piézoélectrique constitué d'un polymère organique dont l'absorption est plus voisine de celle des tissus que celle du quartz ou des céramiques habituellement utilisées. On retrouve donc, dans un domaine pourtant très éloigné du leur, le souci de "l'équivalence aux tissus" bien connu des dosimétristes des rayonnements ionisants.

EFFETS BIOLOGIQUES

Il serait trop long de passer ici en revue les effets biologiques très variés des différents types de RNI et les très nombreux problèmes qui se posent encore à leur sujet. Je n'en résumerai donc simplement que les principaux caractères.

Bien que l'effet fondamental ne soit plus ici l'ionisation, les altérations subies par les molécules du fait des différentes interactions que nous venons de voir peuvent, comme pour les rayonnements ionisants, conduire à des modifications fonctionnelles ou structurelles, à des mutations et à la mort des cellules. La restauration sera plus ou moins importante selon l'ampleur et la nature des dégâts produits et l'effet final pourra soit passer inaperçu, soit donner lieu à une lésion décelable.

Activites du Comite International de l'IRPA

La nature, l'ampleur et l'importance physiologique des effets produits par les différents types de RNI sont extrêmement variables et dépendent de facteurs encore plus nombreux, semble-t-il, que dans le cas des rayonnements ionisants. Si l'énergie des photons incidents, autrement dit la longueur d'onde ou la fréquence du rayonnement, est primordiale puisqu'elle conditionne leur profondeur de pénétration dans les tissus et leur absorption préférentielle par tel ou tel type de structure moléculaire, l'intensité du champ, le mode d'émission du rayonnement (cohérent ou non cohérent, continu ou pulsé), les dimensions et la configuration géométrique du sujet, la constitution cellulaire des tissus irradiés et souvent même certaines caractéristiques physiologiques, comme l'irrigation sanguine des tissus, seront des facteurs non moins importants. Or, bien que depuis quelques années l'intérêt pour les effets biologiques des RNI se soit beaucoup développé et que des recherches de plus en plus nombreuses soient entreprises sur ce sujet à travers le monde, le rôle précis joué par ces différents paramètres, leur interférence et plus particulièrement leur relation quantitative avec l'effet biologique produit sont encore très mal connus.

Les problèmes posés et les difficultés rencontrées par les biologistes sont tout à fait analogues à ceux que l'on connaît dans le domaine des rayonnements ionisants. En dehors des paramètres mentionnés plus haut, la gravité des effets dépendra de l'étendue de l'irradiation (irradiation partielle ou irradiation totale) de la sensibilité et de l'importance des organes atteints. D'une manière générale, on retrouve ici les deux types d'effets : stochastiques et non-stochastiques. Certains effets non-stochastiques sont relativement bien connus, comme par exemple érythème et brûlures au niveau de la peau ainsi qu'effets oculaires (conjonctivite, brûlures rétiniennes, cataracte) pour les rayonnements optiques, effets résultant de l'échauffement des tissus, couramment appelés effets thermiques, dus aux microondes à débit de fluence énergétique élevé. Dans quelques cas, il a même été possible d'en déterminer les seuils d'apparition, dose érythème pour les ultraviolets, par exemple.

Par contre, de grandes incertitudes subsistent sur les effets non thermiques qui pourraient être produits par les microondes en particulier aux faibles débits de fluence énergétique. Les résultats obtenus par divers chercheurs en France, aux Etats-Unis et en URSS notamment, semblent en effet montrer que dans certaines gammes de fréquences les ondes électromagnétiques peuvent induire des effets biologiques, qui ne sont pas dus à un simple échauffement, au niveau des systèmes nerveux, endocrinien et immunitaire, par exemple. D'autre part, il semble maintenant établi qu'il existe un intervalle de débit de fluence énergétique, appelé "fenêtre", au-dessous duquel peuvent apparaître certains effets produits par les microondes à faible débit de fluence ; à l'intérieur de cet intervalle, ces effets disparaissent et au-delà on verra se manifester les effets thermiques. Toutefois, ni le mécanisme d'induction de ces effets non-thermiques, ni leurs conséquences pathologiques éventuelles pour l'homme n'ont été éclaircis. Les difficultés pour extrapoler à l'homme les résultats obtenus sur des animaux de laboratoire relativement petits sont en effet encore bien plus grandes qu'avec les rayonnements ionisants, aussi bien en ce qui concerne l'aspect physique de l'interaction entre le champ de rayonnement et le sujet exposé que du point de vue des effets biologiques produits.

Quant aux effets stochastiques (induction de cancers, effets génétiques), si l'on sait bien qu'ils existent dans certains cas (cancer de la peau dû au rayonnement ultraviolet), on n'en est encore qu'au stade des recherches préliminaires pour les autres types de RNI. Ainsi, il semble qu'au cours de quelques expériences faites avec des ultrasons on aurait observé l'induction d'aberrations chromosomiques dans des cellules ces résultats toutefois doivent encore être confirmés.

Enfin, il faut noter que, sauf dans le cas de certains effets dus aux rayonnements optiques, on manque à peu près totalement de données sur les relations quantitatives entre l'exposition aux différents types de RNI et la réponse de l'organisme.

NORMES DE PROTECTION : TENDANCES ACTUELLES

Méconnus, comme la plupart des autres nuisances, les rayonnemen non-ionisants, à l'exception peut-être du rayonnement ultraviolet, n'avai pas jusqu'à ces 15 dernières années suscité beaucoup d'intérêt chez tous ceux qui pouvaient être concernés par la protection sanitaire des travail leurs et du public. Aussi, contrairement à ce qui s'est passé pour les rayonnements ionisants, l'essor de leurs applications scientifiques médicales et industrielles n'a-t-il pas été accompagné dès le début de mesures de protection appropriées formulées sur un plan national et encore moins sur un plan international. S'il est vrai que des informations supplémentaires sont encore nécessaires pour établir scientifiquement des limites d'exposition couvrant à la fois les effets aigus et les effets à long terme pour les travailleurs comme pour les personnes du public, la situation a néanmoins changé et des réglementations sont adoptées dans un nombre croissant de pays, en particulier dans le domaine des microondes et des hautes fréquences. En l'absence de recommandations faites par un organisme dont l'autorité serait reconnue sur le plan international, les valeurs utilisées comme limites sont souvent très différentes d'un pays à l'autre.

Rayonnements optiques

Dans le cas des rayonnements optiques, on s'est préoccupé depuis assez longtemps de l'exposition au rayonnement ultraviolet. Certaines limites ont été proposées pour l'exposition professionnelle dont les plus complètes sont celles recommandées par l'American Conference of Governmental Industrial Hygienists (ACGIH), lesquelles comprennent des valeurs différentes selon la longueur d'onde des UV et la durée d'exposition. Elles ont été adoptées depuis par le National Institute for Occupational Safety and Health (NIOSH) des Etats-Unis et également recommandées par le National Radiological Protection Board (NRPB) en Grande-Bretagne. Toutefois, ces limites ne sont fondées que sur les effets non-stochastiques et, en l'absence de données sur la relation dose-réponse, l'on n'a pas tenu compte du risque d'induction de cancers.

Des normes pour la protection contre les lasers ont été recommandées par un certain nombre d'organismes notamment aux Etats-Unis et en Grande-Bretagne. S'appuyant sur les mêmes données, la Commission des Communautées Européennes prépare actuellement une recommandation à ce sujet. Les limites proposées ne sont fondées que sur les effets immédiats et irréversibles au niveau de la peau et de l'oeil considérés comme étant les organes critiques. Elles forment un tableau complexe de valeurs qui, pour chacun des deux organes critiques, varient en fonction de la longueur d'onde utilisée ainsi que de la durée et de la fréquence de répétition des impulsions. De grandes incertitudes subsistent sur les facteurs correctifs

à introduire pour tenir compte d'éventuels effets cumulés lors de la
répétition des impulsions ou en cas d'exposition chronique ou répétée
à des lasers de faible puissance.   Un effort important pour harmoniser la protection au niveau de la source est actuellement fait par la
Commission Electrotechnique Internationale qui cherche à réaliser un accord
international sur une classification des différents appareils lasers en
fonction de leurs risques et des spécifications techniques appropriées.

Microondes et hautes fréquences

Dans un nombre restreint de pays, il existe déjà une réglementation au plan national pour limiter l'exposition des travailleurs, et parfois du public, aux microondes et aux hautes fréquences. Dans d'autres, des limites ont été recommandées par des organismes spécialisés ou sont en vigueur dans l'armée. Mais les valeurs adoptées sont très disparates, car les critères fondamentaux qui ont servi à les établir sont en général très différents d'un pays à l'autre.

Aux Etats-Unis et dans un certain nombre de pays occidentaux, les limites utilisées sont fondées uniquement sur les effets dits thermiques qui résultent d'un échauffement excessif des tissus, alors qu'en U.R.S.S. et dans d'autres pays de l'Europe de l'Est on tient compte également de la possibilité d'existence de certains effets neuro-végétatifs. Il en résulte des différences jusqu'à un facteur 1000, puisque pour l'exposition continue des travailleurs, par exemple, elles sont comprises entre 0,01 et 10 $mW.cm^{-2}$, et même plus dans le cas de l'exposition du public (0,001 à 10 $mW.cm^{-2}$) (tableau 1). De même, des différences importantes existent dans les valeurs d'exposition admises pendant des temps courts et les plafonds absolus introduits par quelques pays (10 à 55 $mW.cm^{-2}$) pour une irradiation de l'ordre de 2 minutes environ.

Actuellement, cependant, on assiste à une certaine évolution. D'une part, du fait des résultats obtenus au cours de certaines recherches, il apparaît de plus en plus probable qu'à certaines fréquences, tout au moins, les ondes électromagnétiques peuvent induire des effets biologiques qui ne sont pas dus à un simple échauffement, au niveau, par exemple, des systèmes immunitaire, endocrinien et de la barrière hémato-encéphalique.

D'autre part, aux Etats-Unis, un effort de quantification et de modélisation mathématique a amené à la définition d'un nouveau concept, le taux d'absorption spécifique (TAS), également appelé parfois débit de dose, destiné à rendre compte du transfert d'énergie à l'organisme vivant sous forme soit d'énergie cinétique, soit d'énergie potentielle. Bien que le détail de ces mécanismes et l'interprétation même de la notion de TAS fassent encore l'objet de discussions, un groupe d'experts du sous-comité spécialisé de l'American National Standards Institute (ANSI) a proposé un projet révisé de normes de protection fondées sur le taux d'absorption spécifique. Les principales caractéristiques en sont que la limite recommandée pour le débit de fluence énergétique varie en fonction de la fréquence, passe par un minimum égal à 1 $mW.cm^{-2}$ pour les fréquences comprises entre 30 et 300 MHz et ne dépasse pas 5 $mW.cm^{-2}$ pour les fréquences supérieures à 1500 MHz (tableau 2).

Bien que l'on observe une certaine tendance à l'abaissement des limites d'exposition les plus élevées, comme on peut le voir en particulier d'après les limites adoptées en Suède en 1976 et tout récemment au Canada, l'accord pour l'instant est loin d'être unanime et un certain nombre de pays, en particulier au sein des Communautés Européennes où un projet de directive est en préparation, sont encore partisans du maintien de la limite à la valeur de 10 $mW.cm^{-2}$.

Il faut cependant noter un point positif dans le domaine de la normalisation technique des appareils mis à la disposition du public. La limite de fuite de 5 $mW.cm^{-2}$ à 5 cm des surfaces externes de l'appareil et les prescriptions techniques proposées par la Commission Electrotechnique Internationale pour les fours à microondes sont en effet adoptées et mises en vigueur par un nombre toujours croissant de pays.

Ultrasons

Malgré l'utilisation de plus en plus importante des ultrasons dans l'industrie et en médecine, les dispositions réglementaires quant à leur emploi semblent pratiquement inexistantes. Des normes relatives aux performances des appareils, en particulier médicaux, existent dans quelques pays, mais aucune limite d'exposition n'a été établie. Aux Etats-Unis, cependant, l'ACGIH commence à se préoccuper du problème et a soumis à discussion certains niveaux d'exposition considérés comme admissibles.

PROTECTION OPERATIONNELLE

Enfin, pour l'ensemble des RNI, peu d'efforts ont été faits pour systématiser la protection dans son ensemble. Sauf dans le cas des lasers, les recommandations d'ordre pratique concernent surtout la protection individuelle (port de lunettes ou de vêtements de protection, durée d'exposition, par exemple) et ne s'attachent guère aux mesures que l'on pourrait prendre au niveau de la source en vue de la protection collective des travailleurs.

Il est probable qu'un grand nombre des principes utilisés pour la protection opérationnelle contre les rayonnements ionisants pourraient également rendre de grands services dans le domaine des rayonnements non-ionisants. Ce serait le cas, entre autres, de la notion de zone contrôlée ou de la définition des personnes exposées. Dans le cas d'un radar, il arrive que ces dernières n'aient aucun rapport avec la source et soient simplement occupées dans une zone balayée par le faisceau à une certaine distance; on a constaté, par exemple, que sur un aéroport ce pouvait être le personnel de passerelle ou le balayeur d'un parking proche.

Bien des problèmes aussi restent posés en ce qui concerne la surveillance médicale des travailleurs exposés, parmi lesquels on peut citer : difficulté d'établir la relation de cause à effet, en particulier pour les effets chroniques et tardifs; incertitude sur les signes cliniques à rechercher à la suite d'exposition continue ou répétée à des rayonnements de faible intensité (lasers, microondes, ultrasons); examens difficiles et coûteux, par exemple pour le dépistage des lésions oculaires, etc...

CONCLUSION

En résumé, la réponse de l'organisme à une irradiation par les différents rayonnements non-ionisants dépend d'un grand nombre de paramètres physiques et biologiques dont l'interférence et l'importance relative ne sont pas toujours bien éclaircies. Qu'il s'agisse des recherches biologiques ou des dispositions réglementaires, on peut dire que la protection contre les rayonnements non-ionisants présente un retard d'environ une génération par rapport aux moyens mis en oeuvre pour la protection contre les rayonnements ionisants. Bien que la tâche soit difficile, le Comité espère qu'après avoir analysé les données de base actuellement disponibles, il sera en mesure dans les quelques années qui viennent de proposer pour les différents RNI un système de limites qui soit acceptable au plan international. Actuellement, un peu partout à travers le monde, on constate un intérêt croissant pour la protection contre les nuisances de toute nature et les RNI semblent y occuper une place privilégiée. Les recherches s'intensifient et un nombre toujours plus grand d'organismes nationaux et internationaux inscrivent les RNI au rang de leurs préoccupations. Parmi eux, l'IRPA, organisme scientifique indépendant qui regroupe les associations nationales les plus spécialisées en protection contre les rayonnements, peut jouer un rôle de premier plan pour la promotion de la protection contre les RNI et l'harmonisation des mesures de protection.

TABLEAU 1

LIMITES D'EXPOSITION AUX MICROONDES

PAYS		TRAVAILLEURS			PUBLIC
		Débit de fluence énergétique (irradiance) en $mW.cm^{-2}$	Durée d'exposition	Fluence énergétique (exposition radiante) $mW.h.cm^{-2}$	Débit de fluence énergétique $mW.cm^{-2}$
ETATS UNIS ANSI (1974)		10	illimitée	1 par 0,1 h	
ETATS UNIS Army + Air Force 1965		10 10-55	illimitée $t(\text{min par h}) = \frac{6\,000}{p^2}$ ×		
ETATS UNIS ACGIH (1978)		10 10-25	8 h	1 par 0,1 h	
CANADA (1979)	10 MHz-1 GHz 1 GHz-300 GHz	1 5	illimitée illimitée		1
	10 MHz- 1 GHz 1 GHz-300GHz	1-25 10-25 1-10	$t(\text{min par h}) = 60/p$ × " $= 300/p$ ×		
FRANCE Min. des Armées (1968)		10 10-55 1	$t \geq 1$ h $t(\text{min par h}) = \frac{6\,000}{p^2}$ × en dehors h de trav.		
POLOGNE (1972)	stationnaire	0,2 0,2-10	10 h $t(\text{h par j}) = \frac{0,32}{p^2}$ ×		0,01
	tournant	1 1-10	10 h $t(\text{h par j}) = \frac{8}{p^2}$ ×		0,1
SUEDE Min. du Travail (1976)		1	8 h	0,1 par 0,1 h	
TCHECOSLOVAQUIE (1970)	continu	0,025 1,6	8 h 1 h par jour		0,0025
	pulsé	0,01 0,64	8 h 1 h par jour		0,001
U.R.S.S. (1977)	stationnaire	0,01 0,1 1	10 h 2 h par jour 20 min par jour		0,001
	tournant	0,1 1	10 h 2 h par jour		0,005

× p = débit de fluence énergétique (irradiance) en $mW.cm^{-2}$

Activites du Comite International de l'IRPA

TABLEAU 2

LIMITES D'EXPOSITION AUX RADIOFREQUENCES
(300 kHz-300 Ghz)
Propositions 1979 de l'ANSI (E.U.)

Fréquence (f) MHz	Débit de fluence énergétique mW.cm^{-2}	E^2 V^2.m^{-2}	H^2 A^2.m^{-2}
0,3 - 3	100	400 000	2,5
3 - 30	900/f^2	4000 (900/f^2)	0,025 (900/f^2)
30 - 300	1,0	4000	0,025
300 - 1500	f/300	4000 (f/300)	0,025 (f/300)
1500 - 300 000	5	20 000	0,125

THE WORK OF THE ILO IN THE FIELD OF PROTECTION OF WORKERS AGAINST IONISING AND NON-IONISING RADIATIONS

G. H. Coppée

One of the basic aims and objectives of the International Labour Organisation, as defined in its Constitution, is "the protection of workers against sickness, disease and injury arising out of his employment". The protection of the workers against the risks due to ionising and non-ionising radiations thus forms part of the field of action of the ILO.

Already, in 1934, the ILO adopted an international instrument providing that persons sustaining occupational injuries caused by ionising radiation would receive compensation and Convention No 121 (1964), concerning benefits in the case of employment injury, includes, under its Schedule I, the compensation of diseases caused by ionising radiations in all work involving exposure to the action of ionising radiations.

In 1949 the ILO published what is probably one of the first sets of practical international standards on radiation protection, which were incorporated into the *Model Code of Safety Regulations for Industrial Establishments*. These provisions were revised and considerably extended in 1957 (Part II of the *Manual of Industrial Radiation Protection*). The Model Code contains also a section on the protection of workers against occupational hazards due to infra-red and ultra-violet radiations. This Code has no binding force but it is intended as a guide to government and industry; it is now being revised.

In June 1960 the International Labour Conference adopted a Convention (No 115) and a Recommendation (No 114) on the protection of workers against ionising radiations. The Convention applies to all activities involving exposure of workers to ionising radiations in the course of their work and provides that each Member of the ILO who ratifies it shall give effect to its provisions by means of laws or regulations, codes of practice or other appropriate methods. To date this Convention has been ratified by 35 countries. The Recommendation further develops the principles stated in the Convention and provides for various measures not specifically mentioned in it. The Conference has also adopted a Resolution requesting the Office to ensure the continued study of the protection of female workers against ionising radiations.

Two other Conventions and Recommendations are also relevant to the protection of workers against the risks due to ionising radiations: the Convention (No 139) and Recommendation (No 147), 1974, concerning prevention and control of occupational hazards caused by carcinogenic substances and agents, the Convention (No 148) and

The Work of the ILO

Recommendation (No 156), 1977, concerning the protection of workers against occupational hazards in the working environment due to air pollution, noise and vibration.

The International Labour Office has published a number of guides on radiation prevention dealing with the principles of radiation protection in industrial operation, particularly concerning the use of industrial X-ray and gamma-ray radiography and fluoroscopy equipment and the use of luminous compounds. Together with the IAEA and the WHO, the ILO has taken part in the production of a number of further guides published by WHO on radiation protection in hospitals and general practice. A manual on the medical supervision of radiation workers has been published by the IAEA under the auspices of the three organisations. A publication on *Mutual Emergency Assistance for Radiation Accidents* has been published by the IAEA in co-operation with FAO, ILO and WHO; it is now being revised in co-operation with UNDRO. A Symposium on the assessment of radioactive body burdens in man was convened by the IAEA, the WHO and the ILO in 1964.

The ILO has published jointly with the IAEA a *Code of Practice on Radiation Protection in the Mining and Milling of Radioactive Ores*, which is currently being revised in co-operation with the WHO. In 1963 a Symposium on radiation protection in mining and milling of radioactive ores was convened by the IAEA, the WHO and the ILO. In 1947 another Symposium on radiation protection in mining and milling of uranium and thorium was organised by the ILO and the French Atomic Agency Commission in co-operation with the WHO and the IAEA. The IAEA and the ILO have also prepared a Manual on radiological safety in uranium and thorium mines and mills, published by the IAEA, which follows and supplements the ILO/IAEA Code of Practice.

The ILO has published an *Encyclopedia on Occupational Health and Safety* which contains, together with other information, a series of articles on the medical and technical aspects of radiological protection and on occupational risks due to non-ionising radiations. This important ILO publication is now being updated. Within the framework of its technical co-operation programme, the ILO has contributed to the training of qualified personnel in radiation protection; it has organised courses for labour inspectors, safety engineers and medical officers. The International Occupational Safety and Health Information Centre - CIS - is in a position to provide comprehensive information and selected bibliographies on the various aspects of occupational safety and health, including the protection of workers against the risks due to ionising and non-ionising radiations.

The IAEA, the WHO, the ILO and the OECD Nuclear Energy Agency are co-sponsoring the revision of the 1967 edition of the *IAEA Basic Safety Standards for Radiation Protection* in the light of the new *ICRP Recommendations* (1977). This revision is carried out by an Advisory Group of Experts which held two meetings in Vienna in 1977 and in 1978; it is to hold a further meeting in 1980. The revised *Basic Safety Standards for Radiation Protection* is expected to become joint standards of the four organisations after submission for approval to their respective competent bodies. The IAEA, in co-operation with the ILO, the WHO, and NEA-PECD and the International Commission on Radiological Protection, has organised a Topical Seminar on the Practical Implications of the *ICRP Recommendations* (1977) and the revised *IAEA Basic Standards for Radiation Protection*, which was held in Vienna from 5-9 March 1979.

The ILO Programme and Budget for 1980-81 provides for a study on the protection of workers against non-ionising radiation. It is proposed to analyse and compare knowledge on the biological effects of non-ionising radiations and their prevention. This study will be done in co-operation with the WHO.

G. H. Coppee

Following the adoption of the Resolution concerning the future action of the ILO in the field of the working conditions and environment by the International Labour Conference in 1975, the International Programme for the Improvement of the Working Conditions and Environment, PIACT, was launched in 1976. This programme seeks to give governments, employers' and workers' organisations, as well as research and training institutes, the necessary help in drawing up and implementing programmes for the improvement of working conditions and environment. The protection of workers against ionising and non-ionising radiations falls naturally within the scope of this consolidated programme, which uses in a co-ordinate manner the various means of action available to the ILO.

Geneva
5 March 1980

THE BEIR-III REPORT AND ITS IMPLICATIONS FOR RADIATION PROTECTION AND PUBLIC HEALTH POLICY

J. I. Fabrikant

INTRODUCTION

My assignment today is to try to give some sort of general background of the implications the current Report (1) of the Committee on the Biological Effects of Ionizing Radiation, National Academy of Sciences-National Research Council (The BEIR-III Report) may have on societal decision-making in the regulation of activities concerned with the health effects of low-level radiation (Table 1). I shall try to discuss how certain of the areas addressed by the present BEIR Committee attempt to deal with the scientific basis for establishing appropriate radiation protection guides, and how the Report (1) may not necessarily serve as a comprehensive review and evaluation of existing scientific knowledge concerning low-level radiation exposure to human populations. Whatever I may consider important in these discussions, I speak only as an individual, and in no way do I speak for the BEIR Committee whose present deliberations are soon to become available. It would be difficult for me not to be somewhat biased and directed in favor of the substance of the BEIR Reports, (1-3) since as an individual I have been sufficiently close to the ongoing scientific deliberations of agreement and disagreement as they developed over the past 10 years.

I think the best thing for one to do is to discuss very briefly why we have advisory committees on radiation, and why the BEIR

[1] Presented as Invited Lecture, Plenary Session, Review of the Activities of the BEIR Committee, Fifth International Congress of The International Radiation Protection Association, Jerusalem, Israel, March 9-14, 1980.

[2] Supported by the Office of Health and Environmental Research of the U.S. Department of Energy under Contract W-7405-ENG-48 and the Environmental Protection Agency.

[3] Professor of Radiology, University of California School of Medicine, San Francisco.

[4] Mailing Address: Donner Laboratory, University of California, Berkeley, California 94720

Committee, and its current Report, (1) may be somewhat different than the others. To do this, I shall review what we know and what we do not know about the health effects of low-level radiation, particularly as these may highlight the controversy which has led to scientific dispute within the Committee. Further, I shall comment on how the risks of radiation-induced cancer in man have been estimated, the sources of the epidemiological data, the dose-response models used, and the uncertainties which limit precision of estimation of excess risks from radiation. And finally, I should like to conjecture with you on what lessons we have learned or should have learned from the BEIR-III Committee experience, and especially on what the implications might be of numerical risk estimation for radiation protection and public health policy.

WHY DO WE HAVE ADVISORY COMMITTEES ON RADIATION?

For more than half a century, responsible public awareness of the potential health effects of ionizing radiations from medical and industrial exposure, from nuclear weapons and weapons testing, and from the production of nuclear energy has called for expert scientific advice and guidance. And, advisory committees on radiation of national and international scientific composition have for these many years met and served faithfully and effectively to deliberate and to report on three important matters of societal concern (Table 2): (1) to place into perspective the extent of harm to the health of man and his decendants to be expected in the present and in the future from those societal activities involving ionizing radiations; (2) to develop quantitative indices of harm based on dose-response relationships in order to provide a scientific basis to be applied to concepts of acceptable risk and protection of human populations exposed to radiation related primarily to somatic and genetic risks; (3) to identify the extent of radiation activities which could cause harm, to assess their relative significance, and to provide a framework on how to reduce unnecessary radiation exposure to human populations.

To a greater or lesser extent, each advisory committee on radiation---such as the UNSCEAR, the ICRP, the NCRP, and the BEIR Committee---have dealt extensively with these matters. But significant differences occur in the scientific reports of these various bodies, and we should expect differences to occur, because of the charge, the scope, and the composition of each Committee, and most important, public attitudes existing at the time of the deliberations of that particular committee, and at the time of the writing of that particular report. The BEIR Report (1) is different; however, the main difference is not so much from new data or new interpretations of existing data, but rather from a philosophical approach and appraisal of existing and future radiation protection resulting from an atmosphere of constantly changing societal conditions and public attitudes.

WHY IS THE BEIR REPORT (1) DIFFERENT?

The Report (1) of the Committee on the Biological Effects

of Ionizing Radiations of the National Academy of Sciences-
National Research Council is the record of the deliberations of a
standing expert scientific advisory committee (the BEIR Committee)
and deals with the scientific basis of the health effects of human
populations exposed to low levels of ionizing radiation. The
current Report (1) broadly encompasses two areas (Table 3):
(1) it reviews the current scientific knowledge---epidemiological
surveys and laboratory experiments---relevant to radiation
exposure of human populations and the delayed or late health
effects of low-level radiation; (2) it evaluates and analyzes
these late health effects---both somatic and genetic effects---in
relation to the risks from exposure to low-level radiation. The
BEIR Committee is an advisory committee to the National Academy of
Sciences-National Research Council. It presently consists of 22
members, selected for their special scientific expertise in areas
of biology, biophysics, biostatistics, epidemiology, genetics,
mathematics, medicine, physics, public health, and the radiological
sciences. The reports (1-3) of this advisory committee have, in
the past, become a reference text as a scientific basis for the
development of appropriate and practical radiation protection
standards.

The 1972 BEIR-I Report (2) and the forthcoming BEIR-III
Report (1) differ from one or more of the other radiation advisory
committee reports of the UNSCEAR, (4,5) the ICRP, (6,7) the
NCRP, (8,9) and of the other national councils and committees,
in four important ways (Table 4):

(1) The BEIR Report (1-3) is intended to be a readable, usable
document for all activities concerned with radiation health. The
conclusions, recommendations, and scientific appendices are purposefully written in a straightforward manner, to be read and understood by physicists and physicians, by congressmen and counsellors,
by unions and utilities, and by engineers and environmentalists.

(2) The BEIR Report (1-3) does not set radiation standards
or public health policy. However, the Report (3) is purposefully
presented so that it will be useful to those responsible for
decision-making concerning regulatory programs and public health
policy involving radiation in the United States. There is no
intent to make the task any easier or to set a firm direction for
those decision-makers who must take into account those considerations of science and technology, the relevant societal and economic
matters, and the development and execution of such regulatory programs. In this regard, the BEIR Report (3) suggests that those
responsible for setting radiation protection standards must always
take into account societal needs at that time, so that such
standards are established on levels of radiation exposure which
are not necessarily absolutely safe, but rather those which are
considered to be appropriately safe for existing circumstances at
the time to fulfill society's needs, particularly in the areas of
general population and occupational exposure from medical
radiation and nuclear energy.

(3) The experimental data and epidemiological surveys are carefully reviewed and assessed for their value in estimating numerical risk coefficients for the health effects in human populations exposed to low-level radiation. Such deliberations require scientific judgment and assumptions based on the available epidemiological and experimental data only, and have necessarily and understandably led to disagreement not only outside the committee room, but among committee members as well. But such dispute and disagreement center not on the scientific facts and not on the existing epidemiological and experimental data, but rather on the assumptions, interpretations, and analyses of the available facts and data. Therefore, the BEIR Report (3) uses a particularly practical format for decision-makers, namely, the numerical risk coefficients estimated are presented in probabilistic terms, within most likely upper and lower boundaries, derived solely from the scientific facts, the epidemiological data, and the scientific hypotheses and assumptions on which they are based.

(4) The BEIR Report (1-3) addresses the continued need to assess and evaluate the benefits from those activities involving radiation as well as the risks. In our resource-limited society, such benefit-risk assessment is essential for societal decision-making for establishing appropriate and achievable radiation protection standards. Decisions can and must be made on the value and costs of technological and societal programs for the reduction of risk by reducing the levels of radiation exposure. This would include societal choices centered as well on alternative methods involving nonradiation activities available through a comparison of the costs to human health and to the environment. (3)

WHAT ARE THE IMPORTANT BIOLOGICAL EFFECTS OF LOW-LEVEL RADIATION?

My remarks here will be restricted primarily to those delayed or late health effects in humans following exposure to low-LET radiation, x-rays and to gamma rays from radioactive sources, and to a much lesser extent to high-LET neutron and alpha radiations, since these are the ionizing radiations most often encountered in medicine and in the nuclear industry. Briefly, low-level radiation can affect the cells and tissues of the body in three important ways (Table 5). First, if the macromolecular lesion occurs in one or a few cells, such as these of the hematopoietic tissues, the irradiated cell can occasionally transform into a cancer cell, and after a period of time, there is an increased risk of cancer developing in the exposed individual. This biological effect is called carcinogenesis; and the health effect, cancer. Second, if the embryo or fetus are exposed during gestation, injury can occur to the proliferating and differentiating cells and tissues, leading to abnormal growth. This biological effect is called teratogenesis; and the health effect, developmental abnormality in the newborn. Third, if the macromolecular lesion occurs in the reproductive cell of the testis or the ovary, the hereditary genome of the germ cell can be altered, and the injury can be expressed in the descendants of the exposed individual. This biological effect is called mutagenesis; and the health effect, genetically-related ill-health.

There are a number of other biological effects of ionizing radiation, such as cataracts of the lens of the eye, or impairment of fertility, but these three important late effects---carcinogenesis, teratogenesis and mutagenesis---stand out as those of greatest concern. This is because a considerable amount of scientific information is known from epidemiological studies of exposed human populations and from laboratory animal experiments. Furthermore, we believe that any exposure to radiation, even at very low levels of dose, carries some risk of such deleterious effects. And, as the dose of radiation increases above very low levels, the risk of these deleterious health effects increases in exposed human populations. It is these latter observations that have been central to the public concern about the potential health effects of low-level radiation, and to the task of establishing standards for protection of the health of exposed populations. Indeed, all reports of expert advisory committees on radiation are in close agreement on the broad and substantive issues of such health effects.

WHAT DO WE KNOW ABOUT THE HEALTH EFFECTS OF LOW-LEVEL RADIATION?

A number of very important observations on the health effects of low-level radiation have now convincingly emerged, and about which there is firm general agreement (Table 6). These observations are based on careful statistical evaluation of epidemiological surveys of exposed human populations, in conjunction with extensive research in laboratory animals, and on analysis of dose-response relationships of carcinogenic, teratogenic and genetic effects, and on known mechanisms of cell and tissue injury in vivo and in vitro.

1) _Cancer_ induction is considered to be the most important late somatic effect of low-dose ionizing radiation. Solid cancers arising in the various organs and tissues, such as the female breast and the thyroid gland, rather than leukemia, are the principal late effects in individuals exposed to radiation. The different organs and tissues vary greatly in their relative susceptibility to cancer induction by radiation. The most frequently occurring radiation-induced cancer in man include, in decreasing order of susceptibility (Table 6): the female breast; the thyroid gland, especially in young children and in females; the hematopoietic tissues; the lung; certain organs of the gastrointestinal tract; and the bones. There are influences, however, of age at the time of irradiation, of sex, and of the radiation factors and types---LET and RBE---affecting the cancer risk.

2) The effects on _growth and development_ in the irradiated embryo and fetus are related to the gestational stage at which exposure occurs. It appears that a threshold level of radiation dose may exist below which gross teratogenic effects will not be observed. However, these dose levels would vary greatly depending on the particular developmental abnormality.

3) It has been necessary to estimate _genetic risks_ based mainly on laboratory mouse experiments because of the paucity of data from exposed human populations. Our knowledge of fundamental

mechanisms of radiation injury at the genetic level is far more complete, thereby permitting greater assurance in extrapolating from laboratory experiments to man. Mutagenic effects are related linearly to radiation dose, even at very low levels of exposure. With new information on the broad spectrum and incidence of genetically-related ill-health in man, such as mental retardation and diabetes, the risk of radiation mutagenesis in man affecting future generations takes on new and special consideration.

WHAT DO WE NOT KNOW ABOUT THE HEALTH EFFECTS OF LOW-LEVEL RADIATION?

In spite of a remarkable understanding of the health effects in exposed human populations, there is still a considerable amount we do not know about the potential health hazards of low-level radiation (Table 7):

1. We do not know what the health effects are at dose rates as low as a few hundred millirem per year. It is probable that if any health effects do occur, they will be masked by environmental or other competing factors that produce similar effects.

2. The vast epidemiological data on exposed human populations are nevertheless highly uncertain in regard to the forms of the dose-response relationships for radiation-induced cancer in man. This is especially the case for low-level radiation. Therefore, it has been necessary to estimate human cancer risk at low doses primarily from observations at relatively high doses, frequently greater than 100 rads and more. However, it is not known whether the cancer incidence observed at high dose levels also applies to cancer induction at low dose levels.

3. We have no reliable method at the present time of estimating the repair of injured cells and tissues of the body exposed to very low doses and dose rates. And further, we do not know how to identify those persons who may be particularly susceptible to radiation injury.

4. Analyses of the numerous epidemiological surveys of irradiated populations exposed in the past demonstrate that we have very limited information on the precise radiation doses absorbed by the tissues and organs. Furthermore, we do not know the complete cancer incidence in each study population, since new cases of cancer continue to appear with the passing of time. Accordingly, any estimation of excess cancer risk based on such limited dose-response information must necessarily be incomplete, until the entire study population has died from natural causes.

5. We do now know the role of competing environmental and other host factors---biological, chemical, or physical factors---existing at the time of exposure, or following exposure, which may influence and affect the carcinogenic, teratogenic, or genetic effects of low-level radiation.

WHAT ARE THE UNCERTAINTIES IN THE DOSE-RESPONSE RELATIONSHIPS FOR RADIATION-INDUCED CANCER?

The present BEIR-III Committee, in its earliest deliberations, recognized that there was great uncertainty in regard to the shapes of the dose-response curves for cancer induction by radiation in humans, and this was especially the case at low levels of dose. Estimates of excess cancer risk at low doses appear to depend more on what is assumed about the mathematical form of the dose-response function than on the available epidemiological data. Accordingly, in estimating the excess cancer risk from low-dose low-LET radiation, the BEIR-III Committee chose to use a linear-quadratic dose-response model felt to be consistent with epidemiiological and radiobiological data in preference to more extreme dose-response models. In this regard, the current BEIR-III Report[1] differs substantially from the 1972 BEIR-I Report[2]. I should like to examine the deliberations of this decision more closely.

In recent years, a general hypothesis for estimation of excess cancer risk in irradiated human populations, based on theoretical considerations, extensive experimental animal studies and epidemiological surveys, suggests that complex dose-response relationships between radiation dose and observed cancer incidence[10-15]. Perhaps the most widely accepted model for cancer induction by radiation, based on the available information and consistent with both knowledge and theory, takes the complex linear-quadratic form: $I(D) = (\alpha_0 + \alpha_1 D + \alpha_2 D^2) \exp(-\beta_1 D - \beta_2 D^2)$, where I is the cancer incidence in the irradiated population at radiation dose D in rad, and α_0, α_1, α_2, β_1 and β_2 are non-negative constants (Figure 1). The multicomponent dose-response curve contains (1) an initial upward-curving linear and quadratic functions of dose which represents the process of cancer induction by radiation; and (2) a modifying exponential function of dose which represents the competing effect of cell killing at high doses. α_0 is the ordinate intercept at 0 dose, and defines the natural incidence of cancer in the population. α_1 is the initial slope at 0 dose, and defines the linear component in the low dose range. α_2 is the curvature near 0 dose, and defines the upward-curving quadratic function of dose. β_1 and β_2 are the slopes of the downward-curving function in the high dose range, and define the cell killing function.

Analysis of a large number of dose-incidence curves for cancer induction in irradiated populations, both in humans and in animals, has demonstrated that for different radiation-induced cancers only certain of the parameter values of these constants can be theoretically determined. However, the extent of the variations in the shapes of the dose-response curve does not permit direct determination from the data of any of these parameter values with precision, or of assuming their values, or of assuming any fixed relationship between two or more of these parameters. In the case of the epidemiological surveys of irradiated human populations, this complex multicomponent general dose-response form cannot be universally applied. Therefore, it has become necessary to simplify the model by reducing the number of parameters which would have the least effect on the form of the dose-response relationship in the dose range of low-level radiation. Such simpler models, with increasing complexity, include the linear, quadratic, linear-quadratic, and finally, the multicomponent linear-quadratic form with an exponential modifier (Figure 2).

The BEIR-III Committee recognized three compelling situations which seriously limit precise numerical estimation of the excess cancer risk of low-level radiation in human populations (Table 8). (1) We lack an understanding of the fundamental mechanisms of cancer induction by radiation in man. (2) The dose-response information from human data is highly uncertain, particularly at low levels of dose. (3) Experimental and theoretical considerations suggest that various and different mathematical forms of dose-response relationships may exist for different radiation-induced cancers in exposed human populations. Nevertheless, these limitations do not relieve decision-makers of the responsibility for determining public health policy based on appropriate radiation protection standards. Accordingly, not only did the BEIR-III Committee consider it essential that quantitative risk estimation be determined, based on the available epidemiological and radiobiological data, but that in addition, it was equally essential that precise explanations and qualifications of the assumptions and procedures involved in the determination of such risk estimates are to be provided. This has been done explicitly in the current BEIR-III Report[1] containing the estimates of excess cancer risk. The Committee recognized that some experimental and human data, as well as theoretical considerations, suggest that for exposure to low-LET radiation, such as x-rays and gamma rays, at low doses, the linear model probably leads to overestimates of the risk of most radiation-induced cancers in man, but that the model can be used to define the upper limits of risk. Similarly, the Committee believes that the quadratic model may be used to define the lower limits of risk from low-dose low-LET radiation. For exposure to high-LET radiation, such as neutrons and alpha particles, linear risk estimates for low doses are less likely to overestimate risk and may, in fact, underestimate risk.

WHAT IS THE CONTROVERSY OVER LOW-LEVEL RADIATION?

The estimation of the cancer risk of exposure to low-level radiation is said to be clouded by scientific dispute. In particular, there appears to be disagreement among some scientists as to the effects of very low levels of radiation, even as low as our natural radiation background. While there is no precise definition of low-level exposure, most scientists would generally agree that low-level radiation is that which falls within the dose range considered permissible for occupational exposure. According to accepted standards (16), 5 rem per year to the whole body would be an allowable upper limit of low-level radiation dose for the individual radiation worker.

In this context, and with this as the boundary condition for occupational exposure, then it could very well be concluded that most of the estimated delayed cancer deaths which may be associated with a so-called hypothetical nuclear reactor accident, for example, are therefore considered by some scientists to be caused by exposures well below the allowable occupational limits.

The BEIR-III Report

Furthermore, if it is assumed that <u>any</u> extra radiation above natural background, however small, causes additional cancer, then if millions of people are exposed, some extra cancers will inevitably result. Other scientists strongly dispute this, and firmly believe that low-level radiation is nowhere near as dangerous as their adversarial colleagues would insist. Central to this dispute, it must be remembered that cancers induced by radiation are indistinguishable from those occurring naturally; hence, their existence can be inferred only on the basis of a statistical excess above the natural incidence. Since such health effects, if any, are so rarely seen under low-level radiation because the exposures are so small, the issue of this dispute may never be resolved--it may be beyond the abilities of science and mathematics to decipher.

It is just this type of controversy that was at the root of the division within the present BEIR-III Committee. There is little doubt that the Committee's most difficult task has been to estimate the carcinogenic risk of low-dose low-LET whole-body radiation. Here, emphasis was placed almost entirely on the human epidemiological studies, since it was felt that little information from animal studies could be applied directly to man. Therefore, as the earlier 1972 BEIR-I Report (2) had done, some members of the present BEIR-III Committee chose it necessary to adopt a linear hypothesis of dose-response to estimate the cancer risk at very low-level radiation exposure where no human epidemiological data are available. Here, it was assumed the same proportional risks are present at low levels as at high levels of radiation. This position implies that even very small doses of radiation are carcinogenic, a finding that could force the Environmental Protection Agency to adopt stricter health standards to protect against occupational and general population exposure. Other members of the Committee do not accept this position, and believe this is an alarmist approach. When there is no human epidemiological evidence at low doses, these scientists prefer to assume that the risks of causing cancer are proportionally lower.

Let us look at some of the problems. In its deliberations, the present BEIR-III Committee concluded two important points: (1) It is not yet possible to make precise low-dose estimates for cancer induction by radiation because the level of risk is so low it cannot be observed directly. (2) There is great uncertainty as to the dose-response function most appropriate for interpolating in the low-dose region. In studies of exposed animal and human populations, the shape of a dose-response relationship at low doses may be practically impossible to ascertain statistically. This is because the population sample sizes required to estimate or test a small absolute cancer excess are extremely large; specifically, the required sample sizes are approximately inversely proportional to the square of the excess. For example, if the excess is truly proportional to dose, and if 1,000 exposed and 1,000 control persons are required in each group to test the cancer excess adequately at 100 rads, then about 100,000 in each group are required at 10 rads, and about 10,000,000 in each group are required at one rad. Thus, it appears that experimental evidence and theoretical considerations are more likely than empirical data to guide the choice of a dose-response function. In this dilemma and after much disagreement among some of its members, the present BEIR-III Committee chose to adopt as a working model for low-LET radiation and carcinogenesis the linear-quadratic dose-response form with an exponential term to account for the frequently observed turndown of the curve in the high-dose region. However, in applying this multicomponent model, only certain of its derivatives, including the linear, linear-quadratic, and pure quadratic, could prove practical.

It should be remembered that in the 1972 BEIR-I Report cancer risk estimates for whole-body radiation exposure were derived from linear model average excess cancer risk per rad observed at doses generally of a hundred or more rads. These estimates have been generally criticized on the grounds that the increment in cancer risk per rad may well depend on dose and that the true risk at low doses may therefore be lower or higher than the linear model predicts (9). In animal experiments, it has been shown, often with considerable statistical precision, that the dose-response curve for radiation-induced cancer can have a variety of shapes. As a general rule, the curve has a positive curvature for low-LET radiation, i.e., the slope of the curve increases with increasing dose. However, at high doses, the slope often decreases and may even become negative. Dose-response curves may also vary with the kind of cancer, with animal species, and with dose rate. On the basis of the experimental evidence and current microdosimetric theory, therefore, the present BEIR-III Committee could quite reasonably adopt as the basis for its consideration of dose-response models the linear-quadratic form with an exponential term for a negative slope in the high dose region.

On the other hand, the Committee recognized that for the most part, the available human data from the vast body of epidemiological studies fail to suggest any specific dose-response model, and are not sufficiently reliable to discriminate among <u>a priori</u> models suggested by the experimental and theoretical work. However, there appears to be certain exceptions; for example, cancer of the skin is not observed at low radiation doses (17), and dose-response relationships observed in the Nagasaki leukemia data appear to have positive curvature (18). The incidence of breast cancer seems to be adequately described by a linear dose-response model (11,19) (Figure 3).

In attempts to apply derivatives of the multicomponent linear-quadratic model to the human data, simplification was required to obtain statistically stable risk estimates in many cases. It is now well known that members of the BEIR-III Committee were divided on this matter; some members of the Committee strongly favor the linear model, others favor the quadratic form. A further modification of the linear-quadratic form was assumed with the linear and quadratic components to be equivalent at some dose, which is consistent with epidemiological data and radiobiological evidence, and avoids dependence on either of the extreme forms (14,15).

WHAT ARE THE UNCERTAINTIES IN ESTIMATION OF THE CARCINOGENIC RISK IN MAN OF LOW-LEVEL RADIATION?

The quantitative estimation of the carcinogenic risk of low-dose, low-LET radiation is subject to numerous uncertainties (Table 9). The greatest of these concerns the shape of the dose-response curve. Others include the length of the latent period, the RBE for fast neutrons and alpha radiation relative to gamma and x-radiation, the period during which the radiation risk is expressed, the model used in projecting risk beyond the period of observation, the effect of dose rate or dose fractionation, and the influence of differences in the natural incidence of specific types of cancer. In addition, uncertainties are introduced by the biological risk characteristics of humans, e.g., the effect of age at irradiation, the influence of any disease for which the radiation was given therapeutically, and the influence of length of observation or follow-up. The collective influence of these uncertainties is such as to deny great credibility to any estimates of human cancer risk that can be made for low-dose, low-LET radiation. It is for these reasons, the present BEIR-III Committee has placed more emphasis on the methods of risk estimation than on any numerical estimates derived thereby.

WHAT ARE THE SOURCES OF EPIDEMIOLOGICAL DATA FOR THE ESTIMATION OF EXCESS CANCER RISK IN EXPOSED HUMAN POPULATIONS?

The tissues and organs involved in radiation-induced cancer in man about which we have the most reliable epidemiological data from a variety of sources from which corroborative risk estimates have been obtained include the bone marrow, the thyroid, the breast, and the lung. The data on bone and the digestive organs are, at best, preliminary, and do not approach the precision of the others. In several of these tissues and organs, risk estimates are obtained from very different epidemiological surveys, some followed for over 25 years, and with adequate control groups. There is impressive agreement when one considers the lack of precision inherent in the statistical analyses of the case-finding and cohort study populations, variability in ascertainment and clinical periods of observation, age, sex and racial structure, and different dose levels, and constraints on data from control groups.

By far, the most reliable and consistent data have been those of the risk of leukemia, which come from the Japanese A-bomb survivors (18), the ankylosing spondylitis patients treated with x-ray therapy in England and Wales (20,21), the metropathia patients treated with radiotherapy for benign uterine bleeding (22), and the tinea capitis patients treated with radiation for ringworm of the scalp (23,24)(Table 10). There is evidence of an age-dependence and a dose-dependence, a relatively short latent period of a matter of a few years, and a relatively short period of expression, some 10 years. This cancer is uniformly fatal.

The data available on thyroid cancer are more complex; the surveys include the large series of children treated to the neck and mediastinum for enlarged thymus (25), children treated to the scalp for tinea capitis (23,24), and the Japanese A-bomb survivors (18) and Marshall Islanders (26) exposed to nuclear explosions (Table 10). Here, there is an age-dependence and sex-dependence--children and females appear more sensitive. Although the induction rate is high, the latent period is relatively short, and it is probable that no increased risk will be found in future follow-up. In addition, most tumors are either thyroid nodules, or benign or treatable tumors, and only about 5% of the radiation-induced thyroid tumors are fatal.

In very recent years, much information has become available on radiation-induced breast cancer in women (13,19) (Table 11). The surveys include primarily women with tuberculosis who received frequent fluoroscopic examinations for artificial pneumothorax (27), post-partum mastitis patients treated with radiotherapy (28), and the Japanese A-bomb survivors in Hiroshima and Nagasaki (18). Here, there is an age- and dose-dependency, as well as a sex-dependency, and the latent period is long, some 20 to 30 years. Perhaps about half of these neoplasms are fatal.

Another relatively sensitive tissue, and a complex one as regards radiation dose involving parameters of the special physical and biological characteristics of the radiation quality, is the epithelial tissue of the bronchus and lung (Table 11). The information from the Japanese A-bomb survivors (18), and uranium miners in the United States and Canada (29,30), and the ankylosing spondylitis patients in England and Wales (20,21) provide reliable risk estimates of lung cancer exposed persons. There is some evidence of age-dependence from the Japanese experience and a relatively long latent period. This cancer is uniformly fatal.

The lifetime risk of radiation-induced bone sarcoma (Table 11), based primarily on radium and thorium patients who had received the radioactive substances for medical treatment, or ingested them in the course of their occupations (31,32) is low. For all other tumors arising in various organs and tissues of the body, values are extremely crude and preliminary estimates.

There is now a large amount of epidemiological information from various comprehensive surveys from a variety of sources; the most extensive, perhaps, include the Japanese A-bomb survivors (18), the patients treated to the spine for ankylosing spondylitis (20,21), the metropathia patients (22) and the early radiologists (33). These data indicate that leukemia is now no longer the major cancer induced by radiation, and that solid cancers are exceeding the relative incidence of radiation leukemia by a factor as high as 5^5. That is, in view of the long latent periods for certain solid cancers to become manifest, it can be estimated that perhaps after some 30 years following radiation exposure, the risk of excess solid cancers may prove to be many times the risk of excess leukemia. These estimates remain very crude, since they do not take into account the obvious lack of precision of certain of the epidemiological studies, particularly as regards radiation dose distribution, ascertainment, latency periods, and other important physical and biological parameters. The BEIR (1,2), the UNSCEAR (4,5) and the ICRP (6,7) Reports have estimated the risk from whole-body exposure in different ways and based primarily on the studies of the Japanese A-bomb survivors (18), and to a much lesser extent, from data on the ankylosing spondylitis patients (20,21), the metropathia patients (22), the tinea patients (23,24), and similar epidemiological surveys carefully followed, many of which now have adequate control study populations, a very crude figure of the total lifetime excess absolute risk of radiation-induced cancer deaths can be derived. This figure for all malignancies from low-LET radiation, i.e., x-rays and gamma rays, delivered at low doses would be an overestimate of the true risk. The actual figure may be much lower in terms of excess cancer cases per million persons exposed per rad total lifetime risk, a large fraction of which would not necessarily be fatal (1,5). Any such estimated figure remains very unreliable, but it does provide a very rough figure for comparison with other estimates of avoidable risks, or voluntary risks, encountered in everyday life.

WHAT ARE THE RISK ESTIMATES OF RADIATION-INDUCED CANCER IN MAN?

The chief sources of epidemiological data used in the current BEIR-III Report (1) are the Japanese populations exposed to whole-body irradiation in Hiroshima and Nagasaki, patients with ankylosing spondylitis and other patients who were exposed to partial body irradiation therapeutically, or to diagnostic x-rays and various occupationally exposed populations, such as uranium miners and radium dial painters. Most epidemiological data do not systematically cover the range of low to moderate radiation doses for which the Japanese atomic bomb survivor data appear to be fairly reliable. Analysis in terms of dose-response therefore rely greatly on the Japanese data. The substantial neutron component of dose in Hiroshima and its correlation with gamma dose limit the value of the more numerous Hiroshima data to the estimation of cancer risk from low-LET radiation. The Nagasaki data, for which the neutron component of dose is small, are less reliable for doses below 100 rads.

For its illustrative computations of the lifetime risk from whole-body exposure, the present BEIR-III Committee chose three exposure situations for low-dose, low-LET radiation:

(1) a single exposure of a representative (life-table) population to 10 rads;

(2) a continuous, lifetime exposure of a representative (life-table) population to 1 rad per year; and

(3) an exposure to 1 rad per year over several age intervals exemplifying conditions of occupational exposure.

The three exposure situations were not chosen to reflect any circumstances that would normally occur, but embrace the areas of concern-- general population and occupational exposure and single and continuous exposure. These were substantially different from the exposure situation chosen for illustrative computation by the 1972 BEIR-I Committee, where 100 mrem per year was selected.

Below these dose levels chosen for the current report, the uncertainties of extrapolation of risk to very low levels were strongly felt by some members of the present Committee to be too great to justify risk estimation. The selected annual exposure, although only one-fifth the maximal permissible dose for occupational exposure, is nevertheless consistent with occupational exposures in the nuclear industry. The U.S. 1969-1971 life-table was used as the basis for the calculations, and all results are expressed in terms of excess cancers per million persons throughout their lifetime after exposure. The expression time was taken as 25 years for leukemia and the remaining years of life for other cancers. Separate estimates were made for cancer mortality and for cancer incidence.

The resulting cancer mortality risk estimates for all forms of cancer differ by as much as an order of magnitude. The uncertainty derives chiefly from the range of dose-response models used, from the alternative absolute and relative projection models, and from the sampling variation in the source data. The lowest estimates are derived from the pure quadratic model; the highest, from the linear model. The linear-quadratic model provides estimates intermediate between these two extremes.

In the absence of any increased radiation exposure, among one million persons of life-table age and sex composition in the United States, about 164,000 persons would be expected to die from cancer, according to present cancer mortality rates. For a situation in which these one million persons are exposed to a single dose increment of 10 rads of low-LET radiation, the linear-quadratic model predicts increases of about 0.5% and 1.5% over the normal expectation of cancer mortality, according to the projection model.

For continuous lifetime exposure to 1 rad per year, the increase in cancer mortality, according to the linear-quadratic model, ranges from about 3% to 8% over the normal expectation, depending on the projection model.

To compare these estimates with those of the 1972 BEIR-I Report (2) and the 1977 UNSCEAR Report (5), it was convenient to express them as cancer deaths per million persons per rad of continuous lifetime exposure. For continuous lifetime exposure to 1 rad per year the linear-quadratic dose-response model for low-LET radiation yielded estimates some 25% to 50% below the comparable linear estimates in the 1972 BEIR-I Report (2), depending on the projection model. Although the present BEIR-III Report (1) uses much more scientific information not available for the earlier 1972 report, the differences mainly reflect changes in the assumptions made by the two BEIR Committees almost a decade apart. The present Committee preferred a linear-quadratic, rather than linear, dose-response model for low-LET radiation, and preferred not to assume a fixed relationship between the effects of high-LET and low-LET radiation. The present risk estimates do not, as in the 1972 BEIR-I Report (2), carry through to the end of life very high relative-risk coefficients obtained with respect to childhood cancers induced in utero by radiation. The present BEIR-III risk estimates do not differ appreciably from those in the 1977 UNSCEAR Report

Cancer-incidence risk estimates were less firm than mortality estimates. The present BEIR-III Committee used a variety of dose-response models and several data sources. The dose-response models produced estimates that differed by more than an order of magnitude, whereas the different data sources gave broadly similar results. For the linear-quadratic model and for continuous lifetime exposure to 1 rad per year, for example, the increased risks expressed as percent of the normal incidence of cancer in males were about 2% to 6%, depending on the projection model. Risks for females were substantially higher than those for males, due primarily to the relative importance of radiation-induced thyroid and breast cancer.

Estimates of excess risk for individual organs and tissues depend in large part on partial-body irradiation and use a wider variety of data sources. Except for leukemia and bone cancer, estimates for individual sites of cancer were made only on the basis of the linear model and were stated in terms of excess cancer cases per year per million persons exposed per rad. For leukemia, the linear-quadratic model yielded about 1.0 to 1.4 excess leukemia cases (or deaths) per year per million persons exposed per rad, for females and males, respectively. For solid cancers, linear-model estimates were, for example: for thyroid in males, about 2, and in females, about 6; for female breast, about 6; and for lung, about 3.5 to 4. These risk coefficients derive largely from epidemiologic data in which exposure was at high doses, and these values may, in some cases, overestimate risk at low doses.

WHAT ARE THE IMPLICATIONS OF NUMERICAL RISK ESTIMATION FOR RADIATION PROTECTION AND PUBLIC HEALTH POLICY?

The present BEIR-III Committee has not highlighted any controversy over the health effects of low-level radiation. In its evaluation of the experimental data and epidemiological surveys, the Committee has carefully reviewed and assessed the value of all the available scientific evidence for estimating numerical risk coefficients for the health hazards to human populations exposed to low levels of ionizing radiation. Such devices require scientific judgment and assumptions based on the available data only, and has led to disagreement not only outside the committee room but among committee members as well. But such disagreement centers not on the scientific facts or the epidemiological data, but rather on the assumptions and interpretations of the available facts and data.

The present scientific evidence and the interpretation of available human data can draw very few firm conclusions on which to base scientific public health policy for protection standards for low-level radiation. However, based on the radiation risk estimates derived, any lack of precision does not minimize either the need for setting public health policy standards nor the conclusion that such risks are extremely small when compared with those available from alternative options, and those normally accepted by society as the hazards of everyday life. When compared with the benefits that society has established as goals derived from the necessary activities of energy production and medical care, it is apparent that society must establish appropriate standards and seek appropriate controlling procedures which continue to assure that its needs and services are being met with the lowest possible risks.

In a third of a century of inquiry, embodying among the most extensive and comprehensive scientific efforts on the health effects of an environmental agent, certain practical information necessary for determination of radiation protection standards for public health policy is still lacking, and may remain so. It is now assumed that exposure to radiation at low levels of dose carries some risk of deleterious effects. However, how low this level may be, or the probability, or magnitude of the risk, still are not known. Our best scientific knowledge and our best scientific advice are essential for the protection of the public health, for the effective application of new technologies in medicine and industry, and for guidance in the production of nuclear energy. Man cannot dispense with those activities which inevitably involve exposure to low levels of ionizing radiation in medicine, where he readily recognizes some degree of risk to health, however small, exists. In the evaluation of such risks from radiation in all other societal activities involving ionizing radiation, including nuclear energy, as is done in medicine, it is also necessary to limit the radiation exposure to a level at which the risk is acceptable both to the individual and to society.

Acknowledgements

The author wishes to acknowledge the many helpful discussions with many of his scientific colleagues, and particularly those members of expert advisory committees on radiation, which have provided the philosophical approach embodied in his presentation. He acknowledges the scientific papers of numerous authors not listed in the bibliography, since this was not intended as a scientific review of the literature. He is continually grateful to Mrs. Barbara Komatsu for her energy, her patience, her good humor, and her expert assistance in the preparation of this manuscript.

Mailing Address

Jacob I. Fabrikant, M.D., Ph.D., Donner Laboratory
University of California, Berkeley, California 94720 U.S.A.

The BEIR-III Report

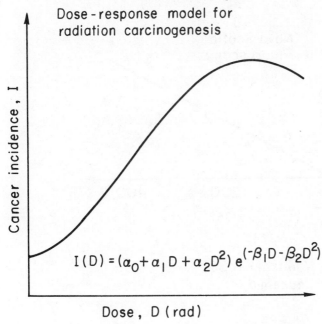

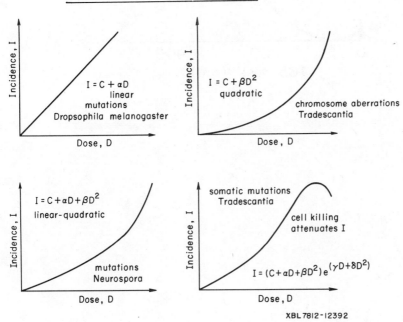

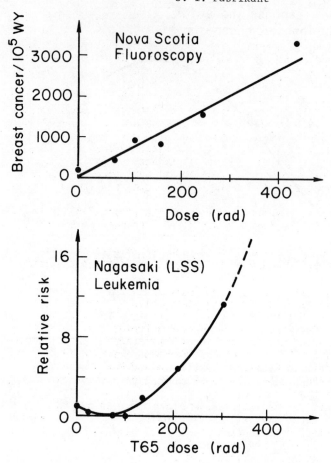

REFERENCES

1. National Academy of Sciences-National Research Council. (1980. To be published): Advisory Committee on the Biological Effects of Ionizing Radiations. The Effects on Populations of Exposure to Low Levels of Ionizing Radiation. Washington, D.C.
2. National Academy of Sciences-National Research Council. (1972): Advisory Committee on the Biological Effects of Ionizing Radiations. The Effects on Populations of Exposure to Low Levels of Ionizing Radiation. Washington, D.C.
3. National Academy of Sciences-National Research Council. (1977): Advisory Committee on the Biological Effects of Ionizing Radiations. Considerations of Health Benefit-Cost Analysis for Activities Involving Ionizing Radiation Exposure and Alternatives. Washington, D.C.
4. United Nations Scientific Committee on the Effects of Atomic Radiation. (1972): Ionizing Radiation: Levels and Effects. New York.
5. United Nations Scientific Committee on the Effects of Atomic Radiation. (1977): Sources and Effects of Ionizing Radiation. New York.
6. International Commission on Radiological Protection. (1966): The Evaluation of Risks from Radiation. ICRP Publication 8. Pergamon Press, Oxford.
7. International Commission on Radiological Protection. (1977): Recommendations of the International Commission on Radiological Protection. ICRP Publication 26. Pergamon Press, Oxford.
8. National Council on Radiation Protection and Measurements. (1971): NCRP Report No. 39. Basic Radiation Protection Criteria. Washington, D.C.
9. National Council on Radiation Protection and Measurements. (1975): NCRP Report No. 43. Review of the Current State of Radiation Protection Philosophy. Washington, D.C.
10. Brown, J.M. (1976): Health Phys., 31, 231.
11. Brown, J.M. (1977): Radiation Res., 71, 34.
12. Upton, A.C. (1977): Radiation Res., 71, 51.
13. Upton, A.C., Beebe, G.W., Brown, J.M., Quimby, E.H., and Shellabarger, C. (1977): J. Natl. Cancer Inst., 59, 480.
14. Fabrikant, J.I. (1979): (In) Epidemiology Studies of Low-Level Radiation Exposure. Report LBL-8667, p. 1, University of California, Berkeley.
15. Fabrikant, J.I. (1980. To be published): (In) Nuclear Reactor Safety: A Current Perspective. Science.
16. International Commission on Radiological Protection. (1970): Protection Against Ionizing Radiation from External Sources, ICRP Publication 15. Pergamon Press, Oxford, 1973. Data for Protection Against Ionizing Radiation from External Sources: Supplement to ICRP Publication 15. ICRP Publication 21. Pergamon Press, Oxford.
17. Shore, R.E., Albert, R.E. and Pasternack, B. (1976): Arch. Environment. Health, 31, 21.
18. Beebe, G.W., Kato, H. and Land, C.E. (1977): Life Span Study Report 8. Radiation Effects Research Foundation Technical Report RERF TR 1-77. National Academy of Sciences, Washington, D.C.

19. Mole, R.H. (1978): Brit. J. Radiol., 51, 401.
20. Court-Brown, W.M. and Doll, R. (1965): Brit. Med. J., 2, 1327.
21. Smith, P.G. and Doll, R. (In press): (In) International Symposium on the Late Biological Effects of Ionizing Radiation, IAEA-SM-224/771, Vienna.
22. Smith, P.G. and Doll, R. (1976): Brit. J. Radiol. 49, 223.
23. Shore, R.E., Albert, R.E. and Pasternack, B.S. (1976): Arch. Environ. Health 31, 21.
24. Modan, B., Baidatz, D., Mart, H., Steinitz, R., and Levin, S.G. (1974): Lancet 1, 277.
25. Hempelmann, L.H., Hall, W.J., Phillips, M., Cooper, R.A., and Ames, W.R. (1975): J. Natl. Cancer Inst. 55, 519.
26. Conard, R.A. (1977): (In) DeGroot, L.J., ed. Radiation-Associated Thyroid Carcinoma, p. 241, Academic Press, New York.
27. Myrden, J.A. and Quinlan, J.J. (1974): Ann. Roy. Coll. Physicians Can. 7, 45.
28. Hempelmann, L.H., Kowaluk, E., Mansur, P.S., Pasternack, B.S., Albert, R.E. and Haughie, G.E. (1977): J. Natl. Cancer Inst. 59, 813.
29. (1976): Health Effects of Alpha-Emitting Particles in the Respiratory Tract. Report of ad hoc Committee on "Hot Particles" of the Advisory Committee on the Biological Effects of Ionizing Radiations. Washington, D.C.
30. Ham, J.M. (1976): Report of the Royal Commission on the Health and Safety of Workers in Mines. Ministry of the Attorney General, Province of Ontario.
31. Mays, C.W. and Spiess, H. (1978): (In) Miller, W.A. and Ebert, H.G., eds. Biological Effects of ^{224}Ra. Benefit and Risk of Therapeutic Application, p. 168, Nyhoff Medical, The Hague.
32. Rowland, R.E. and Stehney, A.F. (1977): (In) Argonne National Laboratory Report ANL-77-65, Part II, 206, Argonne, Illinois.
33. Matanoski, G.N., Seltzer, R., Sartwell, P.E., Diamond, E.L. and Elliott, E.A. (1975): Amer. J. Epid. 101, 119.

THE RADIATION PROTECTION RESEARCH PROGRAMME OF THE COMMISSION OF EUROPEAN COMMUNITIES

F. F. Van Hoeck

The Commission of European Communities performs two types of activities in the field of radiation protection:

- regulatory activity: it lays down "basic safety standards" which, when adopted by the Council of Ministers, are compulsory in the nine Member States. A Commission proposal for standards revised after the last ICRP recommendations is presently under discussion.

- a research activity which organizes a cooperative research effort among the nine Member States.

The Radiation Protection Programme is designed to gain an adequate understanding and control of radiation risks, with two main objectives:

- the improvement of scientific and technical knowledge with a view to updating basic standards for the health protection of the general public and workers against the hazards arising from ionizing radiation;
- the evaluation of the biological and ecological consequences of nuclear activities and of the use of nuclear energy and ionizing radiation, in order to ensure an adequate protection of man and of the environment whenever unacceptable harm could otherwise be caused to them.

The programme is rather arbitrarily divided in the following sectors:

1. Radiation dosimetry and its interpretation
2. Behaviour and control of radionuclides in the environment
3. Short-term somatic effects of ionizing radiation
4. Late somatic effects of ionizing radiation
5. Genetic effects of ionizing radiation
6. Evaluation of radiation risks

Implementation takes place through cost-sharing contracts with universities, laboratories, institutions and national research centers. Actual cooperation between scientists active in the various sectors is fostered through systematic organization of study groups, seminars and symposia. Research results and projects are continuously reviewed by an "Advisory Committee on Programme Management" composed of delegates nominated by the authorities of the Member States.

A new programme decision for the 1980-84 period has just been adopted by the Council of Ministers. This new programme will place more emphasis on, for example, the following problems:

- the development of methodologies for evaluation of radiation risks, including the assessment of individual and collective doses, of the detriment, and of the economic and social consequences of radiation exposure;

- the development of a convergent approach to the problem of biological effects at low doses and at low dose rates;

- the prognosis and treatment of local radiation injuries, a type of accident which is relatively frequent;

- the development of methods to estimate and control a possible (accidental) contamination of the environment by radionuclides and associated pollutants;

- the analysis of radiation induced genetic damage in man directly on human systems or on systems from which extrapolation to man is reliable;

- the assessment of realistic relationships between dosimetric quantities and the genetic and somatic effects and risk, including measurement methods and models for intake, and methods for evaluating low doses;

- the reduction of medical doses without loss of efficacy.

The budget allocation for 1980-84 is 59 million European units of account, corresponding to about 80 million US $, or 16 million $ per year. These funds, mainly used for the Commission's participation in research contracts, are matched by about 22 million $ brought in by the contract partners, bringing the total yearly expenditure to about 38 million $.